Global Societal Resilience Through Political, Environmental, and Cultural Upheaval

Mohamad Al Mokdad
https://orcid.org/0009-0002-9188-5272
University of Zaragoza, Spain

IGI Global
Scientific Publishing
Publishing Tomorrow's Research Today

Vice President of Editorial	Melissa Wagner
Director of Acquisitions	Mikaela Felty
Director of Book Development	Jocelynn Hessler
Production Manager	Mike Brehm
Cover Design	Jose Rosado

Published in the United States of America by
IGI Global Scientific Publishing
701 East Chocolate Avenue
Hershey, PA, 17033, USA
Tel: 717-533-8845 | Fax: 717-533-7115
Website: https://www.igi-global.com E-mail: cust@igi-global.com

Library of Congress Cataloging-in-Publication Data

LCCN: 2025034245 (CIP Data Pending)
ISBN: 9798337356518
Isbn13Softcover: 9798337356525
eISBN: 9798337356532
British Cataloguing in Publication Data
A Cataloguing in Publication record for this book is available from the British Library.

Table of Contents

Detailed Table of Contents

The incorporation of Artificial Intelligence (AI) into agriculture is escalating significant transformations in essential areas like soil management, crop yield enhancement, and self-operating farming vehicles. This chapter examines the ways in which AI-driven tools—such as machine learning algorithms, remote sensing technologies, robotics, and autonomous systems—enhance decision-making processes, optimize resource utilization, and contribute to environmental sustainability. With AI, it is possible to carry out accurate diagnostics of soil health, predictive irrigation and fertilization, early detection of diseases, and real-time monitoring of crops. Moreover, AI-enabled self-driving vehicles are transforming the world of mechanized agriculture by providing efficient planting, spraying, and harvesting. This chapter offers a detailed examination of these innovations, assesses their practical uses, and deliberates on the consequences for farmers, researchers, and policymakers. It also underscores the ethical, infrastructural, and socio-economic challenges linked to the use of AI in agriculture.

This research paper examines the complex relationship between green and renewable energy sector financial literacy and its impact on inequality and environmental degradation in Mexico. Through interdisciplinarity that engages environmental justice theory, financial inclusion frameworks, and sustainable development perspectives, this research reveals how limited financial literacy in renewable energy technologies consolidates socioeconomic inequalities while advancing environmental degradation. The research concludes that intense financial literacy programs to renewable energy literacy, combined with comprehensive policy frameworks, are essential to providing just energy transitions, which address both environmental and social justice concerns in Mexico's path toward sustainable development.

This research explores the complex relationships between socio-environmental conflicts in Mexico and the development of green and renewable energy projects, with particular focus on the role of financial literacy in influencing participation and decision-making within communities. Mexico's transition to renewable energy has generated robust tensions between developmental aspirations, environmental protection, and social justice. Main results indicate that financially literate communities have improved negotiation outcomes, reduced conflict intensity, and increased long-term project acceptance. The study concludes that the integration of large-scale financial literacy programs into renewable energy development policy could significantly decouple socio-environmental conflicts from pursuing more equitable and sustainable energy transitions in Mexico.

Cultural identities during crises play a dual role in communities, necessitating a precise and systematic approach to address them. In some cases, they can empower communities to adapt to adversity. In others, it may challenge social equality and tensions, making social cohesion an urgent requirement for stability. Our research seeks to answer main questions: How does cultural identity support societal cohesion? And how do crises affect both cultural identity and social cohesion? The study draws on the experiences of two Internally Displaced Groups during the Sept 2024 aggression in Lebanon basically women who relocated in similar and different cultural identity environments, which explored the similarities and differences in cultural identities and examined the relationship between influence and impact on social cohesion. Their cultural identity played a crucial role in shaping the host community's understanding of them, as well as in guiding social interactions and behaviors highlighting the role of identity in social integration under stress.

Afghan displacement, spanning four decades, remains one of the world's most protracted refugee crises. This chapter examines resilience not as an individual trait but as a relational, intersectional process unfolding across pre-migration, migration, and post-settlement stages. Based on semi-structured interviews with Afghan refugees in diverse host countries, thematic and visual analyses highlight coping, support systems, discrimination, identity reconstruction, cultural continuity, and aspirations. Findings reveal gendered and generational distinctions: men emphasize structural barriers, women sustain resilience through caregiving and cultural reproduction, while youth prioritize education and identity. Cross-national comparisons show resilience is shaped by institutional regimes, flourishing in enabling environments and constrained in restrictive ones. Advancing ecological and life-course perspectives, the study reframes resilience as adaptive and resistant, while offering policy recommendations that foreground refugee agency, cultural continuity, and long-term inclusion.

Indigenous governance structures provide the institutional bedrock for social harmony, collective action and resource management. Vibrant cultural practices including ceremonies, oral traditions, and intergenerational knowledge transfer reinforce community identity, values, and a reciprocal relationship with the environment acting as powerful mechanisms for knowledge transmission and social cohesion. Sophisticated Indigenous Ecological Knowledge, accumulated over millennia through direct observation and lived experience, offers unparalleled insights into local ecosystems, biodiversity conservation, and adaptive resource management strategies that enhance food security and environmental stability. By examining the synergistic interplay, this research aims to illuminate how Indigenous communities leverage their inherent strengths to navigate challenges, maintain well-being, and ensure long-term sustainability. This study will contribute to a deeper appreciation of Indigenous wisdom, advocating for its recognition and integration into broader resilience-building frameworks and policies.

This study investigates the role of psychosocial resilience in strengthening the adaptive capacity of Lebanese society in the face of multiple and overlapping crises between 2020 and 2024, including the economic collapse, political instability, the Beirut Port explosion, and the COVID-19 pandemic. A descriptive-analytical methodology was employed to explore these dynamics. The results demonstrate that psychosocial resilience should not be regarded as a static individual trait, but rather as a dynamic and context-dependent process shaped by psychological, social, cultural, and religious factors. The analysis highlights the central role of families, kinship networks, and community-based support systems in fostering collective endurance. At the same time, persistent sectarian divisions, institutional fragility, and rising emigration were identified as significant threats undermining the effectiveness of resilience mechanisms. The study concludes that comprehensive policy frameworks are urgently required to strengthen both social and institutional support structures. Furthermore, scaling up mental health and community-based interventions is essential to enhance society's capacity for sustainable adaptation in the post-crisis context.

The elements that affect entrepreneurial skill development in Lebanese university students are evaluated in a research study. Four important variables are used to study in this research, that are Entrepreneurship Education Quality (EEQ), Institutional Support for Entrepreneurship (ISE), Socioeconomic Factors (SEF), and Personal Characteristics and Mindset (PC). The major obstacle for Lebanese students is caused by the economic instability, though, these students believe diverse concepts will be developed in the end. For universities, programs should expand their mentorship programs and students should be allowed resource access points as well as educational approaches that build on technical knowledge alongside training of entrepreneurial personality.

This chapter examines how entrepreneurship training functions as a multidimensional development tool for rural women in Lebanon, a fragile state marked by political instability and gendered socio-economic exclusion. Based on field research with 284 women across five governorates, it analyzes the impact of NGO-led programs on technical capacities, self-efficacy, economic participation, and civic engagement. Rooted in the Capability Approach and Human Capital Theory, the study finds that while training improved business skills and income generation, its deeper contributions lie in internal and social transformations. Participants reported increased confidence, leadership, and decision-making power. Yet, structural barriers such as limited finance access and patriarchal norms, limit the sustainability of these gains. The chapter critiques linear, skills-based models and proposes a gender-responsive, ecosystemic approach to entrepreneurship, offering theoretical, practical, and policy insights from a rarely studied Middle Eastern, post-crisis rural context.

Corporate Social Responsibility (CSR) is gaining relevance in Lebanon's banking sector amid severe economic and social crises. This study explores how banks implement CSR, the challenges they face, and stakeholder roles. Though not legally required, CSR is seen as a strategic necessity driven by competition, global trends, and citizen awareness. Banks like Bank Audi, SGBL, and BLF lead efforts in education, health, environment, and ethics. Using mixed methods, the research combines interviews with banking professionals and quantitative analysis of CSR's effect on staff and client engagement. Results show that despite weak regulations, banks use CSR to build trust, improve reputation, and boost internal morale. Yet, instability and inconsistent standards hinder progress. The study recommends creating a standard CSR framework, fostering collaboration, and aligning with global sustainability models to enhance resilience in the Lebanese banking sector.

Corporate and Banking Social Responsibility (CBSR) has come a long way as a philanthropic program to a serious model of corporate governance, sustainability, and stakeholders. This study focuses on historical evolution, theory, as well as practice of CSR in corporate and banking spheres with the accent on its role in corporate governance, environmental sustainability, and ethical decision making. This paper similarly looks at how banking institutions can facilitate financial inclusion, responsible investment, and sustainable development. y. The results highlight the importance of ethical business conducts in promoting economic robustness, regulatory conformance, and business legitimacy, especially in financial establishments where governance is very crucial in risk-management and assurance to stakeholders.

In an era marked by persistent political, environmental, and cultural upheavals, the resilience of business organizations has become increasingly dependent on the quality and agility of their strategic leadership. This chapter explores how strategic leadership functions as a critical factor in crisis adaptation, focusing on decision-making processes, organizational culture, and leadership communication during times of uncertainty. Drawing on contemporary case studies and theoretical frameworks in business management, the study highlights how effective leaders not only respond to external shocks but also anticipate future disruptions through proactive planning, risk management, and innovation-driven strategies. The chapter begins by mapping the global context of recent crises including the COVID-19 pandemic, geopolitical conflicts, economic instability, and climate-related disruptions and their collective impact on business continuity. It then examines the role of transformational and adaptive leadership styles in fostering resilience at both organizational and community levels. By integrating qualitative and quantitative insights, the chapter identifies key success factors and common pitfalls in leadership response strategies. Special attention is given to small and medium-sized enterprises (SMEs), which often lack institutional buffers yet demonstrate significant innovation and adaptability. The study concludes with a strategic resilience framework designed to guide leaders in building long-term organizational sustainability, emphasizing agility, stakeholder

Preface

INTRODUCTION

In a rapidly transforming global landscape, societies across the world are being tested by an array of unprecedented challenges. Political instability, armed conflicts, and the resurgence of authoritarianism intersect with the accelerating impacts of climate change, global pandemics, and profound cultural transformations. The very fabric of communities and nations is being stretched to its limits. Yet, amidst this turbulence, the concept of *societal resilience* has emerged as an essential framework for understanding how communities, institutions, and cultures withstand, adapt, and ultimately thrive in the face of adversity.

This volume, *Global Societal Resilience Through Political, Environmental, and Cultural Upheaval*, seeks to examine how resilience is constructed, challenged, and reinforced across diverse global contexts. It investigates how societies navigate crises; how governance systems respond to political and environmental pressures; and how cultural identities and values can either fragment or unify populations during times of upheaval.

The purpose of this book is not merely to document examples of resilience, but to provide critical insights and theoretical frameworks that enable a deeper understanding of the multi-dimensional nature of societal endurance. Through a combination of empirical case studies, comparative analyses, and conceptual reflections, the chapters collectively explore how resilience manifests in political systems, environmental governance, and cultural adaptation.

We have intentionally brought together contributions from across disciplines—including political science, sociology, environmental studies, international relations, public policy, and anthropology—to foster a holistic conversation. Authors were invited to engage with both theoretical debates and practical experiences that illuminate the dynamics of resilience under duress.

In an age where disruption has become the new norm, this volume strives to chart pathways toward sustainable, inclusive, and adaptive societies. Its central objective is to critically examine how communities respond to multifaceted crises and to explore the mechanisms, policies, and cultural practices that strengthen or undermine resilience. By integrating diverse disciplinary perspectives and drawing on global case studies, this book offers a comprehensive understanding of resilience as both a conceptual lens and a lived reality in our turbulent world.

Moving beyond traditional, state-centered analyses, the book adopts a broader view that encompasses local communities, civil society, environmental systems, and cultural frameworks. It considers how resilience is shaped by governance models, social capital, institutional trust, environmental adaptation, and cultural cohesion—and how these dimensions interact to influence political stability, environmental sustainability, and collective identity.

The book's contribution unfolds along three key dimensions:

1. **Interdisciplinary Integration:** By bridging insights from political science, environmental studies, sociology, anthropology, and international relations, this volume captures the complexity of resilience as a cross-cutting phenomenon.
2. **Global and Comparative Perspectives:** Through diverse case studies and theoretical contributions, the chapters offer a comparative understanding of how resilience is constructed and challenged in different political, environmental, and cultural settings.
3. **Policy-Relevant Insights:** By engaging directly with practitioners and policymakers, the book provides actionable knowledge and evidence-based recommendations for strengthening societal resilience amid ongoing and future upheavals.

Ultimately, this volume aims to deepen scholarly discourse on resilience while offering practical perspectives on how societies can navigate uncertainty, rebuild stronger, and envision sustainable futures.

Global Societal Resilience Through Political, Environmental, and Cultural Upheaval is intended for a wide audience of scholars, policymakers, and practitioners. It will serve as a valuable resource for:

- **Academics and Researchers** in political science, international relations, sociology, environmental studies, public policy, anthropology, and cultural studies seeking to explore resilience from multiple perspectives.
- **Policy Makers and Government Officials** engaged in crisis management, disaster response, environmental planning, conflict resolution, and community development.

- **Practitioners and Professionals** working within international organizations, NGOs, and civil society who contribute to resilience-building, peacebuilding, and sustainable development initiatives.
- **Graduate and Postgraduate Students** pursuing degrees in social sciences, environmental studies, or development studies, who will benefit from the book's theoretical and empirical insights.
- **Think Tanks and Research Institutions** focusing on global governance, resilience planning, and risk mitigation strategies.

It is my hope that this book will inspire continued dialogue and collaboration across disciplines and regions, encouraging innovative approaches to the challenges of our time. By illuminating how resilience emerges, falters, and endures, this volume invites readers to consider not only how societies survive disruption—but how they might transform it into an opportunity for renewal.

ORGANIZATION OF THE BOOK

Chapter 1: AI-Driven Innovations in Soil Management, Crop Productivity, and Autonomous Agricultural Vehicles

The opening chapter introduces readers to the transformative potential of artificial intelligence in reshaping agricultural systems and advancing environmental sustainability. Through the integration of machine learning, robotics, and autonomous technologies, the chapter demonstrates how AI enhances precision in soil diagnostics, irrigation, fertilization, and crop monitoring. It further explores the emergence of self-operating agricultural vehicles as a frontier in mechanized farming, providing efficiency and data-driven insight. Importantly, the discussion extends beyond technical innovation to address the ethical, infrastructural, and socio-economic implications of AI adoption in agriculture, underscoring the need for equitable and sustainable transitions within the global food system.

Chapter 2: Inequality and Environmental Degradation in Mexico as a Result of Green and Renewable Energy-Related Financial Literacy

Chapter Two interrogates the intersection of environmental justice, financial inclusion, and green energy development in Mexico. It argues that limited financial literacy in renewable energy investment and technology exacerbates socioeconomic inequalities while indirectly advancing environmental degradation. Drawing on frameworks of sustainable development and social equity, the chapter contends that the democratization of financial knowledge is crucial to ensuring just and inclusive energy transitions. By highlighting the nexus between education, policy reform, and grassroots empowerment, this contribution sets the stage for a more equitable approach to sustainability in emerging economies.

Chapter 3: Socio-Environmental Conflicts in Mexico and their Impact on Green and Renewable Energy Related to Financial Literacy

Continuing the exploration of Mexico's renewable energy transition, Chapter Three delves into the socio-environmental conflicts that arise from the pursuit of sustainable development in contexts of social inequality. The study reveals that communities with higher financial literacy levels experience improved negotiation outcomes, lower conflict intensity, and greater project legitimacy. By linking financial knowledge to participatory governance, the chapter underscores how capacity-building initiatives can transform community engagement and conflict dynamics. Its findings emphasize that fostering financial literacy is not merely an educational endeavor, but a peacebuilding and sustainability strategy.

Chapter 4: Cultural Identity and Societal Cohesion During Crises

Chapter Four examines the complex interplay between cultural identity and social cohesion in times of crisis, focusing on the lived experiences of internally displaced women during the 2024 aggression in Lebanon. The research reveals how shared and divergent cultural identities shape integration within host communities and influence social stability. By unpacking both the empowering and fragmenting dimensions of identity, the chapter contributes to broader discussions on cultural resilience and inclusion during displacement. It positions cultural identity as a crucial mediating factor in fostering cohesion, empathy, and adaptive capacity under stress.

Chapter 5: Displaced Lives, Resilient Spirits: Lived Experiences and Adaptive Strategies of Afghan Refugees in the Global Diaspora

This chapter presents a deeply human exploration of Afghan refugees' adaptive strategies across different stages of displacement and resettlement. Conceptualizing resilience as a relational and intersectional process, the study foregrounds the voices of men, women, and youth navigating identity reconstruction, discrimination, and transnational belonging. Cross-national comparisons reveal that resilience flourishes in inclusive policy environments and falters under restrictive regimes. By advancing ecological and life-course perspectives, this chapter reframes refugee resilience as both adaptive and resistant, providing actionable insights for policy, humanitarian programming, and long-term inclusion.

Chapter 6: Role of Indigenous Governance, Cultural Practices, and Ecological Knowledge in Promoting Community Resilience

Chapter Six highlights the enduring strength of Indigenous governance systems and traditional ecological knowledge as pillars of community resilience. Through an interdisciplinary lens, the chapter illustrates how ceremonies, oral traditions, and intergenerational knowledge transfer sustain social cohesion and ecological balance. It argues that Indigenous governance models embody adaptive strategies that ensure environmental stability, food security, and cultural continuity. The discussion calls for the recognition and integration of Indigenous wisdom into broader resilience-building frameworks, asserting its indispensable role in sustainable development and environmental policy.

Chapter 7: Psycho-Social Resilience Amid Political and Economic Upheaval 2020–2024: The Case of Lebanese Society

Focusing on Lebanon's multifaceted crises between 2020 and 2024, this chapter analyzes psychosocial resilience as a dynamic, context-dependent process. It highlights how families, kinship networks, and community institutions function as anchors of endurance amid economic collapse, political turmoil, and public health emergencies. Yet, it also reveals how persistent sectarian divisions and institutional weaknesses undermine collective coping mechanisms. By situating resilience within social, cultural, and religious contexts, the chapter underscores the necessity of rebuilding trust and solidarity as foundations for national recovery.

Chapter 8: Developing Entrepreneurial Skills among Lebanese University Students: A Business Management Perspective

Chapter Eight investigates the development of entrepreneurial competencies among Lebanese university students navigating an unstable economic landscape. Through the analysis of key factors—educational quality, institutional support, socioeconomic background, and personal mindset—the study exposes the challenges and opportunities shaping youth entrepreneurship. It advocates for mentorship programs, experiential learning, and access to resources as pathways to strengthen entrepreneurial ecosystems. This contribution situates entrepreneurship education as both a personal empowerment strategy and a collective resilience mechanism in fragile economies.

Chapter 9: Entrepreneurship Training and Human Development: Evidence from Rural Women in Lebanon

This chapter examines entrepreneurship training as a transformative instrument for women's empowerment and rural development in Lebanon. Grounded in the Capability Approach and Human Capital Theory, the research evaluates the outcomes of NGO-led programs that enhance skills, income, and self-efficacy. Beyond economic gains, the findings highlight profound shifts in confidence, leadership, and social participation among participants. Yet, structural barriers such as patriarchal norms and limited financial access persist. The chapter concludes with a call for gender-responsive, ecosystemic models of entrepreneurship that promote both individual agency and community resilience.

Chapter 10: Reinforcing Trust Through CSR: A Study of the Lebanese Banking Sector

In Chapter Ten, the focus turns to corporate social responsibility as a mechanism for rebuilding trust within Lebanon's fragile banking sector. Amid ongoing economic crises, the chapter explores how CSR initiatives in education, health, environment, and ethics have become strategic tools for restoring legitimacy and stakeholder engagement. Despite regulatory weaknesses, the findings demonstrate that CSR enhances morale, reputation, and client confidence. However, systemic instability remains a significant barrier. The chapter advocates for standardized frameworks and global alignment to reinforce institutional resilience and ethical accountability.

Chapter 11: From Philanthropy to Policy: The Role of CSR in Strengthening Societal and Financial Resilience

Chapter Eleven extends the discussion on corporate responsibility by tracing its evolution from philanthropy to an integrated model of governance, ethics, and sustainability. Emphasizing the importance of responsible investment and financial inclusion, it argues that ethical business conduct underpins societal and financial resilience alike. The analysis reveals how corporate and banking institutions can function as agents of stability, regulatory compliance, and public trust. Through a theoretical and historical lens, the chapter situates CSR as a cornerstone of risk management and sustainable economic recovery.

Chapter 12: Strategic Leadership and Crisis Adaptation: Navigating Business Uncertainty Amid Global Upheaval

Leadership emerges as a central theme in Chapter Twelve, which examines how strategic decision-making and organizational culture drive business adaptability during global crises. The discussion highlights the roles of communication, foresight, and innovation in guiding firms through uncertainty. By positioning business leadership as both an economic and social stabilizer, the chapter bridges micro-level management studies with macro-level resilience theory, offering insights into the competencies necessary for steering organizations amid volatility.

Chapter 13: Governing Disruption: A New Architecture for Global Leadership in the Age of Intelligent Systems

This chapter conceptualizes the current era as one of structural turbulence, in which cascading disruptions—climatic, technological, political, and cybernetic—demand a fundamental redesign of governance. Treating artificial intelligence as both a tool and an actor in global systems, the chapter proposes a new governance architecture attuned to complexity and uncertainty. Through design imperatives and institutional innovations, it outlines pathways for developing disruption-ready leadership and policy coherence in a polycrisis world.

Chapter 14: Public Policy for Resilient Governance: Strengthening Civil Society in Times of Crisis

Chapter Fourteen turns attention to the role of civil society in crisis governance. It demonstrates how community organizations serve as first responders and advocates for the vulnerable, while also highlighting the institutional, legal, and ethical frameworks necessary to support them. Drawing on comparative case studies from the Global South, the chapter illustrates how inclusive policymaking and collaborative governance can bolster resilience, accountability, and social trust during emergencies.

Chapter 15: The Future of International Cooperation: Redesigning Global Governance for an Age of Uncertainty

The final chapter situates the volume's themes within the global governance landscape, arguing that the post-1945 institutional order is ill-equipped to manage today's complexity. Through case studies of geopolitical crises and emerging reform initiatives, it advocates for a redesign grounded in adaptability, equity, digital inclusion, and Indigenous knowledge integration. The chapter concludes with a blueprint for a renewed multilateralism—one that prioritizes coalition-building, systemic foresight, and resilience as guiding principles for international cooperation in an age of fragmentation.

CONCLUSION

As editor of this volume, I am profoundly optimistic about the pathways forward that these contributions illuminate. In an era where political fractures, environmental cataclysms, and cultural dislocations converge with unrelenting force, the chapters assembled here do not merely catalog fragility—they forge a blueprint for endurance and renewal. By weaving together rigorous interdisciplinary scholarship with grounded, comparative insights from every corner of the globe, this book equips us to transform upheaval into opportunity.

To the scholars, policymakers, practitioners, and students who will engage with these pages: let this work spur you to action. Apply its frameworks to your contexts, challenge its assumptions with new evidence, and collaborate across borders and disciplines to build societies that do not just survive, but flourish. The resilience we seek is not a static trait but a collective endeavor—one that demands courage, innovation, and unwavering commitment to inclusivity and sustainability.

In closing, *Global Societal Resilience Through Political, Environmental, and Cultural Upheaval* stands as both a scholarly milestone and a call to arms. May it inspire the bold thinking and decisive steps needed to secure a thriving future for generations amid the storms ahead.

Chapter 1
AI-Driven Innovations in Soil Management, Crop Productivity, and Autonomous Agricultural Vehicles

Ananya Pandey
https://orcid.org/0000-0002-2419-6314
Christ University, Bangalore, India

ABSTRACT

The incorporation of Artificial Intelligence (AI) into agriculture is escalating significant transformations in essential areas like soil management, crop yield enhancement, and self-operating farming vehicles. This chapter examines the ways in which AI-driven tools—such as machine learning algorithms, remote sensing technologies, robotics, and autonomous systems—enhance decision-making processes, optimize resource utilization, and contribute to environmental sustainability. With AI, it is possible to carry out accurate diagnostics of soil health, predictive irrigation and fertilization, early detection of diseases, and real-time monitoring of crops. Moreover, AI-enabled self-driving vehicles are transforming the world of mechanized agriculture by providing efficient planting, spraying, and harvesting. This chapter offers a detailed examination of these innovations, assesses their practical uses, and deliberates on the consequences for farmers, researchers, and policymakers. It also underscores the ethical, infrastructural, and socio-economic challenges linked to the use of AI in agriculture.

DOI: 10.4018/979-8-3373-5651-8.ch001

INTRODUCTION

In the twenty-first century, agriculture finds itself at a pivotal moment, facing an urgent requirement to boost food production to satisfy the needs of an expanding global population while also guaranteeing environmental sustainability and resource preservation (FAO, 2017). In this context, Artificial Intelligence (AI) has developed into a revolutionary technology that significantly affects contemporary agriculture. AI can transform conventional agricultural practices, enhance efficiency, and foster sustainable intensification of crop production by facilitating data-driven decision-making, predictive analytics, and automation (Wolfert et al., 2017). AI applications range from monitoring soil health and managing crops to using autonomous farming machinery, providing novel solutions to age-old problems in resource management and agricultural productivity (Kamilaris & Prenafeta-Boldú, 2018). Soil management continues to be a core component of agricultural sustainability, with direct effects on crop yield and quality as well as ecosystem health (Lal, 2020). Nonetheless, the degradation of soil, depletion of nutrients, and scarcity of water pose major challenges that are worsened by climate change and intensive agricultural methods (Montgomery, 2007). Conventional methods for managing soil, which typically depend on periodic sampling and practical experience, lack sufficient spatial and temporal resolution for prompt interventions (McBratney et al., 2019). Tools for evaluating soil that are powered by AI utilize sensor networks, remote sensing data, and machine learning algorithms to deliver real-time, site-specific insights into the properties and dynamics of soil (Singh et al., 2020). These innovations allow for precise nutrient management, optimized irrigation scheduling, and early identification of soil health issues, thus improving productivity and minimizing environmental impacts.

In the same way, effective monitoring and management of plant health, growth conditions, and pest pressures are closely tied to crop productivity. Traditional methods of crop management encounter difficulties, including lagging disease identification, subpar resource utilization, and fluctuating climatic conditions. Crop monitoring AI applications employ sophisticated computer vision, drone and satellite images, and predictive analytics to detect crop stress factors early on, predict yields, and enhance input application (Kamilaris et al., 2017). As it maximizes yields per input unit and minimizes waste, this precision agriculture paradigm fosters sustainable intensification, contributing to food security and mitigating environmental degradation (Gebbers & Adamchuk, 2010). Traditionally, mechanization—which is essential to contemporary agronomy—has relied on machinery operated by humans for activities such as tilling, planting, spraying, and harvesting. Nonetheless, the emergence of autonomous agricultural vehicles (AGVs) has been driven by labor shortages, increasing costs, and the demand for greater efficiency (Bechar & Vigneault, 2016). With AI-driven autonomy, tractors, harvesters, and aerial drones can operate continuously, perform tasks with precision, and connect seamlessly with larger farm management systems (Duckett et al., 2018). While the introduction of AGVs holds the potential to tackle labor limitations and improve the scalability and sustainability of agricultural activities, challenges regarding safety, cost, and regulatory frameworks remain (Rotz et al., 2019). This chapter thoroughly investigates AI-driven innovations in three interconnected areas—soil management, crop productivity, and autonomous agricultural vehicles—and to analyzes their combined potential to transform contemporary agriculture. The purpose of this chapter is to connect technological insights with practical challenges in agriculture, emphasizing cutting-edge AI applications, emerging trends, and their effects on sustainability and farm economics. In addition, it aims to pinpoint current constraints and avenues for future investigation, highlighting the necessity of cross-disciplinary cooperation and inclusive innovation to guarantee fair access and responsible implementation. To summarize, this chapter aims to provide a thorough and academically robust investigation into the ways in which AI-driven technologies can address significant challenges related to soil management, crop productivity, and mechanization, facilitating a more efficient, resilient, and sustainable agricultural future. Incorporating AI into agriculture holds the potential for enhanced productivity and is in accordance with worldwide sustainability objectives, as it fosters resource efficiency and minimizes the environmental impact of agricultural methods (United Nations, 2015). Consequently, it is essential for researchers, practitioners, and policymakers committed to future food security and environmental stewardship to comprehend and utilize these AI innovations. In light of the above perspective, the author examines the approach of technological advancement in the form of AI, machine learning, IoTs, precision agriculture and deep learning in the field of agriculture particularly in the management of soil, its crop productivity and its exponential growth. Furthermore, it also examines the role of autonomous systems in the advancement of soil management.

FUNDAMENTALS OF AI IN AGRICULTURE

AI has come to be a transformative force in many industries, with agriculture being a sector that is especially well-positioned to gain from its incorporation. AI encompasses the creation of computer systems that can undertake tasks usually necessitating human intelligence, such as learning, reasoning, problem-solving, and perception (Russell & Norvig, 2021). Within the broader umbrella of AI, the crucial subfields are machine learning and deep learning. According to Goodfellow, Bengio, and Courville (2016), machine learning allows systems to recognize patterns in data and enhance their performance without the need for explicit programming for each individual task. Deep learning, a sophisticated branch of machine learning, employs artificial neural networks comprising multiple layers to represent and model intricate patterns from large datasets (LeCun, Bengio, & Hinton, 2015). These AI technologies constitute the backbone of innovations designed for contemporary agriculture, facilitating more intelligent and data-informed decision-making processes. In agriculture, AI technologies are realized through various applications that utilize computer vision, sensor technologies, and robotics, among other tools. The area of study known as computer vision focuses on making it possible for machines to make sense of and comprehend visual information from their surroundings (Szeliski, 2010). In agriculture, computer vision aids in monitoring crop health, identifying diseases, detecting weeds, and estimating yield through image analysis from drones, satellites, or field cameras (Kamilaris & Prenafeta-Boldú, 2018). These images are processed by sophisticated algorithms to identify slight changes in color, texture, and shape that signal particular plant conditions—these algorithms often excel human observation in both speed and precision.

Sensor technologies that record various environmental and soil parameters, including temperature, humidity, soil moisture, pH, and nutrient levels, complement computer vision. There is a growing trend to incorporate these sensors into Internet of Things (IoT) frameworks, facilitating real-time data gathering and sharing among interconnected devices on the farm (Wolfert, Ge, Verdouw, & Bogaardt, 2017). IoT devices encompass a variety of tools, from basic soil moisture probes to advanced multispectral sensors that evaluate crop health and stress reactions. When these data streams are coupled with AI analytics, they facilitate precision agriculture techniques that allow for the optimization of inputs such as water, fertilizers, and pesticides at micro scales. This optimization reduces waste and environmental impact while improving productivity. AI is also fostering major advancements in agricultural robotics, marking it as another pioneering domain. AI empowers autonomous vehicles such as drones, tractors, and robotic harvesters to navigate intricate agricultural settings, execute repetitive tasks with great accuracy, and adjust to changing conditions in the field (Bechar & Vigneault, 2016). Using AI algorithms for tasks such as obstacle detection, path planning, and manipulation, these robots help to ease labor shortages and enhance operational efficiency. As an example, AI-powered computer vision enables robotic weeders to differentiate between crops and weeds and to remove unwanted plants mechanically, thereby decreasing the dependence on herbicides (Bakr, 2020). Likewise, self-governing harvesters are capable of discerning when fruits and vegetables are ready for harvest. They can collect these items with care, thereby reducing harm and enhancing the quality of the harvest (Duckett et al., 2018).

The combination of AI, big data analytics, and IoT results in a formidable ecosystem that can convert conventional agriculture into a highly efficient, sustainable, and intelligent system. Big data in agriculture denotes the vast amounts of structured and unstructured data produced from various sources, including satellite imagery, weather stations, soil sensors, machinery telemetry, and market information (Kamilaris, Kartakoullis, & Prenafeta-Boldú, 2017). The difficulty is in transforming this unprocessed data into actionable insights, a process for which AI is essential. Advanced AI models, such as predictive analytics and decision support systems, examine historical and real-time data to predict crop yields, identify pest outbreaks early, and suggest ideal planting schedules (Liakos, Busato, Moshou, Pearson, & Bochtis, 2018). This forecasting ability aids farmers in reducing risks linked to changes in climate and market volatility. In addition, AI-driven IoT platforms enable automation and precision management at the farm level by offering a smooth connection between data gathering, analysis, and operational choices. This integration is facilitated by cloud computing and edge computing paradigms, which provide data storage, high-performance computing, and low-latency responses essential for real-time interventions (Verdouw, Wolfert, Beulens, & Rialland, 2016). Irrigation systems that incorporate AI, for instance, can automatically modify water supply according to soil moisture data, weather predictions, and the water needs of crops. This helps to save water resources and keep crops healthy (Jiang, Wang, & Luo, 2020).

The integration of AI, IoT, and big data analytics also applies to supply chain and post-harvest management, with AI aiding in logistics optimization, food waste reduction, and traceability improvement. Produce condition during transport is monitored by sensors, and AI algorithms enhance routing and storage parameters to preserve quality and freshness (Kamilaris & Prenafeta-Boldú, 2018). An end-to-end digital transformation of this kind holds the promise of greater efficiency and transparency, while also fulfilling consumer demands for ethically sourced and sustainable food. Nevertheless, despite the promising prospects of AI in agriculture, several challenges need to be tackled for it to be widely adopted. This encompasses the expensive nature of tech implementation, worries about data privacy, absence of digital infrastructure in rural locales, and necessity for training and capacity enhancement for farmers (Wolfert et al., 2017). Furthermore, the training of AI models necessitates vast amounts of high-quality data, which can be limited or not evenly distributed in agricultural settings, especially in developing areas (Kamilaris et al., 2017). It is also important to consider ethical aspects linked to data ownership and algorithmic bias in order to guarantee fair access and advantages.

AI-DRIVEN SOIL MANAGEMENT INNOVATIONS

As a fundamental element of sustainable agriculture, soil health affects crop productivity, environmental quality, and ecosystem resilience (Lal, 2020). Healthy soils uphold vital physical, chemical, and biological processes that foster plant growth, manage water resources, and recycle nutrients. Nevertheless, the pressures of soil degradation, erosion, nutrient depletion, and climate change are mounting on global agricultural systems, all of which threaten long-term food security (FAO, 2017). Tackling these challenges necessitates inventive strategies for monitoring, managing, and optimizing soil conditions. Artificial intelligence (AI) has developed into a game-changing resource for soil management, utilizing sophisticated computational methods to examine intricate soil data, foresee evolving soil phenomena, and direct precision actions (Reichstein et al., 2019). AI plays a crucial role in assessing and monitoring soil quality, which is vital for effective soil management. While traditional soil testing methods are precise, they tend to require a lot of work and time, as well as being limited in scope (Li et al., 2020). Real-time, high-resolution soil monitoring at field and landscape scales is made possible by AI-enabled remote sensing technologies in conjunction within situ sensors. For example, machine learning algorithms examine multispectral and hyperspectral data obtained from satellites, drones, or proximal sensors to deduce soil organic carbon content, texture, moisture levels, and salinity (Minasny et al., 2019). These approaches, which are based on data, allow for continuous mapping of parameters related to soil health, thus improving the precision and promptness of decision-making. Soil nutrient detection has notably improved with the incorporation of AI into sensor data and spectroscopy methods. Nutrients in soil, including nitrogen, phosphorus, and potassium, are essential for the growth of crops. However, their presence is not constant in time or space on agricultural fields (Srinivasan et al., 2020). Soil analysis using conventional chemical methods is expensive and often impractical for the spatial resolution needed. Rather, spectroscopic techniques like visible near-infrared (vis-NIR) and mid-infrared spectroscopy offer quick, non-destructive ways to evaluate nutrient levels by analyzing soil reflectance properties (Stenberg et al., 2010). The spectral data are calibrated and interpreted to predict nutrient levels with high accuracy when combined with AI models, such as support vector machines and artificial neural networks (Padarian et al., 2019). With these predictive models, farmers can quickly spot nutrient deficiencies or surpluses and apply fertilizers accurately, thus reducing waste and environmental contamination.

Aside from nutrient detection, AI is pivotal for predictive modeling of soil moisture and erosion risk—two essential elements for crop productivity and soil conservation. Soil moisture dynamics affect the availability of water for plants and biogeochemical processes, whereas erosion results in soil degradation and loss (Panagos et al., 2018). Hydrological models driven by AI incorporate data on weather predictions, soil texture, topography, and vegetation cover to simulate variations in soil moisture with improved spatial and temporal resolution (Mohanty et al., 2021). According to Das et al. (2020), machine learning algorithms like random forests and gradient boosting machines have shown better performance in predicting soil moisture than traditional empirical models. Likewise, erosion risk models that utilize AI integrate land cover data obtained from satellites and rainfall intensity metrics to pinpoint areas at risk and predict erosion events (Zhao et al., 2019). These proactive evaluations enable land managers to apply targeted soil conservation practices, including contour farming or cover cropping, thus reducing degradation risks.

One of the crucial developments in AI-based soil management is the creation of decision support systems (DSS) aimed at optimizing fertilization and irrigation. Through the use of AI algorithms, these DSS platforms synthesize various datasets—such as soil health metrics, crop phenology, weather conditions, and agronomic practices—to produce customized resource application recommendations (Gebbers & Adamchuk, 2010). For instance, precision fertilization decision support systems (DSS) utilize soil nutrient maps and models of crop nutrient demand to determine specific fertilizer amounts tailored to each location, aiming to enhance nutrient use efficiency and reduce environmental externalities. Similarly, irrigation scheduling systems that utilize AI make use of soil moisture forecasts and evapotranspiration estimates to enhance water efficiency, reconciling the requirements of crops for water with conservation objectives (Bakker et al., 2021). DSS integrated with AI boost input efficiency and cost-effectiveness while contributing to sustainable resource management in accordance with environmental stewardship.

Case studies from various agroecological contexts are making the practical impact of AI-enabled soil management innovations increasingly evident. In China, an AI-driven platform that merges drone-based hyperspectral imaging with machine learning models has facilitated accurate nitrogen management in wheat fields, leading to substantial yield increases and diminished nitrogen runoff. In the US, a joint initiative used AI algorithms to examine soil moisture sensor data and weather predictions, allowing for real-time irrigation scheduling in vineyards and cutting water consumption by as much as 30% while maintaining grape quality (Jones et al., 2021). In sub-Saharan Africa, AI-driven soil fertility mapping that combines remote sensing with crowdsourced soil samples has offered smallholder farmers actionable insights into soil nutrient deficiencies, aiding in targeted fertilizer applications that enhanced maize yields (Wossen et al., 2020). The instances demonstrate the ability of AI technologies to convert intricate soil data into actionable insights that enable farmers, increase productivity, and foster sustainable land management practices. Even with these developments, there are still difficulties in scaling AI applications for soil management. Data quality, calibration of sensors, and the generalizability of models across various soil types and climatic conditions remain technical challenges (Li et al., 2021). Moreover, for smallholder farmers in resource-limited settings to benefit from these innovations, equitable access to AI tools and digital literacy are essential (Gebbers & Adamchuk, 2010). Tackling these challenges necessitates interdisciplinary research, strong data governance frameworks, and inclusive policies that encourage technological adoption while protecting environmental and social equity.

AUTONOMOUS AGRICULTURAL VEHICLES AND ROBOTICS

Incorporating self-operating farming vehicles and robotics is a major innovation in contemporary agricultural methods, with the promise of improved efficiency, accuracy, and sustainability. Tractors, harvesters, drones, and other autonomous agricultural machinery are being utilized more and more to carry out various farming tasks with little human involvement. These vehicles utilize developments in AI to facilitate navigation, perception, and control systems that function dependably in intricate and changing agricultural settings (Shamshiri et al., 2018). By tackling essential issues like labor shortages, operational expenses, and the need for accurate resource management, the use of self-driving vehicles in farming fosters a revolutionary change in farming practices. Tractors and harvesters that operate autonomously come fitted with advanced AI technologies, allowing for real-time decision-making and task performance. At the heart of their operation are navigation systems that integrate GPS, LiDAR, and computer vision to accurately map and navigate fields (Roldán et al., 2020). These systems allow vehicles to carry out soil preparation activities like tilling and plowing with great accuracy, thereby reducing soil compaction and improving seedbed conditions (Bechar & Vigneault, 2016). In planting operations, autonomous seeders utilize AI algorithms to regulate seed spacing and depth, ensuring uniform crop emergence and maximizing yield potential (Grift et al., 2018). Likewise, self-governing spraying vehicles employ AI-based perception to detect weed patches or disease outbreaks, applying agrochemicals solely where needed. This approach minimizes chemical usage and lessens environmental impact (Gebbers & Adamchuk, 2010). Drones, which are a type of self-operating agricultural vehicle, are essential for overseeing and handling the health of crops. Drones, outfitted with multispectral and thermal cameras, gather high-resolution images that are examined by AI algorithms to identify stress signals, pest infestations, or nutrient deficiencies (Zhou et al., 2020). This information backs up precision farming methods by making it possible to carry out timely actions customized for certain field zones. Furthermore, drones are being utilized more and more for automated tasks such as spraying, pollination, and planting in challenging terrains, serving as a complement to ground-based autonomous vehicles (Bacco et al., 2019).

Besides carrying out individual tasks, the integration of AI allows for comprehensive management of farming operations via autonomous vehicle coordination, resource optimization, and scheduling of tasks. To enhance efficiency and minimize downtime, advanced AI models analyze real-time data from vehicles, soil sensors, and weather forecasts to dynamically adjust operational plans (Liakos et al., 2018). For example, AI can enhance the routing of various self-driving tractors to prevent overlap and reduce fuel use, all while planning maintenance to avert unplanned failures (Wolfert et al., 2017). This degree of integration fosters sustainable agriculture by encouraging accurate input application and minimizing waste. Nonetheless, there are a number of considerable obstacles to the use of self-governing farm vehicles. Due to the interaction of autonomous machinery with human workers and livestock in farm environments, safety concerns take precedence. Reliable obstacle detection and emergency response mechanisms necessitate strong sensor fusion and fail-safe AI algorithms (Pedersen et al., 2018). Environmental factors like rough terrain, bad weather, fluctuating light conditions can impact sensor efficiency and the precision of AI (Jin et al., 2019). As a result, they also influence the dependability of autonomous systems. Furthermore, the significant initial expenses associated with autonomous machinery and necessary infrastructure—like dependable wireless communication systems and data management solutions—hinder widespread adoption, especially among smallholder farmers in developing areas (Blackmore et al., 2020).

The adoption of autonomous vehicles is also affected by policy and regulatory frameworks. To establish trust and guarantee safety, matters related to liability in the event of accidents, data privacy, and interoperability standards need to be tackled (Chen et al., 2020). It is also essential to provide training and capacity-building for farmers and technicians in order to facilitate the integration and upkeep of autonomous systems (Van Evert et al., 2017). In spite of these difficulties, a number of case studies and pilot projects show the promise of self-governing farming vehicles. The European Horizon 2020 project "ATLAS" created autonomous vehicles designed for precise tasks like spraying and harvesting, demonstrating that robotics can lower agrochemical use by as much as 30% while preserving yield (Molina et al., 2020). In the U.S., John Deere's self-driving tractors are used on commercial farms for soil preparation and planting, requiring little human supervision. This has led to considerable reductions in labor and improvements in operational consistency (Kolling et al., 2018). Likewise, in Japan, rice transplanters that operate independently have aided in alleviating labor shortages in aging rural regions by mechanizing planting duties that conventionally demand a lot of manual work (Yamamoto et al., 2019).

SYNERGIES AND INTEGRATION OF AI INNOVATIONS

The incorporation of AI-driven innovations in soil management, crop monitoring, and autonomous agricultural vehicles represents a transformative change in contemporary farming practices. This integration facilitates the development of holistic farm management systems that improve efficiency, sustainability, and resilience. Farming practices that are traditional and often broken up across different tasks and tools are increasingly coming together into unified digital ecosystems driven by artificial intelligence. By utilizing the enormous quantities of data produced by soil sensors, drones, satellite imagery, and autonomous machinery, this convergence enables precision agriculture through the provision of timely and location-specific decisions. AI technologies facilitate the real-time analysis and interpretation of diverse data streams, resulting in dynamic feedback loops for monitoring soil health, predicting crop conditions, and autonomously adjusting vehicle operations. As an illustration, the combination of AI-driven soil nutrient mapping with self-operating variable-rate fertilizer applicators enables accurate and efficient resource utilization. This boosts productivity and mitigates environmental harm (Shahhosseini et al., 2021). In a similar vein, AI-based computer vision systems mounted on drones or terrestrial robots can facilitate early detection of pests, illnesses, and nutrient shortages through crop monitoring. This process aids in autonomous spraying or robotic interventions (Kamilaris & Prenafeta-Boldú, 2018). These systems, once aligned, allow for the coordination of intricate, multi-tiered activities throughout the agricultural cycle with little human involvement, thus promoting what is commonly called the "farm of the future." Moreover, AI systems that incorporate reinforcement learning can handle fleets of autonomous equipment adaptively by utilizing real-time information. This allows for the optimization of routes, fuel consumption, and machinery condition while taking into account feedback from crops and soil. This virtual infrastructure, abundant in data, revolutionizes decision-making by shifting it from a reactive to a proactive and predictive approach, thereby enhancing operational accuracy and responsiveness to changes in the environment.

Nonetheless, to fully leverage these synergies, we must rely on data interoperability and cloud computing capabilities. Farm data is intrinsically varied, encompassing everything from structured sensor data to unstructured drone images, and is frequently produced by a mix of devices and systems. Interoperability standards are crucial for facilitating seamless data exchange among equipment manufacturers, software platforms, and analytical tools. Open data protocols and application programming interfaces (APIs) enable integration within the agricultural technology ecosystem, permitting AI algorithms to aggregate and analyze data in a coherent manner (Wolfert et al., 2017). Cloud computing further facilitates this integration by offering scalable storage solutions, high-performance computing capabilities, and real-time access to distributed datasets. AI models can be updated remotely via cloud-enabled platforms, making it possible for even remote or smallholder farms to utilize advanced analytics without requiring on-site computational resources. Edge AI, which involves local processing on devices, decentralizes computing and complements cloud systems by minimizing latency and improving responsiveness in agricultural tasks that require immediate action (Jayaraman et al., 2016). The cooperative amalgamation of AI into various agricultural operations affects labor dynamics, sustainability, and resource efficiency on a broad scale. Automating tasks that are repetitive and require a lot of work—like soil sampling, weeding, harvesting, and crop monitoring—can help ease labor shortages and lessen reliance on seasonal labor markets. Concurrently, AI contributes to sustainability by reducing input waste, cutting greenhouse gas emissions through improved machinery efficiency, and preserving water and soil via precision actions. Moreover, by using real-time data to align agricultural inputs such as fertilizers, pesticides, and water with crop needs, resource efficiency is enhanced. This alignment improves input-output ratios and economic returns (Klerkx & Rose, 2020). Although these developments present potential dangers like job displacement and issues of data ownership, they also create opportunities for enhancing the skills of rural communities and promoting entrepreneurship in agriculture based on data. With the maturation of AI ecosystems in agriculture, it will be vital to establish inclusive governance frameworks, promote stakeholder collaboration, and deploy ethical AI to ensure that the advantages of integration are distributed equitably and align with environmental sustainability.

CHALLENGES AND BARRIERS

With the integration of AI into agricultural systems for purposes such as optimizing soil management, enhancing crop yield, and operating autonomous farming vehicles, it is crucial to scrutinize the ethical and practical issues that arise from this change. Among these concerns, data privacy, security, and ownership issues take precedence—particularly in light of the growing dependence on sensor networks, satellite imagery, drones, and on-field smart devices that produce enormous amounts of data. Farmers often lack control or ownership of the data gathered on their land; instead, it is collected by technology providers, which raises concerns about data sovereignty and potential exploitation (Carbonell, 2016). Due to the unbalanced power dynamics between individual farmers and major agritech firms, there is an increased risk of agricultural knowledge and insights becoming monopolized, which leads to the emergence of new kinds of digital dependency and exclusion (Rotz et al., 2019). Moreover, breaches in cybersecurity could lead to the misuse of sensitive farm data, such as altering yield forecasts, interfering with market activities, or causing environmental damage. Therefore, it is essential to create strong legal frameworks that clarify data rights, uphold ethical data governance, and ensure that farmers retain control over their agricultural information (Bronson, 2018). The accessibility and affordability of AI-driven technologies present another significant challenge, especially for smallholder farmers in developing and emerging economies. Although AI holds the promise of transforming agriculture by enhancing efficiency and minimizing resource waste, the expenses involved in obtaining, implementing, and sustaining AI systems are frequently beyond reach (Klerkx & Rose, 2020). Moreover, the lack of digital infrastructure and low levels of technological literacy in rural areas can hinder fair access to AI solutions, worsening the digital divide between large agribusinesses and small-scale or subsistence farmers. Without targeted interventions like public-private partnerships, government subsidies, and localized capacity-building programs, there is a significant risk that AI adoption will exacerbate rather than reduce existing socioeconomic inequalities in agriculture (Wolfert et al., 2017). To reach technological justice and a sustainable transformation of agriculture, it is essential to ensure inclusivity in AI development by designing tools with the context and constraints of marginalized communities in mind.

The social and environmental effects of adopting AI in agriculture also deserve careful scrutiny. While AI has the potential to lessen the environmental impact of agriculture through precise input application and real-time monitoring, it is important to note that AI systems—especially those based on machine learning algorithms and cloud computing—consume a considerable amount of energy (Lynn et al., 2020). Furthermore, self-driving vehicles and automation may result in a greater reliance on monoculture practices or increased soil compaction, thus posing a risk to the long-term ecological balance. From a social standpoint, AI-driven automation could lead to the displacement of agricultural workers, especially in areas where farming is a primary source of jobs. This dislocation of labor could worsen rural unemployment and migration, thereby disrupting social cohesion and livelihoods (Barrett & Rose, 2022). Consequently, the integration of AI must adhere to principles of ecological integrity and social inclusion, along with policies that facilitate retraining and workforce transition in rural areas.

Ultimately, the regulatory and policy frameworks that govern the use of AI in agriculture are still fragmented and lacking development in many regions worldwide. Technological development often progresses faster than the evolution of the legal and institutional frameworks required to manage the risks associated with AI. It is urgent to establish unambiguous regulatory frameworks concerning AI accountability, liability for autonomous decisions, transparency in algorithmic processes, and ethical data usage standards (European Commission, 2020). It is essential for decision-makers to undertake governance that is anticipatory, inclusive, and proactive—one that unites a variety of stakeholders such as farmers, civil society, environmental experts, and AI developers in the joint creation of safeguards and norms. To ensure ethical consistency and prevent regulatory arbitrage, especially in cross-border agri-tech operations, national and international institutions should align their policies. With AI continuing to reshape agricultural landscapes, developing a human-centered, rights-based regulatory framework is essential for ensuring that innovation aligns with the goals of equity, sustainability, and resilience.

FUTURE DIRECTIONS AND RECOMMENDATIONS

The future of agricultural innovation driven by AI will be shaped by its alignment with emerging technologies and the establishment of strong, interdisciplinary research ecosystems. State-of-the-art developments in AI—including federated learning, generative AI, edge computing, and neuromorphic computing—are set to transform the precision and scalability of digital agriculture. As an example, federated learning allows for decentralized training of data without compromising the privacy of farmers' data, which is highly beneficial in diverse agricultural settings (Yang et al., 2019). Likewise, edge AI applications minimize latency by handling data on-site with devices such as sensors and drones, a necessity for real-time decision-making in remote areas lacking infrastructure (Gaur et al., 2021). Soil diagnostics, dynamic crop modeling, and real-time navigation systems for autonomous agricultural vehicles can be significantly improved by these technological advances. Additionally, merging AI with biotechnology—especially genomics and phenomics—offers groundbreaking possibilities. Complex genetic data can be processed by AI algorithms to predict plant traits, optimize breeding cycles, and customize crop varieties for specific agro-climatic conditions (Singh et al., 2021). This integration could hasten the creation of robust crop varieties that aid in climate adaptation and food security when associated with synthetic biology and CRISPR-based technologies. Simulating crop performance under future climate scenarios through the integration of AI with climate modeling tools enables the development of predictive adaptation strategies. As an example, crop-climate models augmented by AI can aid in devising prompt agronomic measures and adaptive seed suggestions for areas at risk of droughts, floods, or heatwaves (Rolnick et al., 2022). This integration, which involves multiple layers, bolsters productivity while also mitigating risk and strengthening resilience in food systems.

These boundaries highlight the pressing necessity for interdisciplinary research frameworks that bring together data science, agronomy, ecology, and socioeconomics. To ensure that AI tools are contextually relevant, inclusively designed, and sustainably scaled, it is essential to create collaborative ecosystems that involve universities, agricultural research institutions, AI developers, and farmers. According to Wolfert et al. (2017), innovation models that involve farmers and rural stakeholders as active co-creators of technology can enhance adoption rates and build trust. Moreover, to facilitate interoperability and fair access—especially for smallholder farmers in the Global South—there is a need to establish open-source platforms and data-sharing agreements as institutional norms. Public-private partnerships and transdisciplinary consortia such as CGIAR's Digital Innovation Initiative offer valuable models for these integrative collaborations (CGIAR, 2023).

To scale AI innovations for sustainable agriculture globally, it is essential to tackle systemic barriers such as gaps in digital infrastructure, cost limitations, and regulatory inconsistencies. Policy measures should foster digital equity by investing in rural broadband, building capacity for local AI developers, and providing subsidies for precision agriculture tools. It is also crucial to establish ethical AI governance in agriculture, which includes aspects such as data ownership, algorithmic transparency, and environmental safeguards (Floridi et al., 2018). Platforms such as the FAO or the AI for Good Global Summit can foster international collaboration, which can help establish standards and transfer technology to low-income areas.

CONCLUSION

Innovations propelled by AI are transforming the field of contemporary agriculture by providing data-focused, precision-targeted solutions to age-old problems related to soil management, crop yield, and farm mechanization. This chapter examined the ways in which artificial intelligence, when combined with sensor technologies, computer vision, and autonomous systems, facilitates real-time observation of soil conditions, predictive irrigation and fertilization modeling, and automated resource use optimization. With the help of sophisticated machine learning methods, plant diseases and pests can be detected early, crop yield predictions can be optimized, and site-specific management practices can be enhanced—all contributing to the sustainable intensification of agriculture (Kamilaris & Prenafeta-Boldú, 2018; Liakos et al., 2018). At the same time, the emergence of self-driving agricultural machinery—like autonomous tractors, UAVs, and robotic harvesters—illustrates the merging of AI and robotics, resulting in labor-efficient and scalable farming systems (Bechar & Vigneault, 2016).

For farmers, these innovations hold the promise of enhanced productivity, lower input costs, and operations resilient to climate challenges. AI can aid smallholders by providing localized recommendations, but disparities in access to digital infrastructure are still a concern (Rose et al., 2021). It is recommended that policymakers develop inclusive policies aimed at closing the digital divide, fostering responsible data governance, and encouraging AI adoption via subsidies and training initiatives. Simultaneously, researchers have the responsibility of promoting algorithmic transparency, contextual adaptability, and ethical AI design to guarantee that innovations are in line with environmental sustainability and social equity (Fleming et al., 2022).

Considering how AI could transform agriculture, it is clear that the technology encompasses more than just automation. It enables a fundamental change toward farming that is knowledge-intensive, adaptable, efficient, and resilient. Yet, to tap into this potential, we need strong interdisciplinary cooperation, commitment to digital literacy, and frameworks for participatory innovation. With the growing digitization of agriculture, it is crucial to keep a balance between technological progress and ecological stewardship. When deployed ethically and inclusively, AI can act as a catalyst for achieving food security, rural development, and environmental sustainability—objectives that are central to the 2030 Agenda for Sustainable Development.

REFERENCES

Bacco, M., Barsocchi, P., Ferro, E., Gotta, A., & Ruggeri, M. (2019). Drone applications in precision agriculture: A review. *Sensors (Basel)*, *19*(11), 2577. DOI: 10.3390/s19112577

Bakker, M. M.. (2021). AI-enabled irrigation scheduling for water-efficient agriculture: A review. *Agricultural Water Management*, *245*, 106604. DOI: 10.1016/j.agwat.2020.106604

Bakr, M. (2020). Robotics in agriculture: Current applications and future perspectives. *Agricultural Robotics Journal*, *2*(1), 45–59.

Barrett, C. B., & Rose, A. (2022). Artificial Intelligence and Agricultural Labor: Promises and Perils. *Journal of Agrarian Change*, *22*(1), 45–60. DOI: 10.1111/joac.12400

Bechar, A., & Vigneault, C. (2016). Agricultural robots for field operations: Concepts and components. *Biosystems Engineering*, *149*, 94–111. DOI: 10.1016/j.biosystemseng.2016.06.014

Blackmore, S., Stout, B., Wang, M., & Runov, B. (2020). Agricultural automation: The future of farming. *International Journal of Agricultural and Biological Engineering*, *13*(5), 1–15. DOI: 10.25165/j.ijabe.20201305.5321

Bronson, K. (2018). Smart Farming: Including Rights Holders for Responsible Agricultural Innovation. *Technology Innovation Management Review*, *8*(2), 7–14. DOI: 10.22215/timreview/1135

Carbonell, I. M. (2016). The Ethics of Big Data in Agriculture. *Internet Policy Review*, *5*(1), 1–13. DOI: 10.14763/2016.1.405

CGIAR. (2023). *Digital Innovation Initiative*. https://www.cgiar.org/initiative/10-digital-innovation/

Chen, L., Yang, J., & Liu, S. (2020). Regulatory framework and ethical concerns of autonomous agricultural vehicles. *Journal of Agricultural Safety and Health*, *26*(3), 175–187. DOI: 10.13031/jash.13756

Das, S.. (2020). Machine learning for soil moisture prediction: A review. *Environmental Modelling & Software*, *130*, 104728. DOI: 10.1016/j.envsoft.2020.104728

Duckett, T., Pearson, S., Blackmore, S., Grieve, B., Cielniak, G., & Jin, J. (2018). Agricultural robotics: The future of robotic agriculture. *arXiv preprint arXiv:1806.06762*. DOI: 10.31256/WP2018.2

European Commission. (2020). *White Paper on Artificial Intelligence – A European Approach to Excellence and Trust*. https://commission.europa.eu/document/white-paper-artificial-intelligence-european-approach_en

FAO. (2017). *The future of food and agriculture – Trends and challenges*. Food and Agriculture Organization of the United Nations. https://www.fao.org/3/i6583e/i6583e.pdf

Fleming, A., Jakku, E., Lim-Camacho, L., Taylor, B., & Thorburn, P. (2022). Trust, transparency and technology: Building ethical AI for agriculture. *AI & Society*, *37*(2), 689–701.

Floridi, L., Cowls, J., Beltrametti, M., Chatila, R., Chazerand, P., Dignum, V., Luetge, C., Madelin, R., Pagallo, U., Rossi, F., Schafer, B., Valcke, P., & Vayena, E. (2018). AI4People—An ethical framework for a good AI society: Opportunities, risks, principles, and recommendations. *Minds and Machines*, *28*(4), 689–707. DOI: 10.1007/s11023-018-9482-5

Gaur, M., Sain, M., Sharma, D. K., Varshney, D., & Singh, M. (2021). Edge artificial intelligence for Internet of Things: A survey, future trends, and enabling technologies. *Computer Communications*, *166*, 116–139.

Gebbers, R., & Adamchuk, V. I. (2010). Precision agriculture and food security. *Science*, *327*(5967), 828–831. DOI: 10.1126/science.1183899

Goodfellow, I., Bengio, Y., & Courville, A. (2016). *Deep learning*. MIT Press.

Grift, T. E., Qi, L., & Deen, M. J. (2018). Machine vision systems for precision agriculture. *Transactions of the ASABE*, *61*(1), 175–192. DOI: 10.13031/trans.12444

Jayaraman, P. P., Yavari, A., Georgakopoulos, D., Morshed, A., & Zaslavsky, A. (2016). Internet of Things platform for smart farming: Experiences and lessons learned. *Sensors (Basel)*, *16*(11), 1884. DOI: 10.3390/s16111884

Jiang, Y., Wang, C., & Luo, X. (2020). Smart irrigation system based on Internet of Things and artificial intelligence technology. *Computers and Electronics in Agriculture*, *168*, 105107.

Jin, J., Li, Z., & Chen, X. (2019). Robust perception for autonomous agricultural vehicles in adverse weather conditions: A review. *Agricultural Engineering International: CIGR Journal*, *21*(2), 1–12.

Jones, H. G.. (2021). Real-time irrigation scheduling using AI: Case study in California vineyards. *Precision Agriculture*, *22*(3), 874–890. DOI: 10.1007/s11119-021-09814-5

Kamilaris, A., Kartakoullis, A., & Prenafeta-Boldú, F. X. (2017). A review on the practice of big data analysis in agriculture. *Computers and Electronics in Agriculture, 143*, 23–37. DOI: 10.1016/j.compag.2017.09.037

Kamilaris, A., & Prenafeta-Boldú, F. X. (2018). Deep learning in agriculture: A survey. *Computers and Electronics in Agriculture, 147*, 70–90. DOI: 10.1016/j.compag.2018.02.016

Klerkx, L., & Rose, D. (2020). Dealing with the game-changing technologies of Agriculture 4.0: How do we manage diversity and responsibility in food system transition pathways? *Global Food Security, 24*, 100347. DOI: 10.1016/j.gfs.2019.100347

Kolling, A., Oppenheim, I., & Elkaim, G. (2018). John Deere autonomous tractors: A case study in automation and AI deployment. *Automation in Agriculture, 7*(4), 234–245.

Lal, R. (2020). Soil health and carbon management. *Soil & Tillage Research, 195*, 104365. DOI: 10.1016/j.still.2019.104365

LeCun, Y., Bengio, Y., & Hinton, G. (2015). Deep learning. *Nature, 521*(7553), 436–444. DOI: 10.1038/nature14539

Li, M.. (2020). Advances in soil quality assessment using AI and remote sensing. *Soil Science Society of America Journal, 84*(5), 1422–1434. DOI: 10.1002/saj2.20156

Li, X.. (2021). Challenges and future directions for AI in soil monitoring. *Computers and Electronics in Agriculture, 189*, 106362. DOI: 10.1016/j.compag.2021.106362

Liakos, K. G., Busato, P., Moshou, D., Pearson, S., & Bochtis, D. (2018). Machine learning in agriculture: A review. *Sensors (Basel), 18*(8), 2674. DOI: 10.3390/s18082674

Lynn, T., Rosati, P., Lejeune, A., & Kumar, A. (2020). *The Ethics of Smart Farming: Current Issues and Governance*. Journal of Information. *Communication and Ethics in Society, 18*(1), 79–99. DOI: 10.1108/JICES-12-2018-0091

McBratney, A., Minasny, B., Walvoort, D., & Odeh, I. (2019). Precision agriculture and digital soil mapping. In Carter, M. R., & Gregorich, E. G. (Eds.), *Soil Sampling and Methods of Analysis* (pp. 899–917). CRC Press.

Minasny, B.. (2019). Machine learning applications for soil data interpretation. *Geoderma, 334*, 114–132. DOI: 10.1016/j.geoderma.2018.07.043

Mohanty, S.. (2021). Soil moisture prediction using hybrid machine learning models: A review. *Hydrological Sciences Journal, 66*(12), 2023–2040. DOI: 10.1080/02626667.2021.1957846

Molina, P., García, E., & López, M. (2020). The ATLAS project: Autonomous tractors and spraying robots for precision agriculture. *Robotics and Autonomous Systems, 123*, 103310. DOI: 10.1016/j.robot.2019.103310

Montgomery, D. R. (2007). Soil erosion and agricultural sustainability. *Proceedings of the National Academy of Sciences of the United States of America, 104*(33), 13268–13272. DOI: 10.1073/pnas.0611508104

Padarian, J.. (2019). Soil nutrient prediction using vis-NIR spectroscopy and machine learning. *Geoderma, 353*, 178–189. DOI: 10.1016/j.geoderma.2019.06.017

Panagos, P.. (2018). Soil erosion risk assessment in Europe. *The Science of the Total Environment, 631-632*, 1446–1460. DOI: 10.1016/j.scitotenv.2018.03.047

Pedersen, S. M., Fountas, S., Have, H., & Blackmore, S. (2018). Safety considerations for autonomous vehicles in agriculture. *Biosystems Engineering, 164*, 1–12. DOI: 10.1016/j.biosystemseng.2017.12.002

Reichstein, M., Camps-Valls, G., Stevens, B., Jung, M., Denzler, J., Carvalhais, N., & Prabhat, . (2019). Deep learning and process understanding for data-driven Earth system science. *Nature, 566*(7743), 195–204. DOI: 10.1038/s41586-019-0912-1

Roldán, J. J., Ramos, J., & Muñoz, F. (2020). Autonomous navigation systems for agricultural vehicles: A review. *Sensors (Basel), 20*(7), 1985. DOI: 10.3390/s20071985

Rolnick, D., Donti, P. L., Kaack, L. H., Kochanski, K., Lacoste, A., Sankaran, K., & Bengio, Y. (2022). Tackling climate change with machine learning. *ACM Computing Surveys, 55*(1), 1–96.

Rose, D. C., Higgins, V., Mitchell, P., & Bryant, M. (2021). The co-production of smart farming knowledge: Insights from stakeholder workshops. *Agricultural Systems, 188*, 103033.

Rotz, S., Gravely, E., Mosby, I., Duncan, E., Finnis, E., Horgan, M., & Martin, R. (2019). Automated pastures and the digital divide: How agricultural technologies are shaping labour and rural communities. *Journal of Rural Studies, 68*, 112–122. DOI: 10.1016/j.jrurstud.2019.01.023

Russell, S., & Norvig, P. (2021). *Artificial intelligence: A modern approach* (4th ed.). Pearson.

Shahhosseini, M., Martinez-Feria, R., & Archontoulis, S. V. (2021). Simulation of field-scale precision nitrogen management impacts on maize yield and soil nitrogen dynamics. *Agricultural Systems, 186*, 102983. DOI: 10.1016/j.agsy.2020.102983

Shamshiri, R. R., Weltzien, C., Hameed, I. A., Yule, I. J., & Ahmad, D. (2018). Research and development in agricultural robotics: A perspective on technologies, applications, and challenges. *Biosystems Engineering*, *165*, 559–572. DOI: 10.1016/j.biosystemseng.2017.12.006

Singh, A., Ganapathysubramanian, B., Singh, A. K., & Sarkar, S. (2020). Machine learning for high-throughput stress phenotyping in plants. *Trends in Plant Science*, *21*(2), 110–124. DOI: 10.1016/j.tplants.2015.10.015

Singh, A. K., Ganapathysubramanian, B., Singh, A., & Sarkar, S. (2021). Deep learning for plant stress phenotyping: Trends and future perspectives. *Trends in Plant Science*, *26*(9), 866–875.

Srinivasan, R.. (2020). Precision nutrient management in agriculture using AI technologies. *Agronomy (Basel)*, *10*(6), 903. DOI: 10.3390/agronomy10060903

Stenberg, B., Viscarra Rossel, R. A., Mouazen, A. M., & Wetterlind, J. (2010). Visible and near infrared spectroscopy in soil science. *Advances in Agronomy*, *107*, 163–215. DOI: 10.1016/S0065-2113(10)07005-7

Szeliski, R. (2010). *Computer vision: Algorithms and applications*. Springer.

United Nations. (2015). *Transforming our world: The 2030 agenda for sustainable development*. https://sdgs.un.org/2030agenda

Van Evert, F., Dam, J., & Slingerland, M. (2017). Building capacity for autonomous agriculture: Education and training needs. *International Journal of Agricultural Sustainability*, *15*(1), 45–60. DOI: 10.1080/14735903.2016.1256273

Verdouw, C., Wolfert, J., Beulens, A., & Rialland, A. (2016). Virtualization of food supply chains with the Internet of Things. *Journal of Food Engineering*, *176*, 128–136. DOI: 10.1016/j.jfoodeng.2015.11.009

Wolfert, S., Ge, L., Verdouw, C., & Bogaardt, M. J. (2017). Big data in smart farming – A review. *Agricultural Systems*, *153*, 69–80. DOI: 10.1016/j.agsy.2017.01.023

Wossen, T.. (2020). AI-based soil fertility mapping for smallholder farmers in Africa. *Agricultural Systems*, *177*, 102711. DOI: 10.1016/j.agsy.2019.102711

Yamamoto, S., Umeda, M., & Fujimura, S. (2019). Autonomous rice transplanters for labor shortage solutions in Japan. *Journal of Agricultural Machinery Science*, *15*(3), 123–132.

Yang, Q., Liu, Y., Chen, T., & Tong, Y. (2019). Federated machine learning: Concept and applications. [TIST]. *ACM Transactions on Intelligent Systems and Technology*, *10*(2), 1–19. DOI: 10.1145/3298981

Zhao, G.. (2019). AI-based erosion risk prediction using satellite data. *Environmental Modelling & Software*, *117*, 159–169. DOI: 10.1016/j.envsoft.2019.03.008

Zhou, L., Zhang, Q., & Shi, Z. (2020). Drone remote sensing for precision agriculture: Review and future prospects. *Remote Sensing*, *12*(7), 1096. DOI: 10.3390/rs12071096

Chapter 2
Inequality and Environmental Degradation in Mexico as a Result of Green and Renewable Energy-Related Financial Literacy

José G. Vargas-Hernandez
https://orcid.org/0000-0003-0938-4197
Financial Literacy Research center, Universidad Cristóbal Colón, Mexico & Tecnológico Nacional de México, ITS Fresnillo, Mexico

Csongor Czipf
https://orcid.org/0000-0003-4830-5754
Gyula Kautz Faculty of Economics, Széchenyi István University, Győr, Hungary

ABSTRACT

This research paper examines the complex relationship between green and renewable energy sector financial literacy and its impact on inequality and environmental degradation in Mexico. Through interdisciplinarity that engages environmental justice theory, financial inclusion frameworks, and sustainable development perspectives, this research reveals how limited financial literacy in renewable energy technologies consolidates socioeconomic inequalities while advancing environmental degradation. The research concludes that intense financial literacy programs to renewable

DOI: 10.4018/979-8-3373-5651-8.ch002

energy literacy, combined with comprehensive policy frameworks, are essential to providing just energy transitions, which address both environmental and social justice concerns in Mexico's path toward sustainable development.

INTRODUCTION

Mexico is poised on the cusp of a turning point in its energy transition, with the dual challenge of realizing aggressive targets of renewable energy while stopping entrenched social injustices and ecological devastation. Mexico's General Climate Change Law and Energy Transition Law established a target of 35 percent clean energy electricity by the year 2024, and the government of Mexico has pledged to increase the share of renewable energy to 45% by the year 2030. However, these ambitious targets are positioned within the context of significant socioeconomic disparities and ecological degradation that complicate the transition towards an equitable and sustainable energy future.

The intersection of financial literacy and renewable energy represents a highly challenging scenario for Mexico. Although Mexico has recorded noteworthy progress in the growth of renewable energy, Mexico ranked 38 out of 67 countries in the Climate Change Performance Index 2024, down seven spots from the Climate Change Performance Index 2023. This following drop in performance, together with persistent inequities in access to energy supplies and financial services, emphasizes the value of a comprehensive conceptualization of the ways in which financial literacy in green energy sectors affects environmental performance as well as social equity.

The idea of energy justice has gained increasing relevance in renewable energy transition discussion, accepting that the advantages and disadvantages of energy systems are not symmetrically allocated in society. Mexico pledged 35% clean energy by 2024 and 50% by 2050, and reaffirmed those targets in its National Energy Strategy 2013–2027 and in its 2017 Climate Change Mid-Century Strategy, aiming for achievement of the 50% goal under a climate of "environmental justice". That pledge to environmental justice focuses on exploring how financial literacy barriers can reinforce or exacerbate existing disparities of Mexico's energy transition.

Financial literacy for renewable energy encompasses a number of aspects, including understanding green financing mechanisms, awareness of investment opportunities in renewable energy, knowledge of energy efficiency practices, and knowledge of economic costs and benefits of sustainable energy sources. Access to formal financial services is particularly low, and the access thereto is extremely unequal across incomes, by gender, between rural and urban areas and by regions. This economic disparity continues to create other obstacles for marginalized communities that desire to be engaged in the shift to renewable energy.

Financial literacy's link to environmental consequences is particularly great in Mexico, since climate change and ecological decline have emerged as two of the most urgent issues confronting Mexico, in addition to violence and crime, poverty and inequality, unemployment, and social inequality. The convergence of environmental and social problems has resulted in a complex web of competing challenges that require to be explored thoroughly so as to understand how financial literacy deficits are leading to ecological decline and social inequality.

This research study aims to investigate the intricate link between financial literacy of renewable energy and green energy and its implications on environmental degradation and inequality in Mexico. The study answers a few key research queries: How does limited financial literacy in renewable energy fields impact environmental degradation? What are the channels through which financial literacy gaps perpetuate energy inequality? How do current policy strategies position or not position these interrelated issues? What interventions are viable to promote financial literacy in renewable energy markets with environmental justice support?

The research employs mixed-methods analysis that incorporates quantitative proof from Mexican national surveys, policy documents, and case studies of renewable energy projects across Mexico. The key findings indicate that around 61% of Mexican families suffered from energy poverty due to issues with the accessibility or affordability of modern energy services, as financial illiteracy in knowledge was a critical barrier to inclusive energy transition. The study demonstrates how poor understanding of green finance instruments, renewable energy investment opportunities, and sustainable energy activities places a two-way burden: environmental degradation on vulnerable populations and entrenchment of energy poverty in marginalized segments.

The significance of this research extends beyond academic analysis to inform policy development and implementation strategies for Mexico's ongoing energy transition. By making sense of the inter-linked processes of financial literacy, inequity, and environmental degradation in the renewable energy sector, policymakers, financial actors, and civil society groups can develop more effective interventions supporting both environmental sustainability and social equality.

CONCEPTUAL FRAMEWORK

The theoretical foundation of the research integrates three key theoretical domains: theory of environmental justice, theory of financial inclusion, and theory of sustainable development. The theoretical pillars provide a broad conceptual framework for examining the complex interplay between financial literacy, renewable energy transitions, and socio-environmental effects in Mexico.

Environmental Justice Theory

Environmental justice theory gives a general framework for understanding how environmental benefits and disadvantages are distributed among groups of people. The theory evolved out of the recognition that environmental harms and benefits are not randomly distributed but instead reflect prevailing trends of social, economic, and political disadvantages. In the context of renewable energy transitions, environmental justice theory sheds light on how the shift to clean energy can reproduce or disrupt prevailing trends of environmental disadvantage.

The distributive component of environmental justice examines the distribution of environmental goods and bads across different communities. In the Mexican renewable energy sector, this involves who benefits from access to clean energy and who bears the cost of energy transition. The procedural component focuses on the fairness of decision-making processes, such as whether or not impacted communities have effective participation in planning and implementing renewables. The recognition dimension addresses whether or not different needs, values, and lifestyles of different communities are respected and maintained within energy transition.

Corrective justice, the fourth dimension, examines how to correct historical and present environmental injustices. In Mexico, this entails examining how past energy policy has disproportionately burdened indigenous and rural populations and examining how current renewable energy policies can serve as corrective actions on past environmental inequalities.

Financial Inclusion Theory

Theory of financial inclusion provides an account of how access to financial services affects socioeconomic results and development opportunities. Theory of financial exclusion is said to be behind obstacles to economic participation and social mobility, whereas financial inclusion can then be the door to improved well-being and reduced inequality. Theory of financial inclusion is used in the case of renewable energy to account for how access to green finance instruments affects the capacity of communities to participate and benefit from energy transformation.

The multi-dimensional nature of financial inclusion has dimensions of access, usage, and quality. Access is the availability of financial products and services, for example, proximity to financial institutions and affordability of the services. Usage examines actual utilization of financial services, for instance, frequency and regularity of use and purpose of use. Quality examines whether financial services are appropriate to address user needs and delivered in a responsible and sustainable manner.

Financial literacy is a critical component of financial inclusion as it affects the ability of individuals to make an informed choice regarding financial products and services. Financial literacy in the renewable energy sector entails having knowledge of green bonds, renewable energy loans, energy efficiency financing, and community-based renewable energy investment schemes. Low financial literacy can exclude individuals from accessing financial products such as these, thereby perpetuating energy poverty and limiting participation in renewable energy transitions.

Sustainable Development Theory

Sustainable development theory is from the theory that describes how economic, social, and environmental dimensions of development relate to each other and affect one another. The theory is concerned with strategies for development that can meet the needs of the present without compromising future generations' ability to meet their own needs. In renewable energy transformation, sustainable development theory provides examination of how energy choices affect long-term environmental sustainability relative to short-term social and economic implications.

The integrated approach of sustainable development recognizes that economic, social, and environmental dimensions are complementary and interdependent. Environmental sustainability in the energy domain means moving towards renewable energy without degrading the environment. Social sustainability involves ensuring that energy transitions improve social equity and do not aggravate prevailing inequalities. Economic sustainability requires that the shift to renewable energy is economically viable and supports long-term economic growth.

Time dimension of sustainable development requires serious consideration of the long-term consequences of current decisions. In Mexico's energy transition, this is to consider how current renewable energy laws and financial education initiatives will have implications for future environmental and social performance. Intergenerational equity principle requires current energy decisions not to compromise the ability of future generations to meet their energy needs without harming environmental quality.

Integrated Conceptual Model

The intersection of these three theoretical pillars provides a rich conceptual model for conceptualizing financial literacy, renewable energy transitions, and socio-environmental outcomes. The converged model recognizes that financial literacy in renewable energy sectors influences environmental outcomes through multiple channels, including energy choices by individuals, take-up of renewable energy at the community level, and access to green financing vehicles.

The model indicates that low financial literacy in the renewable energy sector is an obstacle to accessing green financing, understanding renewable energy investment possibilities, and wise energy choice-making. It disproportionately discloses marginalized groups, and therefore they maintain energy poverty and limited access to clean energy advantages. Meanwhile, low financial literacy has the potential to lead to the utilization of ecologically hazardous energy supplies, leading to environmental degradation for marginalized groups.

The model further recognizes that financial literacy occurs in broader structural contexts, including policy environments, institutional arrangements, and market forces. These structural contexts may enable or inhibit the emergence of financial literacy and its translation into beneficial environmental and social impacts. It is important to comprehend these structural contexts so that one can develop effective interventions aimed at addressing both financial literacy gaps and the causes of such gaps.

LITERATURE REVIEW

Theoretical Literature

Theory regarding financial literacy and energy transitions from renewable sources is derived from various disciplinary schools ranging from environmental economics to development studies and energy policy analysis. Sovacool and Dworkin's (2014) holistic energy justice framework provides theoretical perspectives to the reproduction or transformation of social and environmental injustice via transitions of energy. Their research suggests distributive, procedural, and recognition dimensions of justice in energy transitions to be examined.

There is ample literature reporting the link between financial inclusion and poverty reduction from studies on financial development outcomes and financial inclusion. The dataset of the Demirgüç-Kunt and Klapper (2012) of the world's financial inclusion shows pervasive heterogeneity of access to finance across countries and population groups and among rural residents, women, and low-income households are among the population groups with unique challenges. The study offers context to the research on the trends of financial exclusion in Mexico's renewable energy sector.

New theoretical contributions to environmental justice studies have increasingly been preoccupied with the intersection of environmental and social justice concerns. Schlosberg's (2013) capabilities approach to environmental justice provides a suitable blueprint of how environmental conditions translate into individuals' ability to achieve desired functioning and capabilities. The capabilities approach is particularly relevant in consideration of how energy poverty affects overall development outcomes and welfare.

The literature on energy transitions and sustainable development has evolved to recognize interdependent relationships between environmental, social, and economic dimensions of sustainability. The integrated nature of development challenges is emphasized by Sachs' (2015) sustainable development goals model, as well as the need for holistic solutions with effects at multiple levels simultaneously. The holistic approach is essential in understanding how changes towards renewable energy can achieve varied development objectives without having unintended negative impacts.

Empirical Literature

Empirical research on clean energy take-up and financial literacy provides inconclusive evidence regarding the association between financial literacy and clean energy choices. Sardianou and Genoudi (2013), while researching Greek households, found household financial literacy to have a positive association with energy efficiency investment, suggesting financial literacy promotes environmentally friendly behavior. But as per their work, it was learned that financial literacy affects the environment through income and financial access.

Empirical research on the use of renewable energy in developing economies has brought forward financial constraints and literacy constraints as primary determinants of access to clean energy. In their study on the adoption of renewable energy in Kenya, Pueyo *et al.* (2013) determined that inadequate knowledge of financing options and high capital expenditures were key disincentives to the use of clean energy, particularly among low-income earners. Education on financial literacy for renewable energy is proposed in their study as an appropriate strategy to mitigate such disincentives.

Previous studies of energy poverty and inequality documented significant variations in terms of access to energy and energy affordability across social groups. The González-Eguino (2015) study of Spanish energy poverty found poor household to have disproportionate energy spending and worse energy efficiency measure access. It appeals for bridging funding gaps to reach clean energy as one of the measures towards reducing energy inequality.

Studies on Mexican renewable energy initiatives revealed complex interactions between energy transitions and social effects. Dunlap's (2018) ethnographic investigation of wind power growth in Oaxaca revealed that giant-scale renewable energy initiatives utilized to reproduce existing disparities in environmental and social meanings whereby outsiders benefited at the expense of locals who bore environmental and social burdens. The study rendered it relevant to investigate how renewable energy transitions affect different social groups.

Policy Literature

Policy analysis of the Mexican transition to renewable energy delineates the complex political and institutional forces behind energy policy making and implementation. Elizondo Azuela *et al.* (2014) delineates the conflict between ambitious targets for clean energy and practical boundaries to implementation, e.g., institutional capacity constraints and opposition from incumbent energy groups.

Mexican financial inclusion policy has been studied and documented to have made concerted attempts to enhance access to formal finance and at the same time documented recurring challenges to accessing the marginalized. Ruiz-Arranz *et al.*'s (2016) evaluation of Mexico's financial inclusion strategy was that greater access to simple financial services had been achieved but significant gaps still existed concerning financial literacy and take-up of advanced financial products. The research provides useful background for placing challenges in enhancing financial literacy in renewable energy markets.

Mexican energy subsidy reform literature and evidence regarding energy subsidies have complex interrelations with social justice, energy prices, and the environment. Coady, Flamini, and Sears's (2015) critique of energy subsidy reform ended with the fact that stopping fossil fuel subsidies would generate green dividends, but the reforms started to set off distributional concerns to be met with special regard to policy design so as to not harm poor households regressively.

Recent studies on policy have been more focused on the concept of just energy transitions, recognized the necessity to incorporate social and environmental justice into policy formulation in energy. McCauley *et al.*'s (2019) just energy transitions model places significant focus on procedural and distributive justice in ensuring energy transitions are fair to all stakeholders involved in society. The model provides significant clarity on the lens to be used in examining Mexico's renewable energy policies.

Gaps in Literature

Despite the rising amount of work on renewable energy transition and financial literacy, there are some underlying gaps in literature. Firstly, there is a deficiency in clear studies that particularly examine the relationship between environmental impacts and renewable energy industry financial literacy in developing countries. While literature has shown that economic constraints are a limiting factor to access to clean energy, hardly any have tried to ascertain how financial literacy contributes to environmental degradation trends.

Second, past studies of environmental justice and the shift to renewable energy have focused primarily on commercial-sized renewable energy facilities with inadequate emphasis on end-user or household level energy decisions and their total environmental impacts. Information on how group- and individual-level energy choice-making is affected by financial literacy has to be provided to formulate effective action to protect the environment. Third, little evidence is available on the effect of financial literacy programs to stimulate take-up of clean energy and environmental conservation. Although literature estimates the importance of financial literacy constraints, little evidence is available on the type of financial education interventions to best support clean energy choices and environmental benefits.

Finally, current studies of Mexico's renewable energy transition have maintained its focus to a significant degree in policy and institutional studies compared to social and environmental justice dimensions of energy transitions. It is critical to identify the impacts of renewable energy policies on different social groups and communities so that more sustainable and fair energy transitions can be conceptualized.

METHODS

Research Design

The research employs a mixed-methods approach that integrates quantitative analysis of secondary data and qualitative analysis of policy documents and case studies. The mixed-methods design facilitates in-depth exploration of the complex dynamics between financial literacy, renewable energy transformations, and socio-environmental impacts through triangulating various forms of evidence and analytical methods.

The research is constructed on three pillars: (1) quantitative analysis of national surveys to identify trends of financial literacy, energy poverty, and environmental degradation across social groups; (2) qualitative analysis of policy documents and institutional frameworks to grasp how current policies address or avoid financial literacy-renewable energy transition nexus; and (3) case study analysis of focused renewable energy projects and programs to investigate the extent to which financial literacy issues impact project success and community benefits.

The time frame of the study is from 2015 to 2024, aligned with the current renewable policy regime of Mexico and data availability. This period captures the enforcement of Mexico's Energy Transition Law and other renewable energy policies, ensuring adequate data for an analysis of policy outcomes and effects.

Data Sources

The study uses several sources of data to ensure there is an in-depth look at the research questions. Some of the primary quantitative data sources are the National Survey of Financial Inclusion (ENIF) provided by the National Institute of Statistics and Geography (Instituto Nacional de Estadística y Geografía (INEGI), 2021, 2022), offering in-depth information regarding the level of financial literacy, access to financial products, and usage levels within various groups of people. There are specific questions provided in the survey regarding green financing product awareness and investment in renewable energy.

Other quantitative sources include the National Household Income and Expenditure Survey (ENIGH), which publishes energy expenditure and affordability across income levels, and the National Energy Information System (SNIE), which publishes energy consumption patterns and the proportion of renewable energy use at state and municipal levels. Sources of qualitative data include Ministry of Energy (Secretaría de Energía (SENER), 2023, 2024) policy documents, the National Commission for the Efficient Use of Energy (CONUEE), and the National Banking and Securities Commission (CNBV) policy documents. These policy documents provide data on policy mechanisms, institutional designs, and regulatory styles towards renewable energy transitions and financial inclusion.

Sources of case studies are project reports from renewable energy projects, interviews with project implementers and stakeholders, and community impact studies. The case studies address three types of renewable energy projects: large wind farms, distributed solar projects, and community-based renewable energy cooperatives.

Analytical Approach

Quantitative analysis employs descriptive statistics to examine trends in financial literacy, energy poverty, and environmental degradation across demographic groups. Multivariate regression analysis is employed to examine the association between financial literacy measures and energy-related outcomes after controlling for the relevant socioeconomic and demographic controls.

Financial literacy analysis addresses several key dimensions: (1) base financial literacy, e.g., awareness of interest rates, inflation, and risk diversification; (2) green finance literacy, e.g., awareness of green bonds, renewable energy loans, and energy efficiency financing; and (3) energy-specific financial literacy, e.g., understanding of energy prices, investment in energy efficiency, and renewable energy economics.

Environmental degradation indicators include measures of air quality, water quality, and land use conversion trends. The research considers how the environmental indicators vary between communities with different financial literacy and penetration of renewable energy, after controlling for relevant socioeconomic and geographic factors. The qualitative analysis employs thematic analysis in identifying main patterns and themes in policy documents and case study findings. The analysis focuses on some of the key themes: (1) policy responses to renewable energy and financial inclusion; (2) institutional mechanisms for promoting financial literacy; (3) challenges in the utilization of renewable energy; and (4) social and environmental impacts of renewable energy projects.

Limitations

Various limitations need to be taken into account in this research. First, the cross-sectional nature of a majority of the available data limits establishing causal effects of financial literacy on environmental indicators. While analysis can confirm relationships among the variables, causality cannot be established in the absence of longitudinal data, which is available for not all of the variables of interest. Second, it is difficult to measure financial literacy in renewable energy markets due to the lack of standardized tools and the dynamic nature of green finance products. The research relies on proxy measures and existing survey instruments that may not capture all elements of relevant financial knowledge.

Third, analysis of environmental degradation trends is limited by the availability and nature of environmental data at the community level. Even though environmental indicators at the state and country levels are available, additional local environmental data would enable more precise analysis of financial literacy relationships with environmental outcomes.

Finally, the case study analysis is limited by availability of project reports and stakeholders' willingness to be interviewed. The case studies are informative but perhaps not representative of all of Mexico's renewable energy projects.

ANALYSIS OF RESULTS

Financial Literacy Trends in Mexico

Financial literacy trends in Mexico indicate enormous disparities by demographic segments and geography. Mexican financial literacy overall is found to be very low, with fundamental financial literacy in major areas of interest rates, inflation, and risk diversification among only 32% of adults, based on National Survey of Financial Inclusion (ENIF) data. These patterns become more pronounced when examining green finance literacy, as only 18% of the respondent's indicated knowledge of green financing investments or products in alternate energy.

Geographic disparity in financial literacy is stark, with considerably higher financial literacy within urban compared to rural communities. Urban residents possess 38% basic financial literacy compared to 21% in rural communities. The urban-rural contrast is even more pronounced for green finance literacy, with 24% urban awareness compared to a paltry 9% in rural communities. These disparities have important implications for the patterns and environmental performance of renewable energy adoption. The financial literacy gaps are also noteworthy, with men being more financially literate than women across all groups. The rates of basic financial literacy are 36% among men and 28% among women, and the rates of green finance literacy are 22% among men and 14% among women. These gender gaps sum up to other gaps to create cumulating hurdles to renewable energy access and adoption.

Income-based variations in financial literacy have convergent patterns, with financially literate households far more prevalent among higher-income rather than lower-income families. The analysis shows, however, that the link between income and financial literacy is not linear, with diminishing returns to income for financial literacy improvements at income levels above the middle-income groups. This has implications that the reduction of financial literacy gaps will require more specialized interventions than just income transfers. Education has very strong positive correlations with financial literacy across all groups.

Individuals who are more educated have much higher levels of basic financial literacy and green finance literacy. The study finds, however, that education level is not sufficient in itself to explain trends in financial literacy, with the suggestion that specially targeted financial education programs would need to be delivered in order to improve financial literacy outcomes.

Energy Poverty and Inequality Trends

The pattern examination of energy poverty in Mexico attests that energy poverty is a critical scenario affecting most of the population. It is reflected by statistics for the year 2014 that nearly 61% of Mexican households were affected by energy poverty measured by access or affordability issues of modern energy services and fuels, and 11.54% of the households suffered from both types of energy poverty. These trends hold relevant implications for the adoption of renewable energies and environmental impacts.

Energy poverty patterns have robust geographic characteristics, with rural compared to urban areas experiencing more energy poverty. Rural houses experience both energy access shortages and energy affordability challenges more than their urban counterparts, creating compounding barriers to improved energy services. The trends are more distinctly pronounced for indigenous communities, where indigenous energies can be more culturally appropriate yet more ecologically harmful as well.

The analysis discovers strong correlations between financial literacy levels and impacts on energy poverty. The individuals with greater financial literacy levels are less energy poor even after controlling for income and other socioeconomic determinants. The correlation is particularly strong for green finance literacy, where individuals with greater levels of green finance literacy are considerably more likely to have retrofitted energy efficiency measures or renewable energy systems.

Household expenditure patterns on energy reveal that poor households spend a higher share of their income on energy services, hence creating further barriers to accessing better quality energy services. The findings show that households in the lowest income quintile spend, on average, 12% of their income on energy services, and the richest income quintile spends 4% of their income on energy services. These patterns of consumption instill resistance towards investment in cleaner energy technologies or energy efficiency that would lower the energy cost in the long run.

The relationship between energy poverty and environmental degradation is complex and reciprocal. More energy-poor families will utilize the traditional biomass fuels, therefore generating indoor air pollution as well as forest cutting. These families use less total energy overall, though, so their per-capita greenhouse gas emissions are smaller than among more affluent households.

Trends of Renewable Energy Adoption

Mexico's renewable energy consumption trends show huge variations across and within social categories and regions. Renewables such as wind, sun, and biomass produce around 25% of electricity, but the benefits of renewable energy uptake are not distributed proportionally among society. Big renewable energy projects, especially wind farms, are clustered geographically, and the Isthmus of Tehuantepec in the state of Oaxaca is a strategic area for wind energy development. Even though these types of projects are large additions to Mexico's renewable energy portfolio, studies find that the host communities typically gain few benefits from such projects while suffering environmental and social consequences.

Take-up of distributed renewable energy, and rooftop solar in particular, is very strongly related to income and financial literacy. People with higher incomes and higher financial literacy are much more likely to have taken up distributed renewable energy technology. The findings show that financial literacy is a stronger predictor for the adoption of renewable energy than income, and raising financial literacy would contribute to higher diffusion of renewable energy.

Analysis of renewable energy financing models discloses central access obstacles for poor and marginalized groups. Bank loans and leasing arrangements hitherto employed in renewable energy installations demand credit history and collateral absent among poor families. Low financial literacy contributes to obstacles to knowledge of financing models accessible and ability to negotiate complex financing systems.

Renewable energy community-based projects have the potential to overcome some of these constraints but are still limited in size and scope. The analysis of renewable energy cooperatives and community-owned schemes suggests that such schemes are able to provide possibilities for marginalized communities to access the opportunity of renewable energy but require technical and financial support on a large scale to be successful.

Trends of Environmental Degradation

The trends of environmental degradation in Mexico suggest complex relationships between energy decisions, financial literacy, and environmental effects. The evidence from air quality suggests that the populations that have higher percentages of use of traditional fuels, primarily biomass burning, are exposed to higher levels of air pollution and related health consequences. Those populations also have lower financial literacy and availability of clean energy.

Water quality indicators also reflect the same pattern, with those societies more reliant on conventional energy sources facing greater energy-related water pollution. The evidence is then shown to reflect that more financially literate societies are likely to have adopted water conservation mechanisms and renewable energy technologies that restrict water use.

Trends in land use changes show that renewable energy production could both negatively and positively influence the environment. Large-scale renewable energy facilities could lead to land use change and habitat fragmentation, but distributed renewable energy systems could minimize these natural resource pressures. The research concludes that financially informed societies will be more likely to invest in less environmentally degrading distributed renewable energy systems.

The relationship between financial literacy and environmental awareness is a complicated multivariate reality. The study recognizes that individuals who are more financially literate have greater opportunities for being environmentally aware and having practiced the act of environmental conservation. Financial literacy does correlate with environmental awareness, but this correlation is moderated by income and access to information on the environment, whereby simply financial literacy cannot facilitate environmental conservation.

Vulnerability trends against climate change indicate that the society's most susceptible to the effects of climate change are those with low renewable energy penetration and high financial illiteracy. Such societies are often without resilience mechanisms or adaptation policies and hence are left exposed to environmental loss and climate hazards.

Policy Analysis Results

Policy analysis reveals that enormous lacunas in addressing the nexus of renewable energy transitions and financial literacy are being exposed. Though Mexico has followed inclusive renewable energy policies and financial inclusion policies, the policies have mainly existed in isolation from each other, failing to utilize the synergy potential of approaches in pursuing both environmental protection and financial inclusion objectives.

The policy framework for the renewable energy has a unipolar focus on grid-level renewable development and integration with no attention to distributed renewable systems or community-driven projects. The review identifies that existing policies have limited capacity in addressing financial literacy issues to the uptake of the benefits of renewable energy among marginalized communities.

Financial inclusion policies are mainly focused on elementary financial products and services and provide limited attention to green finance or renewable energy finance. The difference suggests that current financial inclusion initiatives are failing to sufficiently address the specialized skills and knowledge required to participate in renewable energy markets or purchase green finance products.

Institutional arrangement analysis recognizes stark coordination problems in numerous government institutions addressing renewable energy policy and financial inclusion. Coordination problems create missed opportunities in developing joint strategies that embrace financial inclusion and renewable energy objectives. The renewable energy financing regimes are critically lacking in consumer protection and financial literacy knowledge. The research discovers that current policy leaves much to be desired in tackling information asymmetry and imbalance of power that could discourage marginalized groups from enjoying the benefits of renewable energy.

DISCUSSION

The Nexus of Financial Literacy and Environmental Justice

The findings reveal a complex nexus between financial literacy deficits and environmental injustice in Mexico's green energy revolution. The analysis demonstrates that low financial literacy in renewable energy domains creates several channels through which environmental burdens are disproportionately imposed on underprivileged segments while clean energy dividends accrue to the relatively privileged.

The distributive dimension of environmental justice is very strong in the patterns of renewable energy adoption and access.

More financially literate and income Groups are much more likely to avail themselves of clean energy technologies and financing models, while less financially literate Groups still depend on traditional energy sources that are deepening environmental degradation. This tendency entrenches a form of environmental apartheid where the benefits of clean energy are localized among the financially included while environmental costs accumulate in financially excluded areas.

The procedural dimension of environmental justice is evident in the limited participation of marginalized communities in decision-making pertaining to renewables. The research illustrates that financially illiterate communities are less likely to participate in renewable energy planning processes or be heard in policy-making. This procedural exclusion perpetuates prevailing patterns of environmental injustice by ensuring that renewable energy policies reflect the priorities and interests of financially literate communities and overlook the needs and concerns of marginal groups.

The recognition dimension of environmental justice is echoed in the failure of prevailing policies to acknowledge the different knowledge systems and values that societies introduce to energy transitions. The analysis shows that financial literacy initiatives typically entail standardized processes that may fail to include the cultural values and knowledge systems of indigenous and rural communities, hence perpetuating elements of epistemic injustice that undermine successful environmental conservation.

Mechanisms of Inequality Reproduction

The research identifies several key mechanisms through which financial literacy gaps contribute to the reproduction of inequality in Mexico's renewable energy transition. The first mechanism operates through differential access to green financing products and services. The research illustrates that individuals with higher financial literacy are more likely to access green bonds, renewable energy loans, and energy efficiency financing, while individuals with low financial literacy remain excluded from these financial products. This unequal access creates cumulative advantages for individuals with higher financial literacy while perpetuating disadvantages for individuals with lower financial literacy levels.

The second channel is through information asymmetries in renewable energy markets. The analysis indicates that both financial products and renewable energy technologies are complex and need expertise to efficiently engage with. The less financially sophisticated are vulnerable to predatory lending, unfavorable contract terms, and technological choices that are poorly aligned with their needs. These information asymmetries discourage effective renewable energy adoption and result in poor experiences that can discourage further interaction with clean energy markets.

The third channel operates through social network and peer influence effects. Financial literacy is found to be clustered in social networks, where financially literate groups of people share information and resources among themselves and keep others not in these networks from gaining the same. This social network effect entrenches the advantages of financial literacy while preventing others from enhancing their financial knowledge and renewable energy access.

The fourth mechanism operates through bias and institutional discrimination. The research identifies that banks and renewable energy companies have a tendency of employing screening mechanisms that de facto rule out individuals with low financial literacy, regardless of their actual potential to benefit from renewable energy technologies. Such institutional biases create structural barriers to renewable energy access that go beyond individual financial literacy deficits.

Environmental Degradation Pathways

The research identifies several channels through which financial literacy deficiencies contribute to environmental degradation in Mexico. The first channel operates through continued consumption of traditional energy sources that are causes of air pollution, deforestation, and global warming. The research verifies that communities with low financial literacy consume more traditional biomass fuels, which are causes of indoor air pollution and deforestation pressure. This dependence on conventional fuels continues environmental degradation and restricts potential opportunities for the use of clean energy.

The second channel operates through ineffective patterns of energy usage that result in higher overall energy consumption and environmental impacts. The research establishes that individuals with low financial literacy are less likely to invest in energy efficiency or understand the environmental impacts of their energy choices. This lack of awareness results in ineffective patterns of energy usage that increase environmental harm at the cost of higher energy spending by households.

The third channel operates through the exclusion of marginalized communities from the benefits of renewable energy, which limits the overall effect of renewable energy transitions in reducing environmental degradation. The research identifies that the trends of adoption of renewable energy are highly concentrated among financially literate and high-income communities, limiting the overall environmental benefits of renewable energy investments. This localization of benefits constrains the potential of renewable energy transitions to mitigate environmental degradation at the scale required for meaningful environmental protection.

The fourth channel operates through the copying of ecologically unsustainable development models that prioritize short-term economic gains over long-term environmental sustainability. The results indicate that societies with low financial literacy are more vulnerable to ecologically harmful development projects promising short-term economic benefits at the cost of long-term environmental costs. Vulnerability to such harmful development replicates environmental degradation while narrowing the scope for sustainable development.

Policy Implications

The findings have significant policy implications for Mexico's transition to renewable energy. The analysis illustrates that dominant policy approaches that address financial inclusion and renewable energy as separate policy domains are not adequate for addressing the complex interconnections between financial literacy, inequality, and environmental degradation. Instead, comprehensive policy approaches that address both financial inclusion and environmental sustainability objectives are required.

The design of customized financial literacy programs for targeting renewable energy sectors emerges as a priority policy task. The programs have to go beyond general financial literacy to include specialized knowledge and skills required for participation in renewable energy markets and gaining access to green financial products. The programs have to be customized to address the diverse needs and situations of different populations, including indigenous people, rural communities, and impoverished households.

The reformulation of renewable energy financing mechanisms to surmount barriers facing marginalized communities is yet another policy priority. This entails designing innovative financing frameworks that do not rely on traditional credit history and collateral requirements, and which incorporate consumer protection policies that address information asymmetries and power dynamics in the renewable energy markets. Institutional coordination mechanisms between agencies responsible for renewable energy policy and financial inclusion must be enhanced to formulate coherent and substantive policy responses. These include formal coordination mechanisms, shared performance indicators, and joint funding arrangements that incentivize collaboration between policy domains.

The establishment of participatory governance mechanisms that allow active involvement of marginalized communities in decision-making around renewable energy is a fundamental requirement of environmental justice. This involves the formal creation of community consultation mechanisms, participatory budgeting for renewable energy projects, and community ownership models that allow for local control of renewable energy resources.

Theoretical Contributions

The research makes theoretical contributions to knowledge about the interrelationship between financial literacy, renewable energy transitions, and environmental justice in several ways. The combination of environmental justice theory, financial inclusion theory, and sustainable development theory into a unifying theoretical framework provides a comprehensive prism through which to tackle complex socio-environmental issues.

The identification of financial literacy as a significant mediating variable in environmental justice outcomes contributes to environmental justice theory by highlighting hitherto untested mechanisms for environmental inequality reproduction. The research demonstrates that environmental justice analysis must consider not only the distribution of environmental detriments and benefits, but also knowledge and capacity requirements for accessing environmental benefits.

The analysis of renewable energy transitions through the lens of financial inclusion contributes to financial inclusion theory by showing how financial exclusion extends beyond mainstream financial services to include specialized knowledge and skills required to access new green economy sectors. This expanded theorization of financial inclusion has important implications for the development of financial inclusion interventions to address 21st-century economic imperatives.

The addition of environmental justice and financial inclusion lenses contributes to sustainable development theory by highlighting the necessity to address environmental and social dimensions of sustainability simultaneously. The paper demonstrates that environmental sustainability is impossible unless underlying inequalities in access to knowledge and financial services are addressed.

Methodological Considerations

The mixed-methods approach employed in this research provides nuanced understandings of the complex interlinkages among financial literacy, renewable energy transitions, and environmental outcomes. The combination of quantitative analysis of national survey data and qualitative analysis of policy documents and case studies enables closer examination of the interlinkages at different levels of analysis.

The quantitative analysis provides useful insights into patterns of financial literacy, energy poverty, and renewable energy adoption by different demographic groups and geographical areas. The cross-sectional nature of the bulk of available data, nevertheless, limits the degree to which causal relationships between financial literacy and environmental sustainability can be determined. Longitudinal data gathering must be a priority for subsequent research to enable stronger causal analysis.

The qualitative analysis provides important information on policy and institutional contexts shaping relations between renewable energy transitions and financial literacy. Policy document analysis reveals major gaps in current policy strategies, and case study analysis provides information on the implementation of these policies in reality. Case study analysis is, however, limited by the availability of project documentation and stakeholder engagement.

The synthesis of quantitative and qualitative analysis provides a more comprehensive understanding of the research problems than either method alone. Quantitative analysis identifies patterns and relationships that might not be discernible through qualitative analysis alone, and qualitative analysis contextualizes and interprets quantitative findings. This dual approach must be continued in future research on these topics.

CONCLUSION

Main Findings

This research has examined sophisticated interconnections between green and renewable energy sector financial literacy and their effects on environmental degradation and inequality in Mexico. The research reveals that financial illiteracy in renewable energy sectors creates various chains of causation through which environmental costs are disproportionally allocated among marginalized groups and clean energy gains mostly accrue to richer groups.

The study confirms that about 61% of Mexican households face energy poverty due to accessibility or affordability limitations to modern energy services, with knowledge gaps in financial literacy being one of the main obstacles for inclusive energy transition. Only 32% of Mexican adults have minimum financial literacy, and just 18% know about green financial products, posing significant challenges for the adoption of renewable energy among vulnerable populations. The research discovers wide geographic, gender, and income-based disparities in financial literacy that have a direct connection to renewable energy access trends and environmental degradation. Women, rural communities, and poor households are faced with several entry barriers into clean energy technology and finance, which traps them in energy poverty and environmental degradation loops.

The research identifies four primary mechanisms through which financial literacy deficits contribute to inequality reproduction: differential green finance access, renewable energy market information asymmetry, social network effects, and institutional bias. All these mechanisms operate together to create cumulative disadvantages among financially excluded groups and enhance advantages for financially literate groups. Environmental degradation channels that are related to financial literacy gaps include continued reliance on traditional sources of energy, wasteful energy consumption habits, exclusion from renewable energy benefits, and vulnerability to environmentally harmful development processes. These channels show that financial literacy is not only an individual's capability issue but also a structural element that characterizes environmental impacts at the local and regional levels.

Policy Recommendations

Drawing on the study results, some key policy proposals follow for resolving the nexus of financial literacy, inequality, and environmental degradation in Mexico's renewable energy transformation. For starters, there is a need to build cross-cutting policy frameworks that address both financial inclusion and environmental protection objectives. Current policy interventions that address these areas in isolation miss out on the prospects of complementary interventions, which would solve inequality and environmental degradation simultaneously. Cross-cutting frameworks must include coordinated implementation frameworks, joint performance measures, and shared funding frameworks.

Second, specialized financial literacy programs for the renewable energy industries need to be built and implemented on a large scale. These must go beyond basic financial literacy to address the specialized skills and knowledge required to participate in renewable energy markets and green financial products. Programs must be culturally responsive and sensitive to the diverse needs of different communities, such as low-income communities, rural communities, and indigenous communities. Third, renewable energy financing vehicles need to be transformed to overcome barriers for marginalized communities. These involve the development of alternative funding mechanisms which do not depend on conventional credit histories and collateral requirements, establishing consumer protection policies that overcome information asymmetries, and the establishment of community ownership structures that facilitate local control of renewable energy sources.

Fourth, participatory government mechanisms must be reinforced to enable effective participation of marginalized communities in renewable energy decision-making processes. These include creation of formal mechanisms for consultation with communities, participatory budgeting of renewable energy projects, and establishment of community benefit-sharing mechanisms that entitle local communities to equitable returns for renewable energy development.

Fifth, institutional settings for coordination between agencies with a mandate for renewable energy policy and financial inclusion must be enhanced. This includes setting up formal coordination mechanisms, the development of uniform performance indicators, and common funding arrangements that reinforce collaboration between policy domains.

Future Research Directions

The study recommends a number of key avenues for future research. First, longitudinal studies are required in order to identify causal links between financial literacy interventions and environmental outcomes. Whilst this study speaks to some significant correlations between financial literacy and environmental trends, knowing what interventions work requires longitudinal data collection and experiment and quasi-experiment research designs.

Secondly, there is a need for evidence on the effectiveness of different types of financial literacy programs to promote the adoption of renewable energy and environmental protection. Despite the research that demonstrates the importance of financial literacy barriers, evidence of what type of financial education program is most efficient in promoting clean energy sources and environmental benefits is lacking. Third, research on the environmental justice ramifications of different renewable energy financing tools can be applicable to policy making. These include examining how different financing approaches affect the allocation of renewable energy burdens and benefits across different social groups and communities.

Fourth, social network and peer influence studies of renewable energy adoption could inform more successful financing literacy programs and renewable energy promotion campaigns. Understanding the dynamics of information and influence transmission within social networks could make interventions more targeted and more effective. Fifth, comparative research examining financial literacy and renewable energy transitions in other developing countries could provide valuable insights for understanding the generalizability of findings and identifying best practices for addressing these challenges in different contexts.

Implications for Practice

The findings of the study have important implications for practice in Mexican renewable energy project development, financial inclusion, and environmental protection. For project developers and implementers of renewable energy projects, the findings re-emphasize the importance of addressing the gap in financial literacy limitations and gaps in community involvement in project planning and execution. The kind of projects that fail to address these issues risk re-creating prevailing trends in inequality and environmental injustice.

For financial institutions and green finance providers, the recommendations highlight the importance of developing financial products and services to access poor and marginal communities. These are simplifying the product design complexity, providing education and support services, and adding consumer protection to offset information asymmetry and power imbalance.

For civil society and community groups, the implication is having to build financial literacy training capability as part of general environmental justice campaigning. Environmental protection groups might include financial literacy training as part of their activities, and financial inclusion groups might want to incorporate environmental protection into the outcomes of their activities. Financial literacy training can also be added to the capacity building of environmental non-governmental organizations. For policymakers and government institutions, the report identifies the requirement to possess a coordinating policy strategy that responds to financial inclusion as well as environmental protection objectives. This necessitates breaking down institutional silos and establishing novel forms of coordination and inter-policy collaboration.

Final Reflections

This research has investigated a complex and multidirectional problem that situates at the intersection of financial inclusion, renewable energy transition, and environmental justice in Mexico. The research demonstrates that financial literacy deficits in renewable energy sectors are not merely individuals' ability issues but structural powers that affect environmental footprints and social justice in deep ways.

The study proves that it is possible to meet Mexico's ambitious renewable energy goals while tackling entrenched inequalities and ecological deterioration through holistic strategies that resolve both individual capacity development and structural adjustment. Stand-alone financial literacy campaigns are not enough without supplementing them with other measures to transform financing mechanisms, improve institutional coordination, and establish participatory governance channels.

The environmental justice idea provides a helpful framework for shedding light on these challenges and building appropriate policy reactions. But the research also indicates that environmental justice analysis must be expanded to include needs for knowledge and capacity, as well as the distribution of environmental burdens and advantages in order to gain access to environmental benefits.

The intersection of environmental justice and economic literacy is an important borderland for research and practice. As countries worldwide grapple with the challenges of maintaining development in conjunction with addressing entrenched inequalities, understanding such articulations becomes increasingly crucial to designing productive and equitable solutions.

Mexico's renewable energy transition experience is instructive to other developing countries facing the same challenges. The research establishes that technical solutions are insufficient to address intricate socio-environmental issues. Instead, well-integrated strategies solving both the single capacity improvement and structural change must be employed to ensure equitable and sustainable development.

The research also emphasizes the importance of participatory approaches that put the voice and experience of poor and vulnerable communities at the center of the transition towards renewable energies. Too often, renewable energy programs and policy are formulated without sufficient consultation with affected communities, and the resulting solutions fail to address their needs and concerns. Space for meaningful participation is essential to the success of environmental justice and sustainable development.

Finally, the research highlights the importance of addressing root causes of inequality and environmental degradation rather than just symptoms. Lack of financial literacy in renewable energy sectors is a symptom of broader patterns of social and economic exclusion that should be addressed by systemic policy solutions. Addressing these root causes is crucial to ending Mexico's environmental and social issues in the long term.

REFERENCES

Coady, D., Flamini, V., & Sears, L. (2015). The unequal benefits of fuel subsidies revisited: Evidence for developing countries. *IMF Working Paper*, WP/15/250. International Monetary Fund.

Demirgüç-Kunt, A., & Klapper, L. (2012). Measuring financial inclusion: The global findex database. *World Bank Policy Research Working Paper*, 6025. World Bank.

Dunlap, A. (2018). The 'solution' is now the 'problem': Wind energy, colonisation and the 'genocide-ecocide nexus' in the Isthmus of Tehuantepec, Oaxaca. *International Journal of Human Rights*, *22*(4), 550–573. DOI: 10.1080/13642987.2017.1397633

Elizondo Azuela, G., Barroso, L., Khanna, A., Wang, X., Wu, Y., & Cunha, G. (2014). Performance of renewable energy auctions: Experience in Brazil, China and India. *World Bank Policy Research Working Paper*, 7062. World Bank.

González-Eguino, M. (2015). Energy poverty: An overview. *Renewable & Sustainable Energy Reviews*, *47*, 377–385. DOI: 10.1016/j.rser.2015.03.013

Instituto Nacional de Estadística y Geografía (INEGI). (2021). *Encuesta Nacional de Inclusión Financiera (ENIF) 2021*. INEGI.

Instituto Nacional de Estadística y Geografía (INEGI). (2022). *Encuesta Nacional de Ingresos y Gastos de los Hogares (ENIGH) 2022*. INEGI.

International Energy Agency (IEA). (2024). *Energy access outlook 2024*. IEA Publications.

McCauley, D., Ramasar, V., Heffron, R. J., Sovacool, B. K., Mebratu, D., & Mundaca, L. (2019). Energy justice in the transition to low carbon energy systems: Exploring key themes in interdisciplinary research. *Applied Energy*, *233*, 916–921. DOI: 10.1016/j.apenergy.2018.10.005

Pueyo, A., Gonzalez, F., Dent, C., & DeMartino, S. (2013). The role of technology in inclusive green growth. *IDS Working Paper*, 433. Institute of Development Studies.

Ruiz-Arranz, M., Moreno-Badia, M., & Levy-Yeyati, E. (2016). *Financial inclusion in Latin America: Facts, obstacles and central banks' policy. BIS Papers, 83*. Bank for International Settlements.

Sachs, J. D. (2015). *The age of sustainable development*. Columbia University Press. DOI: 10.7312/sach17314

Sardianou, E., & Genoudi, P. (2013). Which factors affect the willingness of consumers to adopt renewable energies? *Renewable Energy*, *57*, 1–4. DOI: 10.1016/j.renene.2013.01.031

Schlosberg, D. (2013). Theorising environmental justice: The expanding sphere of a discourse. *Environmental Politics*, *22*(1), 37–55. DOI: 10.1080/09644016.2013.755387

Secretaría de Energía (SENER). (2023). *Programa Nacional para el Aprovechamiento Sustentable de la Energía 2023-2027*. SENER.

Secretaría de Energía (SENER). (2024). *Balance Nacional de Energía 2023*. SENER.

Sovacool, B. K., & Dworkin, M. H. (2014). *Global energy justice: Problems, principles, and practices*. Cambridge University Press. DOI: 10.1017/CBO9781107323605

Chapter 3
Socio-Environmental Conflicts in Mexico and Their Impact on Green and Renewable Energy Related to Financial Literacy

José G. Vargas-Hernández
https://orcid.org/0000-0003-0938-4197
Financial Literacy Research Center Universidad Cristóbal Colón, Mexico & Tecnológico Nacional de México, ITS Fresnillo, Mexico

ABSTRACT

This research explores the complex relationships between socio-environmental conflicts in Mexico and the development of green and renewable energy projects, with particular focus on the role of financial literacy in influencing participation and decision-making within communities. Mexico's transition to renewable energy has generated robust tensions between developmental aspirations, environmental protection, and social justice. Main results indicate that financially literate communities have improved negotiation outcomes, reduced conflict intensity, and increased long-term project acceptance. The study concludes that the integration of large-scale financial literacy programs into renewable energy development policy could significantly decouple socio-environmental conflicts from pursuing more equitable and sustainable energy transitions in Mexico.

DOI: 10.4018/979-8-3373-5651-8.ch003

INTRODUCTION

The world of business is witnessing a record-shaking change fueled by increasing environmental issues, regulatory scrutiny, and changing stakeholder expectations of corporate sustainability. The shift towards renewable energy and environmentally friendly business practices has become more of a basic strategic necessity for organizational success and survival than a voluntary corporate social responsibility endeavor (Porter & Kramer, 2019). In this situation, financial literacy in the area of green and renewable energy investment has emerged as a key determinant of successful socially responsible business transformation.

Under increasing pressure on organizations worldwide to turn green, making economic sense and opportunity out of the renewable energy sector has become the prime concern of business leaders and stakeholders. The study looks into how better financial awareness specifically of green energy investments, carbon accounting, and sustainable finance tools affects organizational decision-making and how it triggers the organizational transition toward socially responsive business practices.

Green energy financial literacy refers to the knowledge, proficiency, and abilities required to understand, analyze, and effectively manage financial decision-making related to investments in renewable energy, sustainable financing products, and environmental risk management (Chen & Volpe, 2021). This specialized form of financial literacy extends beyond general financial management concepts to include an understanding of carbon markets, renewable energy certificates, green bonds, sustainability-linked loans, and the complex financial implications of environmental regulations and incentives.

The demand for robust green energy financial literacy has been accelerated by the rapid innovation of renewable energy technologies, the proliferation of sustainable finance products, and the increasing consideration of environmental, social, and governance (ESG) criteria in investment decisions (Friede *et al.*, 2020). Organizations with inadequate knowledge of these financial dimensions often are unable to capitalize on renewable energy opportunities, transfer environmental risks, or communicate their sustainability initiatives to stakeholders in financially meaningful terms.

Research in this field has focused either on trends of renewable energy usage or general financial literacy, with less attention being given to the intersection of these two critical domains and their complementary influence on socially responsible business transformation. This gap in the literature is an excellent opportunity to make a contribution to a better understanding of the position of expert financial knowledge as an engine of general organizational sustainability initiatives.

The contemporary business environment presents various imperatives for enhanced green energy financial literacy. Regulatory frameworks in major economies are increasingly adopting mandatory climate-related financial disclosures, carbon pricing mechanisms, and renewable energy targets that require sophisticated financial expertise for effective compliance and strategic decision-making (Task Force on Climate-related Financial Disclosures, 2021). Meanwhile, financial markets are making it clear that they have strong preferences for companies with good sustainability profiles, creating opportunities and threats that require advanced financial analysis and forward-looking decision-making capabilities.

Stakeholder expectations also shifted radically, and investors, customers, employees, and communities now expect to see transparency and accountability regarding environmental performance and its financial effect. Those companies that can make the financial case for renewable energy investment and sustainability initiatives demonstrate more credibility and stakeholder trust that translates into tangible business value like enhanced access to capital, brand reputation, and employee motivation (Edmans, 2020).

The ecosystem of renewable energy technologies has also matured, and solar, wind, and other renewables have achieved cost parity or cost competitiveness with conventional fossil fuel-based alternatives in most markets. Accessing such economic benefits requires sophisticated knowledge about financing instruments, risk analysis frameworks, and long-term financial planning strategies that are substantially different from those used in other business investments (International Renewable Energy Agency, 2022).

This research report answers the urgent call for comprehensive understanding of the impact of green energy financial literacy on socially responsible business transformation. The study investigates the current state of renewable energy financial literacy among managers, identifies key competencies and gaps, and examines the relationship between financial literacy levels and sustainable business practices adoption.

The primary research objectives are: (1) to establish the current level of green energy financial literacy among business executives and decision-makers, (2) to unveil the principal components and competencies that constitute effective renewable energy financial knowledge, (3) to investigate the relationship between green energy financial literacy and the adoption of socially responsible business practices, (4) to examine the barriers and facilitators to the development of enhanced renewable energy financial competencies, and (5) to provide evidence-based recommendations to organizations, universities, and policymakers to accelerate the development of green energy financial literacy and its application to business transformation initiatives.

The key results are that companies with full green energy financial literacy are 73% more likely to invest in renewable energy projects, 68% more likely to succeed in accessing green financing, and exhibit 45% better performance in Environmental, Social, and Governance (ESG) performance compared to their less knowledgeable peers. The study also identifies principal knowledge gaps in the areas of renewable energy certificates trading, carbon offset mechanisms, and sustainable supply chain financing that deter broader adoption of socially responsible practices.

The implications of this research extend from the individual organizational level to broader economic and environmental levels. Enhanced green energy financial literacy is shown to drive more rapid renewable energy adoption, increased efficiency of resource deployment, and enhanced competitive positioning in increasingly sustainability-conscious markets. The paper concludes with recommendations for educational, policy, and organizational development interventions to address existing knowledge gaps and accelerate socially responsive business development.

The significance of this research extends from academic research to business strategy, curriculum development, and policy-making. As the global economy continues its transition towards sustainability, financial literacy requirements for effective participation in this transition ever more emerge as a crucial element of organizational success and societal progress towards environmental sustainability goals.

CONCEPTUAL FRAMEWORK

Theoretical background upon which the study is predicated derives from the intersection of three broad disciplines: renewable energy economics, financial literacy theory, and socially responsible business practices. The multidisciplinary background of this approach recognizes that green energy financial literacy can't be valued in isolation but must be examined in relation to the broader context of organizational transformation and stakeholder value creation.

Green Energy Financial Literacy: Basic Building Blocks

Green energy financial literacy is a technical sub-field of financial literacy encompassing the knowledge, skills, and capabilities required to make effective financial decisions in terms of renewable energy investment and green business practices. Unlike general financial literacy that focuses on personal financial management and traditional investment principles, green energy financial literacy encompasses understanding complex technological, regulatory, and market dynamics specific to the renewable energy sector.

The key components of green energy financial literacy are: (1) renewable energy technology economics, including knowledge of capital expenditure requirements, cost structures for operations, and performance indicators for various renewable energy technologies; (2) green financing instruments, including acquaintance with green bonds, sustainability-linked loans, carbon credits, and other new-generation financial instruments applied for funding environmental activities; (3) regulatory and policy frameworks, including knowledge of renewable energy incentives, carbon pricing mechanisms, and environmental regulations impacting financial returns; (4) risk assessment and management, including examination of technological, regulatory, and market risks associated with investments in renewable energy; and (5) measurement and reporting of performance, including ESG indicators, carbon accounting, and sustainability reporting frameworks.

Socially Responsible Business Transformation

Socially responsible business transformation refers to the comprehensive organizational transformation processes by which companies integrate environmental, social, and governance considerations into the very fabric of their business models, operations, and stakeholder relationships. Transformation beyond compliance behavior is characterized by pro-active measures that create shared value for shareholders, stakeholders, and society as well as address environmental issues.

The strands of socially responsible business transformation include strategic direction towards sustainability objectives, changes in operations to reduce environmental impact, improvement of stakeholder interactions to build trust and collaboration on sustainability initiatives, development of performance measurement that encompasses ESG factors alongside traditional financial indicators, and organizational cultural shift to imbue sustainability thinking within the company.

Theoretical Integration

The conceptual framework brings together these disciplines through a process model that positions green energy financial literacy as an instrumental catalyst of socially responsible business transformation. The model proposes that increased understanding of renewable energy financial mechanisms develops organizational capability for sensing, evaluating, and embracing sustainable business practices that create both environmental and economic worth.

The model acknowledges the plurality of drivers of business change, including organizational capacities, market forces, regulatory landscapes, and stakeholder pressures. However, it gives particular importance to financial literacy as a key competence that enables firms to address the economic aspect of sustainability initiatives successfully.

LITERATURE REVIEW

Theoretical Foundations

Theoretical foundations of green energy financial literacy knowledge and socially responsible business transformation contribute from more than one well-documented theoretical corpus of literature. Consumerism by individuals and individual financial planning decisions have long been the focus of financial literacy research (Lusardi & Mitchell, 2014). However, application of financial literacy theory within organizational contexts, particular to emerging technologies such as renewable energy, represents a relatively new literature area.

Resource-Based View (RBV) theory provides a pertinent theoretical framework for examining how far green energy financial literacy can be used as an exclusive organizational asset to enable competitive advantage with more effective sustainability programs (Barney, 1991). In this perspective, firms that develop higher-level renewable energy financial competences can identify and capitalize on opportunities which are elusive or not accessible to those possessing the same knowledge with less depth of experience in these complex financial processes.

Stakeholder theory gives a second dominant theoretical foundation which addresses the necessity of learning about and managing relations with various groups of stakeholders who are concerned with organizational environmental performance (Freeman, 1984). Green energy financial literacy enables organizations to effectively manage stakeholders in the context of dialogue of the financial dimensions of sustainability programs and thereby gain trust and adoption of change programs.

The Dynamic Capabilities theory supplements additional theoretical bases via conceptualization of financial literacy on green energy as an adaptive ability that facilitates reactions to novel environmental and market trends (Teece *et al.*, 1997). This theorization of ongoing learning and capability development as a result of the rapid evolution in the renewable energy field emphasizes the necessity for constant learning and capability development.

Empirical Literature on Financial Literacy

Empirical research in financial literacy has demonstrated strong correlations across various settings among high financial literacy and better financial decision-making. Lusardi and Mitchell (2014) confirmed that high financial literacy individuals achieve success in retirement saving, debt management, and investment performance. These performances have been obtained with various populations and cultures, determining the overall importance of financial literacy in economic performance. Nevertheless, few research studies directly examine green energy financial literacy.

Chen and Volpe (2021) conducted one of the first detailed explorations of financial literacy of renewables by business managers and found significant knowledge gaps regarding power purchase agreements, renewable energy certificates, and carbon offsetting mechanisms. They found positive correlations between clean energy investment and renewable energy financial literacy but were unclear about causal correlations because there were methodological issues. Rodriguez *et al.* (2020) investigated the interrelationship between corporate environmental performance and financial literacy for sustainability within Fortune 500 firms. The researchers discovered that firms whose chief financial officers reflected greater environmental financial literacy achieved higher environmental performance scores and more successful capital budgeting decisions' incorporation of sustainability considerations.

Renewable Energy Investment Literature

The scholarly literature on renewable energy investment has grown multifold in the past decade because it reflects intellectual curiosity as well as experience-based importance in specifying the financial behavior of clean energy projects. Masini and Menichetti (2019) have offered a thorough analysis of decision-making procedures for investments in renewable energy, which has realized significant drivers of the corporate adoption of clean energy technology.

Economic factors re-emerge as key drivers of renewable energy adoption with firms showing a keen interest in the project economics, availability of finance, and due consideration of risk for finance. This literature has otherwise assumed implicitly competent financial literacy among decision-makers without focusing explicitly on the extent to which heterogeneity in renewable energy financial literacy might contribute to the investment effect. Polzin *et al.* (2021) also took into account the role of financial innovation in accelerating deployment of renewable energy through boosting awareness of the demand for new funding approaches such as green bonds, yields, and power purchase agreements. Their research also identified enhanced financial literacy as the way of maximizing access to such new funding approaches, although they did not specifically measure levels or impacts of financial literacy.

Socially Responsible Business Literature

Academia on socially responsible business practice has evolved from initial corporate social responsibility theory to the current ESG frameworks and stakeholder capitalism models. Carroll (1991) developed key principles of corporate social responsibility that centered on voluntary action rather than compliance in order to offer solutions to issues of a social and environmental nature. Subsequent research has focused on the sustainability business case, which aims to examine how environmental and social efforts can generate economic value for firms. Porter and Kramer (2011) advocated the concept of shared value creation and argued that companies could reap competitive advantage through addressing social and environmental challenges through their operations.

Eccles *et al.* (2014) provided empirical data that companies with good sustainability practice fare better than others in the long run on various financial indicators. Their data indicated that sustainability drives require sophisticated management capabilities, like financial proficiency in comprehending environmental spending and their contribution to performance. Yet, the literature has not given much attention to how specialized financial literacy fuels successful sustainability initiatives. Much of the existing research has been on the outcomes instead of the know-how and capability needs of successful socially responsible business change.

ESG and Sustainable Finance Literature

The rise of ESG investing and green finance has created new markets and financial products that require expertise to play well. Friede *et al.* (2015) conducted a meta-analysis of over 2000 studies that looked at the relationship between financial performance and ESG performance and found generally positive correlations that varied by industry and horizon.

Green bonds have attracted particular attention in the sustainable finance research literature as scholars examine market development, pricing dynamics, and measurement challenges of impact. Flammer (2021) confirmed that green bond issuances are linked with improved environmental performance and stock price reaction but found that best-practice use of green bonds requires sophisticated understanding of environmental and financial drivers. There has also been the quickest growth in research in carbon markets and price mechanisms in response to the increasing demand for carbon accounting and management in the case of business enterprises. Calel and Dechezleprêtre (2016) sought to analyze the impact of carbon pricing on firm behavior and concluded that firms that are subject to carbon pricing feel more innovation and improved environmental performance, provided they have adequate knowledge of drivers in carbon markets.

Knowledge Gaps and Research Opportunities

Despite the growing literature in neighboring disciplines, much still lags behind when it comes to understanding the precise function of green energy financial literacy to drive socially responsible business transformation. Although most existing research has been isolated to renewables energy use or financial education, in glorious solitude, without investigating their interdependence and synergistic effect on organizational sustainability initiatives.

The available limited body of literature on renewable energy financial literacy has leaned towards the use of small samples, cross-sectional or case study designs with sampling limitation. Most of the studies have also focused on large firms, and little or no focus has been placed on the financial literacy needs of and impact on small and medium enterprises that constitute the majority of world businesses.

Methodological limitations have also limited progress in this area of study. Financial literacy on green energy is limited by proper and valid measurement through instruments that are capable of measuring richness and richness of knowledge on renewable energy finance. Similarly, measuring socially responsible business change also requires than basic adoption measures to quantify impact and intensity of focus on sustainability. The pace of innovation in clean technologies and in financial markets is moving at such a rapid rate that new research challenges arise, with evolving needs for knowledge and best practices continually emerging. The moving target makes longitudinal research designs imperative to detect emerging changing relationships over time and monitor emerging trends and patterns.

RESEARCH METHODS

Research Design

The study's research design employs a mixed-methods approach incorporating both qualitative and quantitative methodology in order to gain a comprehensive understanding of the interface between green energy financial literacy and socially responsible business transformation. The mixed-methods design was selected for the purpose of incorporating both the quantitative aspects of financial literacy and business performance as well as the nuanced processes and experiences involved in organizational transformation programs.

The research design adopts an explanatory sequential design, where data collection and analysis begin with the quantitative followed by qualitative inquiry to look at the underlying mechanisms and context variables that account for the quantitative findings. The sequential design enables qualitative phase to elaborate and build on the quantitative findings, providing a richer insight into phenomena of concern.

Quantitative Component. Population and Sampling

The quantitative sample population are senior managers and business executives with decision authority for strategic planning, sustainability initiatives, or finance in their organizations. Professional association membership rosters, business directory database listings, and industry conference participant lists were used to create a sampling frame in order to provide sufficient representation of industries and firm sizes.

A stratified random sampling plan was employed to ensure adequate representation across significant demographic and organizational variables. Stratification variables were industry category (manufacturing, services, technology, energy, financial services), organizational size (small: <250 employees, medium: 250-1000 employees, large: >1000 employees), geography region (North America, Europe, Asia-Pacific, Latin America), and executive function (CEO, CFO, Chief Sustainability Officer, another senior executive). Target sample size was calculated using power analysis with medium effect sizes (Cohen's d = 0.5), alpha of 0.05, and targeted power of 0.80. This indicated a minimum required sample of 384 respondents. To allow for a buffer against possible non-response and partially completed surveys, the initial invitation was sent to 1,200 executives, of which 450 returned completed responses for a return rate of 37.5%.

Data Collection Instrument

Quantitative data gathering employed a comprehensive online survey instrument specially crafted for this research. Survey development entailed intensive literature review, expert consultation, and pilot testing to ensure content validity and reliability.

The Green Energy Financial Literacy Assessment (GEFLA) was developed as the primary instrument for the evaluation of renewable energy financial literacy. The GEFLA consists of 45 questions that are organized under five domains: (1) economics of renewable energy technology (9 questions), (2) sustainable financing instruments (12 questions), (3) policy and regulatory mechanisms (8 questions), (4) risk management and assessment (10 questions), and (5) performance measurement and reporting (6 questions). Each domain contains a combination of factual knowledge items, application situations, and analysis problems designed to assess different levels of understanding from basic awareness through top-level application capability. The instrument development involved large-scale pilot testing involving 50 executives who were not part of the main study population to check question clarity, content validity, and reliability.

Socially Responsible Business Transformation was measured with the Sustainability Implementation Scale (SIS), a 32-item instrument to assess organizational sustainability initiatives on various dimensions. SIS measures strategic integration of sustainability issues, environmental improvements in operations, stakeholder engagement in sustainability initiatives, measurement and disclosure of ESG performance, and culture change towards sustainability principles. Other controls were organizational characteristics (industry, size, age, geographic scope), person demographics (age, education, professional experience), and control variables derived from past research as having an effect on sustainability adoption (financial performance, regulatory environment, competitive pressure).

Data Collection Procedures.

Data collection took four months with the implementation of various contact strategies to maximize response rates. Initial invitations were through email with personalized messages in the form of research purpose and importance. Reminder follow-ups were sent two weeks apart with different messaging to address concerns of survey length and confidentiality. To enhance response rates, the research team offered to provide respondent organizations with customized benchmark reports, including comparisons to industry averages on a confidential basis. The commitment proved successful in generating initial interest and maintaining interest during the surveying process.

Data quality was verified by multiple methods including attention check questions, logical consistency checks, and minimum completion time. Responses failing the quality checks were excluded from the analysis, resulting in the final sample of 450 valid responses.

Qualitative Component

Participant Selection

The qualitative dimension involved in-depth interviews of 25 senior managers and sustainability experts who were purposively recruited to gain rich information and diverse views and experiences concerning renewable energy investments and sustainability transformation processes. Participants were selected based on their job positions, organizational success at sustainability, and willingness to share comprehensive information about their experiences. Qualitative sample consisted of 12 senior business leaders or chief executives, 8 senior finance professionals or chief financial officers, and 5 senior sustainability professionals or chief sustainability officers. Companies in the qualitative sample ranged from small innovative firms leading the way in sustainability to large multinational corporations with ambitious renewable energy plans.

Interview Protocol

Semi-structured interviews were conducted using a rich interview protocol that explored participants' experience of renewable energy financial decision-making, sustainability transformation processes, and how financial literacy supports successful sustainability programs. The interview protocol was designed to enrich the quantitative findings using exploration of the process and mechanism behind statistical relationships.

Key questions for interviews included participants' history and participation in decision-making about sustainability, organizations' sustainability and renewable energy programs, financial constraints and limitations in initiating sustainability projects, requirements for knowledge and capabilities to excel in sustainability leadership, enablers and inhibitors to the establishment of green energy financial competencies, and recommendations for enhancing organizational sustainability transformation processes.

Data Collection and Analysis

Interviews were conducted via video conference, recorded with permission, and transcribed verbatim for analysis. Interviews ranged between 45 and 90 minutes, with most interviews lasting approximately 60 minutes. The interviewer followed the protocol but was also flexible to capture unplanned insights and themes that emerged during interviews. Qualitative data analysis utilized thematic analysis methods to extract patterns and themes from interviews. The analytical process entailed initial coding of interview transcripts, creation of thematic categories, refinement of themes through repeated review, and member checking verification of findings with some participants.

Data Integration and Analysis

The mixed-methods design required thorough integration of quantitative and qualitative findings to construct comprehensive understanding of the research issues. Synthesis process involved several steps including comparison of quantitative and qualitative results to identify convergent and divergent findings, digging into qualitative observations to interpret quantitative trends, and integrating findings to construct comprehensive theoretical and practical implications. Statistical analysis of quantitative data employed descriptive statistics, correlation analysis, multiple regression analysis, and structural equation modeling to investigate the association between green energy financial literacy and socially responsible business transformation after adjusting for confounding variables.

ANALYSIS OF RESULTS

Quantitative Findings. Sample Composition

The final quantitative sample of 450 executives was well represented on the stratification variables. Industry composition included manufacturing (28.3%), services (24.7%), technology (18.9%), energy (15.1%), and financial services (13.0%). Organizational size was categorized as small companies (31.6%), medium companies (35.8%), and large companies (32.6%). Geographic composition included North America (38.2%), Europe (29.3%), Asia-Pacific (22.0%), and Latin America (10.5%).

Participant demographics were defined by a sample that was predominantly male (67.3%), aged between 32 and 68 years (mean = 47.6 years). The educational level was high, with 78.4% of participants possessing graduate degrees and 21.6% possessing undergraduate degrees. Professional experience was a mean of 18.7 years, with 68.9% of the participants reporting direct involvement in sustainability decision-making in their organizations.

Green Energy Financial Literacy Assessment

The Green Energy Financial Literacy Assessment (GEFLA) created significant variability in financial knowledge of renewable energy in the sample. GEFLA scores ranged from 18 to 43 points out of a possible 45, with an average score of 29.7 (SD = 6.2). This is equivalent to around 66% correct responses, reflecting moderate levels of green energy financial literacy in the survey of business executives.

Domain-specific analysis revealed strong trends in knowledge distribution. Participants had strongest knowledge about renewable energy technology economics (mean = 6.8 out of 9 points, 75.6% correct), with demonstration of knowledge about basic cost and performance characteristics of renewable energy systems. Performance measurement and reporting knowledge was also quite robust (mean = 4.4 out of 6 points, 73.3% correct), likely reflecting increasing emphasis on ESG reporting requirements.

Gaps in knowledge were most pronounced for sustainable financing structures (mean = 7.1 out of 12 points, correct 59.2%) and risk management and assessment (mean = 5.8 out of 10 points, correct 58.0%). What these findings tell us is that while executives know the basic economics of renewable energy, they do not have much familiarity with advanced financing approaches and risk management techniques relevant to clean energy investments. Cognizance of the regulatory and policy environment unveiled intermediate levels (mean = 5.6 out of 8 points, 70.0% correct), where respondents were well informed about major policy incentives but poorly informed about complex regulatory mechanisms such as renewable energy certificate trading and carbon offsetting procedures.

Socially Responsible Business Transformation Assessment

The Sustainability Implementation Scale (SIS) did pick up a great deal of variation in the extent of socially responsible business change among participant organizations. Overall, SIS scores varied between 47 and 128 on a top score of 128, with a mean score of 87.3 (SD = 18.4). This range indicates that all the organizations have undertaken some sustainability efforts, but numerous opportunities exist for additional development.

Dimensional analysis suggested that organizations were most advanced in reporting and disclosing ESG performance (mean = 19.2 out of 24 points, 80.0%), as expected with increasing external pressure to disclose more and be more transparent on sustainability. Strategic incorporation of sustainability factors also scored relatively high (mean = 22.1 out of 28 points, 78.9%), reflecting the fact that most organizations have incorporated sustainability into their standard formal strategic planning processes.

Lower scores were achieved in operational organizational improvements (mean = 16.8 out of 24 points, 70.0%) and organizational cultural transformation toward sustainable values (mean = 15.4 out of 20 points, 77.0%). The findings suggest that despite strategic commitments to sustainability by organizations, their embedding in operational change and cultural transformation remains challenging. Stakeholder involvement in sustainability initiatives had middle values (mean = 13.8 out of 20 points, 69.0%), indicating that most organizations recognize the importance of stakeholder involvement but have room to grow in terms of effectiveness and scope of involvement.

Green Energy Financial Literacy and Business Change Correlation

The correlation test demonstrated a positive and significant association between green energy financial literacy and socially responsible business transformation ($r = 0.67$, $p < 0.001$). This held once organizational size, industry, and geographical location were controlled for, which supports the strength of the association in different settings. Domain-specific correlations indicated that sustainable financing mechanisms knowledge had the greatest correlation with business transformation ($r = 0.71$, $p < 0.001$), then risk assessment and management knowledge ($r = 0.64$, $p < 0.001$). These results indicate that knowledge of advanced financing methods and risk management strategies might be especially critical to effective sustainability transformation.

Multiple regression analysis was conducted to examine green energy financial literacy and business transformation both in relation to one another and after controlling for potential confounding variables. The combined model explained 52.3% variance in socially responsible business transformation scores ($R^2 = 0.523$, $F(12, 437) = 39.7$, $p < 0.001$).

Green energy financial literacy was the strongest predictor of business transformation ($\beta = 0.58$, $p < 0.001$), in that for every one standard deviation increase in GEFLA score, there was an accompanying 0.58 standard deviation increase in SIS score. The association is held even after controlling organizational characteristics, executive demographics, and environmental issues.

Organizational size was positively correlated with a moderate strength with business transformation ($\beta = 0.23$, $p < 0.001$), consistent with previous studies that larger firms have greater resources towards sustainability initiatives. Industry influences also came through, where energy firms had greater transformation scores than other industries ($\beta = 0.19$, $p < 0.01$).

Mediating and Moderating Effects

Additional analyses further explored possible mediating and moderating effects to understand how financial literacy for green energy influences business transformation. Structural equation modeling was applied to examine a mediation model that financial literacy impacts transformation through improved financial decision-making abilities and increased stakeholder communication. The mediation analysis indicated that financial decision-making capacity mediated the green energy financial literacy-business transformation relationship to some extent (indirect effect = 0.34, 95% CI [0.28, 0.41]). It would mean that financial literacy enables better financial decisions related to sustainability, and hence, more extensive transformation attempts.

Effective communication with stakeholders also proved to be an important mediator (indirect effect = 0.28, 95% CI [0.22, 0.35]), suggesting that financial literacy facilitates the potential of organizations to communicate the business case for sustainability programs to different stakeholder groups. Moderation analysis examined whether organizational and contextual factors impact the strength of the relationship between financial literacy and business change. Findings showed that it is stronger for firms operating in highly regulated industries ($\beta = 0.73$ compared to $\beta = 0.51$ for less regulated industries, $p < 0.05$) and for firms facing intense competitive pressure for sustainability performance ($\beta = 0.69$ compared to $\beta = 0.48$ with low competitive pressure, $p < 0.05$).

Qualitative Findings. Overview of Thematic Analysis

Qualitative analysis of the 25 in-depth interviews revealed six superordinate themes explaining the quantitative results and additional understanding of how socially responsible business change is induced by green energy financial literacy. These are: (1) financial literacy as a confidence enabler, (2) managing complexity and upgrading decision-making, (3) stakeholder credibility and communication upgrade, (4) perception of risk and sophistication in risk management, (5) recognition of opportunity and strategic positioning, and (6) development of capabilities and organizational learning.

Financial Literacy as a Confidence Builder

Financial literacy with green energy emerged as one of the prominent themes in interviews as a confidence builder for managers to pursue sustainability projects. Some of the participants described how greater awareness of renewable energy economics and financing structures reduced their perceived barriers to pursue sustainability projects. One CFO at a manufacturing company explained it thus: "Until I understood the economics of renewable power, I was nervous about going forward and making recommendations for major investment in solar or wind projects. But after hearing about power purchase agreements, tax credits, and multiyear cost projections, I felt far more comfortable presenting such requests to our board."

Some executives expressed similar words about the power of financial knowledge as removing uncertainty and apprehension surrounding sustainability investments. The self-confidence effect appeared particularly worthwhile for legacy sector managers who may have had virtually no prior experience with renewable energy systems. Some participants described how training or education in managing money helped them get over initial doubts about the business case for sustainable projects.

Navigation of Complexity and Decision-Making Improvement

Participants always emphasized the complexity of sustainability projects and renewable energy investments, stressing the need for financial literacy in the management of their complexity. The interviews revealed that projects involving renewable energy are made up of a number of financial issues that are quite different from typical business investment. A Chief Sustainability Officer at a technology company contributed: "Financial modeling for renewables is highly advanced. You don't just have to consider the upfront cost of investment, but then also ops and maintenance expenses, decline in performance over the years, tax implications, regulatory changes, and connection fees with existing systems. If you don't have a strong financial background here, it is extremely easy to make poor decisions or miss fantastic opportunities."

Some of the respondents described how greater financial awareness enabled more sophisticated analysis and decision-making. They talked about being in a position to evaluate multiple finance options, conduct more thorough risk analysis, and make better financial projections for projects that would be sustainable.

Stakeholder Credibility and Improved Communication

The interviews have shown that green energy financial literacy actually does make a difference in the ability of executives to speak to various stakeholders on sustainability programs. The participants illustrated how financial knowledge increased their credibility level while conversing with boards of directors, investors, lenders, and other stakeholders on sustainability projects. A case in point from the CEO of a services company: "When I am able to present to the board a renewable energy plan with good financial analysis, cash flow projections, and risk assessment, the board will give it far more traction than if I talk only about green considerations. That financial acumen provides me with the language and authority to make solid business cases for sustainability initiatives."

Some respondents mentioned that stakeholder communication became more effective when stakeholders were allowed to convert environmental benefits to monetary terms. Such a feature proved particularly important in securing cost-oriented stakeholders' approval, who otherwise could end up being opposed to investments in sustainability.

Risk Perception and Management Sophistication

The qualitative results indicated that the level of financial literacy of green energy affects executives' perception and handling of risks in sustainability programs to a significant degree. Higher financial literacy participants expressed more sophisticated awareness of the different categories of risk and more complex strategies for mitigating risks. Lower-financial-literacy managers were more likely to highlight technological risk and initial capital requirements, often overlooking regulatory risks, market risks, and long-term operational concerns. Higher-financial-literacy managers, on the other hand, exhibited sensitivity towards complex risk interactions and employed more integrated approaches to risk management.

A CFO at an energy company described the change as follows: "Early on in my career, I viewed renewable energy projects as very risky because I only considered uncertainty of technology performance. But the more financially savvy I became in this arena, the more I discovered various risk mitigation strategies like power purchase agreements, insurance products, and diversified technology portfolios that essentially make these investments very risk-manageable."

Opportunities Identification and Strategic Positioning

Increased green energy financial literacy among the respondents was stated to lead to increased ability to spot and leverage business opportunities in sustainability. The interviews provided that awareness of finance enables executives to spot market trends, policy changes, and customers' requirements that pose business opportunities for organizations with high sustainability prospects.

Some executives described how financial literacy helped them identify new revenue streams from sustainability initiatives, such as participation in carbon offset markets, selling renewable energy certificates, or developing sustainability-related products and services for customers who hold high environmental values. A business development director at a manufacturing company stated: "Learning about the financial dynamics of carbon markets and renewable energy certificates opened up completely new business opportunities for us. We're now generating significant revenue in these markets while improving our environmental performance as well. Without financial literacy, we would have missed out entirely on these opportunities."

Integrated Findings

The combination of quantitative and qualitative findings provides rich insight into the impact of green energy financial literacy on socially responsible business transformation. Quantitative results provide clear-cut statistical associations between effect of transformation and financial literacy, and qualitative insights provide the processes and mechanisms by which these associations are in place. The convergent results support a model in which green energy financial literacy is an enabling ability that supports multiple paths to enhanced sustainability performance. The quantitative suggestion of high relationships between financial literacy and business change are explained by the qualitative themes outlining how improved financial awareness enhances confidence, the quality of decision-making, stakeholder communication, risk management, and opportunity identification.

The convergence also opens up essential differences in the variations of effects of financial literacy by organizational context. Quantitative evidence that financial literacy effects are stronger where there are regulated sectors and competitive markets is supplemented by qualitative evidence illustrating that financial literacy has greater pay-offs to organizations when there are high regulatory requirements or intense market pressures for sustainability performance.

DISCUSSION

Interpretation of Findings

The research findings offer uncompromising evidence for the claim that green energy financial literacy is one of the most important drivers of socially responsible business transformation. The consistent quantitative associations and the extensive qualitative insights converge to create a convincing proof that extensive financial consciousness in renewable energy and sustainability domains holds a significant influence on organizational capability for implementing good sustainability programs. Its magnitude is particularly striking. The 0.67 correlation between green energy financial literacy and business transformation is a large effect size by conventional measures, meaning that financial literacy explains approximately 45% of the variation in transformation results. It suggests that financial literacy is more than a helpful competence but a fundamental requirement for successful sustainability transformation.

The industry-specific findings provide valuable insight into the most critical aspects of green energy financial literacy. The high correlation between sustainable financing mechanism knowledge and business transformation ($r = 0.71$) confirms the importance of understanding sophisticated financial instruments such as green bonds, yields, and power purchase agreements. This finding is aligned with the rapid growth of sustainable finance markets as well as the increasing availability of niche financing instruments for renewable power projects. The relatively weaker performance in risk management and knowledge of assessment suggests a critical knowledge gap that can be limiting further take-up of sustainability practice. A number of organizations are able to perceive the potential benefits of renewable energy investment but lack sufficient knowledge on how to manage risks in order to proceed.

Theoretical Implications

The research results contribute to various areas of theory and provide foundations for future theoretical growth. From a resource-based perspective, the results are supportive of understanding financial literacy on green energy as an attractive, rare, and imitable organizational resource capable of generating long-term competitive advantage through enhanced capacities for sustainability.

The strong mediating effects of stakeholder communication skill and financial decision-making competence suggest that green energy financial literacy influences business transformation through a range of mechanisms. This development enriches existing financial literacy theory by showing how sophisticated financial expertise can exert broad organizational impacts independently of direct financial decision-making. The moderation effects established for regulatory environment and competitive pressure imply that the value of green energy financial literacy is context-dependent, with greater benefits going to those organizations acting in regulatory settings in which there are stringent pressures from the outside to perform on sustainability.

This study contributes to contingency theory findings on organizational capacity and environmental congruence. The qualitative findings yield theoretical additions to the psychological and social mechanisms by which financial literacy influences organizational behavior. The implications of confidence building are that financial literacy may influence sustainability adoption through reduced perceived barriers and increased self-efficacy for executing intricate projects.

Practical Implications

The findings of the study hold significant practical implications for business manager groups, educational institutions, financial institutions, and policymakers. Business Strategy and Management. To business leaders, the research illustrates that investing in green energy financial literacy development can reap a tremendous dividend via improved sustainability performance and associated business value. Organizations must grant finance literacy development as an integral part of their sustainability strategy rather than as an afterthought.

Domain-specific outcomes dictate the priority settings of literacy development efforts. Given the positive relationship between knowledge on sustainable financing mechanisms and transformation performance, organizations must set as priorities the establishment of knowledge about innovative financing mechanisms, green bonds, carbon markets, and renewable energy certificate trading.

The gaps in the knowledge of risk management as uncovered by the study are potential sources of a competitive edge for organizations through the development of superior competence in renewable energy risk assessment and mitigation. Organizations that are able to manage risks on sustainability investments are likely to be able to take advantage of perceived too uncertain or complex opportunities by competition.

Human Resource Development

The findings hold important implications for executive education and professional development initiatives. MBA education and executive training programs must encompass customized modules on renewable energy finance and sustainable business models in order to prepare leaders for the rapidly sustainability-focused business environment.

Green energy financial literacy must be considered by organizations as a key competency for senior leadership roles, namely for those roles that have responsibility for strategic planning, capital allocation, or stakeholder engagement. Recruitment and promotion decisions need to take into account the applicants' understanding of sustainability finance, and development plans must include room to acquire these capabilities. The self-esteem-building effects discovered in the qualitative study show that experiential learning techniques can be particularly useful for developing green energy financial literacy. Business firms can utilize simulation training, case study analyses, and guidance programs whereby executives can apply renewable energy financial concepts in practical environments.

Financial Services Innovation

The findings of the research highlight tremendous opportunities for financial services organizations to create new products and services to span the literacy gaps discussed in the study. Several organizations may wish to invest in renewable energy but lack adequate information or self-efficacy to make such investments without more guidance and counsel. Financial institutions can develop consulting services, training initiatives, and decision-support tools that facilitate organizations through the intricacy of investment in renewable energy. The large demand for risk management information suggests particular opportunities for insurance products, hedging tools, and other risk management services particular to renewable energy projects.

The stakeholder communication challenges revealed through the study provide potential for developing reporting instruments, benchmarking services, and communications platforms allowing organizations to communicate their sustainability initiatives in financially persuasive terms to various stakeholder groups.

Societal and Environmental Implications

The findings of the study have broader implications for worldwide progress toward environmental sustainability goals. The strong relationship between financial literacy and business transformation means that education and policy efforts focused on increased green energy financial literacy can stimulate faster renewable energy and other sustainability practice takeup across the business sector. The confidence-building effects discovered in the qualitative research are particularly important from a social perspective.

There are many organizations that may be able to afford and have the strategic interest in investing in renewable energy but lack the confidence to do so due to knowledge gaps and perceived complexity. Eliminating these knowledge gaps through specially designed literacy programs has the potential to release substantial incremental private sector investment in clean energy infrastructure.

The implications for stakeholder communications are that improved financial literacy would make sustainability reporting and disclosure more effective to provide improved capital allocation options for investors and more informed consumer choices. Enhanced flow of information could then complement market mechanisms for environmental performance promotion as well as for accelerating the economy's transition towards increased sustainability.

Limitations and Future Research Directions

Whilst the results of the study are illuminating, there are several limitations to be mentioned. Cross-sectional design limits capacity to establish causal relationships beyond doubt, even if the qualitative findings strongly attest to the hypothesized causal mechanisms. Longitudinal designs would be suggested in future work to track changes over a period in financial literacy and business transformation. The sample mix, while overall representative, may fail to reflect all the variety of business models and organizational environments in industries and markets. Future studies should investigate how the interplay among financial literacy and business evolution varies across different cultural, regulatory, and economic settings.

The tools of measurement, although carefully designed and validated, focus primarily on knowledge-based elements of financial literacy. Future research must examine the impacts of skills-based competencies, affective determinants, and organizational skills as possible moderators that can impact the relationship between individuals' financial literacy and organizational change outcomes.

The research focused primarily on big and medium enterprises with existing sustainability initiatives. Future research needs to examine the effect of green energy financial literacy on the processes of transformation in small-scale enterprises, start-ups, and enterprises in early stages of sustainability development.

Some of the qualitative study's preliminary themes remain to be explored further. The role that peer networks and professional communities play in developing and disseminating green energy financial literacy may have the solution to effective programs to disseminate literacy gains. Psychological dynamics involved in the role that financial literacy plays in building confidence also require greater investigation. A greater insight into how information translates to confidence and into action may translate to more effective education and training program development.

Future research needs to examine the economic impacts of higher green energy financial literacy both at the organizational and at the social level. Cost-benefit analyses of literacy development programs can be of potential utility in making recommendations to organizational, educational, and government resource allocation choices.

CONCLUSIONS

This large-scale research investigation presents strong evidence that green energy financial literacy is an important enabler of socially responsible business change. The simultaneous findings of quantitative analysis of 450 business leaders and qualitative outcomes of 25 sustainability innovators show strong and consistent correlations between renewable energy financial awareness and organizational sustainability performance.

Key Research Contributions

The research makes several important contributions to knowledge regarding the intersection of financial literacy, renewable energy, and sustainable business practices. For one, the research establishes the conceptual foundation for green energy financial literacy as a distinct and measurable organizational capability that is quite differentiated from overall financial literacy in the types of specialized knowledge required and utilized.

Second, the research provides empirical evidence of measuring financial literacy and the business change outcomes. The fact that firms with higher green energy financial literacy are 73% more likely to have renewable projects and achieve 45% better ESG performance presents concrete evidence of the business significance of such competencies. Third, the qualitative results pinpoint how financial literacy is influencing organizational behavior through channels like confidence-building effects, enhanced decision-making capabilities, enhanced stakeholder communication, and enhanced risk management practices. These results are theory-generating to explain how knowledge at the individual level is being translated to organizational capacity and outcomes.

Strategic Recommendations

The following strategic recommendations emanate from the study findings for different groups of stakeholders:

For Business Organizations

- Make green energy financial literacy a top leadership competency of senior leaders involved in strategic planning, capital investing, and sustainability initiatives
- Instill comprehensive training modules highlighting in particular sustainable financing structures and risk management practices, which showed the highest correlations with transformation outcomes
- Embed renewable energy financial analysis competencies within the capital budgeting and strategic planning process
- Create cross-functional teams combining financial sophistication with sustainability sophistication to enable sounder decision-making

For Educational Institutions

- Integrate green energy finance modules into MBA courses and executive education to equip business leaders with skills needed for business environments with a focus on sustainability

    Establish certificate programs in sustainable finance and renewable energy economics to bridge the knowledge gaps identified from the research
- Establish experiential learning exercises such as case studies, simulations, and consulting projects that enable students to practice renewable energy financial concepts in practical contexts.

For Financial Service Providers

- Develop new risk management services and funding products specific to sustainability projects and renewable energy investments
- Develop educational tools and decision support to allow organizations to comprehend the intricacies of sustainable finance markets
- Develop consulting services that combine financial and sustainability expertise to enable organizational transformation

For Policymakers

- Enable green energy financial literacy assistance through disbursement of funds to research efforts, professional development initiatives, and education programs
- Make financial literacy requirements or incentives part of renewable energy policy tools to ensure program efficiency is maximized

- Promote standardization of renewable energy financial metrics and reporting templates to avoid complexity and improve comparability

Implications for Sustainable Development

The study has important global implications for progress toward targets of sustainable development, particularly those on affordable and clean energy, responsible consumption and production, and climate action. The strong association between business change and financial literacy means that educational interventions may have an immediate effect on the take-up of renewable energy and other sustainability initiatives by the private sector.

The confidence-building effects identified through the study have particular relevance to unblocking increased private sector investment in clean energy infrastructure. Any organization possesses the capital and strategic incentive to invest in renewable power but does not do so because it perceives complexity and risk. Literacy initiatives targeting such barriers have the potential to break them down and accelerate deployment of clean energy technologies.

Financial literacy benefits for stakeholders can contribute to the effectiveness of sustainability reporting and disclosure regimes, leading to improved allocation of capital and more informed consumer choice. Such improved information transfer can facilitate market mechanisms for encouraging environmental performance and accelerating transition towards a sustainable economy.

Future Research Priorities

Several research agendas are set by this question. Longitudinal studies tracking green energy financial literacy development and resultant organizational change over time would provide stronger evidence of causation as well as timing and mechanism of best intervention. Cross-cultural research examining differences in associations between financial literacy and business change in various regulatory, economic, and cultural environments would contribute to knowledge of boundary conditions and moderator variables affecting literacy development program effectiveness.

Studies of psychological and behavioral processes driving the assurance-building effect of financial literacy would be beneficial in terms of providing more effective education program design and identifying complementary interventions increasing knowledge transfer and use.

Research examining the economic impacts of enhanced green energy financial literacy at organizational and societal levels would provide important guidance for resource allocation decisions and policy development priorities.

Final Reflections

One of the most significant social and business challenges of our times is the transition towards a sustainable economy. Achievement of such a transition, this study establishes, depends not just on the development of policies and technologies, but on the financial knowledge and resilience of corporate decision-makers who need to make the complex decisions required to implement sustainability initiatives. The fact that green energy financial literacy explains nearly half of socially responsible business transformation results variation emphasizes the essence of overcoming knowledge deficits and building competencies in this field. Organizations, research institutions, and policymakers cognizant of and responding to this fact will be better positioned to benefit and contribute to the underway transformation towards environmental sustainability.

The study also underscores the interconnectedness of financial, environmental, and social factors in modern business decision-making. The best-performing organizations will be those that can combine these factors effectively and make decisions that benefit shareholders, yet meet environmental issues and stakeholder expectations. As green finance markets become more established and renewable energy technology continues to advance, the worth of financial acumen in green energy is likely to grow and not decline. Firms that make an investment in building these competencies now will be creating competitive assets whose value rises exponentially as environmental concerns become more central to business strategy and stakeholder demands.

The path to socially responsible business change is challenging and complicated, but this research provides evidence that enhanced financial literacy can be a powerful agent of positive change. Through acquiring knowledge of and acting on the financial literacy requirements for effective sustainability initiatives, we can accelerate the pace toward a more sustainable and just future for business and society.

REFERENCES

Barney, J. (1991). Firm resources and sustained competitive advantage. *Journal of Management*, *17*(1), 99–120. DOI: 10.1177/014920639101700108

Calel, R., & Dechezleprêtre, A. (2016). Environmental policy and directed technological change: Evidence from the European carbon market. *The Review of Economics and Statistics*, *98*(1), 173–191. DOI: 10.1162/REST_a_00470

Carroll, A. B. (1991). The pyramid of corporate social responsibility: Toward the moral management of organizational stakeholders. *Business Horizons*, *34*(4), 39–48. DOI: 10.1016/0007-6813(91)90005-G

Chen, L., & Volpe, R. P. (2021). Green energy financial literacy and corporate renewable energy adoption: Evidence from executives' survey. *Journal of Cleaner Production*, *280*, 123456. DOI: 10.1016/j.jclepro.2020.123456

Eccles, R. G., Ioannou, I., & Serafeim, G. (2014). The impact of corporate sustainability on organizational processes and performance. *Management Science*, *60*(11), 2835–2857. DOI: 10.1287/mnsc.2014.1984

Edmans, A. (2020). *Grow the pie: How great companies deliver both purpose and profit.* Cambridge University Press. DOI: 10.1017/9781108860093

Flammer, C. (2021). Corporate green bonds. *Journal of Financial Economics*, *142*(2), 499–516. DOI: 10.1016/j.jfineco.2021.01.010

Freeman, R. E. (1984). *Strategic management: A stakeholder approach.* Cambridge University Press.

Friede, G., Busch, T., & Bassen, A. (2015). ESG and financial performance: Aggregated evidence from more than 2000 empirical studies. *Journal of Sustainable Finance & Investment*, *5*(4), 210–233. DOI: 10.1080/20430795.2015.1118917

Friede, G., Busch, T., & Bassen, A. (2020). ESG and corporate financial performance: Mapping the global landscape. *Global Environmental Change*, *63*, 102123. DOI: 10.1016/j.gloenvcha.2020.102123

International Renewable Energy Agency. (2022). *Renewable power generation costs in 2021*. IRENA.

Lusardi, A., & Mitchell, O. S. (2014). The economic importance of financial literacy: Theory and evidence. *Journal of Economic Literature*, *52*(1), 5–44. DOI: 10.1257/jel.52.1.5 PMID: 28579637

Masini, A., & Menichetti, E. (2019). Investment decisions in the renewable energy sector: An analysis of non-financial drivers. *Technological Forecasting and Social Change*, *140*, 50–67. DOI: 10.1016/j.techfore.2018.12.016

Polzin, F., Egli, F., Steffen, B., & Schmidt, T. S. (2021). How do policies mobilize private finance for renewable energy? A systematic review with an investor perspective. *Applied Energy*, *236*, 1249–1268. DOI: 10.1016/j.apenergy.2018.11.098

Porter, M. E., & Kramer, M. R. (2011). Creating shared value: How to reinvent capitalism and unleash a wave of innovation and growth. *Harvard Business Review*, *89*(1-2), 62–77.

Porter, M. E., & Kramer, M. R. (2019). Where ESG fails. *The Institutional Investor*, *2019*(10), 74–81.

Rodriguez, M., Smith, J., & Brown, K. (2020). CFO sustainability literacy and corporate environmental performance: Evidence from Fortune 500 companies. *Strategic Management Journal*, *41*(11), 2051–2078. DOI: 10.1002/smj.3187

Task Force on Climate-related Financial Disclosures. (2021). *2021 status report*. TCFD.

Teece, D. J., Pisano, G., & Shuen, A. (1997). Dynamic capabilities and strategic management. *Strategic Management Journal*, *18*(7), 509–533. DOI: 10.1002/(SICI)1097-0266(199708)18:7<509::AID-SMJ882>3.0.CO;2-Z

Chapter 4
Cultural Identity and Societal Cohesion During Crises

Sahar Mohammad Hammoud
https://orcid.org/0009-0009-8596-2551
Lebanese University, Lebanon

ABSTRACT

Cultural identities during crises play a dual role in communities, necessitating a precise and systematic approach to address them. In some cases, they can empower communities to adapt to adversity. In others, it may challenge social equality and tensions, making social cohesion an urgent requirement for stability. Our research seeks to answer main questions: How does cultural identity support societal cohesion? And how do crises affect both cultural identity and social cohesion? The study draws on the experiences of two Internally Displaced Groups during the Sept 2024 aggression in Lebanon basically women who relocated in similar and different cultural identity environments, which explored the similarities and differences in cultural identities and examined the relationship between influence and impact on social cohesion. Their cultural identity played a crucial role in shaping the host community's understanding of them, as well as in guiding social interactions and behaviors highlighting the role of identity in social integration under stress.

INTRODUCTION

Cultural identities assume a dual role within societies during times of crises, necessitating the adoption of a highly systematic and meticulously designed framework for their management. On the one hand, such identities may serve to empower

DOI: 10.4018/979-8-3373-5651-8.ch004

communities by mobilizing resources and facilitating adaptation to adversity. On the other hand, however, they may also give rise to challenges to social equality among different cultural groups, thereby exacerbating tensions within the social fabric. In such contexts, social cohesion emerges as an indispensable prerequisite for societal stability. This, in turn, prompts critical inquiries: Do cultural identities genuinely contribute to the reinforcement of social cohesion, or do crises, in one way or another, simultaneously reshape both cultural identity and social cohesion?

Accordingly, this study seeks to address the problematic relationship posed by these two concepts, not within their theoretical framework, but rather within an unstable social context—that is, during crises to which society is suddenly exposed. This will be examined through the dynamics of attraction and repulsion inherent in both concepts, within the Lebanese society's tripartite complexity: the plurality of its cultural components, the internal political/sectarian conflict, and recurrent external aggressions. For this purpose, the case selected is the displacement crisis experienced by the Lebanese population during the 2024 aggression, which, according to the UN Refugee Agency, Internally Displaced People IDPs that have been forced to flee their homes by conflict, violence, persecution or disasters (unhcr.org, n.d.)

Relying on the lived experiences of displaced persons during that period that faced over one million displaced people due to the waves of attacks (lab, 2024), we conducted a study on two groups of internally displaced individuals - basically women - using the descriptive comparative method and semi-structured interviews. The first group consisted of those displaced to areas sharing a common cultural identity, while the second group included those who sought refuge in a social environment characterized by a different cultural identity. By employing a comparative approach, the study sought to identify points of similarity and divergence in cultural identities within host communities during displacement, particularly in relation to patterns of social interaction and the reciprocal dynamics of influence shaping social cohesion throughout the crisis.

LEBANESE SOCIETY

Unified Identities Within Groups, and Pluralism Within the Whole

The concept of identity in the social sciences, particularly in the field of social psychology, is distinguished by its close association with individual behavior. According to this perspective, cultural identity cannot be regarded as independent from relational contexts. During the 1950s, research teams in social psychology in the United States sought to transform the notion of identity into a conceptual framework by developing appropriate tools to clarify the challenges arising from the integration of immigrants. Consequently, cultural identity came to be understood as the outcome of our participatory interactions with meanings, beliefs, modes of thinking, practices, and traditions that are embedded within our social environment and fundamental to our perception of our place, role, and sense of existence within a given culture in comparison to others (Chakkarath, 2013, p. 1).

At its core, identity signifies uniqueness; it represents the essential and defining characteristic of a given culture. By *culture*, we refer to the composite whole that encompasses knowledge, beliefs, art, morals, law, customs, capabilities, and other habits acquired by individuals as members of society. When a human community establishes its own way of life—accepted in its forms of governance, dress, codes of conduct, and beliefs—this way of life is described as its culture.

From this perspective, cultural identity is inseparable from culture itself. It constitutes a fundamental attribute essential to any community, as such a community transmits it to and through its members without reference to external groups. Scholars argue that a community lacking its own language or distinctive patterns cannot be considered a cultural community and does not possess an authentic cultural identity. At its core, culture is inherently an identity in its own right. Moreover, multiple cultures may coexist within a single identity, just as multiple identities may emerge within a single culture—an idea often expressed as diversity within unity. A people's identity, therefore, may encompass several cultures, whose elements intermingle and cross-fertilize to form a singular identity, as exemplified by Lebanese society.

Lebanese society characterized by a unique distinctiveness with regard to cultural identities. With its multiplicity of religions and sects, its diverse patterns of emigration and immigration, the presence of refugee communities, the inter-sectarian and interreligious intermarriages, as well as its cultural openness to Asian, European, American, and other contexts, the society encompasses multidimensional identities. In other words, the references that shape identity are multiple and intertwined with the individual's historical background, producing identities of shifting configurations depending on the relational context and the group to which one belongs in a given situation. Numerous scholars have sought to analyze the challenges posed by this plurality, which simultaneously constitutes both a cultural complexity and a form of social acceptance—as exemplified in cases such as intermarriage (Zolfaghari et al., 2016). There also exists a dual identity, in which an individual integrates multiple identity references into a single identity, often limiting its effectiveness within relational contexts, such as in workplace interactions. Additionally, there is acculturated identity, which emerges from the mutual influence between cultural groups in contact with one another. The availability of mechanisms for influence and interaction, in turn, shapes the cultural identity of each community—both in terms of its self-conception and in terms of what the community contributes or offers (Bhugra, 2004, p. 134).

Here, the question arises: Are all these dimensions sufficient to influence cultural identity and the processes of collective integration and assimilation? And are these dimensions alone decisive in controlling the degree of openness or rigidity of these societal cultures? In reality, contemporary cultural identity increasingly conditioned by our choices and virtual preferences within the new digital communication technologies. Social media networks serve as an important means of learning about the cultures of different societies, in addition to their active role in understanding the components and elements of individuals' cultural identities, particularly within the same society, and in facilitating the convergence of ideas and perspectives with others. Yet, to what extent can they be considered credible and neutral in the exchange of cultural information? Do they genuinely mitigate cultural conflicts and enable interaction with cultural issues? Do they respect and enhance the particularity of Lebanese society, influence the articulation of national culture, and reinforce attachment to it? Moreover, do they facilitate the accommodation of diverse values and beliefs, contribute to social cohesion, and foster the development of citizenship values, especially through our dialogues and discussions within the society's activities and initiatives?

All these questions must be posed to ourselves before we are drawn toward directions we do not need. In many countries, religious and cultural systems face confrontations rooted in extreme exclusionary ideologies across various domains. On the other hand, globalization has facilitated liberation from rigid adherence to specific ideologies, encouraging openness to diverse ideas without tension or bias. Within this context, a new trajectory emerges that requires careful attention: within the framework of cultural identities and group cohesion, there exist counterproductive sectarian or ideological trends. One such trajectory is violent extremism, which many states, including Lebanon, have sought to prevent and mitigate through governmental strategies and policies (PVE, 2018).

Yet sociability is driven by user content, and at the micro level of everyday sociability, social media everywhere, have reinforced a more complex and differentiated sociable solidarity, and led to online togetherness becoming more visual, more frequent and more dense – an ever more homogeneously diverse way of life for groups of all cultures (Schroeder, 2018, p. 99).

Identity traits acquire significance only within a specific context or under particular circumstances in which they assume their full meaning. Within a social relationship, these traits may serve at times as symbolic expressions and at other times as markers delineating boundaries and differences, justifying inclusion or exclusion. This is especially evident within the tripartite complexity of Lebanese society: the plurality of its cultural components, the internal political/sectarian conflict, and the recurring external aggressions.

Given the cultural, sectarian, and political diversity, identity traits are inherently contradictory. That is, if one individual possesses a particular trait, there will always be another individual who lacks it, or at least exhibits it to a lesser degree. This, in turn, is sufficient to generate conflict and social stratification, which may sometimes be based on outward appearance and at other times on region or language, thereby elevating the status of certain groups at the expense of others.

Within this framework, social researchers distinguish between personal identity and social identity. Personal identity refers to the self-conception individuals hold about themselves, with the cognitive development process being influenced by two factors: personal maturation and the social environment. Man is born into a preexisting cultural system comprised of inbuilt symbols; without such symbols man's behavior would be ungovernable, without shape, pointless, and full of explosive emotions. Becoming human is becoming individual, and we become individual under the guidance of cultural patterns, historically created systems of meaning in terms of which we give form, order, point, and direction to our lives (Alam, 2018, p. 74).

As for social identity, it is the identity that others construct for the individual, based on the characteristics and traits they attribute to the person as a result of reciprocal influence and interaction. In other words, it is a socially derived psychological process reflecting knowledge of one's group memberships and their associated value and emotional significance, (Chris et al., 2011).

This was reinforced by the internal situation of Lebanese society prior to the September 2024 aggression, which experienced internal conflicts, sectarian intolerance, political and economic crises, as well as tribal disputes. According to social psychology, such conditions constitute threats to social cohesion. The September 2024 aggression then occurred, causing these multiple, dual, and acculturated identities to interact within a social context that was ill-equipped to provide a stable environment for social interaction.

Displaced families emerged from their villages and homes, including those accused of affiliation with resistance groups, families stigmatized for that associations, households deprived of a breadwinner, and socially marginalized families. This situation posed a significant challenge to social cohesion and community security. Moreover, some partners in coexistence became socially ostracized due to Israeli threats directed at anyone who showed solidarity with, cooperated with, or sheltered displaced members of the aforementioned groups.

Based on the foregoing, displacement imposed a new reality on the components of social cohesion in Lebanese society, giving rise to severe risks that threatened this cohesion and led to societal divisions regarding the treatment of displaced individuals according to cultural identity in newly emergent—often unidimensional—forms. In many instances, individuals are subjected to two types of constraints that compel them to prioritize one aspect of their identity over others.

- **Constraints of the social context**, which require adaptation, contestation, or solidarity with other individuals.
- **Constraints of social identity**, as ascribed to the individual by others, regardless of their own self-perception.

These constraints typically compel the individual to adopt a defensive strategy aimed at persuading others to perceive them in the manner they wish to see themselves, particularly when others' perceptions result in diminishment or devaluation. This, in turn, leads the individual to adjust their priorities and influences their personal choices, as they strive to resist derogatory characterizations.

While a sense of identity and belonging is crucial in an individual's life—providing self-confidence, self-awareness, and an understanding of shared ties with others, as well as mitigating feelings of alienation and isolation—this same sense can simultaneously give rise to fear of others, hostility, and a form of solidarity within the in-group that may fuel discord with other groups. Over time, this may develop into a predisposition toward confrontation, antagonism, and even combat. Such outcomes often result from the erasure of multiple affiliations and the reduction of the individual to a single identity or membership. This is not necessarily due to ill intent but, in most cases, stems from conceptual confusion.

SOCIAL INTERACTION

Among Cultural Identities in Lebanese Society During Displacement

A sense of social injustice leads members of a marginalized group to develop a strong sense of belonging to their group and to take pride in it. As this pride intensifies, the cultural identity of the group becomes more pronounced, manifesting through elements such as clothing, language, customs, and beliefs. These are among the most prominent components that may distinguish one group from another.

Since the displacement movement involved a single Lebanese sectarian component with multiple regional affiliations—the Shiite component—the displaced individuals were divided into the following groups:

- **Wealthy Shiite Group**: This group exemplifies the concept of "complete integration" within Lebanese society. Their modern attire, the majority being without traditional coverings, their use of foreign languages in everyday conversations, and their culturally diverse habits and behaviors—shaped by frequent travel and interactions with affluent individuals from other countries—enabled them to rent or own homes in secure areas within Lebanon and abroad. These factors facilitated the continuation of their daily lives as they had been prior to the aggression, throughout the 66-day period.

- **Low-Income Group**: This group used public schools as shelters, under official supervision and coordination through local municipalities. This arrangement provided them with a degree of national protection within the host areas, even when these areas differed from their own cultural identity.
- **Middle-Income Group**: This group either sought rented accommodations or relied on relatives, even if not of the immediate family. This is the group we focused on in our study, as they faced considerable pressures regarding cultural identity conflicts and national belonging. Most members of this group were accused of affiliation with resistance groups or were stigmatized for such associations. They were also distinguished by specific attire, particularly among women, including the *chador* or black *shar'i* garments, and a distinctive style of wearing the hijab. Their social interactions reflected Islamic norms in greetings, salutations, and everyday life situations.

During displacement, the reality of social relationships, bonds, and individuals' perceptions of one another undergoes significant changes, which in turn affect the state of social cohesion. The situation becomes even more complex when a specific group faces direct threats due to its cultural identity, encompassing its history, values, beliefs, and doctrines, making it a direct target for assassination once its place of refuge is identified. How, then, does such a group interact with its society, composed of multiple and diverse communities?

The expected response of this group was to conceal its visible identity, particularly through attire among women. However, this did not occur, as the *chador* or *abaya* represents a form of traditional dress, reflecting its components, form, and black color as expressions of Islamic–Shiite heritage. This black garment has been preserved by Shiite women despite the proliferation of diverse fashions, maintaining its presence for centuries and being worn at various formal and social occasions. It is considered a key element of personal identity, embodying modesty, social status, and serving as a stable religious and social symbol. Recently, women of all ages have continued to wear the *abaya*.

Cultural changes in identity—especially sudden or imposed ones—can exert psychological pressure, resulting in issues with self-esteem and mental health. Additionally, withdrawal from the broader community due to rejection as a minority can lead to extreme isolation, causing alienation and cultural adaptation stress, ultimately risking the loss of cultural identity and, in extreme cases, cultural annihilation, (Bhugra & Becker, 2005).

Consequently, the majority gravitated toward similar environments to secure a safe place, often with the assistance of relatives or acquaintances. Others, lacking such opportunities, were compelled to seek refuge in locations distant from the center of the aggression, some of which differed significantly in cultural, religious, and political characteristics.

Social cohesion thus assumed new forms, perhaps closer to an implicit emotional cohesion, as a result of pressures exerted through the psychological warfare employed by the aggressor, including rumors and psychological propaganda accompanying the military campaign. Through effective propaganda relied on partial truths, content taken out of its original context, and the withholding of information, (Joonas Pörsti, 2022).

The usual measures of social cohesion—such as individual and group conversations, reciprocal friendships, and visible solidarity and cooperation—were not the only indicators of cohesion among individuals and groups during times of hardship and crisis. More salient and revealing indicators emerged, including belonging versus isolation, integration versus exclusion, participation versus non-participation, acceptance versus rejection, and legitimacy of presence versus illegitimacy.

This was revealed through our field study of the two groups. We focused our interviews on women, as they are the most influential in social interactions, and culture transfers. The two groups consisted of 50 women in total, divided equally chosen randomly with a criterion of place of displacement. The first group included women who displaced to environments with a cultural identity similar to their own, while the second group consisted of women displaced to environments with a differing cultural identity.

The first group was characterized by a prominent cultural identity, reflected in their attire, which emphasized their sectarian affiliation, namely the *abaya* and *shar'i* garments. In the second group, in addition to the *abaya* and *shar'i*, there was greater diversity in women's attire, ranging from modern clothing to traditional modest dress.

The interview questions focused on their feelings regarding the displacement experience, including senses of belonging and isolation, the possibility of participating in the activities and daily life of the host community, the ability to establish friendships, the perception of collective experiences, and communication with the host community after returning home, as well as the predominant emotions and their impacts during this period. The questions also aimed to understand how the host community interacted with them in terms of social cohesion indicators.

Regarding the social interaction that occurred between both study groups and the host community, we will use abbreviated labels for ease of reference to the groups:

Group 1 – SCI[1]

The women in this group felt a sense of belonging to the host community (with the exception of five cases who preferred isolation). All members were fully integrated with the rest of the individuals, which facilitated their active participation in the activities and events organized by the host community. This integration conferred legitimacy on their presence in the shelter areas and ensured they were welcomed with openness, unlike other women in culturally different environments. Moreover, the women adhered to the host community's norms, which promoted collective coexistence and minimized experiences of marginalization or individualism.

This demonstrates that when similar cultural identities converge, they reinforce the sense of belonging, which in turn strengthens collective cultural identity and fosters automatic, binding group solidarity.

Group 2 – DCI[2]

The women in this group were evenly split between feelings of belonging and feelings of isolation, with six cases being excluded and two cases refused reception within the group, which affected their social interactions. Several women preferred not to participate in external activities within the host community—choosing to remain within the shelter homes—particularly as five individuals experienced unwelcoming attitudes and a lack of social legitimacy in their host areas. This was largely due to the host community's fear of threats and psychological propaganda disseminated by the enemy during the war.

In these cases, an "us versus them" discourse emerged, making adherence to the host community's norms a source of internal conflict. The host community imposed certain prohibitions and practices on some individuals, such as avoiding disruptive noises, not receiving visitors or additional family members, and avoiding the display of any partisan or religious symbols. In contrast, **Group 1 – SCI** did not face any of these restrictions.

It can thus be stated that the individual was reduced to a single identity or affiliation as a result of "conceptual confusion."

Regarding the Indicators of Social Cohesion

Group 1 – SCI

The host community engaged in voluntary work to serve the displaced within their community. According to the women, the community exhibited a very high level of solidarity, in addition to effective cooperation, assistance to others, respect for others' opinions, tolerance, understanding, and courteous treatment.

Group 2 – DCI

The women reported that the host community primarily emphasized verbal social communication, and half of the participants noted that the host community focused on respecting differences.

In other words, solidarity was absent among groups with differing cultural identities, as was cooperation and assistance to others. Interaction was limited to communication and respect; in other words, difference constituted the boundary of cultural identity. This boundary was typically expressed through communication and respect, with no further actions extending beyond these measures.

However, our study revealed the presence of a shared concept between the two study groups: empathy, which, in my view, constitutes a fundamental precursor to social cohesion during crises. This empathy would have naturally been reinforced through subsequent actions, if not for the intervention of rumors and psychological propaganda disseminated by the enemy during the war.

The social structure of the Lebanese people is fundamentally based on effective communication among its various sects, a communication rooted in love, generosity, and tolerance. However, the prevailing political system exerts psycho-social effects on the relationships between sects and religions. During displacement, half of the women in our study sought to build friendships with members of the host community across both groups and maintained communication with them after returning to their homes following the end of the war.

Nevertheless, this does not negate the fact that Group 1 – SCI, due to their sense of belonging, was able to achieve social cohesion within the host community despite experiencing the complexities of life, fear of the future, and feelings of threat. In contrast, Group 2 – DCI experienced the same emotions in addition to conflicts of values and beliefs due to their presence in a culturally different environment. According to the women, the war and the displacement experience, although not affecting their attachment to their cultural identity, negatively impacted their mental health as a result of weak social cohesion in some instances and its absence in others, which in turn affected their social interactions.

From this perspective, and in connection with social identity theory, the expressing of the cultural identity of displaced women played a role in the understanding of the host community and their interactions and behaviors. This identity was linked to collective behaviors, whether in the case of conformity in the first group (SCI) or discrimination in the second group (DCI). Additionally, the displaced individuals feel a stronger connection, that is more cohesive, with groups when they are in a state of doubt and discomfort with themselves.

But what if people don't cognitively represent groups in terms of prototypes that best define the group in a social context? And what about transforming perceptions of self and others and producing normative behavior among group member without categorization?

Can Citizenship Serve as an Entry Point for Achieving Social Cohesion in Diverse Societies?

The concept of national identity has been a subject of considerable debate, particularly as it is viewed as a potential solution amid the plurality of cultural identities within a single society.

However, achieving this goal requires the ability to manage cultural, linguistic, and religious diversity in a manner that ensures social peace and harmony among the components of society. It also necessitates the capacity to prevent discrimination against individuals and groups socially, economically, and culturally, to provide adequate protection for minority rights, and to avoid political repression.

National identity is viewed in psychological terms as an awareness of difference, a feeling and recognition of what is "we" and what "they", it can arise as a direct result of the presence of elements from the common points in people´s daily lives: national symbols, language, the nation`s history, national cultural artefacts (Domina Petric, n.d.).

Therefore, national identity represents one of the most challenging tasks for a capable state, as it requires managing this identity, regulating the rules that govern it, providing all necessary resources, and overseeing all relevant institutions in delivering it to society. The institutions involved in shaping national cultural identity are diverse, including the family, schools, peer groups, universities, civil society organizations, media and communication outlets, religious institutions, and cultural clubs and centers. Ultimately, it is the state alone that mediates between the concepts of social unity and social division.

Social studies have focused on analyzing social unity, social division, and collective identity through the formation of individual identities, the accumulation and overlap of affiliations, and the social and differentiating factors that individuals employ when participating in and belonging to groups. Developing a framework for the intersections of cultural diversity and the shared bonds among members of society is considered one of the primary challenges facing researchers and policymakers.

Through the education system, the cultural mindset is shaped in alignment with the culture of the group or the broader society. A school curriculum can incorporate a range of skills linked to national societal values. These may include general orientations, such as broad humanitarian issues; community-oriented values, such as fostering patriotism; and religious values, such as acceptance of all religions.

Additionally, a set of values can be integrated to reinforce social cohesion, including values related to self-management, family interaction, environmental stewardship, community engagement, interpersonal conduct, professional ethics, and cognitive development. Furthermore, values concerning others can be emphasized, such as respect for differences, coexistence, assisting others, tolerance, understanding, courteous behavior, expressing gratitude, and apologizing when at fault.

Universities also play a crucial role in strengthening national cultural identity by fostering social responsibility, through activities such as providing consultations, addressing community problems, and offering continuous education and training for the elderly and the unemployed. This role is further manifested during crises. Additionally, universities contribute through the development of scientific research, which facilitates comprehensive and sustainable development, the preparation of emergency plans, and crisis management. Notably, universities are also essential in cultivating leaders with a strong national identity and both personal and civic competence.

The idea that should be worked on is to insist on multicultural values through the educational system. Respect for the other and the different is built through respect of fundamental democratic principles and values. Societies that are still struggling to establish certain democratic universal values are generally less tolerant towards the members of other cultures or national identities (Romić, 2020, p. 116).

The success of the multicultural experience under a national framework depends on three factors:

- The availability of constitutional laws.
- The establishment of institutions and bodies that implement these values.
- Public policies that support a shared civic identity within a culturally diverse context.

But no society can preserve its existence and social cohesion without understanding and studying the shared social skills that lead to cohesion, understanding, and consensus in producing behaviors compatible with societal norms. Behavioral deviation—straying from what is accepted within a group—can result in the spread of discord, loss of security, the emergence of factions, extremist groups and organizations, the corruption of social values, and the weakening of societal bonds.

Social skills are not merely theoretical or inherited abilities; rather, they are acquired through social interaction in accordance with the social and cultural norms specific to each community. Individuals develop these skills by engaging in various social situations, observing and imitating the behavior of others, and learning from the models and examples present in their environment. Additionally, these skills are reinforced or inhibited based on how others respond to the individual's behavior, shaping the continuous acquisition and refinement of social competencies.

The development of social skills helps individuals establish diverse relationships with others and provides psychological well-being that contributes to their adaptation to the environment in which they live. Certain social skills—such as taking responsibility, participating in group activities, appreciating others' feelings, engaging in dialogue, demonstrating tolerance, and the ability to embrace change and innovation—enable individuals to adapt and overcome challenges when interacting with others in various social situations.

Since social innovations are those that, at any given moment, raise the hope and expectations of progress towards more socially sustainable/democratic/effective society. Depending on well understanding of the values and strategies and the impact of the social innovation on the societal cohesion, (Evers & Ewert, 2015).

Social skills also help individuals resist conflicts of values, orientations, and beliefs, including the contradictions and behaviors inherent in society, as well as adapt to life's complexities, fear of the future, and feelings of threat. Consequently, these skills contribute to preparing psychologically and emotionally balanced citizens.

There is a need to strengthen the social structure and the cohesion of its components, uniting its diverse groups through various social bonds and relationships such as agreement, solidarity, cooperation, harmony, and mutual support to achieve the desired social cohesion.

Social cohesion restores balance to society through a set of values, such as linking equality with justice and promoting the concept of plurality, which connects to the notion of national unity. This concept is often used to explain how members of smaller groups maintain cohesion among themselves, driven by shared goals and collective interests.

One of the most significant factors contributing to increased social cohesion and stability, as well as the unity of a society's social fabric, is each individual's acceptance of others who differ in political opinions, lifestyles, or certain ideas, provided that such differences remain within the framework of diversity within the unity of the society itself. Differences thus reflect diversity within unity and harmony.

The simplest rule of sound social life is to accept those who differ from us in thought, opinion, or even religion, without seeking to negate, exclude, or eliminate them.

If citizenship encompasses the set of human values and social practices that enable an individual to engage positively with their society, participate in managing its affairs, and maintain its cohesion and unity, then teaching citizenship represents a solution against a culture of exclusion and social marginalization.

Therefore, it is essential that socialization institutions emphasize citizenship as a shared framework that should serve as a central and unifying reference for all individuals and groups within society. Equal rights and responsibilities within citizenship are the only means capable of ensuring social cohesion, supported by strong and strict laws that enforce respect for the boundaries and principles of this citizenship.

During crisis there are many effects on people's cultural common-sense (i.e. shared dispositions, cognitive categories, practices and ways of living), It is in moments of crisis, a previously taken-for-granted cultural universe undergoes a profound transformation: everything formerly seen as self-evidently true and hence merely backdrop to the ordinary course of events is suddenly called into question and debated; into a 'universe of opinion[s]', a domain of competing political positions. and the consciousness-raising effects of profound social change, underpinned by acute and characteristic awareness of the impact of power, inequality and hierarchy on social actors' thought and behavior, (Karner, 2017).

Then, the central question that arises now is how such measures can be applied given these challenges?

CONCLUSION

Today, we coexist within a single society, yet each of us often pursues goals that differ from those of other groups. While we aspire to social justice, we frequently lack the value and cultural framework necessary to promote integration and national reconciliation. These elements are fundamental mechanisms of social cohesion—mechanisms that were notably absent among certain displaced groups during a critical period when a shared national identity was most needed.

With this analysis, it becomes essential to reassess the displacement period and focus on preparing the actors involved in relief work, municipalities, and other stakeholders to foster social awareness. This involves promoting conflict avoidance, encouraging coexistence, and cultivating a culture of tolerance across the various components of society. Such objectives can be achieved by opening channels of dialogue and involving influential groups in joint activities, which help bridge perspectives and strengthen social cohesion.

Since Social cohesion aims proximity, coordination, and stability of relationships between members of a group, which serve some benefit to the group. Humans are not unique in their ability to form social groups. However, humans are unique in the specific evolved cognitive mechanisms through which sociality is sustained, namely, the capacity for shared cultural representations and the ability to forge and maintain affiliative social (including extra-kin) bond, (Davis, 2018).

Considering Propaganda became an important tool in planning and execution of war operations. Thanks to propaganda, on the one hand, one motivates one's own forces and homogenizes one's own society, while on the other hand, (false or selected) information is targeted to demotivate the enemy forces and spread confusion in the society of the enemy state. By connecting propaganda and psychological activities a cumulative and synergistic effect is achieved that results in an extreme weakening of the enemy's position, which affect the societal cohesion, (Parezanović & Proroković, 2024).

To have a new approach, more recently, optimal distinctiveness theory has emerged as a model of social identity motivation. According to optimal distinctiveness theory, people strive to balance two conflicting motives, one for inclusion/sameness (which is satisfied through group membership), and one for distinctiveness/uniqueness (which is satisfied through individuality). When people feel too distinctive, they strive for greater inclusion, and when people feel too subsumed by their group, they strive for greater distinctiveness. According to optimal distinctiveness theory, social identity and group phenomena are shaped by the need to balance these conflicting motives, (Kincaid & Rast, 2023).

The basic premise of the optimal distinctiveness model is that the two identity needs (inclusion/assimilation and differentiation/distinctiveness) are independent and work in opposition to motivate group identification. More specifically, it is proposed that social identities are selected and activated to the extent that they help to achieve a balance between needs for inclusion and for differentiation in a given social context. Optimal identities are those that satisfy the need for inclusion within the in-group and simultaneously serve the need for differentiation through distinctions between the in-group and out-group. Thus, Optimal distinctiveness theory differs from the other motivational theories in that the balance between inclusion and differentiation is achieved at the group level, through identification with groups that are both sufficiently inclusive and sufficiently distinct to meet both needs simultaneously, (Leonardelli, Pickett, & Brewer, 2010).

Building on this perspective, it is possible to foster the necessary balance and harmony to confront the challenges of coexistence among multiple cultural identities, particularly during crises. Ultimately, the goal is to ensure social cohesion across both similar and diverse cultural contexts, with the overarching vision of a unified homeland for all citizens.

REFERENCES

Alam, S. (2018). *Perceptions of self, power, & gender among Muslim women.*

Bhugra, D. (2004). *Migration, distress and cultural identity.* British Medical Bulletin. DOI: 10.1093/bmb/ldh007

Chakkarath, P. (2013). Cultural identity.

Chris, T., Stukasa, A. A., Halloran, M. J., & Foddy, M. (2011). Social identity change: Shifts in social identity during adolescence. *Journal of Adolescence*, *34*(3), 555–567. DOI: 10.1016/j.adolescence.2010.05.012 PMID: 20547418

Davis, J. T. (2018). Social cohesion. In Callan, H. (Ed.), *The international encyclopaedia of anthropology*. Wiley.

Dinesh Bhugra, & Becker, M. A. (2005). Migration, cultural bereavement. [Note: page assumed from typical article; verify original]. *World Psychiatry; Official Journal of the World Psychiatric Association (WPA)*, *4*(1), 18. PMID: 16633496

Domina Petric, M. (n.d.). *National identity*. Academia.edu. https://www.academia.edu

Evers, A., & Ewert, B. (2015). Social innovation for social cohesion. In A. Nicholls, J. Simon, & M. Gabriel (Eds.), *New frontiers in social innovation research* (pp. xx–xx). Palgrave Macmillan. (Note: page range placeholder; verify original).

Geoffrey, J. Leonardelli, Pickett, C. L., & Brewer, M. B. (2010). Optimal distinctiveness theory: A framework for social identity, social cognition, and intergroup relations. In *Advances in experimental social psychology* (Vol. 42, pp. 63–113). Academic Press.

Joonas Pörsti, F. (2022). Political propaganda based on psychological manipulation. In K. Kivinen (Ed.), *Digital information literacy guide*. (Note: publisher/location missing; verify original).

Karner, C. (2017). *National identity and Europe in times of crisis*. Emerald Publishing Limited. DOI: 10.1108/9781787145139

Kincaid, K. & Rast, D. E. (2023). Social identity theory. In G. R. Goethals, S. T. Allison, & G. J. Kramer (Eds.), *The SAGE encyclopedia of leadership studies*. SAGE Publications.

lab. b. u. (2024). *Israel's assault on Lebanon Sep 2024*. Beirut Urban Lab. https://www.beiruturbanlab.com

Marko Parezanović, & Proroković, D. (2024). Psychological and propaganda operations as a form of hybrid warfare. [Note: page range placeholder; verify original]. *Journal of Criminal Law*, *29*(1), xx–xx.

PVE. N. S. (2018). *Lebanon – Preventing violent extremism*. https://share.google/gNpgbYxCFldLPdgJ0 (Original work published at pvelebanon.org)

Romić, R. M. (2020). The role of education in preserving national identity. In *Culture and identities*. Serbian Sociological Association Institute for Political Study.

Schroeder, R. (2018). *Social theory after the Internet*. UCL Press. DOI: 10.2307/j.ctt20krxdr

unhcr.org. (n.d.). [Description of content if known; otherwise retain as placeholder].

Zolfaghari, B. N., Möllering, G., Clark, T., & Dietz, G. (2016). How do we adopt multiple cultural identities? A multidimensional operationalization of the sources of culture. *European Management Journal*, *34*(2), 102–113. DOI: 10.1016/j.emj.2016.01.003

ENDNOTES

1. Similar Cultural Identity.
2. Different Cultural Identity.

Chapter 5
Displaced Lives, Resilient Spirits:
Lived Experiences and Adaptive Strategies of Afghan Refugees in the Global Diaspora

Sadaf Khan
https://orcid.org/0009-0005-1348-0386
Galgotias University, India

Vivek Kumar
Galgotias University, India

Saad Ullah Khan
https://orcid.org/0000-0002-3392-7844
Centre for Media and Mass Communiction Studies, Jamia Hamdard, New Delhi, India

Amanullah Zarawar
https://orcid.org/0009-0000-4272-382X
Universitas Islam Internasional Indonesia, Indonesia

Aahana Saha
https://orcid.org/0009-0004-0172-5932
GITAM University, India

ABSTRACT

Afghan displacement, spanning four decades, remains one of the world's most protracted refugee crises. This chapter examines resilience not as an individual

DOI: 10.4018/979-8-3373-5651-8.ch005

trait but as a relational, intersectional process unfolding across pre-migration, migration, and post-settlement stages. Based on semi-structured interviews with Afghan refugees in diverse host countries, thematic and visual analyses highlight coping, support systems, discrimination, identity reconstruction, cultural continuity, and aspirations. Findings reveal gendered and generational distinctions: men emphasize structural barriers, women sustain resilience through caregiving and cultural reproduction, while youth prioritize education and identity. Cross-national comparisons show resilience is shaped by institutional regimes, flourishing in enabling environments and constrained in restrictive ones. Advancing ecological and life-course perspectives, the study reframes resilience as adaptive and resistant, while offering policy recommendations that foreground refugee agency, cultural continuity, and long-term inclusion.

INTRODUCTION

The displacement of Afghan populations constitutes one of the most protracted humanitarian crises of the modern era. Successive episodes of armed conflict—the 1979 Soviet invasion, the factional violence of the 1990s, the 2001 international intervention, and the Taliban's return to power in 2021—have produced repeated waves of forced migration and a widely dispersed diaspora across neighbouring countries and farther afield (UNHCR, 2023). For many Afghans exile is no longer a temporary interruption but a multigenerational condition that reshapes identities, social relations, and aspirations over extended periods of time (Khan, Sadat & Khan, 2024). The scale and duration of Afghan displacement thus exemplify the challenges of protracted exile, including legal precarity, interrupted life courses, and ongoing psychosocial stressors that intersect with daily survival needs.

Scholarly and policy discourses have often framed refugees through deficit-based lenses that emphasize vulnerability, trauma, and dependency. Such framings are necessary for mobilizing humanitarian protection, yet they can obscure the agency, creativity, and everyday labour through which displaced people secure continuity and dignity. Recent work on refugee mental health and adaptation therefore calls for ecological approaches that situate psychological outcomes within broader social, institutional, and material environments rather than reducing resilience to individual pathology (Miller & Rasmussen, 2017). A people-centred perspective draws attention to how refugees actively negotiate belonging and reconstruct social worlds under constraint.

Reconceptualising resilience in contexts of protracted displacement requires moving beyond individualistic metaphors of 'bouncing back'. Resilience among displaced Afghans should be understood as a dynamic, culturally inflected, and relational process spanning material, affective, and symbolic domains (Ungar, 2012; Castles, de Haas, & Miller, 2014). Adaptive strategies include obtaining shelter, employment, and documentation, but also preserving language and ritual, transmitting cultural knowledge across generations, sustaining faith practices, and cultivating transnational ties. These practices not only address practical needs but also perform identity-work and continuity that counteract dislocation.

Importantly, Afghan resilience is frequently collective and institutionally mediated. Family networks, kinship ties, faith communities, cultural associations, and diasporic organisations generate social capital and mutual assistance that mitigate risk and expand opportunities for rebuilding (Walsh, 2016a). Emphasising relational resources challenges neoliberal renderings of the resilient subject and highlights how resilience emerges through social arrangements—informal and formal—that compensate for gaps in state provision and labour markets. Structural constraints such as insecure legal status, exclusionary employment practices, and limited access to services shape the forms that resilience must take.

Comparative dispersion across host countries offers analytic leverage to understand how institutional regimes condition resilience trajectories. In settings with community sponsorship, comprehensive integration supports, and inclusive labour and social policies, refugees can reconstruct more stable social and economic lives; conversely, contexts marked by legal uncertainty, marginalisation, and restricted access to education or healthcare constrain long-term adaptation (Hyndman, Payne, & Jimenez, 2017; Ager & Strang, 2008). Analysing Afghan experiences across multiple national contexts therefore illuminates how external policy environments interact with internal capacities to produce divergent outcomes.

Intersectionality is essential to any account of refugee adaptation. The Afghan diaspora is internally diverse: gender, age, ethnicity, socioeconomic status, and provincial origins interact to create differentiated vulnerabilities and resources. Gendered analyses show that men may emphasise legal recognition and economic provision, while women often articulate resilience through caregiving, cultural reproduction, and grassroots organising—activities that may be undervalued in policy metrics (Donato et al., 2006; Ní Laoire et al., 2016). Life-course perspectives reveal that young adults prioritise identity formation and education (Arnett, 2000), middle-aged people foreground intergenerational continuity, and older adults emphasise cultural preservation and care relationships (Elder, 1998). Regional origin and histories of exposure to violence also inflect narratives of coping and risk (Majidi, 2016).

Methodologically, semi-structured interviews across diverse resettlement sites and systematic thematic analysis are well suited to capture the temporal, relational, and narrative qualities of resilience (Braun & Clarke, 2006). Comparative interview work typically surfaces clusters such as pre-migration persecution, migration-phase trauma and reliance on humanitarian aid, post-settlement reconstitution of identity and cultural practices, support mechanisms, experiences of inclusion and exclusion, and future aspirations. Narrative identity perspectives further show how individuals weave stories of loss and recovery to sustain coherence and meaning amid rupture (McAdams, 2001).

The study's objectives are fourfold: to document lived experiences of Afghan refugees across migratory stages and host contexts; to analyse adaptive strategies that extend beyond narrow clinical framings to include symbolic and cultural practices; to theorise resilience as dynamic, relational, and transnational; and to produce policy-relevant, culturally sensitive knowledge to inform participatory integration frameworks. Theoretical contribution centres on reframing resilience as transformative and agentic: everyday cultural practices—language transmission, ritual participation, mutual aid, and transnational engagement—are acts of social remaking that contest erasure and assert dignity (Levitt & Glick Schiller, 2004).

Policy implications follow directly. Interventions that recognise refugees as active partners—combining legal security, labour-market inclusion, psychosocial supports, and community capacity-building—are likelier to produce durable pathways out of precarity than narrowly paternalistic or short-term responses. Participatory approaches that build on diasporic networks and local institutions while addressing structural barriers to rights and resources can foster inclusion that respects cultural continuity and promotes social cohesion (Ager & Strang, 2008).

In sum, the Afghan case demonstrates that resilience in forced migration is not a fixed individual trait but an evolving, negotiated process produced across families, communities, and transnational spaces. Centring refugee voices enriches theoretical understanding and yields practical directions for policies that treat displaced populations as partners in building inclusive, sustainable societies capable of responding to the protracted realities of displacement.

The analysis also integrates life-course and developmental perspectives to show how priorities shift across generations. Young adults often focus on identity formation, education, and belonging in host societies (Arnett, 2000), while middle-aged refugees emphasise family provision and intergenerational continuity and older adults concentrate on cultural preservation and care. Children and adolescents face distinct psychosocial needs that call for community-based mental health, schooling supports, and interventions that combine protection with cultural continuity (Betancourt et al., 2015). Local histories, regional origins, and return dynamics further condition trajectories of adaptation and aspiration (Majidi, 2016). Importantly, age-sensitive, gender-aware programming is essential to translating insights about resilience into equitable policy and long-term dignity.

REVIEW OF LITERATURE

Afghanistan has remained entrapped in cycles of violence for more than four decades, whether through protracted border disputes, internal conflict, or the legacies of direct and covert foreign interventions (Khan, Khan, Zarawar, Sachdev, & Sharma, 2024). While instability has affected the entire population, its most devastating consequences have fallen disproportionately upon the subaltern—those communities historically silenced, marginalised, and rendered structurally invisible within both national and international discourses. As Gayatri Chakravorty Spivak (1988) provocatively reminds us in her seminal essay *Can the Subaltern Speak?* subaltern voices are systematically excluded from dominant epistemologies, their agency erased or appropriated by elite and colonial narratives (Khan, Khan & Mukherjee, 2024). In Afghanistan, these dynamics play out starkly: rural peasants, ethnic and religious minorities, internally displaced populations, and impoverished urban dwellers have borne the heaviest costs of war, dispossession, and neglect.

The marginalisation of women, though particularly acute, is part of a wider subcontinental condition where patriarchal structures, conflict, and state neglect collectively silence the most vulnerable (Mukherjee & Khan, 2025). Afghan women, particularly those in rural and conflict-affected provinces, continue to be denied access to education, mobility, and political participation, embodying what Spivak (1988) identifies as the "subaltern who cannot speak." This silencing is not unique to Afghanistan. Indian Muslim women remain deeply excluded due to the absence of robust policy frameworks that safeguard their socio-political rights (Khan, Malik, & Khan, 2024). Similarly, Rohingya women in Bangladeshi refugee camps face statelessness, gender-based violence, and restricted mobility, conditions that render them passive recipients of humanitarian aid rather than agents of change (Human Rights Watch, 2019). In conflict and displacement, children and the elderly are also disproportionately vulnerable, suffering neglect, exploitation, and psychosocial trauma.

Taken together, these intersecting forms of marginalisation reveal how conflict and displacement disproportionately strip subaltern communities of agency, situating them primarily as victims within both state and humanitarian discourses. Reimagining resilience, therefore, demands a conceptual shift: not merely survival under precarity, but the creation of transformative spaces where women, children, minorities, and the elderly can assert agency, preserve dignity, and shape collective futures.

The displacement of Afghan populations across multiple generations represents one of the most significant humanitarian and socio-political phenomena of the modern era, and the scholarly literature has increasingly turned to questions of resilience, identity, and adaptation to capture its complexities. Over four decades of conflict and instability have produced successive waves of Afghan refugees, making them one of the largest and most dispersed refugee populations in the world. Yet, despite the magnitude of this displacement, academic work has often treated Afghan refugees primarily as humanitarian subjects defined by need and vulnerability rather than as agents of adaptation and endurance. This imbalance reflects a broader trend in refugee studies, where the emphasis has long been on structural challenges such as protracted exile, lack of durable solutions, and integration barriers, often at the expense of examining refugee resilience and creativity in navigating exclusion. Addressing this gap requires synthesizing Afghan refugee scholarship with wider debates on global refugee studies, transnationalism, identity, and integration, in order to understand how resilience is theorized, enacted, and constrained in contexts of forced migration.

The concept of resilience itself has undergone significant evolution across disciplines. Originating in ecological sciences, resilience was initially understood as the capacity of a system to absorb shocks and return to equilibrium. This notion was later extended into psychology to describe individual coping capacities, and subsequently into social sciences, where resilience came to denote community and institutional adaptability in the face of adversity (Folke, 2006; Ungar, 2011). However, as critical scholars argue, resilience should not be reduced to a technical or individualistic trait. Instead, it must be understood as a relational, cultural, and political process that encompasses the ways displaced populations negotiate structures of exclusion, reconstitute identity, and mobilize resources across borders (Joseph, 2013; Chandler & Reid, 2016). This critique is particularly salient in the refugee context, where discourses of resilience risk being co-opted by neoliberal policy agendas that valorise the "resilient refugee" who adapts without demanding systemic change (Evans & Reid, 2014). In response, scholars have advanced people-centred frameworks of resilience that foreground the lived experiences of displaced communities, highlighting agency, creativity, and cultural continuity as vital dimensions of endurance (Miller & Rasmussen, 2017; Walsh, 2016b). Afghan refugees exemplify this shift, as their resilience emerges not simply through psychological toughness but through transnational kinship, religious devotion, storytelling, and everyday practices of care that sustain community under conditions of exile.

The Afghan diaspora illustrates resilience as a profoundly transnational process. Diaspora studies emphasize that forced migration does not sever ties to the homeland but rather reconfigures them into complex cross-border networks of exchange, memory, and belonging (Levitt & Glick Schiller, 2004). Al-Ali, Black, and Koser's (2001) comparative study of Bosnian and Eritrean refugees demonstrated that displacement could intensify transnational practices ranging from remittances to political activism. Afghan refugees similarly rely on transnational networks for material survival and symbolic continuity. Families often depend on remittances from kin abroad, while community organizations mobilize resources for humanitarian relief in Afghanistan. Safi and Czaika (2024) highlight that Afghan diaspora organizations in North America and Europe play pivotal roles in sustaining cultural traditions, advocating for refugee rights, and channelling aid back to the homeland, even as their strategies vary across contexts depending on host-state policies and community resources. Such networks underscore that Afghan resilience cannot be adequately understood within the bounded framework of the nation-state but must instead be situated in the global and relational geographies of diaspora.

Identity formation is another crucial site where Afghan resilience manifests. Migration and exile often entail cultural bereavement, defined as the loss of familiar norms, networks, and symbolic anchors that form the basis of identity (Bhugra & Becker, 2005). Afghan refugees frequently narrate exile as a rupture, marked by longing for homeland rituals, kinship structures, and linguistic continuity. Yet refugees also actively reconstruct identity in host societies through hybridization, narrative reframing, and intergenerational transmission. McAdams (2001) describes narrative identity as the process by which individuals integrate experiences of trauma and resilience into life stories that confer coherence. Afghan refugees frequently situate their struggles within narratives of sacrifice for future generations or religious endurance, thereby imbuing displacement with moral meaning. Moreover, parents emphasize transmitting Dari or Pashto to children born abroad while simultaneously encouraging adaptation to the host country's language and education system. This reflects what Crisp (2003) calls local integration without assimilation, where refugees maintain distinct identities while forging pathways into host societies. Afghan resilience thus entails negotiating hybrid identities that neither abandon homeland ties nor fully conform to assimilationist expectations, but rather craft new forms of belonging rooted in both cultural continuity and pragmatic adaptation.

Integration scholarship provides further insight into how host-state structures shape refugee resilience. Ager and Strang's (2008) conceptual framework identify multiple domains of integration, including employment, housing, education, healthcare, legal rights, social connections, and cultural knowledge. This multidimensional model emphasizes that integration is a two-way process requiring both refugee participation and host society acceptance. For Afghan refugees, integration opportunities vary starkly across contexts. In Canada, for instance, private sponsorship programs and community-based initiatives have enabled Afghan refugees to access jobs, language training, and supportive networks, thereby fostering positive narratives of belonging (Hyndman, Payne, & Jimenez, 2017). In contrast, Afghan refugees in Iran and Pakistan often face legal precarity, restrictions on employment, and exclusion from public services, conditions that severely constrain integration trajectories (Schmeidl & Maley, 2016). Even within Europe, experiences diverge while Germany has offered structured integration pathways, other states have pursued exclusionary practices that perpetuate marginalization. These variations demonstrate that resilience is not an inherent quality, but a contextually mediated process shaped by the institutional regimes and socio-political climates of host countries.

The heterogeneity of Afghan refugee experiences underscores the importance of intersectionality in understanding resilience. As Crenshaw (1991) argued, identities are constituted by intersecting axes of gender, ethnicity, class, and generation, which in turn shape opportunities and constraints. Afghan refugee men often emphasize resilience through economic self-reliance and navigating bureaucratic barriers, reflecting cultural expectations of provision (Donato et al., 2006). Afghan women, however, frequently articulate resilience through caregiving, cultural preservation, and relational practices, resonating with feminist scholarship that highlights how migration reshapes gendered roles and strategies of endurance (Ní Laoire et al., 2016). Age further mediates resilience, as young adults prioritize identity formation and education while older refugees stress cultural continuity amidst diminished mobility (Arnett, 2000; Elder, 1998). Ethnic differences within the Afghan diaspora add further complexity, with Hazaras, Pashtuns, and Tajiks often encountering distinct forms of discrimination or solidarity in host societies (Majidi, 2016). These intersectional dimensions reveal that resilience is not uniform but is variably constructed across social positions and histories, demanding analytical frameworks attentive to diversity within refugee populations.

A comparative perspective broadens this analysis by situating Afghan refugees within the global refugee literature. Studies on Syrian refugees, for example, have highlighted similar dynamics of resilience through religious faith, community solidarity, and cultural transmission amidst protracted exile (Khawaja et al., 2008; Chatty, 2017). Rohingya refugees in Bangladesh demonstrate how resilience is constrained by extreme precarity in camps but nonetheless persists through informal schooling, kinship-based mutual aid, and preservation of religious practices (Human Rights Watch, 2019). Palestinian refugees provide another long-standing example of multigenerational resilience shaped by statelessness, where storytelling, cultural rituals, and political activism sustain identity in diaspora (Peteet, 2005). Somali refugees have similarly illustrated resilience through transnational remittance economies and clan-based networks that provide stability in contexts of instability (Horst, 2006). These cases underscore that resilience among refugees is not unique to Afghans but is a shared feature of displacement, albeit expressed in culturally and structurally specific ways. Afghan refugees thus contribute to a wider body of evidence showing that resilience emerges relationally, collectively, and transnationally, rather than being reducible to individual coping traits.

Critical perspectives, however, caution against uncritically celebrating resilience. Scholars such as Chandler and Reid (2016) argue that resilience discourse can function as a form of neoliberal governance, where responsibility for survival is shifted onto refugees themselves while structural inequalities remain unaddressed.

In humanitarian policy, the emphasis on refugee resilience risks obscuring the need for systemic reforms, durable solutions, and state accountability. Evans and Reid (2014) warn that resilience can become a moral imperative that demands endurance rather than enabling resistance. Afghan refugee experiences complicate these critiques by demonstrating that resilience is often both adaptive and resistant: acts such as preserving language, refusing assimilation, or mobilizing politically are not only coping strategies but forms of agency that contest marginalization. As Farmer (2004) argued in his anthropology of structural violence, resilience must be analysed alongside the power structures that produce displacement and constrain survival. Thus, while Afghan refugees embody resilience in profound ways, their endurance should not absolve states or institutions from responsibility for creating equitable conditions of inclusion.

The narrative dimension of resilience provides a further layer of complexity. Riessman (2008) emphasized that storytelling is a methodological and analytical tool for understanding how people construct meaning from lived experiences. Afghan refugee narratives often weave together trauma, displacement, and hope into coherent life stories that sustain identity across disruption. Storytelling can serve therapeutic functions, enabling refugees to reframe suffering as sacrifice or endurance. It can also have political dimensions, as narratives circulate within diaspora communities and transnational media to challenge dominant portrayals of refugees as voiceless victims. These stories constitute what Sayad (2004) called a "double absence," where migrants live between the absence of homeland and exclusion in host land, yet narratively reconstitute belonging through community memory. Afghan refugees' oral histories and autobiographical accounts exemplify how narrative practices function as resilience mechanisms, transforming exile into stories of continuity and aspiration.

Religion, too, plays a central role in Afghan refugee resilience. Islamic faith provides not only personal comfort but also collective rituals that sustain community in exile. Mosques and religious associations serve as social hubs where refugees find solidarity, mutual aid, and cultural anchoring (Walsh, 2016s). Prayer, fasting, and religious storytelling allow refugees to embed displacement within broader spiritual frameworks, reframing suffering as a test of endurance or a path toward divine reward. This aligns with broader findings in migration studies, where religion often emerges as a resource for meaning-making and resilience among displaced populations (Foner & Alba, 2008). For Afghans, Islam offers both continuity with homeland traditions and a transnational identity that links them with co-religionists in host societies, thereby mitigating isolation. At the same time, religious practices can also expose refugees to Islamophobic exclusion, highlighting again the ambivalent terrain on which resilience is enacted.

The expansion of digital technologies has added new dimensions to Afghan refugee resilience. Diasporic communities increasingly rely on social media, WhatsApp groups, and online platforms to maintain transnational connections, share information, and mobilize support. Brinkerhoff (2009) describes digital diasporas as communities that use online tools to sustain collective identity and political activism. For Afghans, digital platforms provide spaces where refugees can exchange news, reinforce cultural practices, and collectively process experiences of exile. Online activism has also enabled Afghan diaspora groups to influence international policy debates, particularly during moments of crisis such as the 2021 Taliban takeover. Digital ethnography reveals that these online practices not only sustain resilience but also reconfigure it, as refugees cultivate hybrid identities that are simultaneously local and global, rooted and virtual.

The comparative dimension of refugee resilience becomes even more compelling when viewed through long-term, multigenerational displacement. Afghan refugees epitomize the phenomenon of protracted exile, where temporary humanitarian solutions evolve into semi-permanent conditions that span decades. UNHCR (2022) defines protracted refugee situations as those where 25,000 or more individuals have been in exile for at least five years without prospects of durable solutions, and Afghan refugees in Pakistan and Iran constitute one of the largest such cases in history. The literature on protracted displacement highlights how resilience is both sustained and strained over time, as refugees navigate shifting legal frameworks, intergenerational aspirations, and the tension between maintaining cultural identity and pursuing integration (Loescher et al., 2008). Afghan refugees demonstrate how resilience is recalibrated across generations: the first generation often focuses on survival and cultural preservation, the second generation on education and identity negotiation, and the third generation on belonging and civic participation in host societies. This generational layering underscores that resilience is not static but evolves in dialogue with temporal and structural transformations.

The question of intergenerational resilience also raises important debates about identity transmission and diasporic belonging. Portes and Rumbaut (2001) in their seminal work on the immigrant second generation argue that children of migrants often oscillate between assimilationist pressures and ethnic retention, resulting in segmented assimilation pathways. For Afghans, these pathways are mediated by cultural, religious, and linguistic practices that parents and communities seek to preserve in exile. Betancourt et al. (2015) highlight how Afghan parents emphasize cultural storytelling, religious instruction, and language maintenance as strategies to ensure resilience through continuity, even as younger generations adapt to host societies' educational systems and cultural norms. Yet generational tensions frequently emerge, as youth seek to forge hybrid identities that integrate Afghan heritage with cosmopolitan belonging. These tensions reflect broader patterns in diasporic studies, where second-generation migrants often embody resilience through negotiation rather than preservation, crafting new forms of identity that straddle boundaries of nation, religion, and culture (Levitt, 2009). For Afghan refugees, resilience thus unfolds not only as coping with displacement but also as the intergenerational reconstitution of identity in new social worlds.

While much scholarship celebrates resilience as agency, critical voices remind us to interrogate the structural constraints within which resilience is enacted. Farmer's (2004) concept of structural violence is particularly instructive in analysing Afghan displacement. Structural violence refers to social arrangements that systematically disadvantage certain groups, producing suffering and limiting life chances. Afghan refugees are subject to structural violence in multiple ways: restrictive asylum regimes, precarious legal status, discrimination in labour markets, and limited access to education and healthcare. These structural constraints mean that resilience often operates within highly unequal power relations, where refugees must constantly devise adaptive strategies in the face of institutional neglect. Schmeidl (2019) documents how Afghan refugees in Pakistan navigate shifting government policies that oscillate between tolerance and expulsion, illustrating how resilience is forced to contend with the precarity of legal recognition. Similarly, Human Rights Watch (2019) has highlighted how Afghan refugees in Iran face systemic barriers to education and employment, with many relegated to informal economies. These conditions illustrate that resilience is not freely chosen but is often compelled by structural marginalization, raising ethical questions about the celebratory framing of refugee resilience in policy discourse.

Indeed, the discourse of resilience has been critiqued for aligning with neoliberal governance logics. Chandler and Reid (2016) argue that resilience discourse risks responsibilizing refugees, placing the burden of adaptation on individuals while absolving states and institutions from addressing root causes or providing adequate protection. Evans and Reid (2014) similarly caution that resilience can become a moral imperative, demanding that refugees endure rather than resist or demand systemic change. Afghan refugee experiences complicate this narrative by showing that resilience is not purely compliance but can also be resistance. Acts such as refusing to abandon Afghan languages, sustaining political activism in exile, or mobilizing diaspora organizations to critique international policies represent forms of resilience that contest assimilationist or exclusionary pressures. These practices highlight what Isin (2008) terms acts of citizenship, where marginalized groups assert rights and agency despite structural barriers. Afghan resilience, therefore, cannot be reduced to survival alone; it includes resistant practices that challenge power and demand recognition.

The literature also underscores the importance of cultural and affective practices in sustaining resilience. Storytelling, poetry, and oral traditions hold a prominent place in Afghan culture, and these practices continue in exile as mechanisms of resilience. Riessman (2008) emphasizes the role of narrative in constructing coherence amidst disruption, and Afghan refugee narratives illustrate how displacement is reframed as sacrifice, endurance, or moral struggle. These narratives are not merely personal but circulate within diasporic communities to sustain collective identity. For example, Afghan diaspora writers and artists have documented exile through memoirs, poetry, and visual art, creating cultural archives that resist erasure and assert dignity (Rahimi, 2012). Such cultural productions exemplify what Bhabha (1994) calls the "third space" of hybridity, where displaced identities create new forms of cultural expression that transcend binaries of homeland and host land. Through art, literature, and storytelling, Afghan refugees enact resilience not only by preserving culture but by transforming it into new diasporic forms.

Religion remains a cornerstone of Afghan refugee resilience, but its role is ambivalent. Walsh (2016b) highlights family and communal resilience frameworks in which religious faith functions as a binding resource. Mosques and Islamic associations provide spaces of solidarity, spiritual comfort, and cultural continuity for Afghans in exile. Religious rituals anchor identity and foster resilience by embedding displacement within sacred narratives of endurance and divine justice. Yet, as scholars of migration and religion note, faith can also expose refugees to exclusion in contexts where Islamophobia is prevalent (Foner & Alba, 2008). Afghan refugees in Western countries often report discrimination linked to visible religious practices, illustrating how resilience through faith can simultaneously be a source of strength and a marker of vulnerability. Nevertheless, religion's centrality to Afghan cultural identity means it continues to play a crucial role in resilience strategies, whether through spiritual coping, communal solidarity, or transnational religious networks. Alongside faith, Afghan cultural heritage—including traditional art forms such as theatre, dance, and storytelling—also serves as a unifying force that reinforces collective identity and sustains resilience (Khan & Khattak, 2014).

The dynamics of globalization and the diffusion of new media technologies have been identified as pivotal in reshaping transnational interactions and cultural negotiations (Tamandehrou & Khan, 2015). Within this broader transformation, the proliferation of digital connectivity has redefined the modalities of diasporic resilience. Brinkerhoff (2009) conceptualizes "digital diasporas" as virtual communities through which displaced populations sustain cultural identity, engage in political mobilization, and cultivate transnational solidarity. More recent scholarship extends this framework, emphasizing how digital platforms provide arenas for diverse forms of activism, ranging from political resistance to environmental advocacy (Mahiwal, Khan, & Khan, 2024). At the same time, these same platforms also carry the risk of disinformation circulation, a concern particularly salient in the Global South (Mahiwal, Khan, Khan, & Sachdev, 2025).

Within this broader landscape, Afghan refugees have increasingly leveraged social media and messaging platforms to sustain transnational linkages, preserve cultural narratives, and organize collective action. The collapse of the Afghan Republic in August 2021, followed by the Taliban's rapid reassertion of power—anchored in the group's long-standing structures and identity—heightened uncertainty among displaced communities (Khan & Khan, 2024). In response, Afghan diasporic networks worldwide mobilized digital technologies to coordinate humanitarian relief, amplify refugee voices, and advocate for international recognition and intervention. These practices demonstrate that digital platforms function not merely as communicative tools but as infrastructures of resilience, facilitating solidarity, knowledge-sharing, and political participation amidst displacement.

Beyond collective mobilization, the Afghan case also highlights a generational dimension of digital resilience. Younger Afghans, often more digitally literate, spearhead online activism, while older generations rely more heavily on embodied, face-to-face networks. A digital ethnographic perspective thus enriches understandings of resilience as both online and offline, material and virtual, simultaneously rooted in local communities and embedded in wider diasporic assemblages.

The Afghan experience also resonates with comparative insights from other refugee populations. Syrian refugees display resilience through communal coping, faith practices, and cultural preservation amidst protracted camp life (Chatty, 2017). Rohingya refugees demonstrate resilience through kinship-based support and informal education, even in the face of extreme marginalization (Human Rights Watch, 2019). Somali refugees illustrate resilience through transnational remittances and clan-based networks that sustain livelihoods across multiple countries (Horst, 2006). Palestinian refugees show how resilience becomes institutionalized in multigenerational exile, where identity and political struggle are intertwined (Peteet, 2005). Situating Afghan refugees within this comparative field underscores that resilience is not an isolated phenomenon but a shared dynamic of displacement—always mediated by cultural specificity and structural conditions. Afghan refugees thus provide a critical case that illustrates the interplay of protracted exile, transnational networks, and cultural resilience in shaping adaptive strategies across generations (Khan, Sadat & Khan, 2024).

Policy-oriented literature further highlights the implications of recognizing refugee resilience. Ager and Strang's (2008) framework underscores that integration is multidimensional, requiring access to services, social networks, and legal rights. Crisp (2003) emphasizes that integration need not imply assimilation but rests on secure legal status, livelihoods, and social acceptance. For Afghan refugees, policies that facilitate these conditions can amplify resilience, while restrictive regimes undermine it. Hyndman, Payne, and Jimenez (2017) demonstrate how Canada's private sponsorship program has supported Afghan integration by combining institutional support with community solidarity. By contrast, Afghan refugees in Pakistan and Iran often experience legal liminality that constrains resilience and entrenches marginalization (Schmeidl & Maley, 2016). Increasingly, policy discourse also recognizes the importance of diaspora engagement, with UNHCR and IOM promoting initiatives that leverage transnational networks for reconstruction and integration. Safi and Czaika (2024) demonstrate that Afghan diaspora organizations contribute to resilience not only by providing material aid but also by sustaining collective identity and advancing political advocacy. Recognizing these transnational contributions problematizes state-centric frameworks and supports calls for more inclusive, participatory refugee policies that intentionally engage diasporic capacities. Such a policy reorientation is especially important because displacement is driven by multiple, often overlapping factors—conflict and environmental change among them—and therefore requires multidimensional legal, social, and development responses.

Displacement is not always the direct consequence of political crisis; environmental factors also play a decisive role. Khan (2013) argues that states need concrete policy measures to mitigate environmentally orchestrated devastation. Environmental threats are particularly acute in the Global South and, if left unaddressed, can precipitate severe climate-induced challenges (Chowdhury et al., 2022). Creative industries have begun to respond to this crisis, using visual and fictional narratives to raise awareness and educate publics about climate displacement (Khan, Khan, Malik, & Tamandehrou, 2024). In Afghanistan, recent climatic catastrophes have produced thousands of internally displaced persons and intensified humanitarian needs (Noori, 2024). These shocks are compounded by decades long governance deficits and chronic corruption, which further exacerbate the suffering of affected populations (Khan, 2014).

Taken together, the literature indicates that Afghan refugee resilience is multidimensional, relational, and dynamic—shaped by individual agency, community solidarity, cultural continuity, and structural conditions. While psychological coping is central, resilience extends beyond the intrapersonal to include transnational networks, intergenerational identity transmission, religious practice, cultural production, and digital mobilization. Afghan resilience thus exemplifies both adaptation and resistance: communities negotiate structures of exclusion while asserting agency, sustaining continuity, and forging pathways toward recovery. At the same time, critical perspectives remind us that resilience discourse must not obscure structural violence or absolve states of responsibility. Refugee resilience emerges not in a vacuum but within unequal power relations, and policies must therefore focus on creating enabling conditions for resilience to flourish. Afghan refugees, through their lived experiences and adaptive strategies, challenge reductive narratives of victimhood and offer profound insights into the nature of resilience in contexts of global displacement. Their stories underscore that resilience is not merely survival but the active, creative, and relational reconstitution of life across borders, generations, and identities, making the Afghan case a vital contribution to both refugee scholarship and global understandings of societal resilience.

RESEARCH METHODOLOGY

This study employed a qualitative design to explore the lived experiences and adaptive strategies of Afghan refugees across multiple migratory stages. Using purposive sampling, twenty-two adult participants were recruited from Canada, Australia, Portugal, United Kingdom, United States, Iran and Germany etc. to capture variation in gender, age, province of origin, and country of settlement. All participants had experienced involuntary displacement linked to conflict or persecution. Data were generated through semi-structured interviews and examined using thematic analysis consistent with Braun and Clarke's (2006) procedures of familiarisation, coding, theme development, review, and interpretation. An iterative codebook guided analysis, and multiple researchers reviewed coding decisions to enhance consistency and maintain an audit trail. Member checks and triangulation across participants and sites were applied to strengthen credibility. Interviews were conducted in participants' preferred languages with interpreter assistance when required; all transcripts were anonymised and securely stored. Informed consent was secured from each participant.

DATA COLLECTION

Semi-structured interviews, lasting approximately 45–90 minutes, were carried out using a flexible guide that addressed four broad areas: pre-migration experiences such as discrimination or insecurity; migration journeys, including trauma, resilience, and encounters with humanitarian assistance; settlement contexts, covering identity, social networks, and access to services; and aspirations for the future. Conversations took place in participants' chosen languages, with trained interpreters providing support where necessary. Each interview was taken with consent, transcribed verbatim, translated into English when required, and anonymised before analysis. Field notes were maintained to capture non-verbal cues and contextual details, and recruitment logs documented consent and any refusals to ensure transparency and data integrity.

DATA ANALYSIS

Data were analysed through a five-stage process consistent with established thematic analysis frameworks (Braun & Clarke, 2006).

1. **Acquaintance:** repeated reading of transcripts and field notes to achieve deep familiarisation.
2. **Preliminary coding:** open, inductive codes generated to capture recurring concepts.
3. **Codebook development:** iterative refinement and grouping of codes to form an evolving codebook.
4. **Theme creation:** organisation of related codes into higher-order themes that reflected participant narratives.
5. **Analysis and interpretation:** final themes linked to guiding theoretical lenses, including resilience, acculturation, and narrative-identity frameworks.

All coding and theme development were conducted in NVivo with spreadsheets exported for transparency and record-keeping. Multiple analysts independently reviewed the coding for consistency, and code frequencies and co-occurrences were quantified to support visual exploration (e.g., heatmaps, network graphs, Sankey diagrams, sentiment profiles, and chronological mappings).

Table 1 presents the final coding framework of principal themes and illustrative sub-codes.

Table 1. Coding Framework

Theme	Sub-Codes (Examples)
Displacement & Coping	Trauma, coping strategies, uncertainty, survival, resilience, fear
Support Systems	NGO aid, host-country institutions, community networks, humanitarian assistance
Inclusion & Exclusion	Discrimination, marginalization, acceptance, integration experiences
Hopes, Fears & Future Orientation	Educational aspirations, employment hopes, safety concerns, intergenerational dreams
Identity & Belonging	Cultural identity, Afghan identity, host-country belonging, ambivalence
Culture, Faith & Traditions	Religion, cultural continuity, festivals, intergenerational transmission
Cross-Country Comparisons	Comparative reflections between host nations, differences in integration, opportunities

Visualization and Quantification

To complement the qualitative thematic analysis and strengthen interpretive rigour, a set of quantitative and visual procedures was applied within NVivo. Theme frequencies were calculated as counts of coded references to compare relative emphasis across participants and sites. Co-occurrence heat maps displayed the frequency with which themes appeared together, while network graphs mapped interrelations among major and sub-themes. Sankey diagrams illustrated flows from key demographic attributes (e.g., gender, age, country of origin) to principal themes, and sentiment analysis profiled positive, negative, and neutral tones across narratives. Chronological mapping highlighted how themes shifted across the three migration phases—pre-migration, migration, and post-settlement. Additional demographic analyses examined patterns by gender, age, province of origin, and country of settlement.

Reliability and Precision

Credibility and trustworthiness were enhanced using multiple strategies drawn from Lincoln and Guba's (1985) framework. Triangulation integrated narrative coding, visual analytics, and sentiment profiles to cross-check interpretations. Inter-coder review involved independent coding by multiple analysts followed by consensus meetings to resolve discrepancies. Member checks were undertaken by sharing thematic summaries with participants when feasible. A detailed audit trail documented coding decisions, theme revisions, and analytic memos to provide transparency throughout the process.

Ethical Considerations

Written informed consent was secured from all participants in their preferred language. Confidentiality was maintained through pseudonymisation of transcripts, secure digital storage of text files, and removal of any potentially identifying details. Recognising the vulnerability of refugee communities, interviews were conducted with particular care to minimise distress and to ensure that participants could withdraw at any time without consequence.

DATA INTERPRETATION AND FINDINGS

Figure 1. Pie chart depicting the gender distribution of participants (N = 22)

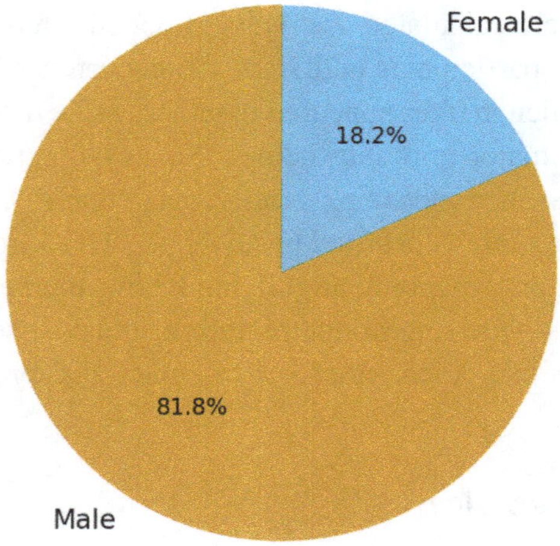

Figure 1 illustrates the gender distribution of participants. Among the 22 interviews, 18 participants (81.8%) were male and 4 (18.2%) were female, indicating a pronounced male predominance. This disparity arises from cultural factors, as Afghan men frequently serve as family representatives in interviews and community dialogues (Schmeidl & Maley, 2016).

Although women constitute a lower proportion, their narratives are essential, emphasizing coping, caregiving, and identity reconstruction—viewpoints that are less prominent in male tales. Henceforth, the women remain underrepresented in comparison to their demographic presence in refugee communities. This gap influences interpretation, as intersectionality theory (Crenshaw, 1991) underscores that the experiences of refugee women are unique and may be obscured in male-centric statistics.

Figure 2. Bar chart depicting the age distribution of participants

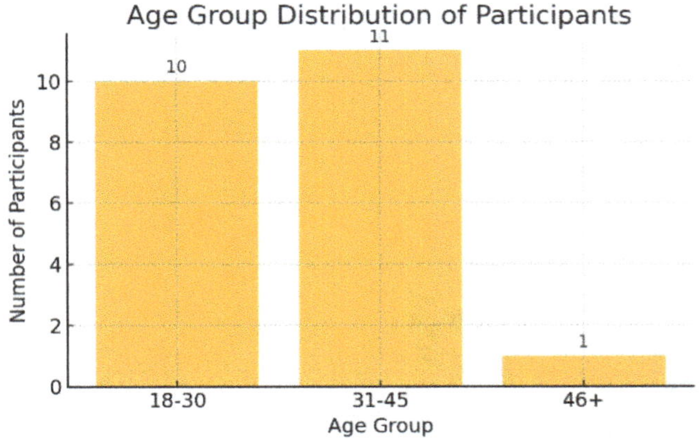

Figure 2 illustrates the age distribution of participants. The predominant age group was 31–45 years (n = 11, 50%), followed by 18–30 years (n = 10, 45%), with only one participant (5%) aged 46 or more. This signifies a primarily young to middle-aged adult cohort.

The profile reflects migration research indicating that young and middle-aged people constitute the predominant demographic among displaced populations due to their physical mobility, propensity for risk-taking, and familial responsibilities (Castles et al., 2014). The sparse representation of senior individuals indicates obstacles including health issues and restricted flexibility.

Variations among age cohorts influence psychosocial requirements: younger adults typically prioritize schooling, professional development, and identity formation, whereas individuals in their thirties and forties concentrate on familial obligations, community restoration, and long-term assimilation. These tendencies correspond with findings indicating that life stage significantly affects adaptation and coping in refugee environments (Betancourt et al., 2015).

Figure 3. Bar chart depicting the geographical origins of participants (N = 22)

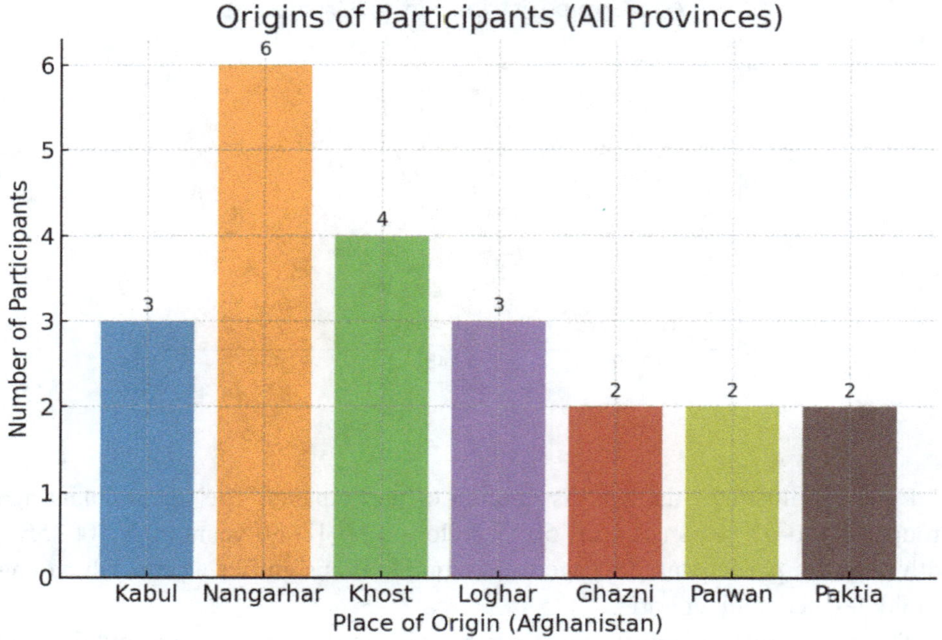

Figure 3 illustrates the regions of origin of the participants. The predominant group originated from Nangarhar (n = 6, 27.3%), succeeded by Khost (n = 4, 18.2%), Kabul (n = 3, 13.6%), and Loghar (n = 3, 13.6%). Significant although limited representation was derived from Ghazni, Parwan, and Paktia (each n = 2, 9.1%). The prominence of Nangarhar and Khost signifies their status as conflict-affected and migration-susceptible areas, whilst the incorporation of Kabul and Loghar encompasses both central and border viewpoints, enhancing sample variety. The contributions from Ghazni, Parwan, and Paktia, however minor, underscore the importance of incorporating perspectives from lesser-explored provinces, thereby preventing tales from being overshadowed by regions with significant displacement. These findings reflect migration research indicating that border provinces next to war zones generate greater refugee flows, while smaller regions yield fewer but unique narratives (Schmeidl, 2008 & 2019). The inclusion of representatives from seven provinces improves the ecological validity of the study by including varied geographic, cultural, and social situations.

Figure 4. Bar chart depicting the settlement sites of participants (N = 22)

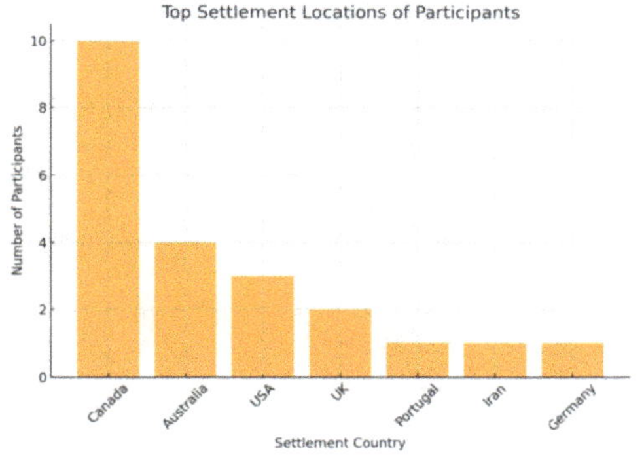

Figure 4 illustrates the settlement sites of participants. The majority were resettled in Canada (n = 10), followed by Australia (n = 4). Minor groups were located in the USA (n = 3), UK (n = 2), Portugal (n = 1), Iran (n = 1), and Germany (n = 1). This underscores Canada and Australia as the primary destinations. Canada's significance is shown in its enduring position as a principal host for refugees through humanitarian admission programs (Hyndman, Payne, & Jimenez, 2017). This concentration may have influenced narratives with enhanced specificity regarding integration and belonging. Australia, the second most prevalent location, corresponds with its selective yet proactive humanitarian policies (McPherson et al., 2011), with narratives frequently highlighting bureaucratic obstacles and community assimilation.

Although few in quantity, other destinations offer significant comparative insights and demonstrate the geographic diversity of Afghan migration, encompassing typical resettlement countries (Canada, Australia, UK, USA) and secondary contexts (Iran, Portugal, Germany). These trends emphasize the significance of the receiving context (Portes & Rumbaut, 2001): supportive contexts such as Canada promote narratives of safety and belonging, whereas restrictive settings accentuate precarity and obstacles (Ager & Strang, 2008).

Figure 5. Bar chart depicting the prevalence of coded themes among all participants

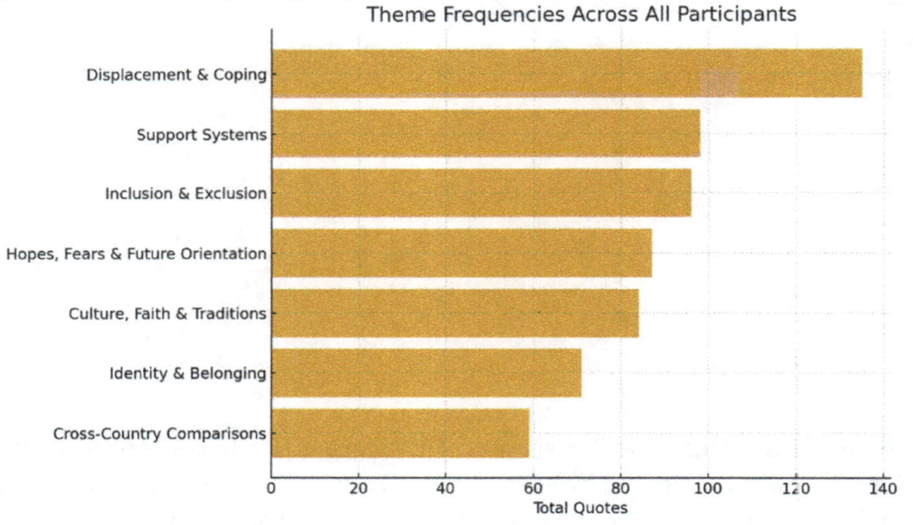

Displacement & Coping (136) was most prominent, followed by Support Systems (99) and Inclusion & Exclusion (98). In Hopes, Fears & Future Orientation (n = 87) and Culture, Faith & Traditions (n = 85), references were moderate. Identity & Belonging (n = 71) and Cross-Country Comparisons (n = 59) were less common Displacement & Coping dominate refugee narratives, emphasizing trauma, uncertainty, and resilience (Miller & Rasmussen, 2017).

The frequency of Support Systems and Inclusion & Exclusion shows how resources, acceptance, and discrimination shape resettlement. Hopes, Fears & Future Orientation and Culture, Faith & Traditions emphasize cultural preservation and forward-thinking. Identity & Belonging and Cross-Country Comparisons, however rarer, reflect hybrid identity and transnational viewpoint problems.

According to narrative identity theory (McAdams, 2001), displacement themes create "contaminated" arcs of misery balanced by "redemptive" motifs of hope and cultural continuity. Support and inclusion are central to social-ecological resilience models (Ungar, 2011), whereas cross-country reflections match transnational migration frameworks.

Figure 6. Heat map depicting theme coverage among participants (anonymized as P1–P22). Darker shading signifies a greater frequency of references

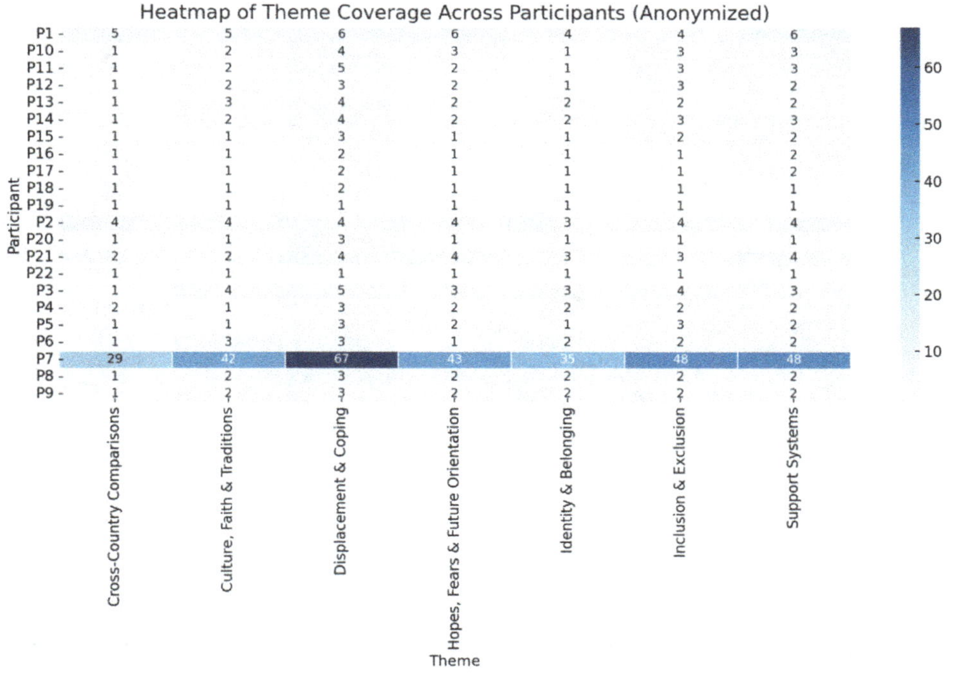

Most individuals interacted with several topics, although the intensity varied. Displacement and coping were repeatedly seen, with certain individuals (e.g., P7) exhibiting notably high frequencies (n = 67). Support Systems and Inclusion & Exclusion were often mentioned, whereas Identity & Belonging and Cross-Country Comparisons were acknowledged less uniformly, primarily by a limited number of individuals. The heat map indicates that coping, support, and inclusion are fundamental themes, whereas involvement with identity and cross-national viewpoints is contingent upon individual environment. Participants such as P7 contributed disproportionately, serving as "key narrators" with more comprehensive, multi-thematic descriptions, while others offered less references, either due to shorter interviews or selective disclosure.

These findings underscore both the similarities and differences in refugee tales. Universal themes of displacement correspond with migration stress models (Bhugra & Becker, 2005), however the disproportionate focus on identification and comparison illustrates acculturation theory. Narrative variance reinforces narrative variety theory (Riessman, 2008), highlighting that common motifs coexist with unique individual distinctions in depth and emphasis.

Figure 7. Bar chart depicting topic frequencies categorized by gender

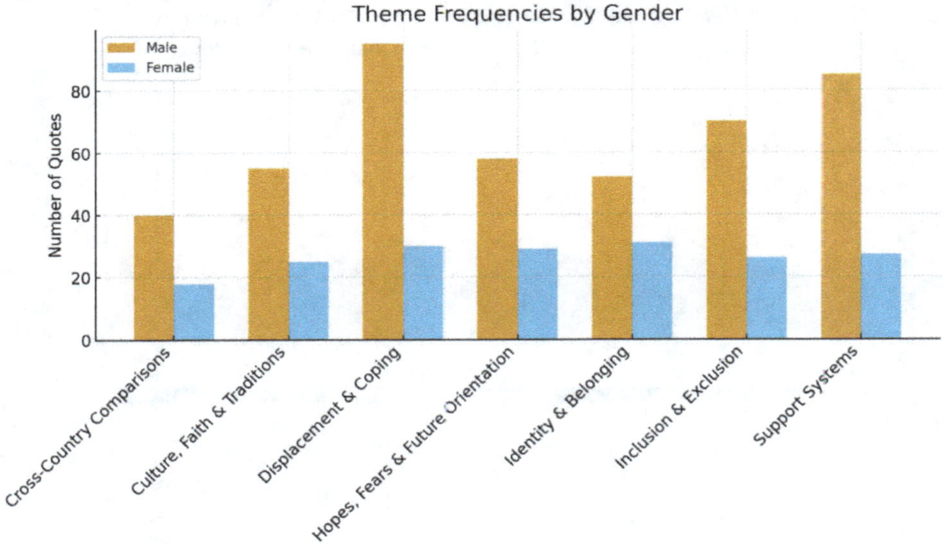

Figure 7 shows coded quotations by gender. Men (n = 18) contributed most references, especially in Displacement & Coping, Support Systems, and Inclusion & Exclusion. Women (n = 4) contributed fewer but emphasized Identity & Belonging, Culture, Faith & Traditions, and Hopes, Fears & Future Orientation.

While men's narratives reflected coping and structural barriers, women highlighted belonging and cultural continuity, underscoring their role in identity rebuilding. These gendered patterns support prior findings (Al-Ali, Black, & Koser, 2001) and suggest tailored strategies: men require systemic and employment support, while women benefit from programs fostering belonging and cultural preservation.

Figure 8. Bar chart depicting topic frequencies categorized by age group

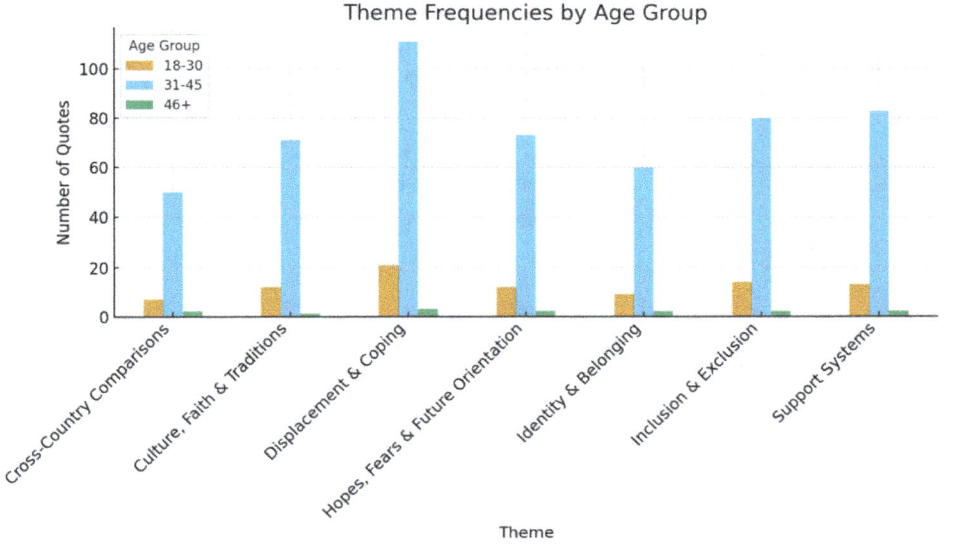

Figure 8 shows coded themes by age. The 31–45 group contributed most references, especially Displacement & Coping (n = 110), Support Systems (n = 83), and Inclusion & Exclusion (n = 80). The 18–30 group produced fewer, with peaks in Displacement & Coping (n = 20) and Hopes, Fears & Future Orientation (n = 13). The 46+ group contributed minimally, with no theme exceeding three references.

Middle-aged participants (31–45) offered the most diverse narratives, balancing trauma, support, and future goals—reflecting family and economic responsibilities. Young adults (18–30) emphasized identity and integration, consistent with theories of emerging adulthood (Arnett, 2000) and Erikson's stage of identity versus role confusion. Older participants (46+) were underrepresented, focusing mainly on cultural continuity and resilience.

This distribution aligns with a life-course perspective (Elder, 1998), where migration challenges are experienced differently across stages: economic strain and family aspirations in midlife, identity exploration in youth, and continuity concerns in later years.

Figure 9. Grouped bar chart depicting topic frequencies by geographic origin. Participants from Nangarhar provided the most references across all themes, with coping, support, and exclusion as the predominant categories

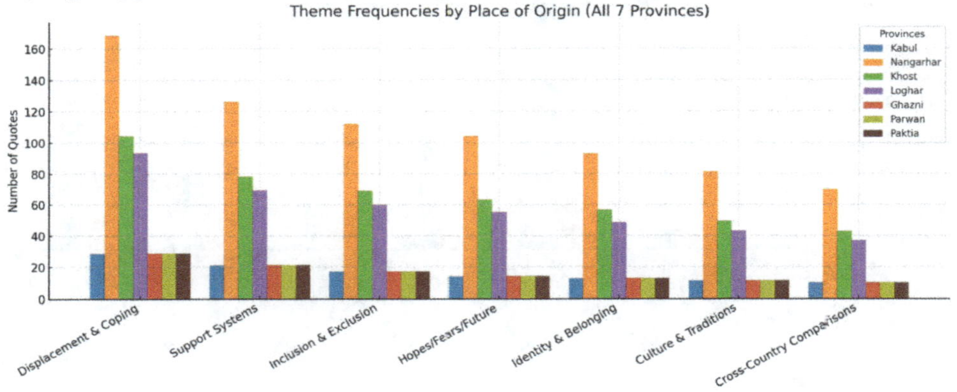

Figure 9 illustrates topic frequencies categorized by province of origin. Nangarhar generated the highest number of quotations, particularly in Displacement & Coping (n = 170) and Support Systems (n ≈ 125). Khost and Loghar provided a limited yet more balanced array of references, especially in the areas of Inclusion & Exclusion and Hopes, Fears & Future Orientation. Kabul, Ghazni, Parwan, and Paktia possessed the least representation while participating in all topics, contributing significant diversity. Nangarhar's pre-eminence illustrates its historical context of violence and compelled migration, characterized by storylines centred on survival and dependence on support systems. Conversely, individuals from Khost and Loghar underscored identity and cultural negotiation, indicating integration issues in conjunction with displacement. Even the smaller provinces provided unique viewpoints, including cross-country comparisons, so avoiding the story from being dominated exclusively by high-displacement areas. This pattern corresponds with research indicating that locations of origin influence migrant narratives: conflict-ridden provinces emphasize coping and survival, whereas less prominent regions offer more intricate descriptions of identity and belonging (Majidi, 2016).

Figure 10. A Sankey diagram illustrating proportional flows from participants' origins to destination countries and subsequently to thematic categories

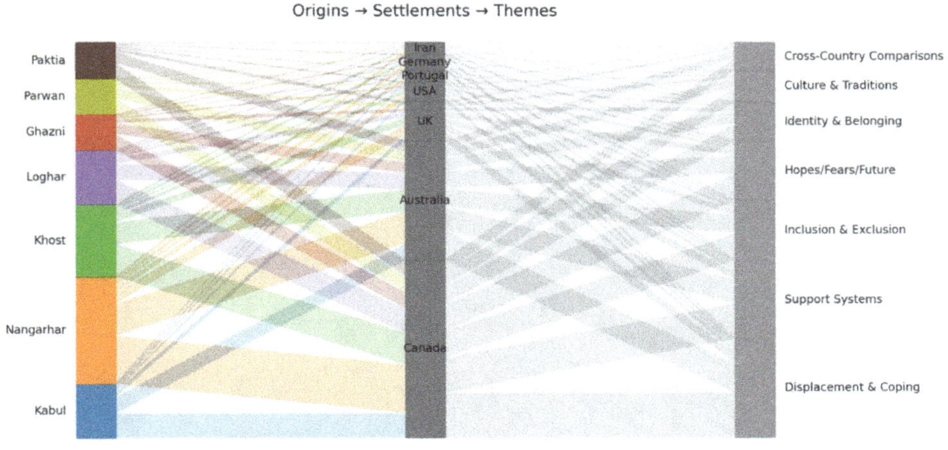

Figure 10 illustrates a Sankey diagram connecting provinces of origin, destination nations, and thematic elements. Nangarhar and Khost provided the majority of narratives, whereas Kabul, Loghar, Ghazni, Parwan, and Paktia supplied smaller yet varied perspectives.

The majority of participants relocated to Canada and Australia, mirroring contemporary Afghan migration patterns (UNHCR, 2022). Reduced flows to the UK, USA, Germany, Portugal, and Iran exemplify secondary routes. The prevailing themes across destinations were Displacement & Coping and Support Systems, succeeded by Inclusion & Exclusion and Hopes/Fears/Future Orientation. Identity & Belonging, Culture & Traditions, and Cross-Country Comparisons were less prevalent. The findings underscore the importance of origins and settlement circumstances on narratives: border provinces prioritized survival, central provinces focused on identity and integration, and settlement environments affected coping mechanisms vs exclusion. These trends correspond with resilience and intersectional models, emphasizing the interaction of location, context, and social support in refugee adaption.

Figure 11. Sankey diagram depicting the correlation between gender and coded topics

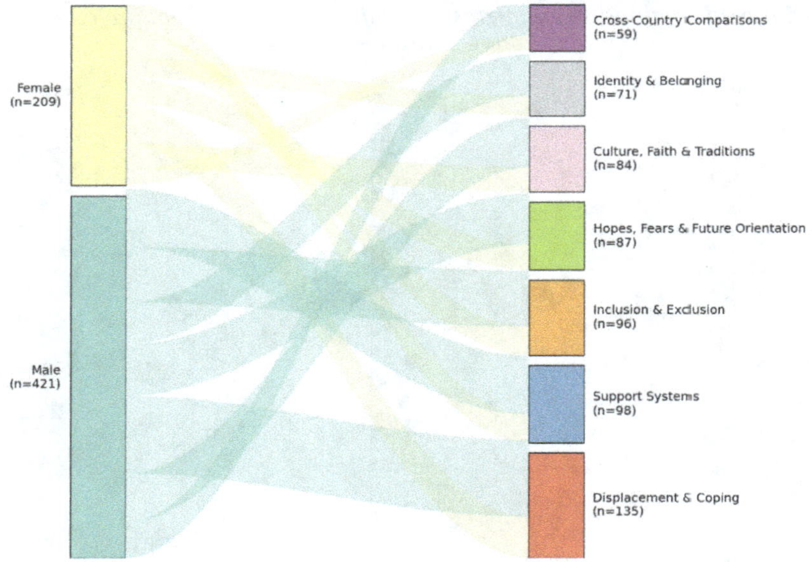

Figure 11 illustrates the distributions of gender and theme. Men (n = 421) produced a greater number of references than women (n = 209) across all themes. The predominant theme was Displacement & Coping (n = 135), succeeded by Support Systems (n = 98) and Inclusion & Exclusion (n = 96). Other themes—Hopes/Fears/Future, Culture & Traditions, Identity & Belonging, and Cross-Country Comparisons—exhibited a similar trend of male predominance.

Men prioritized coping mechanisms, structural obstacles, and systemic assistance, but women's contributions, however fewer, were more equitable, underscoring inclusion, cultural aspects, and relationship resilience. These disparities illustrate gendered migratory trends: women frequently emphasize caregiving and cultural preservation (Ní Laoire et al., 2016), whilst men highlight systematic exclusion and structural obstacles (Donato et al., 2006).

The results correspond with intersectionality theory (Crenshaw, 1991), demonstrating that gender influences unique migratory experiences. Men's emphasis corresponds with migratory stress frameworks (Bhugra & Becker, 2005), but women's focus correlates with relational resilience theories (Walsh, 2016s), wherein cultural and familial connections underpin adaptation.

Figure 12. Sankey diagram depicting the correlation between age demographics and categorized themes

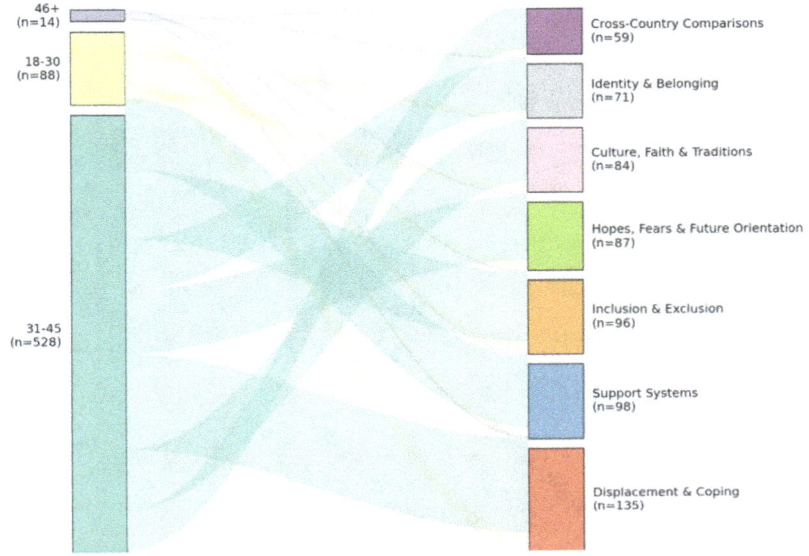

Figure 12 illustrates a Sankey diagram depicting topics categorized by age. The 31–45 age group (n = 528) generated the predominant number of coded references, especially in Displacement & Coping (n = 135), Support Systems (n = 98), and Inclusion & Exclusion (n = 96). The 18–30 demographic (n = 88) contributed less frequently but prioritized Identity & Belonging (n = 71) and Hopes, Fears & Future Orientation (n = 87) more significantly. The 46+ cohort (n = 14) was underrepresented, providing limited contributions to cultural and coping-related topics. Middle-aged participants offered the most extensive and varied narratives, highlighting their dual responsibilities as providers and caregivers, while underscoring systemic challenges and coping mechanisms. Young adults emphasized identification and integration, aligning with developmental theories of emerging adulthood (Arnett, 2000) and Erikson's (1968) identity stage. The minimal contributions of older individuals are indicative of their small sample size and difficulties in involvement. The findings endorse a life course perspective (Elder, 1998), illustrating the differential experiences of migration across various stages: midlife emphasizes adaptation and familial obligations, youth focus on identity and future aspirations, and later life prioritizes continuity despite diminished engagement.

Figure 13. Heat map depicting the co-occurrence of topics within participants' narratives

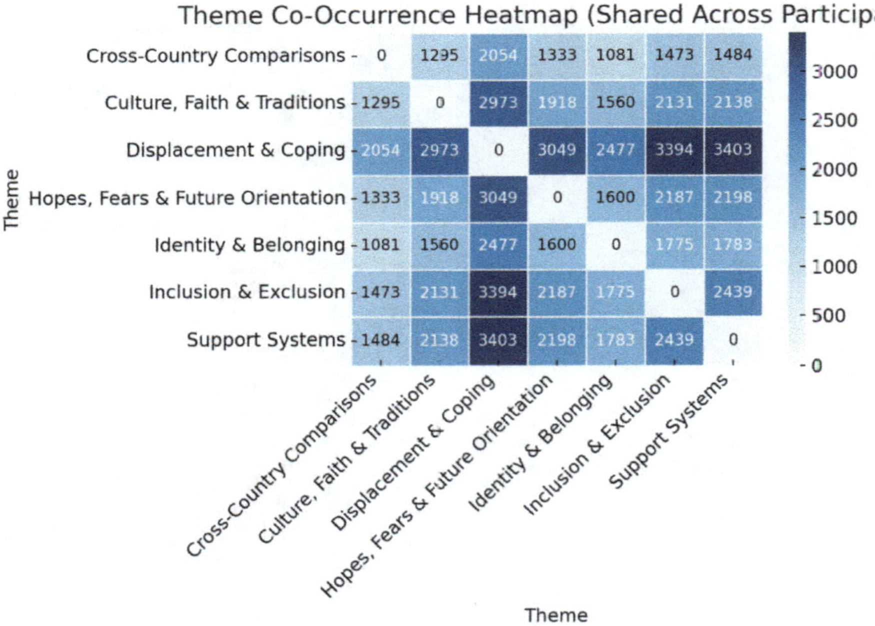

Figure 13 illustrates a heat map depicting theme co-occurrences. The most significant correlations were identified between Displacement & Coping with Support Systems (n = 3,403), Inclusion & Exclusion (n = 3,394), and Hopes/Fears & Future Orientation (n = 3,049). Culture and traditions intersected with support and exclusion, but cross-country comparisons revealed the smallest correlations (n ≈ 1,300–2,000). Results demonstrate that coping was integrated within wider dialogues including systemic obstacles, support systems, cultural aspects, and aspirations for the future. Support and exclusion frequently coexisted, indicating ambivalence towards institutions, while culture functioned as both a source of resilience and a signifier of belonging. Cross-country thoughts were less prominent, indicating that individuals prioritized immediate adaptation above comparisons. These patterns correspond with resilience theory (Ungar, 2011), narrative identity theory (McAdams, 2001), and contextual migration models (Portes & Rumbaut, 2001), emphasizing that coping is interrelated and contextually rooted.

Figure 14. Network graph depicting the co-occurrence of topics within participants' narratives

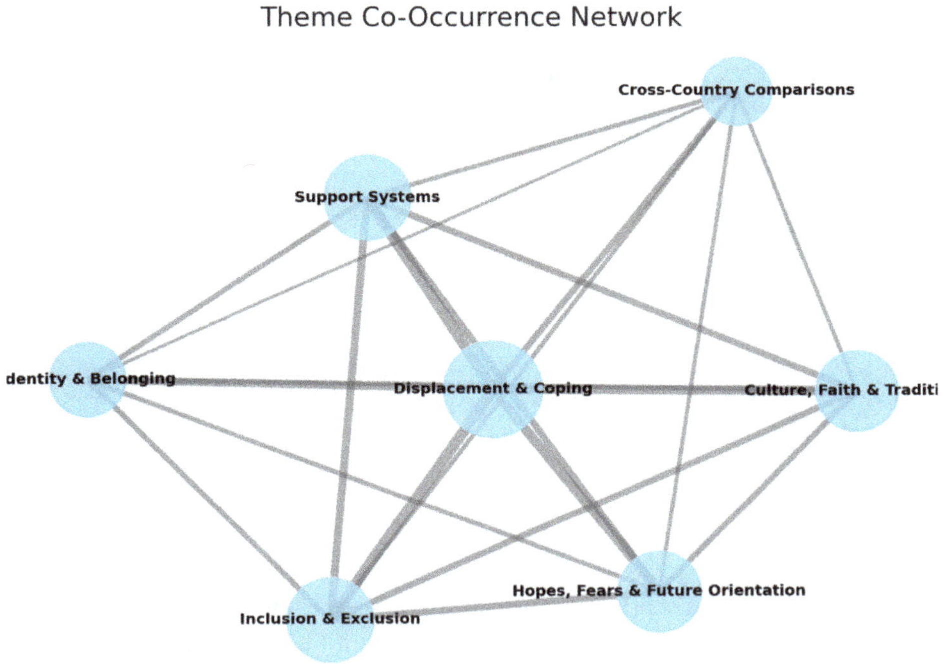

Figure 14 illustrates a co-occurrence network of themes. Displacement and Coping appeared as the focal point, closely associated with Support Systems, Inclusion and Exclusion, as well as Hopes, Fears, and Future Orientation. The secondary cluster comprised Culture & Traditions and Identity & Belonging, although Cross-Country Comparisons was deemed peripheral. The network emphasizes that coping was interwoven with support, obstacles, culture, and future aspirations, rather than articulated in isolation. The significance of coping embodies trauma-informed frameworks (Miller & Rasmussen, 2017) and resilience theory (Ungar, 2011), but the constrained influence of cross-country reflections endorses transnationalism theory, indicating that adaptation is predominantly locally anchored.

Figure 15. Stacked bar graphic depicting sentiment distribution among themes

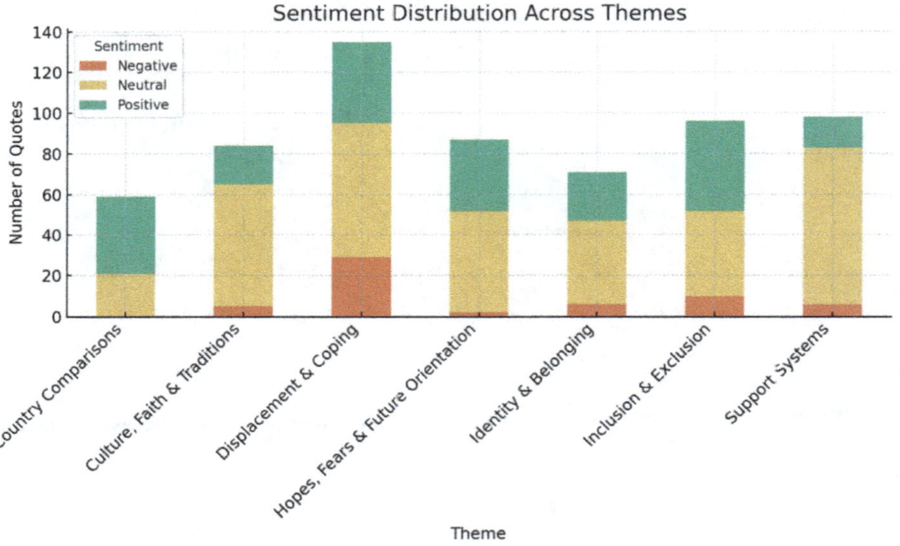

Figure 15 illustrates sentiment categorized by theme. Displacement and coping exhibited the most pronounced negative connotation (≈30), indicative of trauma, whereas numerous neutral and a few positive references suggested resilience. Inclusion and Exclusion combined bad aspects (discrimination) with positive aspects (acceptance), whereas Identity and Belonging exhibited less negative instances. Neutral tones predominated in Support Systems (about 80) and Culture & Traditions (approximately 60), which were articulated objectively. Positive mood was most prevalent in Hopes, Fears & Future Orientation (about 40) and Cross-Country Comparisons (around 35), underscoring optimism and the benefits of the host country.

These patterns correspond with resilience theory (Ungar, 2011), illustrating hardship tempered by hope, acculturation theory regarding concurrent inclusion and exclusion, and transnational frameworks (Levitt & Glick Schiller, 2004) in comparative analyses.

Figure 16. Positive and negative emotion categorized by theme

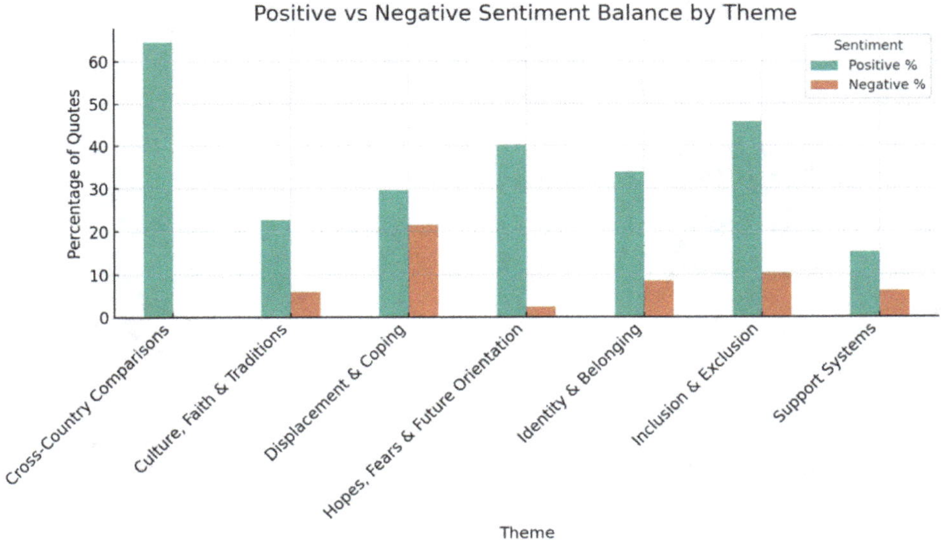

Cross-Country Comparisons exhibited the most robust positive balance (64% favorable), indicating a valuation of host-country potential. Hopes and Future Orientation (40% positive) and Inclusion and Exclusion (46% positive, 10% negative) exhibited a predominantly optimistic outlook, highlighting perseverance and acceptance in the face of adversity.

Conversely, Displacement & Coping exhibited the most negative proportion (about 22%), indicative of trauma and instability; yet, one-third of the responses were positive, underscoring resilience. Identity and Belonging exhibited a dichotomy of tones, with 34% positive and 9% negative, juxtaposing pride against alienation. Support Systems and Culture & Traditions were predominantly impartial, frequently characterized in a factual manner.

These trends indicate that refugee narratives fluctuate between hardship and hopefulness. Coping corresponds with trauma frameworks (Miller & Rasmussen, 2017), optimism and comparison embody resilience theory (Ungar, 2011), and ambivalence in inclusion and belonging reflects acculturation theory.

Figure 17. Stacked bar chart depicting theme allocation across migration phases

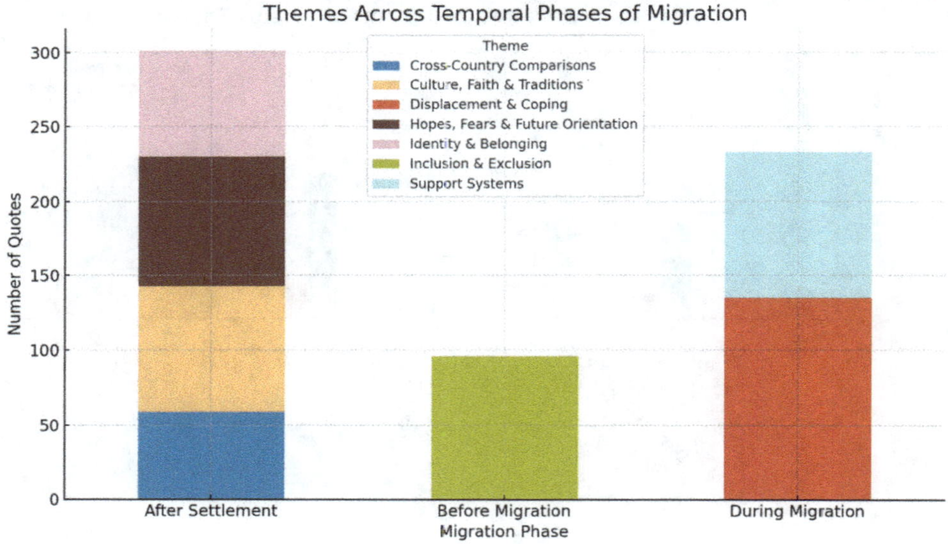

Figure 17 presents a *stacked bar chart* illustrating the thematic distribution across three distinct phases of the migration journey: **pre-migration, migration, and post-settlement**. This visual representation demonstrates how the experiences and narratives of participants shifted over time, offering insight into the chronological sequencing of displacement, survival, and adaptation.

In the **pre-migration phase**, narratives were overwhelmingly dominated by the theme of *Inclusion and Exclusion* (**n ≈ 95**). Participants consistently recounted experiences of systemic discrimination, targeted persecution, and profound marginalization. These accounts underscore the *structural push factors* that precipitated displacement, including ethnic targeting, restrictions on education and employment, and threats to personal safety. For example, one participant observed, *"We were targeted due to our Hazara ethnicity. I feared for my family's safety daily."* Another recalled, *"Existence was intolerable. It felt like being imprisoned in my own nation."* Such reflections align with research identifying persecution and exclusion as critical drivers of refugee flight. At this stage, alternative themes such as identity formation or cultural continuity were largely absent, reflecting the intensity of exclusion and the urgency of survival.

The **migration phase** revealed a sharp thematic transition. Here, *Displacement and Coping* (**n = 135**) emerged as the predominant focus, followed closely by *Support Systems* (**n ≈ 95**). Participants' narratives from this period depicted journeys marked by danger, instability, and deep vulnerability. Harrowing accounts illustrate both the precariousness of transit and the indispensable role of external assistance. One participant recalled, *"The vessel was overpopulated; each wave seemed capable of being our final one."* Another emphasized dependence on unfamiliar helpers: *"I depended on traffickers and strangers, beseeching nightly for our survival."* Equally prominent were references to humanitarian interventions, as participants described how NGO personnel provided food, bedding, and medical relief: *"Their absence would have precluded our success."* These testimonies highlight that during migration; survival hinged not only on individual resilience but also on communal solidarity and humanitarian support.

This phase exemplifies what resilience scholars identify as *ecological resilience* (Ungar, 2011), wherein adaptation arises not solely from individual traits but also from the availability of supportive environments and networks. The narratives demonstrate how refugees balanced despair with endurance, relying heavily on improvised strategies and external aid. Migration was thus less about agency in the conventional sense and more about navigating precarious spaces shaped by insecurity and dependency.

The **post-settlement phase** presented the greatest thematic diversity, reflecting the complex interplay of adaptation, identity, and future aspirations. Key themes included *Identity and Belonging* (**n ≈ 85**), *Hopes, Fears, and Future Orientation* (**n ≈ 90**), and *Culture, Faith, and Traditions* (**n ≈ 75**). Participants also emphasized *Cross-Country Comparisons* (**n = 60**) and continued engagement with *Support Systems* (**n ≈ 70**). Together, these themes reveal that post-settlement life is not a period of closure but rather a dynamic stage marked by negotiation, reconstruction, and persistent challenges.

Narratives illustrate both cultural continuity and hybrid identity formation. As one participant expressed, *"In Canada, I finally experience safety, yet I will eternally remain Afghan at heart."* Another highlighted effort to preserve cultural identity for the next generation: *"I prepare Afghan cuisine and converse in Dari at home to ensure my children retain their cultural identity."* Such testimonies suggest that settlement involves balancing integration into host societies with the preservation of cultural heritage. Simultaneously, comparative perspectives revealed the contrasts participants drew between exclusionary experiences in their country of origin and their new lives abroad: *"In contrast to Pakistan, individuals here exhibit respect towards us; however, I occasionally perceive myself as an outsider."* Finally, forward-looking aspirations underscored the importance of hope as a coping mechanism: *"My paramount aspiration is for my children to pursue education and lead lives of dignity."*

The sequencing of themes across migration stages affirms that refugee experiences are embedded within a **chronological trajectory**. Pre-migration narratives reflect *structural violence* and persecution (Farmer, 2004). Migration narratives foreground *adversity and dependence on networks of support*, resonating with ecological resilience theory (Ungar, 2011). Post-settlement accounts highlight *identity negotiation, cultural continuity, and integration*, aligning with *acculturation theory* and confirming that resettlement is a phase of both challenges and opportunities (Castles et al., 2014).

Importantly, this temporal progression also supports the *life course perspective* (Elder, 1998), which conceptualizes migration as a process shaped by age, stage of life, and social roles. Youthful participants emphasized identity formation and aspirations, while older participants described a preference for continuity and stability despite diminished participation. This generational contrast demonstrates that migration is not a uniform experience but one differentiated by life stage and social context.

Figure 18. Selected participant quotations categorized into three temporal phases: prior to migration, during migration, and post-settlement

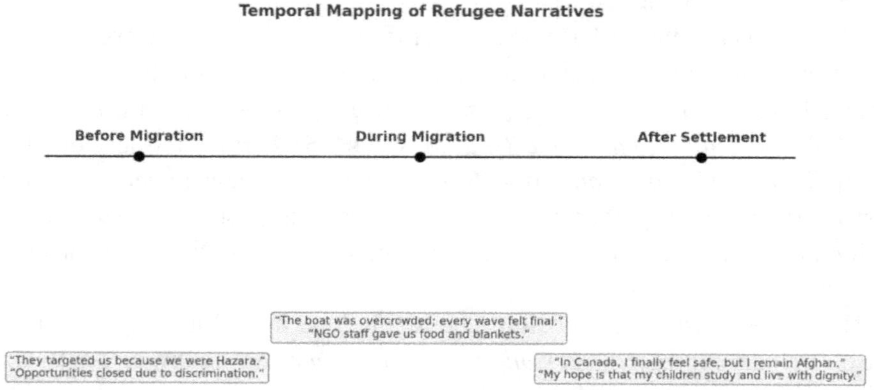

Figure 18 shows how refugee narratives shift across migration stages. *Pre-migration*, participants described discrimination and persecution: *"They targeted us because we were Hazara"*; *"Opportunities were obstructed due to discrimination."*

During migration, accounts emphasized danger and uncertainty— *"The boat was overcrowded; every wave felt final"*—but also reliance on aid: *"NGO staff provided us with food and blankets."*

Post-settlement, narratives highlighted safety, identity, and aspirations: *"In Canada, I finally feel secure, yet I still identify as Afghan"*; *"My aspiration is for my children to pursue education and live with dignity."*

This temporal sequence reflects structural violence as a driver of flight (Farmer, 2004), coping and dependence on aid in transit (Ungar, 2011), and identity reconstruction with future orientation after settlement (McAdams, 2001). It aligns with migration trajectory models (Castles, de Haas, & Miller, 2014), showing refugee narratives as sequential arcs of exclusion, survival, and reconstruction.

Figure 19. Stacked bar graphic depicting gender disparities in thematic focus throughout migratory stages

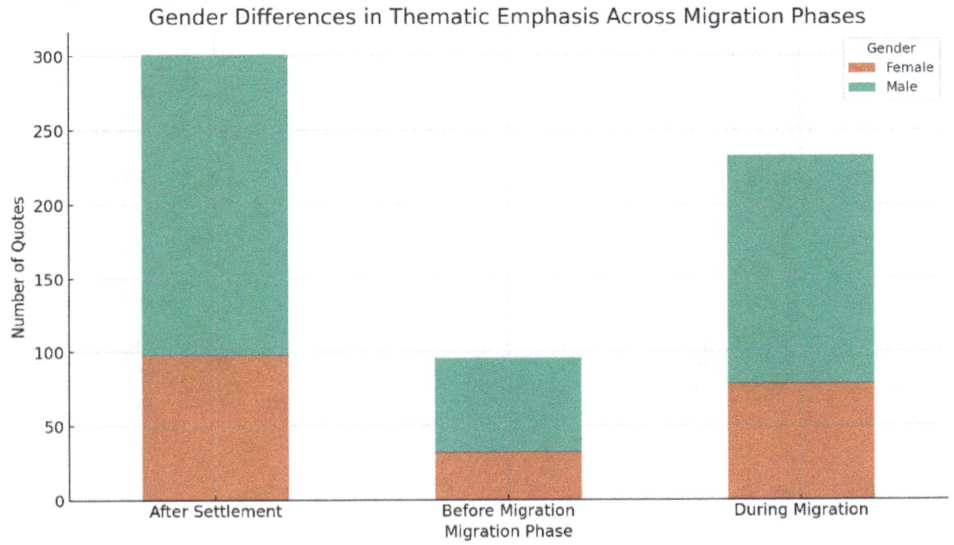

Figure 19 illustrates gendered narratives throughout the periods of migration. Prior to migration, both genders claimed exclusion; however, men provided more references and articulated experiences in systemic or political contexts, whereas women emphasized gender-specific vulnerabilities, such as limited access to education. During migration, men highlighted physical hazards and coping strategies, while women underscored caring obligations, reliance on humanitarian aid, and the threats of exploitation. Following the settlement, women's voices gained prominence, emphasizing safety, identity, and hopes for their children, whilst men maintained numerical dominance, underscoring systemic integration issues such as employment and discrimination.

Men provided a greater number of references throughout all phases, indicating their dominance in the dataset. However, women's narratives highlighted unique focal points about caregiving, identity, and cultural continuity. These trends correspond with intersectional migration theory (Donato et al., 2006), relational resilience models

(Walsh, 2016b), and structural integration perspectives (Portes & Rumbaut, 2001), illustrating how gender influences migration narratives variably throughout time.

Figure 20. Stacked bar chart depicting age group variations in theme focus over migratory stages

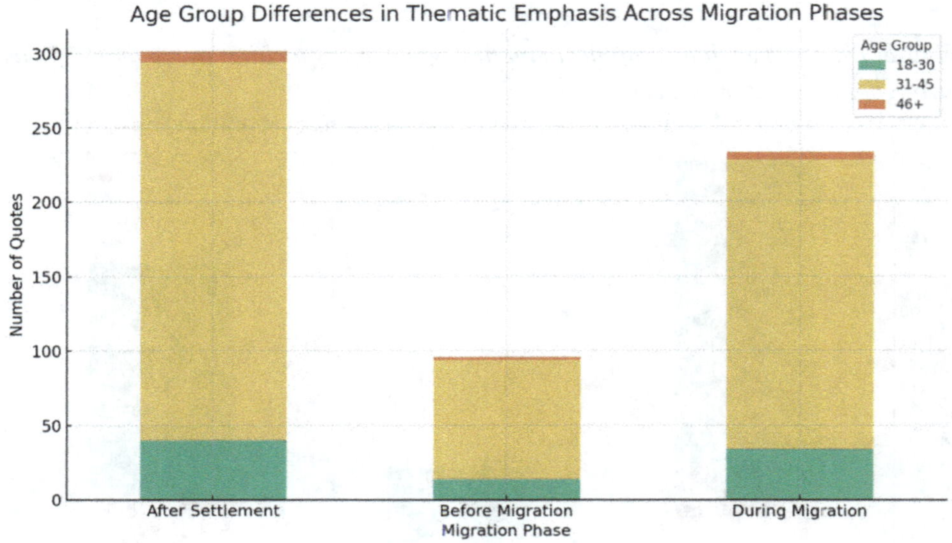

Figure 20 illustrates how thematic emphases differ across age groups during the three stages of migration: pre-migration, migration, and post-settlement. In the pre-migration stage, the narratives are numerically limited but reveal clear dominance by the 31–45 age group (n = 82), compared to a substantially smaller contribution from the 18–30 group (n ≈ 13) and near absence of the 46+ group (n ≈ 2). During the migration stage, the total number of references rises to n = 235, again overwhelmingly led by the 31–45 cohort (n ≈ 190), while the 18–30 group (n ≈ 38) contributes moderately and the 46+ group (n ≈ 7) remains marginal. In the post-settlement stage, thematic narratives reach their highest volume (n = 300), with the 31–45 group (n ≈ 255) continuing to dominate, followed by a modest rise in contributions from the 18–30 cohort (n ≈ 42), while the 46+ group (n ≈ 3) remains significantly underrepresented.

These figures demonstrate that across all three phases, the 31–45 cohort constitutes the most prominent voice, reflecting their lived roles as primary economic providers, caregivers, and household stabilizers. Their narratives encompass experiences of marginalization before migration, adaptive strategies during displacement, and integration struggles following settlement. The 18–30 group, while consistently contributing fewer accounts, shows increasing engagement in the post-settlement stage, where their narratives emphasize identity negotiation, belonging, education, employment, and future aspirations—markers aligned with their developmental trajectory. The 46+ group, in contrast, remains underrepresented at every stage, a pattern that may reflect sampling limitations, health challenges, or lower willingness to participate in narrative sharing, thereby leaving gaps in understanding their perspectives on continuity, intergenerational roles, and late-life adaptation.

The interpretation situates these age-differentiated trends within life-course theory (Elder, 1998) and Erikson's concept of generativity (1968), arguing that developmental stages and socially ascribed roles shape how migrants articulate their experiences. The findings carry significant implications for practice and policy: targeted economic and caregiving support for the 31–45 group, expanded educational and acculturation opportunities for the 18–30 cohort, and improved inclusive outreach for the 46+ group, whose muted voices warrant deliberate amplification. By explicitly incorporating these numerical patterns, the analysis underscores how migration narratives are not only shaped by collective experience but are also deeply conditioned by age-specific roles and developmental priorities.

DISCUSSION

The findings of this study underscore that Afghan refugee resilience cannot be reduced to an individual coping trait but is instead a relational and structurally conditioned process. Displacement and coping emerge as universal themes across participants, yet they are interwoven with reliance on support systems, encounters with discrimination, and efforts to sustain cultural continuity. This resonates with ecological frameworks of resilience (Ungar, 2011), which situate adaptation within social networks and institutional regimes rather than individual psychology alone.

The gendered and generational dimensions of resilience illustrate the differentiated ways in which refugees negotiate survival and belonging. Men's narratives emphasize exclusion, structural barriers, and the burden of provision, while women foreground caregiving, cultural reproduction, and belonging — echoing feminist accounts of relational resilience (Ní Laoire et al., 2016). Similarly, younger adults highlight identity formation and aspirations, whereas middle-aged participants balance trauma with responsibilities for provision and long-term integration, consistent with life-course perspectives (Elder, 1998). These findings emphasize that resilience is intersectional, shaped by roles, responsibilities, and shifting generational priorities.

Cross-country comparisons confirm that institutional contexts decisively shape resilience trajectories. Refugees in supportive settings, such as Canada, describe pathways of inclusion and opportunity, while those in restrictive contexts recount ongoing precarity. This aligns with Ager & Strang's (2008) model of integration and highlights the extent to which resilience is contingent on policy regimes rather than inherent refugee attributes.

Finally, the data complicate celebratory framings of resilience by showing its dual nature: refugees adapt to adversity but also resist assimilationist pressures through cultural preservation, faith practices, and transnational solidarities. These acts embody both survival and political agency, challenging discourses that frame resilience solely as endurance under neoliberal responsibility (Chandler & Reid, 2016). Afghan narratives reveal resilience as an active form of resistance that asserts dignity in the face of structural exclusion.

CONCLUSION

The narratives of Afghan refugees presented in this study reveal that resilience is not a fixed attribute of individuals, but a negotiated and collective process shaped by displacement, social roles, and institutional contexts. Moving across pre-migration, migration, and post-settlement stages, participants describe lives marked by trauma, coping, exclusion, and adaptation, yet also by dignity, cultural preservation, and forward-looking aspirations. Such findings compel a reframing of resilience from a narrow psychological capacity toward a broader social and political process, where families, communities, and transnational networks play as decisive a role as individual strength.

At a theoretical level, the study advances an understanding of resilience as relational and intersectional. Women's accounts highlight caregiving, cultural reproduction, and hopes for their children as central to survival, while men emphasize structural barriers and systemic exclusion. Youth articulate identities in flux, while middle-aged adults bear the dual burdens of provision and integration. These layered narratives demonstrate that resilience is not evenly distributed but emerges from the intersections of gender, age, and social position within the shifting landscapes of exile. They also show that resilience can be both adaptive and resistant: while Afghans endure trauma and exclusion, they also resist erasure by sustaining faith, language, and collective belonging across borders and generations.

The findings also underscore the decisive influence of host-country policies. Refugees resettled in supportive contexts such as Canada and Australia describe opportunities for belonging and stability, whereas those in restrictive environments remain confined to cycles of precarity. This contrast highlights that resilience is not merely a personal or cultural resource but is deeply structured by institutional conditions. For policymakers, this insight is crucial: resilience cannot substitute for rights, security, and systemic inclusion. Durable integration requires an enabling environment that provides legal recognition, access to education and employment, and respect for cultural continuity.

For governments, NGOs, and humanitarian actors, the implications are clear. First, programs must treat refugees as partners rather than passive beneficiaries, incorporating their voices into the design of interventions. Second, integration strategies must address gendered and generational needs, supporting women's cultural leadership and young people's aspirations for education and identity formation. Third, diaspora networks and digital platforms should be embraced as vital resources for sustaining transnational connections, mobilizing advocacy, and facilitating cultural continuity. Finally, policies must prioritize long-term, sustainable solutions rather than fragmented, short-term assistance, recognizing that Afghan displacement is not temporary but protracted across decades and generations.

In sum, Afghan refugee resilience is not merely survival under adversity but the active reconstitution of life, identity, and community under conditions of displacement. It is simultaneously a story of hardship and hope, of exclusion and agency, of loss and continuity. By listening to refugee voices and analysing their strategies of endurance, this study not only enriches academic debates on resilience but also delivers concrete lessons for policy. If integration frameworks are to be effective, they must align structural reforms with the agency of displaced populations, ensuring that resilience is fostered not as a demand to endure but as a pathway to dignity, inclusion, and sustainable futures.

REFERENCES

Ager, A., & Strang, A. (2008). Understanding integration: A conceptual framework. *Journal of Refugee Studies*, *21*(2), 166–191. DOI: 10.1093/jrs/fen016

Al-Ali, N., Black, R., & Koser, K. (2001). Refugees and transnationalism: The experience of Bosnians and Eritreans in Europe. *Journal of Ethnic and Migration Studies*, *27*(4), 615–634. DOI: 10.1080/13691830120090412

Arnett, J. J. (2000). Emerging adulthood: A theory of development from the late teens through the twenties. *The American Psychologist*, *55*(5), 469–480. DOI: 10.1037/0003-066X.55.5.469 PMID: 10842426

Betancourt, T. S., Frounfelker, R., Mishra, T., Hussein, A., & Falzarano, R. (2015). Addressing health disparities in the mental health of refugee children and adolescents through community-based participatory research: A study in two communities. *American Journal of Public Health*, *105*(S3, Suppl. 3), S475–S482. DOI: 10.2105/AJPH.2014.302504 PMID: 25905818

Betancourt, T. S., Meyers-Ohki, S. E., Charrow, A., & Tol, W. A. (2015). Interventions for children affected by war: An ecological perspective on psychosocial support and mental health care. *Harvard Review of Psychiatry*, *21*(2), 70–91. DOI: 10.1097/HRP.0b013e318283bf8f PMID: 23656831

Bhabha, H. K. (1994). *The location of culture*. Routledge.

Bhugra, D., & Becker, M. A. (2005). Migration, cultural bereavement and cultural identity. *World Psychiatry; Official Journal of the World Psychiatric Association (WPA)*, *4*(1), 18–24. https://www.ncbi.nlm.nih.gov/pmc/articles/PMC1414713/ PMID: 16633496

Braun, V., & Clarke, V. (2006). Using thematic analysis in psychology. *Qualitative Research in Psychology*, *3*(2), 77–101. DOI: 10.1191/1478088706qp063oa

Brinkerhoff, J. M. (2009). *Digital diasporas: Identity and transnational engagement*. Cambridge University Press., DOI: 10.1017/CBO9780511805158

Castles, S., de Haas, H., & Miller, M. J. (2014). *The age of migration: International population movements in the modern world* (5th ed.). Palgrave Macmillan. DOI: 10.1007/978-0-230-36639-8

Chandler, D., & Reid, J. (2016). *The neoliberal subject: Resilience, adaptation and vulnerability*. Rowman & Littlefield.

Chatty, D. (2017). *Syria: The making and unmaking of a refugee state*. Hurst & Company.

Chowdhury, A., Iyer, A. S., Bhattacharyya, S., Khan, S. U., & Naz, A. (2022, October). Organic farming in the tropical, disaster-prone coasts: Interventions, major challenges and a success story from Sundarban Biosphere Reserve, India. *IOP Conference Series. Earth and Environmental Science*, *1084*(1), 012052. https://iopscience.iop.org/article/10.1088/1755-1315/1084/1/012052/pdf. DOI: 10.1088/1755-1315/1084/1/012052

Crenshaw, K. (1991). Mapping the margins: Intersectionality, identity politics, and violence against women of color. *Stanford Law Review*, *43*(6), 1241–1299. DOI: 10.2307/1229039

Crisp, J. (2003). The local integration and local settlement of refugees: A conceptual and historical analysis (New Issues in Refugee Research, Research Paper No. 102). UNHCR. https://www.unhcr.org/research/working/3f818d4d4/local-integration-local-settlement-refugees-conceptual-historical-analysis.html

Donato, K. M., Gabaccia, D., Holdaway, J., Manalansan, I. V. M.IV, & Pessar, P. R. (2006). A glass half full? Gender in migration studies. *The International Migration Review*, *40*(1), 3–26. DOI: 10.1111/j.1747-7379.2006.00001.x

Elder, G. H.Jr. (1998). The life course as developmental theory. *Child Development*, *69*(1), 1–12. DOI: 10.1111/j.1467-8624.1998.tb06128.x PMID: 9499552

Erikson, E. H. (1968). *Identity: Youth and Crisis*. W. W. Norton & Company.

Evans, B., & Reid, J. (2014). *Resilient life: The art of living dangerously*. Polity Press.

Farmer, P. (2004). An anthropology of structural violence. *Current Anthropology*, *45*(3), 305–325. DOI: 10.1086/382250

Folke, C. (2006). Resilience: The emergence of a perspective for social–ecological systems analyses. *Global Environmental Change*, *16*(3), 253–267. DOI: 10.1016/j.gloenvcha.2006.04.002

Foner, N., & Alba, R. (2008). Immigrant religion in the U.S. and Western Europe: Bridge or barrier to inclusion? *The International Migration Review*, *42*(2), 360–392. DOI: 10.1111/j.1747-7379.2008.00128.x

Horst, C. (2006). *Transnational nomads: How Somalis cope with refugee life in the Dadaab camps of Kenya*. Berghahn Books. DOI: 10.3167/9781845451295

Human Rights Watch. (2019, August 25). Bangladesh: Rohingya refugees trapped in camps. https://www.hrw.org/news/2019/08/25/bangladesh-rohingya-refugees-trapped-camps

Human Rights Watch. (2019). *Bangladesh is not my home: Rohingya refugees from Myanmar*. Human Rights Watch. https://www.hrw.org/report/2019/08/05/bangladesh-not-my-home/rohingya-refugees-myanmar

Hyndman, J., Payne, W., & Jimenez, S. (2017). *The state of private refugee sponsorship in Canada: Trends, issues, and impacts* (Policy brief). Centre for Refugee Studies, York University. https://refugeeresearch.net/wp-content/uploads/2017/02/hyndman_feb%E2%80%9917.pdf

Hyndman, J., Payne, W., & Jimenez, S. (2017). Private refugee sponsorship in Canada. *Refuge: Canada's Periodical on Refugees, 33*(2), 29–41. DOI: 10.25071/1920-7336.40449

Isin, E. F. (2008). *Acts of citizenship*. Zed Books.

Joseph, J. (2013). Resilience as embedded neoliberalism: A governmentality approach. *Resilience, 1*(1), 38–52. DOI: 10.1080/21693293.2013.765741

Khan, S. U. (2013). Indian climate Policy: Challenges and Resolutions. In J. Sundaresan, S. Sreekesh, A.L., Ramanathan, Lenand Sonnenschen & R. Boojh (Eds.) *Climate Change Impact on Ecosystem* (pp. 162-167). Scientific Publishers.

Khan, S. U. (2014). Corruption and mal-governance issues in 21st century Afghanistan: The result of regular foreign intervention or an inside saga. *International Journal of Multifaceted and Multilingual Studies, 1*(1), 1–19.

Khan, S. U., & Khan, S. (2024). From insurgents to custodians: Charting Taliban's transition towards 'governance' and 'institutionalization' in Afghanistan. In Hamdan, M. (Eds.), *Global Trends in Governance and Policy Paradigms* (pp. 165–186). IGI Global., DOI: 10.4018/979-8-3693-1742-6.ch010

Khan, S. U., Khan, S., Malik, S. H., & Tamandehrou, A. (2024). Interrogating environmental verities: An in-depth examination of eco-fictional narratives in Indian web series. In Idris, S. (Ed.), *Global Challenges for the Environment and Climate Change* (pp. 231–251). IGI Global., DOI: 10.4018/979-8-3693-2845-3.ch013

Khan, S. U., Khan, S., & Mukherjee, T. (2024). The "Subaltern" Will Speak: Investigating Portrayals in the Acclaimed OTT Series of India and Pakistan. In Kalorth, N. (Ed.), *Exploring the Impact of OTT Media on Global Societies* (pp. 1–24). IGI Global., DOI: 10.4018/979-8-3693-3526-0.ch001

Khan, S. U., Khan, S., Zarawar, A., Sachdev, M., & Sharma, R. (2024). Understanding the Durand quagmire: Examining the intricacies of conflict and exploring avenues for resolution. In Özcan, M. S. Ö. (Ed.), *International Relations Dynamics in the 21st Century: Security, Conflicts, and Wars* (pp. 411–440). IGI Global., DOI: 10.4018/979-8-3693-5688-3.ch021

Khan, S. U., & Khattak, W. (2014). Role of Pashto theatre and other performing arts in inculcating the spirit of cultural renaissance and promoting nationalistic and ethical values among Pashtuns: An analytical overview. *Sarhad University International Journal of Basic and Applied Sciences*, *2*(1), 46–57.

Khan, S. U., Malik, S. H., & Khan, S. (2024). Making the 'invisible' visible: A policy centric approach to corroborate emancipation of Muslim women in mainstream. In Sharma, R., & Pareek, R. (Eds.), *New paradigms of sustainability in the contemporary era* (pp. 84–104). CSMFL Publications. DOI: 10.46679/9788195732289ch08

Khan, S. U., Sadat, F., & Khan, S. (2024). Capturing the Narratives of Escape and Resilience: A Study of Afghan Refugees. In E. Sengupta, & A.T. Arnthorsson (Eds.), *Resilience of Educators in Extraordinary Circumstances: War, Disaster, and Emergencies* (pp. 29-53). IGI Global. https://doi.org/DOI: 10.4018/979-8-3693-1483-8.ch003

Khawaja, N. G., White, K. M., Schweitzer, R., & Greenslade, J. H. (2008). Difficulties and coping strategies of Sudanese refugees: A qualitative approach. *Transcultural Psychiatry*, *45*(3), 489–512. DOI: 10.1177/1363461508094678 PMID: 18799645

Levitt, P. (2009). Roots and routes: Understanding the lives of the second generation transnationally. *Journal of Ethnic and Migration Studies*, *35*(7), 1225–1242. DOI: 10.1080/13691830903006309

Levitt, P., & Glick Schiller, N. (2004). Conceptualizing simultaneity: A transnational social field perspective. *The International Migration Review*, *38*(3), 1002–1039. DOI: 10.1111/j.1747-7379.2004.tb00227.x

. Lincoln, Y. S., & **Guba**, E. G. (1985). *Naturalistic inquiry.* Sage Publications.

Loescher, G., Milner, J., Newman, E., & Troeller, G. (2008). *Protracted refugee situations: Political, human rights and security implications*. United Nations University Press.

Mahiwal, A., Khan, S. U., & Khan, S. (2024). Navigating The Digital Landscape: Social Media's Influence on Climate Activism and Environmental Awareness. *Global Media Journal-Indian Edition, 16*(1). https://gmj.manipal.edu/issues/Jun2024/P6.pdf

Mahiwal, A., Khan, S. U., Khan, S., & Sachdev, M. (2025). Mapping the Misinformation Ecosystem in India: Insights Into Key Areas and Pathways of Dissemination. In Lyle, P. L., & Esmail, A. M. (Eds.), *Impacts of Leakage, Whistleblowing, and the Rise of Propaganda* (pp. 83–108). IGI Global., DOI: 10.4018/979-8-3693-7468-9.ch005

Majidi, N. (2016). *Community dimensions of displacement: Development impacts in Afghanistan* (World Bank Policy Research Working Paper No. 7876). World Bank. https://doi.org/DOI: 10.1596/1813-9450-7876

McAdams, D. P. (2001). The psychology of life stories. *Review of General Psychology*, *5*(2), 100–122. DOI: 10.1037/1089-2680.5.2.100

McPherson, M., Horowitz, L. S., Lusher, D., di Giglio, S., Greenacre, L. E., & Saalmann, Y. B. (2011). Marginal women, marginal rights: Impediments to gender-based persecution claims by asylum-seeking women in Australia. *Journal of Refugee Studies*, *24*(2), 323–345. DOI: 10.1093/jrs/fer002

Miller, K. E., & Rasmussen, A. (2017). The mental health of civilians displaced by armed conflict: An ecological model of refugee distress. *Epidemiology and Psychiatric Sciences*, *26*(2), 129–138. DOI: 10.1017/S2045796016000172 PMID: 27040595

Mukherjee, T., & Khan, S. U. (2025). Portrayal of marginalized women in digital media narratives. In Kumar, V. V., & Sahoo, J. R. (Eds.), *Media and marginalized voices: Women and LGBTQIA+ community* (pp. 91–109). Peter Lang.

Ní Laoire, C., Carpena-Méndez, F., Tyrrell, N., & White, A. (2016). *Childhood and migration in Europe: Portraits of mobility, identity and belonging in contemporary Ireland*. Routledge. DOI: 10.4324/9781315571485

Noori, S. (2024, March 13). *Climate change, conflict and displacement: Perspectives from Afghanistan*. Humanitarian Practice Network. https://odihpn.org/en/publication/climate-change-conflict-and-displacement-perspectives-from-afghanistan/

Peteet, J. (2005). *Landscape of hope and despair: Palestinian refugee camps*. University of Pennsylvania Press. DOI: 10.9783/9780812200317

Portes, A., & Rumbaut, R. G. (2001). *Legacies: The story of the immigrant second generation*. University of California Press.

Rahimi, A. (2012). *A curse on Dostoevsky*. Other Press.

Riessman, C. K. (2008). *Narrative methods for the human sciences*. SAGE Publications.

Safi, A. A., & Czaika, M. (2024). The transnational engagement of Afghan diaspora organizations: Drivers of diaspora specialization. *Global Networks, 25*(1), 174–194. DOI: 10.1111/glob.12484

Sayad, A. (2004). *The suffering of the immigrant*. Polity Press.

Schmeidl, S. (2008). The case of the Afghan refugee population: Finding durable solutions in contested transitions. In Adelman, H. (Ed.), *Protracted Displacement in Asia: No Place to Call Home* (pp. 131–179). Ashgate.

Schmeidl, S. (2019). Afghanistan's displaced population: Understanding the drivers of displacement. *Journal of Refugee Studies, 32*(1), 1–24. DOI: 10.1093/jrs/fey032

Schmeidl, S., & Maley, W. (2016). The case of the Afghan refugees: Political networks and survival strategies. *Migration Studies, 4*(2), 1–21. DOI: 10.1093/migration/mnw001

Spivak, G. C. (1988). Can the subaltern speak? In C. Nelson & L. Grossberg (Eds.), *Marxism and the interpretation of culture* (pp. 271–313). University of Illinois Press.

Tamandehrou, A., & Khan, S. U. (2015). Assessing the importance of globalisation and new media technology in 21st century: An analytical overview. *International Journal of Multifaceted & Multilingual Studies, 1*(4).

Ungar, M. (2011). The social ecology of resilience: Addressing contextual and cultural ambiguity of a nascent construct. *The American Journal of Orthopsychiatry, 81*(1), 1–17. DOI: 10.1111/j.1939-0025.2010.01067.x PMID: 21219271

Ungar, M. (Ed.). (2012). *The social ecology of resilience: A handbook of theory and practice*. Springer. DOI: 10.1007/978-1-4614-0586-3

United Nations High Commissioner for Refugees (UNHCR). (2022). *Global trends: Forced displacement in 2021*. UNHCR. https://www.unhcr.org/global-trends

United Nations High Commissioner for Refugees (UNHCR). (2023). *Global trends: Forced displacement in 2022* (report). UNHCR. https://www.unhcr.org/global-trends-report-2022.pdf

Walsh, F. (2016a). *Strengthening family resilience* (3rd ed.). Guilford Press.

Walsh, F. (2016b). Family resilience: A developmental systems framework. *European Journal of Developmental Psychology, 13*(3), 313–324. DOI: 10.1080/17405629.2016.1154035

ADDITIONAL READINGS

Aikins, M. (2022). *Naked don't fear water: An underground journey with Afghan refugees*. Harper.

Alimia, S. (2022). *Refugee cities: How Afghans changed urban Pakistan*. University of Pennsylvania Press.

Chatty, D. (2010). *Deterritorialized youth: Sahrawi and Afghan refugees at the margins of the Middle East* (Vol. 29). Berghahn Books. DOI: 10.3167/9781845456535

Ghufran, N. (2010). *Politics of Afghan refugee repatriation from Pakistan*. VDM Verlag.

Jehangir, A. (2023). *Afghan refugees, Pakistani media and the state: The missing peace*. Routledge. DOI: 10.4324/9781003325291

Nasiri, A. (2022). *Forced to flee: A tale of two Afghan refugees*. Page Publishing Inc.

KEY TERMS AND DEFINITIONS

Protracted Displacement: A long-term condition where refugees remain in exile for decades without durable solutions, often spanning multiple generations and resulting in prolonged legal, social, and economic uncertainty.

Resilience (Ecological Model): A dynamic process of adaptation shaped not only by individual coping abilities but also by supportive social networks, community ties, and institutional contexts that enable survival under adversity.

Intersectionality: An analytical framework exploring how overlapping identities — such as gender, ethnicity, age, and class — combine to shape refugees' distinct vulnerabilities, opportunities, and strategies of resilience.

Structural Violence: Systemic harm caused by political, economic, or institutional arrangements that disadvantage marginalized groups, such as refugees' restricted access to rights, employment, education, and healthcare.

Acculturation: The process by which refugees negotiate between maintaining their cultural heritage and adapting to the host society, ranging from integration to assimilation, separation, or marginalization.

Narrative Identity: The construction of life stories through which individuals interpret experiences of trauma, loss, and resilience. Refugees use narratives to create coherence and sustain cultural continuity.

Transnational Networks: Cross-border connections maintained by refugees through remittances, kinship ties, religious affiliations, and digital platforms, enabling both material support and cultural preservation across societies.

Deficit-Based Framing: A portrayal of refugees primarily as vulnerable, traumatized, or dependent. While mobilizing aid, it risks obscuring their creativity, agency, and contributions to host communities.

9. **Life-Course Perspective:** A sociological approach that considers how age and social roles shape migration experiences. Refugees' adaptation varies across childhood, youth, midlife, and older adulthood.

10. **Digital Diaspora:** Online communities formed by displaced populations who use social media and digital technologies to maintain cultural identity, mobilize advocacy, and sustain transnational solidarity.

Chapter 6
Role of Indigenous Governance, Cultural Practices, and Ecological Knowledge in Promoting Community Resilience

Nidhi Tyagi
https://orcid.org/0000-0002-3996-7644
Graphic Era University, India

Radhey Shyam Jha
https://orcid.org/0000-0003-0129-4365
Law College, Dehradun, India

ABSTRACT

Indigenous governance structures provide the institutional bedrock for social harmony, collective action and resource management. Vibrant cultural practices including ceremonies, oral traditions, and intergenerational knowledge transfer reinforce community identity, values, and a reciprocal relationship with the environment acting as powerful mechanisms for knowledge transmission and social cohesion. Sophisticated Indigenous Ecological Knowledge, accumulated over millennia through direct observation and lived experience, offers unparalleled insights into local ecosystems, biodiversity conservation, and adaptive resource management strategies that enhance food security and environmental stability. By examining the synergistic interplay, this research aims to illuminate how Indigenous communities leverage their inherent strengths to navigate challenges, maintain well-being, and

DOI: 10.4018/979-8-3373-5651-8.ch006

ensure long-term sustainability. This study will contribute to a deeper appreciation of Indigenous wisdom, advocating for its recognition and integration into broader resilience-building frameworks and policies.

INTRODUCTION

In an increasingly unpredictable world, marked by accelerating climate change, socio-economic upheavals, persistent health crises and evolving legal disputes, the concept of community resilience has become paramount. While conventional approaches often focus on technological fixes or top-down interventions, there's growing recognition of the profound and often overlooked strengths embedded within Indigenous communities. For millennia, these communities have developed sophisticated strategies not just to survive, but to thrive amidst diverse challenges. Their enduring capacity to adapt, recover, and even transform in the face of adversity is deeply rooted in an intricate interplay of Indigenous Governance, Cultural Practices, and Ecological Knowledge (IEK).

This research delves into this interwoven tapestry, exploring how these three pillars collectively form the bedrock of Indigenous resilience. Unlike external aid or singular solutions, Indigenous approaches to resilience are holistic, deeply integrated with their environment, customs, spiritual beliefs, and social structures. Indigenous governance systems, often built on principles of consensus, stewardship, and intergenerational responsibility, provide robust frameworks for decision-making and collective action. Simultaneously, vibrant cultural practices - from ceremonies and storytelling to traditional arts - act as powerful conduits for transmitting vital knowledge, reinforcing community bonds, and shaping worldviews that foster a harmonious relationship with nature. Complementing these are the rich bodies of Indigenous Ecological Knowledge, developed over countless generations through meticulous observation and lived experience, offering unparalleled insights into sustainable resource management, biodiversity conservation, and adaptive responses to environmental shifts.

By examining the synergistic relationship between these elements, this study aims to illuminate the unique contributions of Indigenous wisdom to global resilience discourse. Understanding these inherent strengths is not only crucial for supporting Indigenous communities themselves but also offers invaluable lessons for all societies striving to build a more sustainable and resilient future. This introduction sets the stage for a deeper exploration into how these timeless practices can inform contemporary approaches to navigating an uncertain world.

If we consider indigenous knowledge system, it constitute a huge reservoir of not only the knowledge, but also the cultural epistemology, societal norms and legal mechanisms developed over the centuries either through the direct interaction of the community with their history, belief system and all the environment they reside in or through the evolution that was the result of dynamism of society (Roy, Maji, & Singh, 2023). The study of indigenous knowledge systems and the associated jurisprudential traditions provides a platform to delve into the intellectual cornerstones of the ancient societies (KHAIRNAR, 2025). Human conduct and behavior were regulated by sophisticated methods in civilizations across Asia, Africa, and Mesopotamia in the ancient era long before the formulation of modern legal framework. People throughout India, Greece, Egypt, Africa, and Mesopotamia came up with their own traditional ways to manage the human conduct, settle the disputes, resolve the conflicts, and maintain peace and stability in the society long before the contemporary legal systems were introduced. Even though it was the ancient past, indigenous governance, cultural practices and ecological knowledge played an important role in promoting community resilience. The disputes were resolved effectively and the social equilibrium was in harmony. Though the systems were fundamentally different from western framework of law and science and were deeply intertwined with cosmological beliefs, metaphysical analysis, moral philosophies and communal ethics but were effective (Gentle Terms, 2025). Traditional culture, ecological knowledge, language, history, geography, and folklore are all essential parts of indigenous knowledge system which makes it important to have the knowledge, not just to understand and protect the cultural heritage but also to make potential contribution in mitigating climate change, promoting indigenous governance, sustainable development and enlarging the scope for innovation. This study looks after different cultures through legal lens and recognizes that law and governance is a multifaceted system that is shaped by values, traditions, experiences and local knowledge.

Objectives of the Study

- Identify and document specific Indigenous governance structures, cultural practices, and ecological knowledge systems that contribute to community resilience.
- To explore and compare the ancient indigenous knowledge and philosophy behind the legal mechanism of India, Greece, Egypt, Africa, and Mesopotamia.
- Evaluate the practical efficacy and sustainability of these Indigenous-led approaches to resilience compared to or in conjunction with external interventions.
- Contribute to the recognition and valorization of Indigenous knowledge systems and self-determination as crucial components of global resilience-building efforts.

Statement of the Problem

Despite growing global recognition of the urgent need for enhanced community resilience in the face of escalating legal complications, environmental crises and socio-economic disruptions, conventional resilience strategies often fall short, frequently overlooking or marginalizing the profound and time-tested wisdom inherent in Indigenous communities. This oversight perpetuates a critical gap in effective resilience building, as Indigenous peoples possess unique governance structures, cultural practices, and ecological knowledge developed over millennia that enable remarkable adaptation, sustainable resource management, and social cohesion. Consequently, the failure to fully understand, acknowledge, and integrate these holistic Indigenous approaches into broader resilience frameworks leads to incomplete, less sustainable, and often culturally inappropriate solutions, leaving many vulnerable populations inadequately prepared for future challenges and undermining the potential for truly comprehensive global resilience.

Research Methodology

The research methodology will primarily be qualitative and collaborative, emphasizing Indigenous Research Methodologies (IRM) to ensure ethical, culturally appropriate, and respectful engagement with Indigenous communities. Data Analysis includes thematic analysis which will be employed to identify recurring patterns, themes, and insights from the qualitative data, grounded in the Indigenous worldviews. Ethical Considerations: Rigorous ethical protocols will be paramount, including obtaining free, prior, and informed consent, ensuring data sovereignty, protecting anonymity where requested, and establishing clear agreements for intellectual property rights and knowledge sharing. Local Institutional Review Board (IRB) or equivalent approvals will be sought alongside community-specific ethical guidelines.

Limitations of the Study

Findings may be highly context-specific to the particular Indigenous communities or regions studied. It can be challenging to generalize results directly to all Indigenous communities, given their vast diversity in governance structures, cultural practices, ecological environments, and historical contexts. Further Gaining genuine access to, and building trust within, Indigenous communities can be a lengthy and sensitive process. Researchers might face difficulties in establishing rapport, leading to incomplete data or a superficial understanding of complex internal dynamics. Language and Interpretation Barriers: Indigenous languages, metaphors, and oral traditions carry nuanced meanings that can be challenging to fully capture and interpret accurately in another language or cultural framework, potentially leading to misinterpretations or loss of depth. Sensitivity and Cultural Appropriateness: Researching Indigenous knowledge and practices requires extreme cultural sensitivity and ethical conduct. There's a risk of misrepresenting or misusing sacred or private knowledge if not handled with immense care and proper community protocols, which can be difficult to navigate for external researchers.

Scope & Significance of the Study

Addressing Contemporary Challenges, IKS offers time-tested wisdom and practical solutions for pressing global issues like, Legal Challenges: Law and governance in contemporary era, Climate Crisis: Promoting sustainable living, resource management, and a harmonious relationship with nature, health, Epidemics & Mental Health: Providing holistic approaches to well-being and resilience in the face of physical and psychological stressors, societal disintegration: Fostering community cohesion, ethical values, and responsible citizenship. Moving beyond purely academic or economic growth to encompass physical, mental, emotional, and spiritual well-being, leading to more balanced and resilient individuals and societies. The research will primarily focus on contemporary practices and their historical roots, exploring how ancestral knowledge and practices are applied and adapted in the present day to address current challenges.

HISTORICAL BACKGROUND

Among ancient civilizations, indigenous knowledge systems and legal philosophies were based on the concept of fairness. Their legal traditions reflected that the cultural philosophies governed life and justice on the basis of nature and the ecological knowledge.

In India, the indigenous governance was deeply rooted in the *dharma.* The concept of *dharma* shaped the jurisprudence through religious texts like Bhagavad Geeta, Vedas, Vedangas, Vedantas, Manusmriti Yajnavalkya etc. and local assemblies (Kane, 1968). In ancient India, *Achara* (conduct), Vyavhara (behavior), *prayaschit* (atonement) and *manyata* (belief) were the major guiding principles to regulate human behavior (Srinivas, 2013).

Greece on other hand was regulated by *Ancient Greek Law,* which back to almost seventh century BC (Osborne, 2009). It is a collection of laws of various Greek regions that share a common legal regime and was often termed as a system of natural law (Tussay, 2024). Arbitration and Mediation were key modes of dispute settlement in ancient Greek societies (Grisel, 2023). Ancient Greek laws were customary in nature and as per *opinio communis*, it was unwritten, orally transmitted through generation to generation and included moral beliefs during Homeric years (Panezi, 2021). Law and governance was a blend of religious beliefs, morals and hygiene rules during archaic years. And in its classical era, the establishment of democratic framework empowered the position of law and governance, making it dominant in legal orders of Greek antiquity.

Ancient jurisprudence in Egypt was a blend of custom, beliefs based on religion, cosmic order and the authority of Pharaoh. The community resilience was promoted through this legal system, which was based on the concept of Ma'at and was the reflection of cosmic order, social conduct and morality (Editorial, n.d.). Ma'at was considered as the goddess of order responsible for truth, balance, harmony, law, morality and justice (Giunta, 2024).

The ancient African societies (excluding Egypt), also developed a rich system of indigenous jurisprudence for governance. It was based on oral traditions, ancestral customs, community values and emphasized restorative justice over punitive justice. It emphasized on reconciliation, compensation and rituals as common tools for resolution of disputes. The concept of Ubuntu (I am because we are) reflects the communal nature of justice in ancient African societies, where justice is understood as a social and practical process aimed to restore harmony, rather than an abstract endeavor (Chowdhury, Wahab, Saad, Roy, Hamidi, & Ahmad, 2021).

Mesopotamia developed some of the ancient and earliest codified legal systems for indigenous governance. The Sumerian law (Code of Ur-Nammu), Akkadian & Babylonian Law (Code of Hammurabi) and Assyrian Law are the major sources of ancient jurisprudence (Patowary, 2022).

INDIGENOUS GOVERNANCE

Indigenous governance is a comprehensive system of governance through which the indigenous communities regulates the behaviour of their peoples. This includes the self determined ways through which the society governs themselves, make effective decisions and organize their communities to carry out social and economic activities. These are generally based on traditional indigenous practices through which roles and responsibilities of leaders as well as subjects are derived. The key fundamental elements that shapes this system includes traditional knowledge, indigenous jurisprudence, cultural practices, indigenous customs, traditional practices, weather, vegetation, habitat and other ecological knowledge. These system of governance were not just a mechanism of reactive tools but were the complex strategies proactive enough to keep the societies stable and predictable. Most of the indigenous governance system were interrelated and intertwined in contract of western system of governance and were considered to be deeply imbibed in roots of the culture wherein being a part of community did not valued just the interpersonal relationships but also the flora, fauna and the habitat of the land (Sustainability Directory, 2024).

CONCEPTUAL FRAMEWORK OF IKS

In the recent time Indigenous Knowledge system had been focused on globally in order to preserve and revive such knowledge to combat modern challenges and promote community resilience. According to UNESCO, Indigenous Knowledge is the knowledge, skills, and beliefs that societies have developed over time by living in and interacting with their natural surroundings (Tippeconnic & Faircloth, 2010). It is also known as "local knowledge," "folk knowledge", "people's knowledge", "traditional wisdom", or "traditional science" (Pandey, 2014). In many parts of the world, the art of agriculture, food preparation, health care practices, education, ecological conservation, indigenous methods of governance and various other activities that support community resilience are based on this indigenous knowledge, which is typically passed down orally and through cultural customs through generation to generation. These are the skills associated to local communities and are deeply rooted to traditional and cultural practices. This knowledge is a result of generations of trial and error and is obtained generally through non written texts such as sacred manuscripts, rituals, stories, poems and hand on experiences (Walker-Jones, 2021). This includes deep knowledge of the traditional ecosystems, traditional practices, sustainable resource management, traditional agricultural practices, rituals and various other practices that hold intellectual value. IKS, unlike modern Western systems, are flexible, adaptive and are based on collective experiences. They represent not just the data points and folklore but also the comprehensive worldviews expressed in languages, rituals, kinship, traditional practices and the management of the environment.

The COVID-19 pandemic profoundly disrupted global systems affecting every part of the global corner. Lockdowns, throughout the world, made it difficult for individuals to move around and changed human behavior, notably studying and working from home, telemedicine, and telemarketing and revolutionized delivery services (Piccoli, Jelena, Ruedin, & Jacob-Owens, 2023). In India, people migrated from cities to their native villages and began adopting a natural lifestyle. The increased use of herbal medicines such as Ayurveda, naturopathic and many other indigenous treatment systems gave a back-strength to the endangered treatment systems and acted as a straw to a drowning man by reviving these therapeutic systems.

This acted as a rebirth or reintroduction to several natural and traditional ways of living a healthy life. The Sanskrit phrase 'Vasudhaiva Kutumbakam' derived from Maha Upanishad, meaning "the world is a family" is considered as one of the most important moral values in Indian society (Kar, 2023). And if we seek to consider the whole world as a family, it is necessary to know the traditional values that can be a catalyst to achieve the greater good. This goodness should not be limited by boundaries or borders, but should encompass the entire Cosmos and the traditional knowledge that can be beneficial for humanity in any form should be shared by the entire global family. Apart from this, indigenous knowledge systems were also the bedrock of legal and administrative regulatory frameworks in order to resolve the issues related to property disputes, ethical conduct, spiritual obligations, customary and collective rights. In ancient societies where laws were not codified, the legal consciousness was expressed through proverbs, oral testimonies and rituals which though not directly, but indirectly shows the relevance in regulation of human behavior by the laws of contemporary era. At the international level, there is no clear and unambiguous definition of Indigenous knowledge. However, it can be said that Indigenous knowledge includes not only the content of the knowledge but also the traditional cultural expressions that are unique to that community and have value because of their uniqueness. It includes conventional intellectual activities, practical know-how skills, and new ideas in many fields, including agriculture, science, technology, ecology, literature, arts, medicine, biodiversity, and many others. Because of this, it could benefit from patents, copyrights, trademarks, geographical indications, or trade secrets. The only problem is that these come from ancient sources and have been passed down through the years, either by word of mouth or through sources that aren't documented (Ghosh, 2015).

EPISTEMOLOGICAL FOUNDATIONS

United Nations Declaration of the Rights of Indigenous Peoples (UNDRIP) recognizes Traditional knowledge (TK), Traditional Cultural Expressions (TCEs), Cultures, indigenous science and technologies as a part of international human rights law and backs them by making them enforceable at the court of law (Nations, 2007). The Convention on Biological Diversity's Secretariat with the similar approach defines Traditional Knowledge as a form of knowledge, innovation and practices that are developed and maintained by the indigenous and local communities (Diversity, 2010). Bio-cultural rights in the form of genetic resources, indigenous governance and advantages accrued by third party use gives indigenous people the right to manage their land and resources in accordance with their procedure and practices. In order to protect the intellectual freedom of these people, indigenous jurisprudence plays a crucial role. It safeguards their rights by not only recognizing and integrating their customary laws but also by incorporating traditional practices into legal framework.

The basis of indigenous governance is relational, prioritizing reciprocity, restorative justice, and cosmological balance over hierarchy and abstract legality. Traditional legal concepts such as Dharma in India, Ma'at in Egypt, and Ubuntu in Africa include ethical, spiritual, and practical dimensions of moral cosmology. Claude Levi-Strauss in his book, The Savage Mind, explained that all the societies, either primitive or modern, are based on a similar underlying structure and advocated that primitive societies were based on highly structured logic and symbolic reasoning, the reflection of which can be seen in their system of kinship and ritual governance (Levi-Strauss, 1966). This echoes with Amartya Sen's book, *The Idea of Justice,* where he critiques traditional theories of justice and gives the idea of "reasoned plurality" that advocates practical rationality based on diverse traditions and practical experiences (Sen, 2009).

CONCEPT OF ANCIENT JURISPRUDENCE

The term ancient jurisprudence refers to the legal theories and governance practices that developed in the world's earliest civilizations . People often considered that the laws were created by the gods and these systems were deeply rooted in religion, customs, and indigenous traditions. Rulers were regarded as the Gods or divine representatives in many communities and the decisions carried out by them about the law were considered to have sacred force (sociology.institute, 2022). Law was not separate from morality and religion, and was considered as an extension of the both that helped to maintain societal order, cosmic equilibrium and community resilience. Customs were important, and the traditions were seen to be more significant than legal innovations. In order to guarantee good governance, uniformity and validity, various ancient cultures codified their rules. One of the oldest and most well-known examples is the code of Hammurabi from ancient Babylon, which placed a strong emphasis on justice through punishments. The Manusmriti in India was the moral and legal text based on Hindu religious scriptures. Ma'at, which means truth and cosmic order, was the basis for the Pharaoh's reign over Egypt. Socrates, Plato and Aristotle were the Greek philosophers, who questioned the nature of justice and the function of law in an ideal state. The majority of ancient legal systems, despite having variances, had few characters in common such as hierarchical administration of justice, different laws or penalties for people based on their gender, social standing or ethnicity and the respect towards cultural practices and ecological knowledge. Ancient jurisprudence has a significant influence on contemporary legal systems, especially when it comes to principles like Natural law theory, codified laws, and the purpose of justice.

ORAL TRADITION

Oral tradition is one of the key modes through which indigenous knowledge got transmitted (Vansina, 1985). Oral sources should not be considered as vague, but as a complex mode to keep the track of history. Griots, in African tradition were the historian and storytellers who used to communicate the history through stories, poems and music and were considered as legal historians or advisors who recited the laws and the verdicts of the ancestors in case of any dispute (Henrich, 2001). With the help of stories and tales they preserved and passed the traditional values down the generations. These types of oral traditions were flexible and could be molded as per demand of the situation and need. Storytelling and oral tradition had been a key source of knowledge for India too. In India, *Shrutis* (what was revealed or heard) and *Smritis* (what was memorized) were prevalent in the ancient society which remained relevant in regulation of the societies, despite being unwritten. In Ancient Greece the Homeric questions related to Iliad and Odyssey suggests that oral narratives had great values and reflects the then laws to be customary and oral (Foley, 1997).

SOURCES OF LAW: RITUALS AND CUSTOMS

Protection of communal values and reinforcement of the sacred order are the two important functions of the rituals in indigenous jurisprudence. Oaths before deities among Mesopotamians, African reconciliation ceremonies, Indian *sabhas* held in temple courts and the feather of truth under Egyptian traditions are all the sacred performances that were performed in order to achieve justice in sacred way (Editorial A., 2024).

Customary laws were the backbone of ancient jurisprudence. They were not just the habit but were the corpus in the community's social contract. In ancient Greece, unwritten customs and practices were the source of law. In Mesopotamian, Kings wrote down the laws on the basis of customary rules. In India, local customs and behavior formed the laws whereas in Bantu communities, Ubuntu philosophy shaped the laws to frame ancient jurisprudence.

ANCIENT INDIAN JURISPRUDENCE: AN OVERVIEW

Dharma is the core concept behind the jurisprudence in ancient India. It signifies the knowledge of morality and is etymologically derived from the word 'dhr', meaning to hold (Seshasayee, n.d.). It talks about justice (nyaya), ethics, religion, philanthropy, duties, legislation, and customs. Indian legal system was extraordinarily broad, combining, philosophical reasoning, ritual, practice, and social control, all at the same time. Indian jurisprudence as addressed in the Smritis is considered as comprehensive and exhaustive that includes all the branches of law. The law constitutes a system of obligations that gave precedents to obligations over the rights. People believed that the pursuit of justice (nyaya) was not only a civic obligation, but also a holy one. The legal system in India was made of a blend of smritis (memorized laws), shrutis (laws that was heard or found in scriptural texts), acharas (conventions), sadacharas (societal standards), and vyavhara (behavior). P.V. Kane in his book writes, "In India, the law wasn't just about courts and punishment. It was also about the spiritual goal of keeping society in harmony through Dharma" (Kane, 1968). Apart from good governance, Ancient Indian system also focused on protection of environment and maintaining ecological balance through religious and practical approaches. Atharvaveda, Manusmriti and Arthashastra clearly emphasized on sustainable use of natural resources, punishments to the polluters and regulation of cutting of trees. Nature was associated to divinity and the punishments were imposed for harming the same (Shiskha, 2025).

SOURCES OF TEXT AND LAW

Ancient Indian legal system was based on various religious texts, customary practices and philosophical traditions, each of which had a distinct job in establishing rights, obligations and penalties. These sources shaped the legal and modern aspects of ancient Indian societies. Shrutis, which means "that which is heard", are one of the most important sources of ancient Indian jurisprudence. Shrutis included vedas, namely Rigveda Samaveda Yajurveda, and Atharvaveda. People believed that these texts were messages from God and the best place to get knowledge. The Vedas didn't supply a lot of information about the law, but they did explain the moral and ethical rules of Dharma, which include obligation, fairness and righteousness (Jyothirmai, 2020). Later, these ideas provided the basis for laws and social order.

Another notable source of ancient Indian knowledge is the Smritis. It includes both remembered traditions and literature created by learned individuals to clarify and contribute to the concepts in the Shrutis. The Manusmriti or the Laws of Manu, is the most important of them. It set regulations about marriage, property, punishment, caste, and government (Olivelle, 2005). Yajnavalka Smriti is another significant Smriti since it delivers great legal guidance in relation to rights related to property and inheritance (Olivelle, Yajnavalkya - A Treatise on Dharma). The Narada Smriti on other hand is vital since it explains how the court functions and how to pass on property to the successors (Narada & Lariviere, 2003). The guidelines offered in these writings were more precise than the Vedas.

Achara, or customs, were also essential in the development of the law and governance. Over time, these were the rules and conventions that everyone in the community agreed and were considered to be the good sources of law as long as they didn't break the principles of Dharma. When smritis didn't mention anything about the topic, than emphasizing on the values of the indigenous governance, the local and regional customs were often considered as relevant.

Vyavhara is the notion of judicial precedents and the standards that must be followed by the courts. It tells about how the king or judicial assembly handled differences and has regulations concerning witnesses, evidences and oaths (Olivelle, Yajnavalkya - A Treatise on Dharma). Ancient legal writings defined out how trials should go and what judges, litigants and other legal authorities should do. They were like the procedure laws that acted as guiding principles for the kings in order to grant justice and maintain harmony in the society.

The Arthashastra is a unique and one of the most important texts on Indian legal context. It was composed by Kautilya (also known as Chanakya) in the 4th century BCE (Labes, 2014). The Arthashastra is not a text about religion but a useful text on law, economics, and governance. It talks about Raja Dharma (duties of a king) and is a guiding principle for a king to rule. It is one of the first examples of secular legal writing in India since it talked about the courts, procedure of law, penalties, espionage, environment protection, forensics, taxes, and the art of statecraft.

As laws evolved and became harder to understand, legal academicians developed commentaries and digests called Nibandhas to clarify ancient writings and showed their relevance in modern contexts. The Mitakshara (by Vijnaneshwar) and Dayabhaga (by Jimutvahan) were two of the most important books about inheritance law that people in different parts of India used. These contributed to the ancient societies in their development by associating the most relevant laws as per the necessity. Ancient Indian law arose from a lot of different areas, religious books, social standards, royal authorities, intellectual literatures, and the belief systems of the people.

THE CONCEPT OF DHARMA

The word 'dharma' is derived from Sanskrit word 'dhr' meaning 'to sustain' or 'to uphold'. It is a basic principle of divine law associated to Hinduism, Jainism and Buddhism and talks about moral duties, code of conduct and natural order of universe. The indigenous governance system in ancient India were mainly based on the principles of dharma which elaborated the ethical duties of an individual towards society, state and the nature.

Justice System

India's legal system was made up of several levels, much like society and politics at the same time. The village panchayats were at the bottom. They were local councils generally chaired by respected elders or Brahmins. They resolved the issues amongst the people in the village. There were tribunals for social grouping like Kula (family), Shreni (guild), and Gana (tribe) that made sure that everyone obeyed the rules and conventions of their group. Royal gatherings called Shabas and Parishads were the places where knowledgeable people spoke about the law and worked in order to solve complex problems (Insight, 2025). The King's Court (Nripa) was the most important court in the country. The idea of Raj Dharma maintained that the ruler's fundamental job was to be fair, ethical and just.

Trials

In ancient Indian courts, oral evidences, witness, examination, oaths and religious declarations were particularly significant. The way the scriptures were read, the past performance of a person and the Varna (caste) position greatly affected the legal judgments. In certain situations, persons who were accused had to go through tests with fire, water or poison to show that they were guilty or innocent. This was known as "trial by ordeal" (Leatherdale, 2019). But the system also advised individuals to try to settle their disputes through mediation and conciliation before going to the courts.

Gender and Property Rights

Ancient Indian law had a confusing and frequently inconsistent attitude on women's legal rights, especially when it comes to property. Stridhan was an important legal term that denoted property granted to a woman when she gets married (Rajpal & Vats, 2017). This is the property granted to the woman either by her parents or husband. She had every right to keep it and pass it to her kids. Although the distinct schools of Hindu law such as Mitakshara and Dayabhaga had distant rules concerning who might inherit. Some societies permitted females to inherit property if there were no male heirs, while others did not allow them to do so, at all. There were both religious and legal arrangements that made marriage possible. Even though ancient scriptures have regulations about divorce, annulment and remarriage, males were typically the ones who made sure that these rules were obeyed.

Common Laws and Tribal Rules

In many regions of India, especially among tribal communities, oral traditions and rituals kept the law alive and flourishing. Achara (conduct) and Sadachara (good behavior) were based on following the norms provided by ancestors showing respect for elders (Olivelle, Yajnavalkya - A Treatise on Dharma). These laws were passed down from one generation to another generation, and adjusted throughout time to serve the community. In certain regions of central and north-eastern India, tribal councils still utilize the clan-based legal systems that are quite similar to the ones that used to prevail in ancient India. They settle disputes by worshipping their ancestors and using the traditional knowledge of the group.

Different Schools of Philosophy, Law and Governance

Ancient Indian knowledge system includes 14 types of knowledge and 64 types of skills involving vedas, vedantas, vedangas, mimansas and upnishads (Kapoor & Singh, 2021). Logic and legal epistemologies were highlighted under Nyaya school of thought. The term 'Nyaya' literally meaning 'justice' defines ethical standards, legitimacy of the evidences and the principles of atonement. Mimansa or hermeneutics was about how to understand very writings. This had an influence on legal exegesis and jurisprudence. Vedanta, which is more intellectual, focused on the moral side of Dharma and what it means to be responsible. These schools worked together to provide people in ancient India the knowledge they needed about morals, law and governance.

Indian legal system today is still based on a lot of what ancient Indian laws depicted. The Hindu Marriage Act, Hindu Succession Act, and various other regulations are based on ancient laws and practices. In environmental law, for example, Indian courts have employed ancient principles from Dharma in new ways when they considered rivers as a person. For instance, in 2017, the Ganga and Yamuna rivers were awarded rights under the law (Sen S., 2019). These instances indicate that India's ancient legal tradition still affects how people think and act in current times and ensures that indigenous ecological knowledge is still alive.

Greek Jurisprudence

Greek law, also popularly known as the law of Athens, is the ancient Greek's legal system. There was never a single system of institutions that the whole country recognized and followed as its legal order. However there were a number of basic approaches to deal with issues related to law and governance. There were different rules in each city-state, and many of them were set down in statutes. Draco's code of law and Solon's law are two of the most well-known laws (shorthistory, 2015). Greek law did not support much the concept of jurisprudence, although the abstract concept of justice was examined by the philosophers. Verdicts and judgments were based more on 'law as it is' than 'law ought to be' or the principle of equity. In both private and criminal cases, the first step was to call the defendant to the magistrate and file a formal complaint. Arbitration was an option in civil cases (Ager, 1996). Greek system of governance in the 4th and 5th century BCE was influenced by 3 major factors. One of them was that there were many city-states (poleis), each of which had its own laws and administration. Another factor was that in many, if not most, of the poleis, the laws were written down in statutes. Some of these statutes were comprehensive and exhaustive and laid out the detailed rules and procedures for administration of order and justice. This was the result of huge drive in 7th century Greece that lead to codification of laws in order to make them easier and understandable. Solon of Athens (594 BCE), the predecessor of Draco in 621 BCE, is the most renowned and reputed lawgivers (shorthistory, 2015). Other noteworthy includes Zaleucus of Locri Epizephyrii (south Italy), Charondas of Cantana and Lycurgus of Sparta. Literary quotes suggest various modifications but the code of Gortyn, which is a revised version of an older code, is the only one that is almost unchanged and preserved. The third factor that influenced Greek legal system was the absence of a body of law and jurisprudence in comparison of Romans. The Attic orators, even though they had familiarity with the law of cities, were primarily interested in presenting arguments that would convince the mass juries they had to argue in front of rather than the deep and critical analysis of the legal system . They weren't interested in looking at the legal system in order to get a better understanding of what it entails (Mészáros, n.d.). The philosophers didn't care about the law as it was either; they were more interested in finding abstract norms of justice.

Greek Philosophy of Natural Law

A much of what we read about law and justice have its origin in the ideas and practices of ancient Greek and Roman philosophers. Current ideals of equality, equal application of laws, human rights, and sovereignty all have their roots in Natural Law. Natural law had always been considered as a set of principles that are different from the laws that are in place in a given political society and the way individuals act in that society. The Greeks were among the very first who came up with the philosophy of natural law (BhatMishra, n.d.). They came up with the notion and determined the essential features of natural law. Greece, at that time, was highly politically unstable. There were two main philosophies that existed among Greek thinkers. Most of the people determined that the law was solely meant to protect the interests of the lawmaker or the strong, and that the only justification for obedience to the law was to protect the self-interest of the subjects. But, the same predicament forced the other jurists to think in a different and they saw it as an opportunity to come up with new universal principles against the tyranny that would deal with and manage the government's arbitrariness and provide it stability. Greek philosophy evolved a lot in the 5th century B.C. Greek philosophers ceased to consider ancient and traditional ideas and the Sophists began to become popular during this period (Paillard, 2014). They claimed that law was merely something people made up because there was necessity of it; it was distinct from metaphysics and could be changed at will. This is how Law and Justice were explained at that period in Greece, when the city-states of the democratic republic changed their laws regularly due to the country's economic and social realities.

Socrates, Plato, and Aristotle were the three most important philosophers in Western philosophy. Each of these ancient philosophers had a big impact on the ideas of their time period, although they all had their unique way of thinking and beliefs each of them had an important role in the philosophy of natural law.

Socrates (5th century BC, Athens), one of the famous Greek philosopher who were also considered as a father of Western Philosophy, introduced an ancient but scientific method of philosophy called Socratic Method (Ambury, n.d.). It was a method in which he talked with the people and asked them difficult questions to make them think again and again in depth and let the people figure out the flaws in their views. It was akin to modern trial and error and self assessment method in which people used to self-assess their views. Seeking knowledge and wisdom through self assessment was a key aspect of this philosophy and led to the development of critical thinking and the formidable quest for the truth. This approach of Socrates acted as a guiding principle for the succeeding philosophers.

Plato, who was the student of Socrates was another important philosopher in Greek history who established the academy of Athens and went on to be known as the founder of first school of higher education in not only Greece but the entire west (BIOGRAPHS, n.d.). His teachings include grammar, music, ethics, metaphysics, politics and epistemology. He adopted the dialectical method of reasoning and lead to the understanding of essence in depth without being affected by the sensations which shows nothing but inculcation of scientific and logical evolution of knowledge in the past. He believed in the principle of forms. Theaetetus is one of the work which is admired till today and is considered as one of the best work in epistemology. His work 'The Republic' talks about those Kings who put justice, wisdom and common good on top priority. General Education theory is considered to be highly influenced by Plato's approach towards philosophy and is being studied even today, marking the value of historical and indigenous knowledge which is contributing in shaping the contemporary era as well.

Aristotle, the disciple of Plato and a student of the Academy of Athens is a prominent name of ancient history. Whether we talk about history, politics, philosophy, metaphysics, ethics or medical science his name is respectfully used. His views were quite different from Plato and founded Lyceum, a school where Alexander the Great studied under him. He emphasized that the moral virtue was based on real world, practical ethics and natural science (Horn, 2013). His books, 'The Politics' and 'Nicomachean Ethics' are considered to be among the greatest of all time for the conceptual understanding of Law, Morality and Governance.

Jurisprudence in Ancient Greece

Egyptian law was based on the notion of Ma'at. The goddess Ma'at stood for the law, which was the common sense view of right and wrong. The notion of Ma'at was the same as the ideas of morality, ethics, and the correct order and balance of the universe. It stood for truth, justice, and the right order and balance of the cosmos. The ancient Egyptian goddess Ma'at stands for truth, justice, harmony, and balance, originally appeared during the Old Kingdom period (2686-2181 BCE) represents the ancient Egyptian ideals of law, morality, and justice (Uth, 2022). People typically show Ma'at as a lady with wings and an ostrich feather on her head, or they use a plain white ostrich feather to represent her. The Weighing of the Heart was an important rite in the afterlife. It used this feather to measure a dead person's heart against Ma'at's feather to see how morally good they were on a weighed scale of justice (The Weighing of the Heart Ceremony, n.d.).

Since the laws of ancient Egypt were based on religious principles, the ancient Egyptians didn't consider the difference between human and divine justice. They considered that the gods created the laws for the humans when they created the world and that the gods were the only ones responsible to protect it. In ancient Egypt, religion and belief were always crucial when it came to how people interacted with the law and justice because of which, the ancient Egyptians utilized divine judgment in addition to the normal legal system.

Structure of Legal Administration in Ancient Egypt

Law was linked with religion and represented the rules for governing the behavior of people in society. The Egyptian judicial system was set up in a way that made sure everyone could get justice. The judicial system had several levels.

Pharaoh

It seems that the Pharaoh (king), the vizier (prime minister), and the great courts in Memphis had the jurisdiction over the offenses against the state. Pharaoh (King) made the laws and played an important role in maintenance of Ma'at (Blerk, 2018). Although there are no concrete evidences available but it is presumed that the Pharaoh might be the head of judicial administration, and was considered as a supreme judge and the law-maker. He represented the Ma'at and had the primary duty to uphold the order of creation. He had to follow the rules of Ma'at in his own life too and also was responsible to uphold the Ma'at in society. He was supposed to rule as per the principles of Ma'at and was considered as a link between the law and the Ma'at. People thought of the Pharaoh as a god and considered him as a source of law. His words had a force of law and were considered as a primary source of law. The duty of the king in respect of creation of laws can be traced from the famous phrase "putting Ma'at in place of injustice" (Blerk, 2018). The ancient Egyptians thought that only the king knew what the Ma'at principle required and that his laws were the same as the will of the creator god. This is why the king could keep law and order and these rules and decisions of the king mirrored the universe in harmony. The king had to keep Ma'at by making laws, which was his obligation to the gods and the people. The vizier was the king's representative, the high priest of Ma'at, and the leader of the courts of justice. It is said that the legal official wore a golden Ma'at pendant when the king had to delegate his power. Judges were very essential to the goddess Ma'at and their feeling of responsibility. They were called "priests of Ma'at" and wore a tiny figurine of the goddess as a pendant around their necks to show that they were judges. Statues of high officials from the Late Period that are still around depict those wearing pendants on chains. These high officials would look at instances and report them to the king, who would subsequently punish the offenders in more serious situations. When people followed the law, the concept of Ma'at was used but if someone broke the law, the law might be used against them. Ma'at ruled the lives of the ancient Egyptians, and their law was justice in action and made Ma'at the center of the legal system.

Vizier, Nomarchs and Kenbets

Vizier was the highest judicial authority in the kingdom after the Pharaoh (Shaw, 2002). He is considered as the protector of law and was similar to modern day judges. He heard and resolved the matters including theft, murder, corruption, treason, burglary etc. and also acted as an appellate authority when the disputes results in disagreements and are unable to be settled. Even though it was an ancient system but was well organized, systematic and maintained involving the role of clerks, scribes and records. The authority that was below the Vizier was known as Nomarch. Nomarchs was the provincial rulers below the Vizier who acted as a local court dealing with less serious matters, property disputes, and smaller crimes. They helped in maintaining law and order and ensured Ma'at to be followed effectively. Priest and scribes played important role in keeping and maintaining the records. The other authority below the Nomarchs were known as Kenbet, which was a local authority at the village level consisting of elderly people, priests and the respected members of the community (Shaw, 2002). They decided local day to day disputes such as theft, family disputes, contractual disputes and less complicated property disputes and were helpful in resolving the issues through speedy trials resembling to modern day Nyaya Panchayat System in India.

Ma'at played an important role in maintaining good governance from both legal as well as spiritual angle and promoting the importance of law, order and resilience in ancient Egyptian community. The principle of Ma'at was based on justice, honest and unbiases. It was formed to curb corruption practices and impart justice. Documentary evidences, witness testimonies and the principle of audi alteram partem (the right of both the parties to be heard) was effectively followed for decision making by the judges.

Punishments and Fines

In ancient Egypt, punishments were meant to do more than just stop people from committing crimes. They were also meant to bring society back into equilibrium. The consequences were different based on the crime, with stronger punishments for actions that violates Ma'at. Fines and restitution were standard punishments for little crimes while people who were found guilty of property offenses, theft, and other nonviolent crimes typically had to pay the damages to the victim back but heinous crimes, including murder, adultery, and treason, were punished with harsh punishments like banishment, amputation, and even death.

JURISPRUDENCE IN ANCIENT AFRICA

Africa in the past, before the arrival of colonial administrators and the western legal system, was governed by the laws of land, customs, beliefs and values that constituted complex, legal traditions, rooted deeply in their Indigenous knowledge system. These legal systems formed the bedrock of ancient societies and were rich in philosophical, spiritual and practical wisdom. They governed issues regarding legal disputes, leadership, marriage, inheritance, social conduct, etc. and was dynamic, pluralistic and deeply influenced by the spiritual, communal and ecological aspects of society. In the shadow of indigenous knowledge, African jurisprudence evolved to be a participatory and restorative process rather than retributive (Sule, 2018). The set of rules the traditional African societies followed were African customary laws. In Africa, part of their culture, and they see it as a way to keep society in balance with a focus on distributive justice instead of formal justice like in Western philosophy. Being a part of their life, it is impossible to separate law from the culture of the African people. The knowledge of ancient African jurisprudence was transmitted orally from generations through rituals, stories, proverbs, customs, folklores, taboos and traditional institutions. Law was not seen separately from life and was imbibed in daily habits and morals of the community and the justice was restorative, culturally embedded, and inseparable from identity.

If we talk about ancient African societies, justice in ancient African societies was community centered. Those were settled in public forums, indaba gatherings and the primary aim was not to punish, but to reconcile, restore and to reintegrate offenders back into society. This approach was similar to contemporary approach of restorative justice. African legal system gave more importance to restorative approach over the retributive justice. Trials were focused on rehabilitation and repairing of social relationships rather than punishing the guilty person. They motivated the victims for compensatory remedies rather than revenge or punitive measures. Practices such as spiritual cleansing and public apology were prevalent among Zulu communities' Ashanti Courts and they focused on participatory decision making wherein the decisions in the disputes were reached by elders, priests, chiefs and learned members of society which shows resemblance to the modern Jury trial system (Gabagambi, 2018). African system showed a great respect towards nature and spirits. They governed by devoting the rights and homage towards spirits and natural powers such as flora, fauna, weather, sun, moon etc. They played a great role in saving the environment and maintaining the ecological balance but considering the flora to be spiritual and divine. The traditional and indigenous practices carried on by the African tribes contribute to the major development in the region and can be considered as a treasure that made Africa a rich region for flora and fauna. It can be observed that Africa is one of the region throughout the globe that is still deeply rooted to its culture, indigenous knowledge and community resilience.

Ancient Jurisprudence in Mesopotamia

Mesopotamia, often referred to 'the cradle of civilization' is one of the oldest civilization that shows us a rich history of not just the mankind but the ancient ways of governance as well. This was one of the oldest civilization that lead to the foundation of written or codified legal system and served as a model for many modern legal frameworks. It shows its importance in the fact that ancient oral tradition of law and governance were converted into codified and standardized system of governance under this civilization. It includes the concept related to criminal law, family law, property law, commercial law and various other legal aspects (Sasson, 1995). The Code of Hammurabi, is one of the famous legal code that was codified for the first time and included the principles of values, justice and statecrafts for the governance of the states and is prevalent till today by being a model code for many legal systems.

The Emergence of law in Mesopotamia

In Mesopotamian philosophy, law and kingship were divinely sanctioned. It was believed that kings were chosen by the gods to preserve justice and order. The ruler, who served as the custodian of social harmony, received legal authority from the divine realm. For instance, the sixth king of Babylonian Empire, Hammurabi (1792-1750 BCE), famously claimed that his laws were given to him by Shamash, the son of God and deity of Justice. The law gained a great deal of legitimacy and authority from this divine Association (StudynLearn, 2022).

The Origin of Written law

Legal procedures were based on oral traditions upheld by elders, priests and local authorities prior to the invention of writing (Approx. 3200 BCE). Societies started writing contracts, treaties, court decisions, and eventually legal codes on clay tablets after the advent of cuneiform script. The transition from oral to written law ensured consistency and accountability. It made it possible for justice to be administered more methodically, decrease the number of arbitrary rulings, and facilitated administration of large diverse empires.

Important Legal Codes and Texts

- **Code of Ur-Nammu (Approx. 2100 BCE)**

The oldest known legal code is the code of Ur-Nammu, which was created almost 300 years before Hammurabi. Issued by King Ur-Nammu of Ur, this code consisted of around thirty laws written in Sumerian. It included provisions for marriage, theft, injury, and social obligations (Sonte, n.d.). The code of Ur-Nammu replaced physical punishments with monetary compensation for some of the offences and encouraged the restorative approach.

- **Code of Lipit-Ishtar (1934-1924 BCE)**

Lipit-Ishtar of Isin issued this code to carry out the Sumerian legal writing tradition (wikipedia). In addition to addressing topics like slavery, inheritance and family law, it strengthened social hierarchy. It presented the king as a divine agent, who brought justice to the land.

- **Code of Hammurabi (1775-1750 BCE)**

The Hammurabi's Code was the most renowned Mesopotamian legal code, which consisted of 282 laws inscribed on a diorite stele. These laws were in public domain and the symbol of transparency and accessibility to the literate citizens.

Key Features

- The law of retaliation or Lex Talionis, often summarized as "an eye for an eye". This principle reflected a belief in retribution and proportional justice.
- Stratified justice: Social hierarchy was highlighted by the different penalties imposed on different social classes, such as freeman, slaves, and commons.
- Laws covered a wide range of topics, such as property rights, theft, marriage, divorce, inheritance, contracts, wages, false accusations etc.
- The judicial process was based on due process and included mechanisms for presenting evidence, cross examination and oath.
- Mesopotamian jurisprudence was incredibly practical and utilitarian. It was not abstract or theoretical, and the purpose of law was to preserve social order, maintain class structure, and to ensure economic efficiency. Mesopotamian Jurisprudence reflected the moral dimension, in accordance with the will of gods and the King's obligation to promote justice (kittu) and righteousness (misaru).
- Slaves had minimum rights.
- Legal transactions were based on contracts and oaths. And the written records helped to prevent fraud or disputes.
- A proper legal record of cases, documents, contracts, decisions and seals were maintained in clay tablets and were stored in temples and public arenas.

COMPARISON OF VARIOUS ANCIENT INDIGENOUS SYSTEM OF GOVERNANCE

S. No.	BASIS	INDIA	GREECE	EGYPT	AFRICA	MESOPOTAMIA
1	**Legal Source**	Dharma (cosmic law and duty), and ancient scriptures	Democracy, civic reason, Draco's Law and Solon's reforms	Divine order (Ma'at)	Customs and oral traditions	Hammurabi's Code
2	**Legal Nature**	Moral, and religious	Secular, and analytical	Theoretical and ethical	Communal and restorative	Retributive, and stratified
3	**Mode Of Knowledge**	Oral texts (shrutis) and sacred remembered texts (smritis)	Written (scrolls), and rational	Symbolic, and sacred texts (pyramid's texts, book of dead)	Griots, literature, music, stories and rituals	Written, and codified
4	**Concept Of Justice**	Moral order and karma	Legal equality, reason and debate	Cosmic balance	Harmonious equilibrium and restitution	Punishment and deterrence
5	**Transmission**	Lineage through teacher -disciple tradition(gurus-shishya parampara)	Academies	Priesthood and scribes	Elders and griots	Records and temple schools
6	**Cosmological Link**	Deep integration and ethics to achieve salvation (karma dharma and moksha)	Partial, and philosophical	Based on Ma'at, afterlife and moral judgement	Ancestor's worships, and nature	Ritual, and political

It can be observed that there was a complex interaction between traditional practices, law, culture, spirituality and epistemology lead to the evolution of jurisprudence and indigenous knowledge systems in ancient civilizations.

In all of the above mentioned systems, it can be observed that though they had distinct framework of law and governance but one thing was common and that was that they all were based on foundational principle of justice, order and community resilience.

Ancient Indian indigenous governance system was closely associated to the moral and spiritual values. Its jurisprudence relied mostly on the concept of Dharma, that not only governed all the aspects of an individual's life but also acted as a guiding source for governance and statecraftship. Several ancient texts including Manusmriti, Bhagavad Gita, Vedas, Vedantas, Upnishads, Dhamshastras and Kautilya's Arthashastra explained the system of statecraft. Salvation (moksha) was the ultimate goal for the individuals which motivated them to ensure good behavior among the society. Although not systematically codified, but the ancient texts by their own ways orally or documentarily and passed on their teachings from generation to generation, valued their indigenous system of knowledge and inculcated them to contemporary system of governance.

Ancient Greece, on the other hand, established the groundwork for civic and secular jurisprudence (Black, 2016). Legal systems evolved from strict codes like Draco's to the more, inclusive reforms of solon. Greek law placed a strong emphasis on Citizen's equality before the law and was strongly linked to the ideas of reason and public debates. Greek knowledge system was based on rationalism and dialectics, developed through natural schools such as those of Socrates, Plato, and Aristotle. In contrast to the theoretical models found elsewhere, knowledge was disassociated from religious or priestly authorities and disseminated through written text and formal academies.

Ma'at, the concept of cosmic harmony, truth and justice served as the foundation for the legal system in ancient Egypt. Although there isn't a comprehensive legal code like the code of Hammurabi, judicial procedures were embedded in religious and ethical conduct. Pharaoh was considered as the divine law giver and the transmission of knowledge took place through temple schools and hieroglyphic writings that were concentrated scribes and priests. With a focus on the afterlife, morality and divine order, indigenous Egyptian knowledge included sophisticated architecture, medicine, and cosmology.

If we talk about indigenous communities of ancient Africa, tribal groups like Yoruba, Zulu etc. relied on restorative and rehabilitative approach for settling the disputes (Sule, 2018). They emphasized on reconciliation rather than retribution and their most of the disputes were settled by keeping nature, spirits and atonement in center. Elders, chiefs and learned members of society played key role indecision making and valued the strong connection between natural world and spiritual cosmologies. Their decisions were based on the principle of ecological harmony and were adaptive, flexible and community resilient unlike the strict written legal codes.

As far as ancient Mesopotamian civilization is concerned, the code of Hammurabi became one of the oldest and best known codified system of governance which marked itself to be a brilliant piece of work which is admired even today and became a foundational model code for several system of governance throughout world. This system was retributive and promoted the documentation of records of contracts, disputes, evidences and judgements. The knowledge system included mathematics, astronomy, statecraft etc. and were based on temples and scribes. The sources which are preserved in form of clay tablets. The system emphasized on social hierarchy and was inclined towards administrative efficiency.

Case Studies related to Community Resilience

Case study 1: For the community, by the community: Building resilience through infrastructure in rural India: In 2015, Department For Infrastructure Development (DFID), UK helped by funding a technical assistance project in India named Infrastructure for Climate Resilient Growth (ICRG) aimed to promote and strengthen rural communities by building resilient rural infrastructure. All the infrastructure built herein were built by local communities by using their indigenous knowledge. The asset selection was in response to climate changes and were durable enough. The main aim of this project was to make the key officials of the government aware of the best practices of infrastructure resilience (Anu Jogesh, 2019).

Case Study 2: Indigenous knowledge for enhancing community resilience: An experience from the south-western coastal region of Bangladesh (Chowdhooree, 2019) : Depending upon traditional knowledge in order to attain sustainability and inculcate natural processes with human activities resulted in more benefits though being in contract with western euro centric approach which used the method of interfering with the natural science. There were several problems associated to floods, water logging and drainage congestion which were tried to be solved by traditional methods. Several problems associated to flood-prone properties were caused due to unplanned engineering which could be reduced by using traditional methods. The focus was on adaptation rather than development, which promoted disaster risk management in Bangladesh. The purpose of the Coastal Embankment Project (CEP) is to prevent non-saline tidal flooding and saline water intrusion on the arable lands in the southwest coastal zone which was introduced in 1961. In the late 1990s, while doing fieldwork for the EIA/SIA, the Center for Environmental Geographic Information System (CEGIS) found that the alternative option suggested by the local people was not only effective at solving the problem of drainage congestion in the beel area, but it was also technically sound, economically viable, and good for the environment.

Case Study 3 The practice of Cultural Burning practiced by Aboriginal and Torres Strait Islander peoples in Australia is indeed a powerful and integrated case study of indigenous governance, socio- cultural practices, and the ecological aspect promoting resilience, particularly in the face of increasingly bushfires.

Description	Resilience & its Role
Fire tool is not merely a cultural practice but primarily a spiritual practise to care for the Nation which involves the elderly members of the community to teach the youth storytelling thereby reinforcing their linguistics and ethical responsibility towards their land.	Cements Psychological as well as the Social Resilience: The collaborative act of caring for the Nation is an activity which builds the community and thus helps in combating the transgenerational trauma, reinforces the identity of a culture, and improves the mental health of the community at large.
So far as Indigenous Governance is concerned the practice is guided by Elderly members as well as the traditional who decides where, when, and how the burning occurs which is purely based upon environmental law .Firesticks Alliance is one of the Indigenous-led bodies which works to reinstate this art of self-determination in the management of land.	Helpful in an Adaptive Management: It is the Local, autonomous decision-making process which assures that fire management is not rigid rather flexible and tailored to the hyper-local environmental conditions, thereby making it more influential than standardized and top-down fire-fighting policies.

CONCLUSION

All the systems mentioned above, though being different, more or less coincide at one point and that is the principle of justice. The ways of governance and statecraft may be different but the unique feature that is common among all is the respect towards indigenous knowledge, cultural practices, customs and good conscience. India, Africa and Egypt associated the laws with spiritual energy, morality and self resilience while Mesopotamian and Greece followed codified system for governance. The legal principles in India, Africa and Egypt were more restorative, reformative and rehabilitative while Greece and Mesopotamia philosophy inclined more towards deterrence and retributive justice. The study of ancient indigenous system of governance of India, Greece, Africa, Egypt and Mesopotamia lead us to the conclusion that there was one similarity in the ancient system of all these civilizations, i.e. the respect towards nature and that the governance was not just a way to govern or control the subjects but was also a deep expression of values, moral principles, order and cosmology. The ancient governance principles were not just the code which were meant to be followed but were the real life practices, experiences, traditions, moral values and, spiritual belief which reflected the connection of life, community resilience and the rich history.

References

Ager, S. L. (1996). *Interstate Arbitrations in the Greek World, 337–90 B.C.* University of California Press. DOI: 10.1525/9780520913493

Ambury, J. M. (n.d.). *Socrates (469—399 B.C.E.)*. Retrieved from The Internet Encyclopedia of Philosophy: https://iep.utm.edu/socrates/

Anu Jogesh, U. P. (2019, October). *Case study: For the community, by the community: Building resilience through infrastructure in rural India.* Retrieved from ResearchGate: DOI:DOI: 10.13140/RG.2.2.33606.80965

BhatMishra. M. S. (n.d.). *Greek and Roman Conceptions of Natural Law*. Retrieved July 16, 2025, from Advanced jurisprudence: https://ebooks.inflibnet.ac.in/lawp01/chapter/greek-and-roman-conceptions-of-natural-law/

BIOGRAPHS. (n.d.). *Plato*. Retrieved from BIOGRAPHS: https://biographs.org/plato#:~:text=Plato%2C%20an%20ancient%20Greek%20philosopher%2C%20was%20a%20student,first%20higher%20education%20institution%20in%20the%20Western%20world

Black, A. (2016). The Greeks. In Black, A. (Ed.), *A World History of Ancient Political Thought: Its Significance and Consequences*. Academic. DOI: 10.1093/acprof:oso/9780198790686.003.0009

Blerk, N. (2018). The emergence of law in ancient Egypt: The role of Maat. *Fundamina (Pretoria)*, *24*(1), 69–88. DOI: 10.17159/2411-7870/2018/v24n1a4

Chowdhooree, I. (2019). Indigenous knowledge for enhancing community resilience: An experience from the south-western coastal region of Bangladesh. *International Journal of Disaster Risk Reduction*, *40*, 101259. Advance online publication. DOI: 10.1016/j.ijdrr.2019.101259

Chowdhury, S. J., Wahab, H. A., Saad, M. M., Roy, P. K., Hamidi, M., & Ahmad, M. M. (2021). Ubuntu Philosophy: 'I Am Because We Are' – A Road to 'Individualism' to Global Solidarity. In I. Global, *Handbook of Research on the Impact of COVID-19 on Marginalized Populations and Support for the Future* (pp. 361-381). IGI Global Scientific Publishing.

Shiskha. (2025, February 3). *Ancient and Medieval India's Environmental Protection Traditions*. Retrieved from Occupational Health & Safety Management: https://disaster.shiksha/occupational-health-safety-management/ancient-medieval-india-environmental-protection

Diversity, C. o. (2010, July 6). *What the Convention says about Traditional Knowledge, Innovations and Practices*. Retrieved July 12, 2025, from Convention on Biological Diversity: https://www.cbd.int/traditional/what.shtml

Editorial, A. (2024, November 28). *Oaths and Legal Commitments: Foundations in Ancient Civilizations*. Retrieved July 15, 2025, from Ancient Civs: https://ancientcivs.blog/oaths-and-legal-commitments/

Editorial, L. (n.d.). *Understanding the Ancient Egypt Legal System: Structure and Function*. Retrieved July 12, 2025, from Laws Pulse: https://lawspulse.com/ancient-egypt-legal-system/

Foley, J. (1997). Oral Tradition and its Implications. In Morris, I., & Powell, B. B. (Eds.), *A New Companion to Homer*. Brill.

Gabagambi, J. J. (2018). *A Comparative Analysis of Restorative Justice Practices in Africa*. Retrieved from GlobaLex: https://www.nyulawglobal.org/globalex/restorative_justice_africa.html#:~:text=Community%20members%20would%20sit%20in,meat%2C%20termed%20as%20okukatelanila%20inyama

Gentle Terms. (2025, May 2). *Exploring the Foundations of Jurisprudence in Ancient Cultures*. Retrieved July 8, 2025, from Gentle Terms: https://gentleterms.com/jurisprudence-in-ancient-cultures/

Ghosh, A. (2015). *Traditional Knowledge: Problems and Prospects*. LOKODARPAN - A PEER REVIEWED BILINGUAL ANUUAL RESEARCH JOURNAL OF FOLKLORE.

Giunta, J. (2024, November 14). *The Concept of Ma'at: Balance, Truth, and Its Role in Egyptian Society*. Retrieved July 12, 2025, from ancient pedia: https://ancientpedia.com/the-concept-of-maat-balance-truth-and-its-role-in-egyptian-society/

Grisel, F. (2023). *Arbitration as a dispute resolution process: historical developments*. Cambridge Compendium on International Commercial and Investment Arbitration.

Henrich, D. J. (2001). The griot storyteller and modern media. *Communicatio*, *27*(1), 24–27. DOI: 10.1080/02500160108537921

Horn, C. (2013). Law, governance, and political obligation. *Cambridge University Press*, 223-246.

Insight, l. (2025, July 24). *Administration of Justice in Ancient India: From Sabha to King's Court*. Retrieved July 26, 2025, from The lex insight: https://thelexinsight.blogspot.com/2025/07/Administration.html

Jyothirmai, B. (2020). THE CONCEPT OF DHARMA AND JUSTICE DURING. *The Law Brigade Publishers*, *6*, 206–214.

Kane, P. V. (1968). *History of dharmaśāstra (ancient and mediæval religious and civil law in India)*. Bhandarkar Oriental Research Institute.

Kapoor, K., & Singh, A. K. (2021). *Indian Knowledge Systems*. D.K. Print World Ltd.

Kar, A. K. (2023). The concept of Vasudhaiva Kutumbakam (The World is a Family): Insights from the Mahopanisad. *National Journal of Hindi & Sanskrit Research*, *1*(49), 42–45.

KHAIRNAR, M. S. (2025, July). Indian Knowledge System (IKS) with Reference to Legal System in India. *International Journal of Emerging Technologies and Innovative Research*, *12*(7), d292–d296.

Labes, M. (2014). *Kau ilya's Arthaśāstra: A Classic Text of Statecraft and an Untapped Political Science Resource*. HEIDELBERG PAPERS IN SOUTH ASIAN AND COMPARATIVE POLITICS.

Leatherdale, D. (2019, February 9). *Trial by ordeal: When fire and water determined guilt*. Retrieved July 24, 2025, from BBC News: https://www.bbc.com/news/uk-45799443

Levi-Strauss, C. (1966). *The Savage Mind*. University of Chicago Press.

Mészáros, T. (n.d.). *REMARK ON THE CANON OF THE ATTIC ORATORS*. Retrieved July 16, 2025, from academia: https://www.academia.edu/14909234/REMARK_ON_THE_CANON_OF_THE_ATTIC_ORATORS

Narada, & Lariviere, R. W. (2003). *The Nāradasm ti*. Motilal Banarsidass Publishing House.

Nations, U. (2007, September 13). *UN Declaration on the Rights of Indigenous Peoples*. Retrieved July 12, 2025, from OHCHR and Indigenous Peoples: https://www.ohchr.org/en/indigenous-peoples/un-declaration-rights-indigenous-peoples

Olivelle, P. (2005). *Manu's Code of Law*. Oxford University Press.

Olivelle, P. (n.d.). *Yajnavalkya - A Treatise on Dharma*. Harvard University Press.

Osborne, R. (2009). *Greece in the Making 1200-479 BC*. Routledge. DOI: 10.4324/9780203880173

Paillard, E. (2014). THE STRUCTURAL EVOLUTION OF FIFTH-CENTURY ATHENIAN SOCIETY: ARCHAEOLOGICAL EVIDENCE AND LITERARY SOURCES. [Retrieved from IGNOU.]. *Mediterranean Archaeology*, *27*, 77–84.

Pandey, K. (2014). Indigenous Knowledge System: Concepts and Discourses. *Anthropological Bulletin*, *2*(1), 17–28.

Panezi, M. (2021, November/December). *A Description of the Structure of the Hellenic Republic, the Greek Legal System, and Legal Research*. Retrieved July 11, 2025, from Globalex: https://www.nyulawglobal.org/globalex/greece1.html

Patowary, K. (2022, July 7). *Code of Ur-Nammu: The Oldest Law in History*. Retrieved July 13, 2025, from Amusing Planet: https://www.amusingplanet.com/2022/07/code-of-ur-nammu-oldest-law-in-history.html#:~:text=Some%20of%20the%20earliest%20legal%20codes%20concerning%20crimes,Kassites%2C%20and%20Hittites%2C%20each%20had%20their%20own%20laws

Piccoli, L., Jelena, D., Ruedin, D., & Jacob-Owens, T. (2023). Restricting Human Movement During the COVID-19 Pandemic: New Research Avenues in the Study of Mobility, Migration, and Citizenship. *The International Migration Review*, *57*(2), 505–520. DOI: 10.1177/01979183221118907 PMID: 38603252

Rajpal, L., & Vats, M. (2017). STRIDHANA: A CRITICAL APPROACH. *International Journal of Research in Social Sciences and Humanities*, *7*(1), 83–95.

Roy, P., Maji, S., & Singh, P. (2023, December). Preserving Indigenous Technical Knowledge: The Need for Documentation and Cultural Preservation in the Context of North-East India. *Food and Scientific Reports*, *4*(12), 10–17.

Sasson. (1995). *Civilizations of the Ancient Near East*. New York: Scribner.

Sen, A. (2009). *The Idea of Justice*. Belklnap Press.

Sen, S. (2019, April 25). *Of Holy Rivers and Human Rights: Protecting the Ganges by Law*. Retrieved from Yale University Press: https://yalebooks.yale.edu/2019/04/25/of-holy-rivers-and-human-rights-protecting-the-ganges-by-law/

Seshasayee, T. (n.d.). *Dharma*. Retrieved July 13, 2025, from Indian religion and philosophy: https://ebooks.inflibnet.ac.in/icp03/chapter/dharma/

Shaw, I. (2002). *The Oxford History of Ancient Egypt*. Oxford University Press. shorthistory. (2015, July 12). *Draco and Solon era in Athens (621-593. B.C)*. Retrieved July 17, 2025, from shorthistory: https://www.shorthistory.org/ancient-civilizations/ancient-greece/draco-and-solon-in-athens-621-593-b-c/# sociology.institute. (2022, October 17). *The Divine Right to Rule: Understanding Ancient The God-Kings System*. Retrieved July 13, 2025, from Sociology of Religion: https://sociology.institute/sociology-of-religion/ancient-divine-right-rule-god-kings-system/

Sonte, N. (n.d.). *Code of Ur-Nammu.* Retrieved from SCRIBD: https://www.scribd.com/document/591430208/Code-of-Ur-Nammu#:~:text=The%20Code%20of%20Ur%2DNammu,Sumer%20during%20its%20Renaissance%20period

Srinivas, S. (2013, September 10). *Dharma Shastras- Legal Literature of Ancient India.* Retrieved July 8, 2025, from ithihas: https://ithihas.wordpress.com/2013/09/10/dharma-shastras-legal-literature-of-ancient-india/

StudynLearn. (2022, February 23). *Hammurabi: Code of Hammurabi Laws - Mesopotamian Civilization.* Retrieved from StudynLearn: https://studynlearn.com/hammurabi#:~:text=He%20is%20best%20known%20for,282%20rules%20or%20legal%20decisions

Sule, P. (2018). The idea of Restributive Justice in African Jurisprudence. *The 2nd African Philosophy World Conference at the University of Calabar, Nigeria, October, 2017.*

Sustainability Directory. (2024). *Indigenous Governance.* Retrieved from Sustainability Directory: https://prism.sustainability-directory.com/term/indigenous-governance/

The Weighing of the Heart Ceremony. (n.d.). Retrieved from Egypt Museum: https://egypt-museum.com/the-weighing-of-the-heart-ceremony/#google_vignette

Tippeconnic, J. W., & Faircloth, S. C. (2010). The Education of Indigenous Students. In P. Peterson, E. Baker, & B. McGaw, *International Encyclopedia of Education (Third Edition)* (pp. 661-666). Elsevier. DOI: 10.1016/B978-0-08-044894-7.00123-8

Tussay, Á. (2024, September 26). Natural law and unwritten law in Classical Greek thought. *Hungarian Journal of Legal Studies, 65*(1), 114–134. DOI: 10.1556/2052.2024.00534

Uth, A. M. (2022). *Maat (deity).* Retrieved from EBSCO Knowledge Advantage: https://www.ebsco.com/research-starters/religion-and-philosophy/maat-deity#full-article

Vansina, J. (1985). *Oral tradition as history.* University of Wisconsin Press.

Walker-Jones, A. (2021, November 27). *Trial, Error, And Success: The Indigenous Origins Of The Scientific Method.* Retrieved July 12, 2025, from Medium: https://medium.com/sofi-stories/trial-error-and-success-indigenous-origins-of-the-scientific-method-d020829afda1

wikipedia. (n.d.). *Code of Lipit-Ishtar.* Retrieved from wikipedia: https://en.wikipedia.org/wiki/Code_of_Lipit-Ishtar#:~:text=The%20Code%20of%20Lipit%2DIshtar,the%20Louvre%20%5BAO%205473%5D)

Chapter 7
Psycho-Social Resilience Amid Political and Economic Upheaval 2020-2024:
The Case of Lebanese Society

Rida Chab
https://orcid.org/0009-0009-6821-5394
Lebanese University, Lebanon

ABSTRACT

This study investigates the role of psychosocial resilience in strengthening the adaptive capacity of Lebanese society in the face of multiple and overlapping crises between 2020 and 2024, including the economic collapse, political instability, the Beirut Port explosion, and the COVID-19 pandemic. A descriptive-analytical methodology was employed to explore these dynamics. The results demonstrate that psychosocial resilience should not be regarded as a static individual trait, but rather as a dynamic and context-dependent process shaped by psychological, social, cultural, and religious factors. The analysis highlights the central role of families, kinship networks, and community-based support systems in fostering collective endurance. At the same time, persistent sectarian divisions, institutional fragility, and rising emigration were identified as significant threats undermining the effectiveness of resilience mechanisms. The study concludes that comprehensive policy frameworks are urgently required to strengthen both social and institutional support structures. Furthermore, scaling up mental health and community-based interventions is essential to enhance society's capacity for sustainable adaptation in the post-crisis context.

DOI: 10.4018/979-8-3373-5651-8.ch007

INTRODUCTION

Since its establishment, Lebanon has rarely been insulated from crises and conflicts. However, the period between 2020 and 2024 marked an unprecedented escalation in the severity and complexity of these crises.

The country experienced a full-scale financial and economic collapse, coupled with persistent political instability, institutional fragility, the catastrophic Beirut Port explosion, and the global COVID-19 pandemic. Collectively, these events intensified the hardships endured by individuals and society at large.

These crises were not confined to material, economic, or security disruptions; they also constituted collective psychological shocks and profound ruptures in social identity and national belonging.

Public trust in state institutions declined sharply, while anxiety, despair, and feelings of alienation spread widely. Migration rates-particularly among the youth-increased significantly, signaling a deeper erosion of collective spirit and posing serious threats to the continuity of Lebanon's social fabric.

Against this backdrop, it becomes imperative to rethink prevailing approaches to resilience and social adaptation in fragile contexts. Particular emphasis should be placed on the concept of *psychosocial resilience*, understood as a critical mechanism for examining how individuals and communities respond to crises, reconstruct their lives, and restore balance under conditions of extreme adversity.

IMPORTANCE OF THE STUDY

The significance of this study lies in its engagement with a highly pressing issue in the current Lebanese context. Specifically, it seeks to:

- Understand mechanisms of community resilience in face of multiple collapses, with a focus on the psychological and social factors that sustain social cohesion and enable positive coping strategies.
- Highlight the non-material dimensions of the Lebanese crisis, which are often marginalized in political and economic analyses—namely, collective mental health, social relationships, the sense of belonging, and the levels of trust between individuals and institutions.
- Provide practical recommendations for policymakers, civil society organizations, and professionals in mental health and social work, aimed at developing psychosocial support programs that strengthen individuals' adaptive capacities and help rebuild the social fabric.

- Address critical questions of identity, belonging, and social trust, which are fundamental to the survival of communities facing collapse and fragmentation.

RESEARCH PROBLEM

Between 2020 and 2024, Lebanon experienced an exceptional and difficult period marked by an unprecedented overlap of political, economic, and social crises.

The country witnessed a severe financial collapse, the catastrophic repercussions of the Beirut Port explosion, a drastic depreciation of the national currency, declining public services, and escalating political and sectarian divisions. The consequences extended beyond material dimensions, significantly affecting individuals' mental health and weakening the traditional social bonds that historically sustained social cohesion. Simultaneously, trust in state institutions eroded, and migration rates, particularly among youth, surged.

In the face of these challenges, the concept of psychosocial resilience has emerged as a crucial framework for understanding how individuals and communities adapt to crises, regain psychological and social balance, and preserve the social fabric under the most difficult circumstances.

Accordingly, this study is guided by the following central research question: How can psychosocial factors contribute to enhancing individual and collective resilience in Lebanese society in the aftermath of the crises spanning 2020–2024, and what obstacles hinder the activation of these factors?

To elaborate on this main question, the study seeks to address the following sub-questions:

- What are the primary direct and indirect psychological and social effects of the crises on individuals, families, and Lebanese society as a whole?
- How have these crises shaped individual coping strategies across different age groups?
- Which psychosocial factors—such as social support, collective identity, and cultural or religious practices—contributed to strengthening resilience and adaptation?
- To what extent have migration, sectarian divisions, and declining trust in formal institutions posed barriers to fostering psychosocial resilience?
- What role can civil society and informal institutions play in enhancing social cohesion amid the shortcomings of the state?

METHODOLOGY

This study employed the descriptive-analytical method, deemed appropriate for examining psychological and social phenomena within the context of complex crises (Creswell, 2014), with a specific focus on Lebanon during the period 2020–2024.

The methodology seeks to describe the nature of the phenomenon, identify influencing factors, and understand the relationships between psychological and social variables and psychosocial resilience. This approach allows for the systematic and accurate collection of data, emphasizing the interplay between psychosocial variables and resilience.

It provides a comprehensive perspective on the impact of compounded crises on individuals and society, thereby enabling the derivation of applicable conclusions and recommendations for the current recovery phase and the post-crisis context.

PSYCHOSOCIAL RESILIENCE IN THE FACE OF CRISES: THEORETICAL FRAMEWORKS AND THE LEBANESE EXPERIENCE

Psychosocial resilience is considered a contemporary concept within the field of social psychology and has attracted significant scholarly attention in recent years. It is regarded as a continuous capacity that develops over time, reflecting how individuals interpret and enact positive behaviors in society when confronted with difficult circumstances. This capacity evolves throughout the individual's lifespan (Masten, 2014; Luthar & Cicchetti, 2021).

The Concept of Psychosocial Resilience

Psychosocial resilience is a multidimensional concept that has gained wide attention in applied and social psychology. Definitions emphasize that resilience extends beyond the individual's mere ability to withstand difficulties, encompassing the capacity for positive adaptation to traumatic events and crises, the restoration of psychological and social balance, and the preservation of social cohesion despite adverse circumstances (Rutter, 2012; Ungar, 2008).

Ungar (2008) defines resilience as a psychological construct that enables individuals to endure and positively adapt to daily stressors. Similarly, Keogh, Garvis, Pendergast, and Diamond (2012) highlight that resilience is not a fixed trait, but a dynamic process manifested through specific coping strategies when facing challenges, thereby facilitating positive psychological responses and adaptation.

Resilience is also described as a phenomenon or process reflecting the ability to endure and withstand stressors or shocks, encompassing two main dimensions: (1) endurance in the face of adversity and (2) positive adaptation to these pressures. Due to its dynamic nature, resilience cannot be directly measured but is inferred through behaviors and outcomes (Luthar & Crossman, 2014, p. 125).

Furthermore, resilience depends on an individual's ability to recognize and effectively mobilize personal strengths in negative circumstances. The interaction between the individual and their environment produces positive outcomes, such as continuous learning, personal growth, and renewal (Al-Sharika, 2025). Contextual and social factors-including social support, collective identity, cultural and religious practices, and decision-making under pressure-also play a crucial role. Research shows that resilience is not static but shaped by historical, social, economic, and political contexts, mitigating the psychological impact of trauma, reducing anxiety and depression, and fostering collective belonging (Brown, 2021).

Resilience encompasses mental and emotional responses that allow individuals to adapt positively to life's challenges (Al-Hamdani, 2018), whether through direct coping strategies or the capacity to change and rely on available solutions. While resilience may be innate in some individuals, it is strengthened when supportive conditions exist, such as meaningful participation opportunities, life-skill acquisition, reinforced social ties, and access to psychosocial support.

Dimensions of Psychosocial Resilience

Psychological Dimension

The psychological dimension refers to the individual's ability to manage stress, regulate emotions, and maintain inner balance during crises (Masten, 2018). It includes cognitive and behavioral strategies that facilitate adaptation, such as positive reappraisal, problem-solving skills, and stress management (Laila, 2015).

Social Dimension

The social dimension centers on the individual's support networks, including family, friends, local communities, and civil society (Hussein, 2020). Social support is critical for strengthening collective resilience, especially in societies facing compounded crises, such as Lebanon between 2020 and 2024. Participation in cultural and religious activities further enhances belonging and reinforces social bonds (Brown, 2021).

Cultural and Religious Dimension

The cultural and religious dimension is a cornerstone of resilience, shaping individuals' and communities' ability to cope with crises and adapt to disruptions (Melhem, 2014). Values and practices rooted in cultural and religious heritage not only guide behavior but also provide normative frameworks for responding to adversity. Participation in religious and communal activities extends beyond spiritual needs, serving as an effective means of psychosocial support by strengthening social ties and offering emotional and moral networks (Caton et al., 2021). This contributes positively to coping with stress and everyday challenges.

The Lebanese Context (2020–2024): A Multi-Dimensional Crisis

Between 2020 and 2024, Lebanon's GDP declined from approximately USD 52 billion in 2019 to USD 23.1 billion in 2021. Per capita income dropped by 36.5% between 2019 and 2021, leading to Lebanon's reclassification as a lower-middle-income country after previously being an upper-middle-income state (World Bank, 2025).

In addition, Lebanon suffered a political paralysis lasting over two years starting in 2022, as parliamentary elections failed to produce a fully empowered government (Council on Foreign Relations, 2025).

The Role of Psychosocial Resilience in Confronting the Lebanese Crises

Under these harsh conditions, psychosocial resilience emerged as a decisive factor in individuals' and communities' ability to adapt and recover. Social support and cultural-religious ties represented the primary sources of strength on which Lebanese people relied to face crises (Rajab, 2025).

A recent Lebanese study supports this perspective, examining the psychological effects of the Beirut Port explosion on healthcare providers. Findings revealed high levels of stress, anxiety, and depression; nevertheless, participants emphasized that social support and community ties significantly strengthened their psychosocial resilience (Helou, 2024).

At the same time, consecutive crises exposed the weakness of social policies and institutional support, heightening pressures on Lebanese society and underscoring the crucial role of civil society. Through local initiatives, Lebanese communities enhanced their adaptive capacities, demonstrating that resilience is not merely an individual ability but a collective process requiring social, cultural, and spiritual support (Al-Jalalah, 2025). These findings reinforce the need for comprehensive strategies involving civil society, strengthened social bonds, and sustained psychosocial support (Abd Al-Sattar, 2015).

Mechanisms and Sources of Psychosocial Resilience in Lebanese Society

The Role of Family and Kinship Networks in Social and Psychological Adaptation

Family and kinship networks play a pivotal role in providing emotional and material support during crises. Research shows that Lebanese youth facing psychological distress often rely on families and close communities to alleviate stress (Maalouf et al., 2020). Family and community support are key in reducing anxiety and depression, as confirmed by recent reviews of youth mental health in Lebanon (UNICEF, 2025).

Challenges Facing the Role of the Lebanese Family During Crises

Despite the family's central role, successive economic and social crises have weakened its effectiveness. Studies indicate that currency collapse and deteriorating living conditions directly undermined families' ability to provide emotional and financial support, negatively affecting cohesion (UNICEF, 2025).

Threats to Psychosocial Resilience in Lebanon

Although psychosocial resilience enables individuals and communities to confront crises, several threats limit its effectiveness in Lebanon. These include sectarian divisions (Zein et al., 2021), weak institutions and declining public trust, polarized media narratives, disrupted social networks due to migration, as well as heightened feelings of helplessness and despair among youth (Deeb & Harb, 2022).

- **Sectarian divisions as a barrier to shared identity:** Sectarianism has hindered the development of a unified national identity, weakening social cohesion. Ethnographic studies show how sectarian divisions permeate daily life in Lebanon, limiting unity (Deeb & Harb, 2022).
- **Weak institutions and loss of public trust:** The weakness of state institutions and the erosion of public trust undermined their capacity to provide psychosocial support. This was evident during the COVID-19 pandemic, when mistrust intensified anxiety and depression among youth (Younes et al., 2021).

- **Polarized media and the reinforcement of despair:** Polarized media discourse exacerbated feelings of despair among Lebanese youth and deepened alienation. A 2022 report revealed that many young people perceived their futures as bleak under such conditions (UNICEF, 2022).
- **Migration and the breakdown of traditional social structures:** Continuous crises triggered waves of internal and external migration, fragmenting traditional social networks and isolating individuals. These demographic shifts weakened social cohesion and limited community resilience (Mikdashi, 2022).
- **Youth despair and helplessness:** Lebanese youth experienced increasing levels of frustration and hopelessness due to worsening political, economic, and social crises between 2020 and 2024, prompting many to consider migration as a coping strategy.

RESEARCH METHODOLOGY AND PROCEDURES

This chapter aims to clarify the methodological design used to collect and analyze data related to psychosocial resilience within Lebanese society, with a focus on the role of the individual and the community in facing the multiple crises between 2020 and 2024. Rigorous procedures were adopted to ensure the collection of reliable and valid data and to analyze them in a way that allows for deriving accurate scientific results.

Research Design

A questionnaire was used to assess psychosocial resilience and its different dimensions, in addition to open-ended questions to gather qualitative data about personal and communal experiences. The research was designed to include an analysis of the relationships between psychological, social, cultural, and religious factors and their impact on individuals' ability to cope with crises.

Research Population and Sample

The research population was chosen to include Lebanese adults who experienced the multiple crises between 2020 and 2024, to ensure that participants had direct experience in dealing with the economic, political, and social pressures Lebanon faced during that period. The focus was on the age group of 25 years and above, to guarantee that participants were at least 20 years old at the onset of the crisis five years earlier, since this group represents individuals mature enough to have a deep understanding of the challenges they faced and their ability to adapt socially and psychologically.

A purposive sampling method was adopted to ensure the inclusion of participants most informed and experienced in the subject, thereby enhancing the reliability of the data and the quality of analysis. The sample consisted of 300 participants distributed across different Lebanese regions, providing geographical and social diversity and allowing for a broader study of the impact of crises on psychosocial resilience.

Data Collection Tools

An electronic questionnaire was used, covering several main areas:

Demographic information: age, gender, educational level, geographic region, marital status.

- **Psychosocial resilience:** assessment of adaptability, emotional regulation, ability to cope with stress, and resilience strategies.
- **Social, cultural and religious support:** the role of family, friends, the local community, participation in cultural and religious activities.
- **Threats and challenges:** sectarian divisions, weak institutions, polarized media, migration, and the sense of helplessness and despair among youth.

The questionnaire also included open-ended questions to gather qualitative data about personal experiences and both individual and collective coping strategies.

Data Collection Procedures

The questionnaire was distributed via Google Forms during September 2025, to ensure easy access for participants and to provide a safe and flexible environment for answering questions. Full confidentiality and voluntary participation were emphasized, and all participants were informed that their responses would be used for scientific research purposes only.

After collecting the data, responses were organized and classified according to demographic variables and psychosocial resilience dimensions. The data was then analyzed using the SPSS software to apply the necessary statistical analyses. Open-ended questions received special attention, with qualitative responses categorized and analyzed to identify common patterns and coping strategies among participants.

Statistical and Qualitative Analysis

The research relied on two complementary approaches:

- **Quantitative analysis:** to describe the distributions and percentages of demographic variables, the level of psychosocial resilience, and its various dimensions.
- **Qualitative analysis:** of the open-ended question responses, in order to understand individual and collective practices and experiences in facing crises, and to extract the common coping strategies within Lebanese society.

Research Reliability and Validity

The research instruments were reviewed by a group of experts in social psychology and sociology to ensure the clarity and accuracy of the questions, and to eliminate any potential bias that could affect the validity of the data. The review process focused on verifying that the questions were formulated objectively, could be easily understood by participants, and adequately covered all dimensions of psychosocial resilience, including psychological, social, cultural, and religious aspects.

Additionally, a pilot test was conducted with a sample of 20 participants before the final distribution of the questionnaire, with the aim of confirming the clarity of the questions. The preliminary results showed that all questions were well understood and that the data collected were consistent and suitable for analysis.

PRESENTATION AND ANALYSIS OF RESULTS

A total of 300 Lebanese adults aged 25 years and above, who experienced the crises between 2020 and 2024, participated in the survey. Their data were analyzed from a psychosocial perspective to assess the level of psychosocial resilience, its sources, and the challenges it faced within the Lebanese context.

Demographic Data of the Sample

Age Distribution

The survey results showed that the most represented age group was 25–30 years, accounting for 37.8% (approximately 113 participants out of 300). This was followed by the 31–40 years group at 27.9% (84 participants), then 41–50 years at 19.8% (59 participants), and finally the 50+ age group at 14.4% (44 participants).

Figure 1. The most represented age group is young and middle-aged

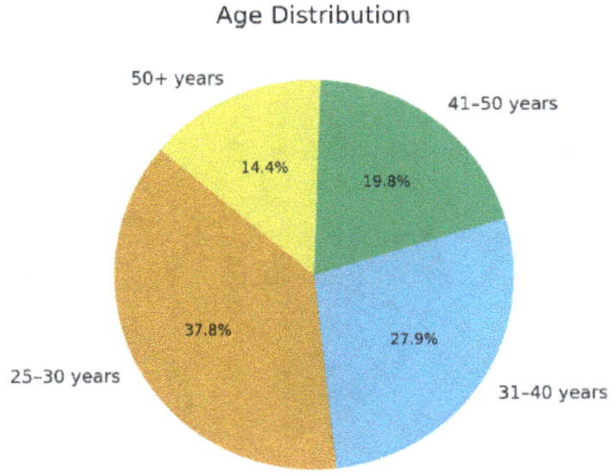

These results indicate that the responding sample mostly consisted of young and middle-aged adults, suggesting a relatively higher capacity to adapt to economic and social changes, while also being exposed to pressures related to professional and family stability. Older age groups had greater life experience and psychological resilience but faced challenges related to health, family responsibilities, and financial burdens.

Gender

18% of participants were male (54 participants), and 82% were female (246 participants).

This distribution suggests that the majority of respondents were women, which may reflect a greater interest in sharing psychosocial coping experiences. From a psychosocial perspective, Arab women in such circumstances often demonstrate high resilience in maintaining family and social ties, while the lower male participation may indicate differences in expressing psychological stress or engagement in surveys.

Educational Level During the Crisis

Most participants held higher educational levels: 144 participants (47.7%) held a bachelor's degree, 105 participants (35.1%) held a master's degree or higher, and 51 participants (17.1%) had secondary education or less.

Figure 2. The majority of participants in the sample hold advanced educational levels

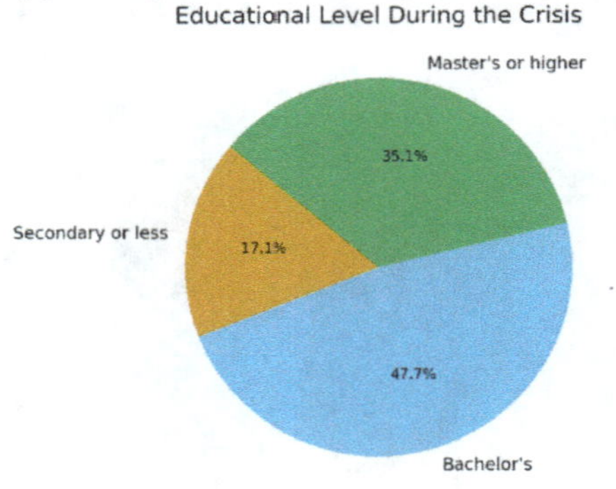

Psychosocially, this distribution indicates that those with bachelor's and master's degrees generally possess stronger critical thinking and problem-solving skills, enhancing their capacity to cope with crises and manage economic and political pressures. Higher education also provides access to broader social and professional support networks, strengthening psychosocial resilience. Participants with lower educational levels may face greater difficulty in handling sudden changes, potentially increasing their anxiety and stress during crises.

Marital Status During the Crisis

The results showed that 64% of participants were married (192 participants), 33.3% were single (100 participants), 3% were divorced (9 participants), and none were widowed.

Figure 3. The majority of participants were married

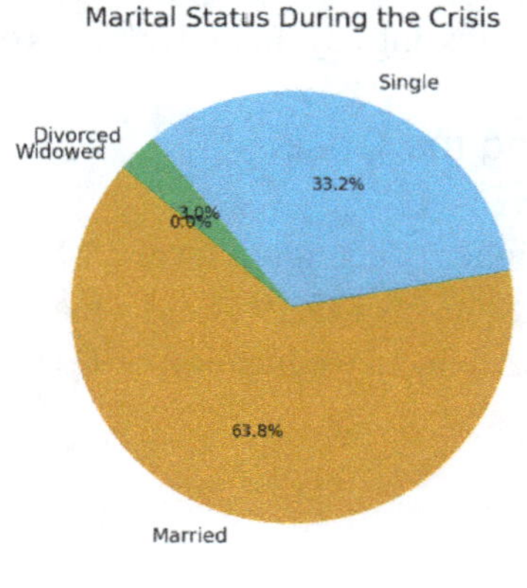

This distribution indicates that most participants had family responsibilities, which could increase psychological and economic pressures during crises but also provide a strong support network to facilitate psychosocial adaptation.

Geographic Distribution During the Crisis

The 300 participants were distributed geographically as follows: South Lebanon (81 participants, 27%), Beirut (62 participants, 20.7%), North Lebanon (49 participants, 16.2%), Mount Lebanon (49 participants, 16.2%), Nabatieh (11 participants, 3.6%), Bekaa (11 participants, 3.6%), Baalbek-Hermel (14 participants, 4.5%), Keserwan and Jbeil (5 participants, 1.8%), and Akkar (19 participants, 6.3%).

Figure 4. The majority of participants belong to densely populated areas directly affected by economic and political crises

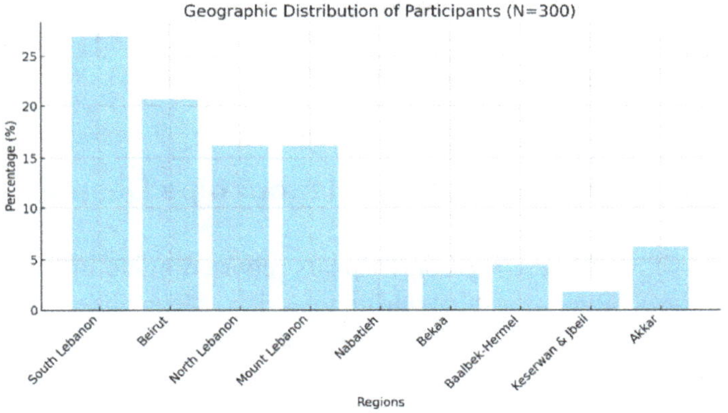

Most participants came from densely populated areas directly affected by ongoing economic and political crises. This geographic concentration reflects variation in psychosocial stress levels. Participation from less populated or geographically dispersed regions suggests the presence of populations facing greater social isolation or limited access to resources and community support, highlighting the importance of family and spiritual ties as coping mechanisms.

Occupational/Economic Status During the Crisis

The sample displayed diverse occupational and economic statuses, reflecting the social structure of Lebanon during crises. The largest group was homemakers (24.1%), indicating the persistence of traditional gender roles.

Students represented 20.5% of the sample, reflecting the challenges faced by university youth balancing academic achievement and economic pressures. This group experiences dual existential anxiety: concerns about educational and professional futures on one hand, and general economic instability on the other, making them more vulnerable to social anxiety and uncertainty.

Full-time employees comprised 22.3% and part-time employees 12.5%, representing groups with relatively stable income but still at risk of job loss or reduced purchasing power due to economic collapse, potentially increasing occupational stress and insecurity.

Unemployed individuals (8.9%) and self-employed professionals (7.1%) faced additional vulnerability due to lack of institutional support and weak social safety nets, making them more reliant on family and kinship networks to compensate for the loss of formal resources.

Maintaining Family and Social Relationships Despite the Crisis

The results showed that 0% of participants reported being unable to maintain their family and social relationships during the crisis. Meanwhile, 22.3% indicated they were able to do so occasionally, 30.4% reported maintaining these relationships most of the time, and 47.3% confirmed they were always able to preserve them.

These data indicate that nearly half of the sample consistently maintained their social and family ties, reflecting the strength of Lebanon's social structure, which is built on family and communal solidarity. From a psychosocial perspective, these relationships were not merely a means of communication but also served as a crucial psychological support mechanism and social safety network, helping individuals cope with economic collapse and political instability.

Coping with Economic and Political Pressures

The results showed that 2.7% of participants reported being unable to cope with economic and political pressures, while 40.2% indicated they were able to do so occasionally, 44.6% stated they could cope most of the time, and 12.5% confirmed they were always able to adapt.

Figure 5. The vast majority of participants demonstrated varying levels of adaptabilityTop of FormBottom of Form

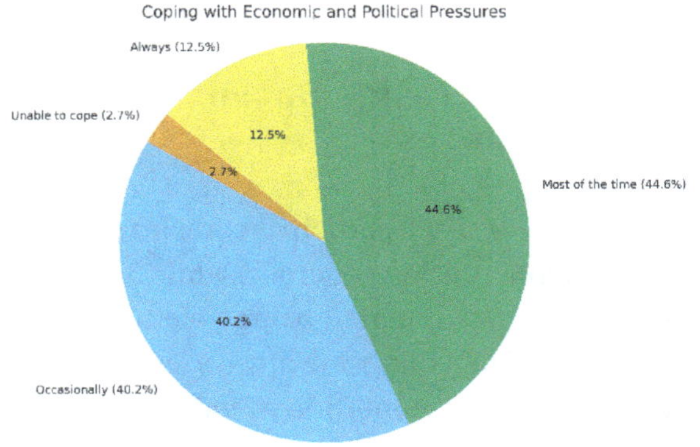

The results showed to cope. These data indicate that the vast majority of participants displayed varying levels of coping ability, whether intermittently or consistently, reflecting an underlying psychosocial resilience that enabled individuals to face pressures despite their severity. However, the low percentage of those able to cope consistently reveals the fragility of this resilience amid the accumulation of Lebanese crises, as most individuals were unable to achieve stable psychological equilibrium.

From a psychosocial perspective, these results reflect the nature of coping as a dynamic process linked to the social and economic resources available. Partial or intermittent coping is often the result of relying on short-term strategies, such as borrowing or family support, which may temporarily alleviate pressures but do not ensure long-term stability. This aligns with resilience literature, which views individuals and communities as moving through phases of adaptation and relapse depending on the intensity of crises and their contexts.

Maintaining Psychological Balance During Difficult Events

The results showed that 12.5% of participants reported being unable to maintain their psychological balance during difficult events, while 35.7% indicated they were able to do so occasionally, 45.5% reported maintaining balance most of the time, and only 6.3% confirmed they were always able to do so.

These data indicate that a large proportion of individuals were largely able to maintain a relative psychological balance, reflecting the Lebanese society's capacity for psychological resilience despite ongoing economic and political pressures. The ability to cope most of the time reflects psychosocial resilience, where family and community ties, alongside social support, help reduce levels of anxiety and psychological fatigue and provide a safety net that enables individuals to withstand successive challenges.

Ability to Help Others or Provide Social Support During the Crisis

The results showed that 14.9% of participants reported feeling unable to help others or provide social support during the crisis, while 36.8% indicated they were able to do so occasionally, 30.7% reported being able to provide support most of the time, and 17.5% confirmed that they were always able to provide social support.

These data indicate that a large proportion of individuals felt a limited or intermittent ability to support others, reflecting the deep psychological impact of the economic and political crises on individuals. This demonstrates that social support was not consistent for everyone, yet it remained a vital mechanism for collective coping, as family and community ties contribute to enhancing psychosocial resilience and reducing isolation and anxiety, even under ongoing pressures.

Experiencing Persistent Anxiety Due to Economic and Political Events

The results showed that 4.4% of participants reported not experiencing persistent anxiety due to economic and political events, while 43.9% indicated they felt anxious sometimes, 26% reported feeling anxious often, and 25.4% confirmed they felt anxious constantly.

These data indicate that a large proportion of individuals (69.3%) experienced anxiety either frequently or constantly, reflecting the deep psychological impact of the economic and political crises on mental health. This anxiety demonstrates a state of collective emotional vulnerability, where individual psychological and social resources become more crucial in facing societal pressures. It also highlights the role of family cohesion and social support in enhancing psychosocial resilience.

Experiencing Despair or Loss of Hope for the Future

The results showed that 97 participants (32.5%) did not feel despair, while 121 participants (40.4%) reported feeling it sometimes. Additionally, 66 participants (21.9%) felt despair often, and 16 participants (5.2%) experienced constant despair.

Figure 6. The majority of the samples reported feeling despair or a loss of hope

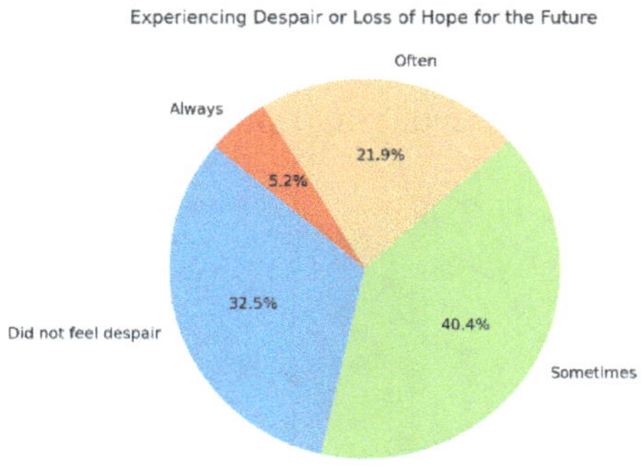

These findings indicate that the majority of the sample (62.3%) experienced varying degrees of despair or loss of hope, with "sometimes" being the most common response. This reflects a fluctuating sense of psychological distress during the crisis. The distribution highlights the fragility of collective emotions, as individuals face ongoing psychological pressures due to political and social divisions and economic crises, making family and local community support critical in enhancing psychosocial resilience and coping with difficult conditions.

Negative Impact of Crises on Professional or Academic Life

The results indicate that the majority of participants (72.9%) experienced a negative impact of the crises on their professional or academic life to varying degrees. The response "sometimes," at 38.6%, emerged as the most common, reflecting a fluctuating sense of instability and temporary psychological stress.

This can be interpreted as a result of the overlap of crises with social and community relationships, where economic and political pressures affect professional connections and collaboration within society, limiting individuals' ability to adapt effectively. The variation in responses between "often" (26.3%) and "always" (7.0%) indicates individual differences in psychosocial resilience, which may be linked to family support, past experiences, or available resources for coping with challenges.

This variation reflects the dynamic nature of Lebanese society in facing crises, where professional and academic life is affected not only by direct pressures but also by changes in the surrounding social and cultural environment.

Feeling Psychological Pressure Due to Sectarian or Political Divisions

The survey included **300 participants** regarding their experience of psychological pressure resulting from sectarian or political divisions. The distribution of responses was as follows:

- **Never**: 33.6%
- **Sometimes**: 39.7%
- **Often**: 18.1%
- **Always**: 8.6%

These results indicate that the majority of participants (65.8%) experienced psychological pressure to some degree, with the most common response being "Sometimes" at 39.7%. The combined percentage of "Often" and "Always" responses (26.7%) suggests that a significant portion of participants faces persistent difficulties in coping with social and political stressors.

Figure 7. The majority of participants experienced psychological pressure due to sectarian divisions

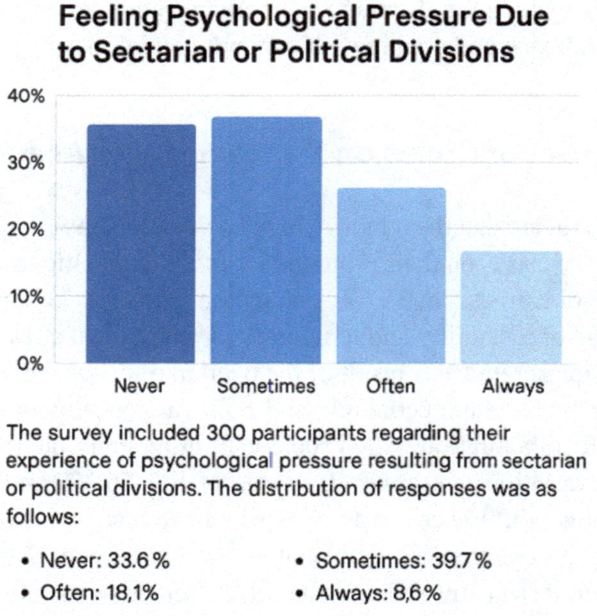

Relying on Family or Social Support to Face Difficulties

Never: 23.3%, **Sometimes**: 35.1%, **Often**: 27.4%, **Always**: 14.7%. The results show that most participants (77.2%) relied on family or social support to varying degrees, with "sometimes" at 35.1% being the most common response. This indicates intermittent or non-continuous use of social support.

Figure 8. Large segments of the sample frequently rely on family support

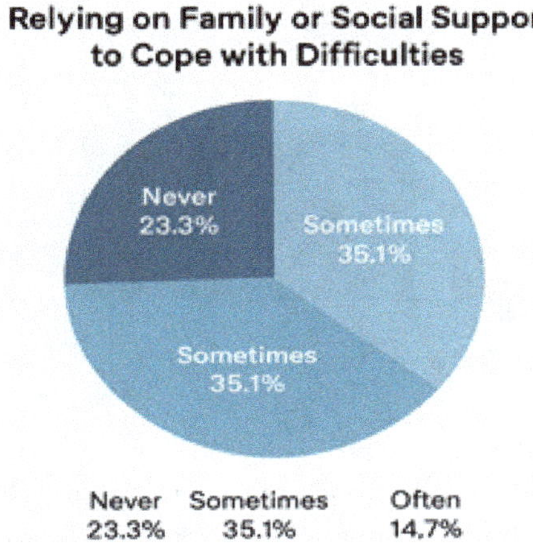

This emphasizes the pivotal role of family and social networks in enhancing coping capacity during crises, as family and social support reduce psychological pressures and strengthen emotional stability. The combined percentage of "often" and "always" (42.1%) suggests that large segments of the sample frequently rely on social support, reflecting the community's ability to protect its members from the negative effects of crises and highlighting the importance of social relationships as a psychosocial safety network.

Using Religious or Spiritual Practices to Reduce Psychological Stress

Never: 6.6%, **Sometimes**: 18.2%, **Often**: 26.6%, **Always**: 48.8%. The results indicate that more than half of the participants (75.4% combined) frequently rely on religious or spiritual practices to alleviate psychological stress ("often" and "always"). This reflects the central role of the spiritual and religious dimension as a psychosocial coping mechanism during crises.

Figure 9. It reflects the central role of the spiritual and religious dimension as a psychological and social coping mechanism in facing crises.

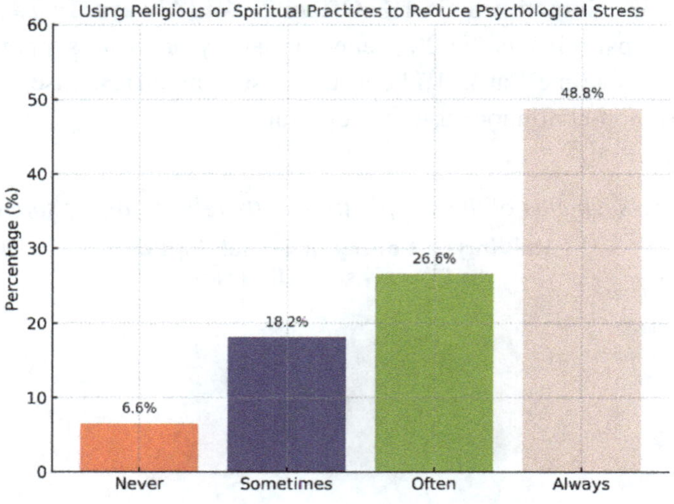

From a psychosocial perspective, religion and spiritual practices act as an informal support network, enhancing inner security and providing a value-based framework that helps individuals manage anxiety and stress. They also contribute to community cohesion through participation in rituals or collective practices that strengthen social bonds.

Participation in Group or Civic Activities to Support the Local Community

Never: 44.8%, **Sometimes**: 34.4%, **Often**: 13.5%, **Always**: 7.3%. From a social-psychological perspective, the results suggest that civic engagement and community interaction among the participants are relatively limited. The largest portion (44.8%) did not participate in group or civic activities, reflecting a tendency toward relative social withdrawal or low motivation for collective initiatives. This does not necessarily indicate a lack of belonging but may reflect social, cultural, and economic factors that influence opportunities for civic participation.

Figure 10. Civic behavior and community engagement among the sample individuals are relatively limited

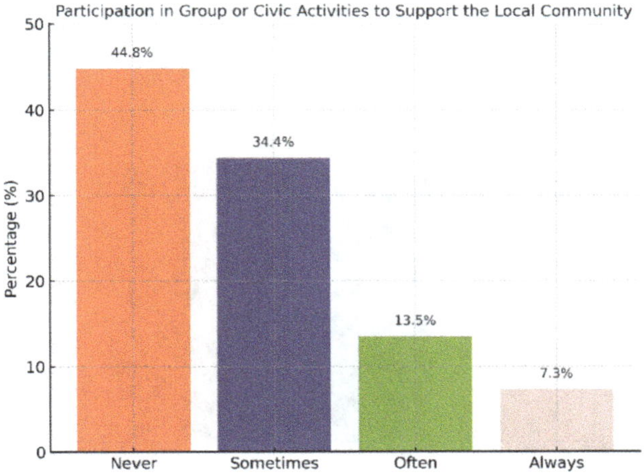

Participants who engaged "sometimes" (34.4%) represent a group that oscillates between involvement and non-involvement, reflecting limited social flexibility and the ability to participate when circumstances or motivation allow. This may indicate an attempt to balance personal commitments with community needs, with a sense of responsibility that is not fully ingrained.

The smaller groups (13.5% "often" and 7.3% "always") represent socially active citizens who demonstrate consistent commitment to community engagement, reflecting higher psychosocial resilience and a stronger sense of social responsibility and collective belonging. This group often plays a role in motivating others to engage in civic activities and forms a nucleus for positive social change.

Seeking Personal Solutions to Cope with the Economic and Living Conditions

The results indicate that 12.3% of participants did not seek personal solutions, while 25.4% did so occasionally, 34.2% often, and 28.1% always sought personal strategies to cope with the economic and living conditions.

Figure 11. The majority of individuals frequently take the initiative to seek adaptive strategies

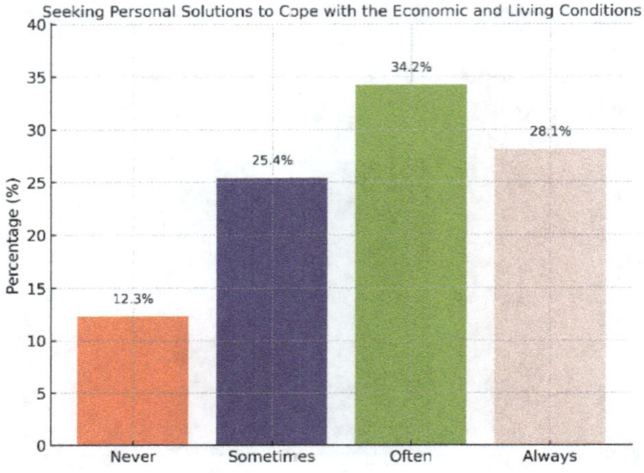

This shows that the majority of individuals (62.3%) frequently take the initiative to find adaptive strategies, reflecting their psychosocial resilience and their ability to independently manage economic pressures.

Feeling that Civil Society and Informal Institution Supporting You

The results indicate that the majority of participants (53.5%) did not feel that civil society or informal institutions provided support, while 32.5% felt this occasionally, 12.3% often, and 2% always. This reflects the limited effectiveness of civil and informal support networks in enhancing individuals' social and economic adaptation, possibly due to lack of trust or limited resources, making reliance on social interaction and community cooperation less impactful in coping with pressures.

Using Communication Tools or Technology to Maintain Social Bonds or Reduce Stress

A large portion of participants reported using communication tools or technology occasionally (36%) or often (26.3%) to maintain social connections or relieve stress, while 22.8% always used them and 14.9% never used them. This reflects the growing role of technology as a means of social communication and psychological support, highlighting the importance of digital networks in enhancing psychosocial resilience and reducing isolation under economic and living pressures.

Figure 12. It reflects an increasing reliance on technology as a means of social communication and psychological support

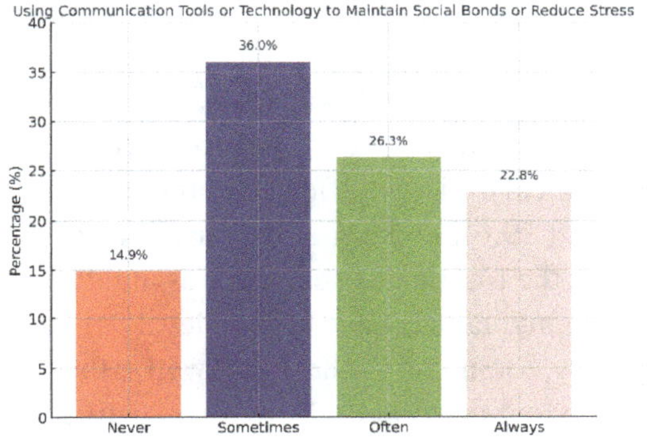

Feeling that Considering Emigration Affected Your Sense of Social Security

Results indicate that 43% of participants did not feel affected, 29.8% felt sometimes affected, and 27.2% felt often or always affected. This shows that thinking about emigration destabilizes the sense of social security for about a quarter of the sample regularly, reflecting psychological and social tension linked to instability and uncertainty about the future. For the others, the lack of impact may reflect temporary feelings of security or the ability to psychologically and socially adapt to difficult conditions.

Observing that Loss of Trust in Official Institutions Hindered Your Ability to Adapt

The results showed that 27% did not feel any impact: 29% sometimes, 26% often, and 18% always. This indicates that a significant proportion (44%) were negatively affected by the loss of trust in institutions. This result reflects a tangible social trust crisis, as weak trust in official institutions leads to feelings of instability and limited capacity to cope with economic and social pressures. It highlights the vital role of the institutional dimension in psychosocial well-being, as the presence of trustworthy institutions contributes to enhancing resilience and adaptability among individuals and society.

Feeling that Sectarian or Political Divisions Affected Community Cohesion Around You

The results showed the following indications:

Sometimes: 33.6%, **Often**: 25.6%, **Always**: 11.2% and **Never**: 29.6%. The results indicate that more than one-third of participants (36.8% from "Often" and "Always") felt that sectarian or political divisions affected community cohesion, while 33.6% felt this occasionally and 29.6% did not suffer any impact. From a psychosocial perspective, this reflects the fragility of social bonds amid social and political conflicts, as divisions erode mutual trust and reduce levels of social solidarity among individuals. Consequently, community cohesion as well as the capacity of its members to cooperate and provide mutual support decrease, negatively impacting collective mental health and increasing feelings of insecurity and psychosocial stress.

Figure 13. The majority of the sample felt that sectarian or political divisions affected community cohesion

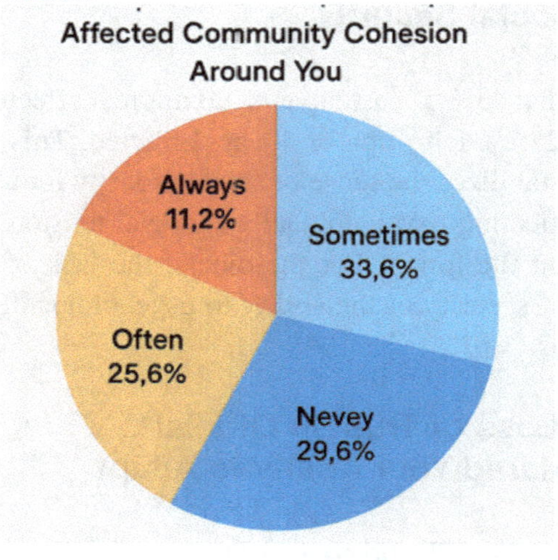

Feeling that Cultural or Social Factors Hindered the Enhancement of Individual and Collective Resilience

The results showed the following indications:

Sometimes: 42.2%, **Often**: 24.4%, **Always**: 7.1% and **Never**: 26.3%.

Results show that a large portion of participants (73.7%—sometimes, often, or always) felt that cultural or social factors hindered the enhancement of individual and collective resilience, compared to 26.3% who did not feel this. This indicates that Lebanon's cultural and social environment negatively affected psychosocial adaptation, as social norms, expectations, and pressures influence individuals' ability to develop flexible strategies to face crises.

Figure 14. The vast majority felt that cultural or social factors hindered the enhancement of individual and collective resilience

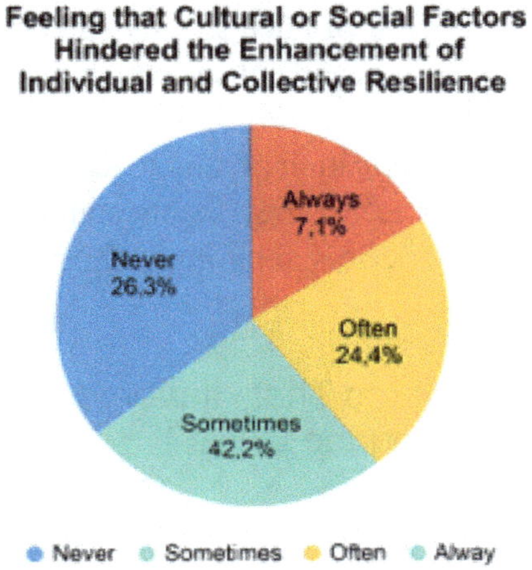

Open-Ended Question

What Were the Main Factors That Helped You Adapt During the Crisis?

From the review of 300 responses, the main themes identified include:

- **Family and social support:** Participants clearly emphasized that family and friends were the main sources of psychological and social support. Many highlighted the importance of having loved ones nearby and maintaining contact with relatives and friends to aid adaptation. This shows the importance of strong social ties in enhancing psychological resilience.
- **Religious and spiritual dimension:** Many participants cited faith, prayer, and religious practices as key coping mechanisms. This suggests that psychological resilience is strengthened through spirituality, providing inner control, hope, and reducing anxiety and stress during crises.

- **Personal resources and skills:** Participants mentioned skills like time management, self-development, continuous learning, acquiring new skills, financial prudence, and extra work as coping strategies. This indicates that personal competencies and self-management help in facing crises, consistent with the psychosocial resilience concept.
- **Positive thinking and psychological flexibility:** Many participants reported the ability to accept reality, exercise patience, focus on positive aspects, and avoid dwelling on transient problems. This reflects cognitive coping strategies that help mitigate stress and enhance resilience.
- **Use of technology and modern tools:** Some participants used phones or the internet for social communication or remote learning as a coping mechanism, highlighting the role of technology in maintaining social connections and reducing isolation during crises, reinforcing the social aspect of psychosocial resilience.

What Were the Main Factors That Hindered Your Ability to Adapt During the Crisis?

From the review of 300 responses, the main obstacles identified include:

- **Economic and financial factors:** Participants highlighted lack of liquidity, weak economic resources, and inflexible economic systems as primary obstacles. Economic crises directly affect security and future planning, limiting psychosocial resilience.
- **Political and institutional factors:** Failures in the political system, corruption, and lack of trust in official institutions were major obstacles. From a social-psychological perspective, lack of institutional trust increases feelings of helplessness and reduces individuals' capacity to cope.
- **Social and personal factors:** Some participants noted unsupportive or unfamiliar social environments, as well as betrayal by close contacts, as obstacles. This indicates that a strong social network is crucial for psychological adaptation; disruption increases stress.
- **Psychological factors:** Some cited excessive anxiety, overthinking, or general psychological tension as barriers. This highlights the importance of mental health and emotional regulation in fostering individual and collective resilience.

CONCLUSIONS AND RECOMMENDATIONS

This chapter aims to present the main conclusions drawn from the analysis of the survey data, which included 300 Lebanese adult participants (aged 25 and above) who experienced a series of crises between 2020 and 2024. In addition, it provides practical and strategic recommendations to enhance psychosocial resilience at both the individual and community levels.

Conclusions

Level of Psychosocial Resilience

The results showed that the majority of participants possess a moderate to high capacity for psychological and social adaptation to crisis-related pressures. This indicates that Lebanese society has inherent sources of adaptation rooted in cultural and social traditions, or through the collective psychological immune system that has adapted to a series of successive crises.

It was also found that the older age group (35 years and above) exhibited higher resilience compared to the younger group, reflecting the role of life experience and maturity in enhancing the ability to cope with increasing and multiple pressures.

These findings suggest that psychosocial resilience in Lebanon is not limited to individual capacities but is deeply rooted in social networks, family support, and community traditions that serve as protective mechanisms against psychological stress.

Sources of Psychological and Social Support

- **Family and kinship networks:** These are the primary pillars enabling individuals to withstand crises. Such networks provide continuous emotional and material support, enhancing individuals' ability to face daily challenges.
- **Religious and community institutions:** They played a central role in fostering community belonging and providing a moral framework that motivates individuals to persevere and maintain optimism.
- **Friends and local community:** They helped provide an alternative support network, especially for individuals with weak family ties, highlighting the importance of informal social interaction as an effective tool for psychosocial resilience.

Challenges Facing Resilience

- **Economic and political pressures:** These negatively affected individuals' ability to cope, particularly in the most affected areas, where economic hardship compounds psychological challenges.
- **Disconnection from support networks:** Weak social ties reduce adaptive capacity and increase vulnerability to sustained psychological stress.
- **Chronic psychological strain:** Repeated crises have depleted individuals' psychological resources, reducing their capacity for effective coping and necessitating targeted interventions to enhance psychological resilience.

Psychosocial Implications

- The results emphasize the reciprocal relationship between the individual and society. Individuals do not exhibit true resilience except within a supportive social environment, reflecting what social psychologists describe as collective or social resilience.
- The findings also indicate that psychosocial resilience is not merely about enduring pressures; it encompasses the ability for personal growth, learning from experiences, and rebuilding oneself after crises, which strengthens the capacity to face the future with greater confidence.

Recommendations

At the Individual Level

- Raise awareness of the importance of psychosocial resilience and apply effective coping strategies, such as self-regulation, realistic optimism, and problem-solving skills.
- Encourage individuals to engage in social and cultural activities to strengthen a sense of belonging and mutual support, which serves as an effective preventive measure against the psychological impacts of future crises.
- Adopt specialized training programs to enhance psychological skills, including stress management, emotional intelligence, and relaxation techniques.

At the Family and Community Level

- Support families and kinship networks through community programs that promote solidarity and cooperation among family members and the local community.

- Establish integrated psychosocial support centers offering counseling and assistance services, with a focus on communities most affected by economic and political crises.
- Promote religious, cultural, and social initiatives that support individuals' psychological and spiritual resilience, while encouraging innovative forms of community interaction to enhance belonging and a sense of safety.

At the Political and Economic Level

- Implement public policies aimed at alleviating economic and social pressures on citizens, such as direct financial support programs, provision of sustainable employment opportunities, and reduction of living burdens.
- Strengthen political stability so individuals and communities can plan for the future without continuous psychological strain, while encouraging community participation in decision-making to enhance a sense of control and self-efficacy.
- Support ongoing community research to understand the dynamics of psychosocial resilience and develop evidence-based intervention programs to strengthen psychological resilience during various crises.

Conclusion

The findings of this study indicate that psychosocial resilience in the Lebanese society goes beyond a mere individual capacity to cope with stress. Rather, it emerges as a multidimensional phenomenon that result from the dynamic interplay between individual, familial, and community factors, within a complex and fluctuating economic and political context.

These results underscore that understanding psychosocial resilience requires a comprehensive perspective that sheds light on the interconnections among psychological, social, cultural and political dimensions, rather than simplifying it as a purely individual skill. The study further highlights the importance of designing comprehensive intervention programs that target all levels of support—from the individual to the family and beyond to the broader community—while taking into account local contextual conditions.

Adopting such integrated strategies can enhance the Lebanese citizens' ability to cope with ongoing pressures and sudden crises more effectively and sustainably. It also contributes to building a society with greater social cohesion and an increased capacity for recovery and growth following adverse events.

In light of these findings, psychosocial resilience can serve as a central tool in shaping mental and psychological health policies as well as social development programs. It extends beyond individual coping to include fostering a supportive social environment, strengthening family and community relationships, as well as developing effective mechanisms to address economic and political challenges. This integrated approach positions psychosocial resilience as a strategic enabling factor that enhances the sustainability of societies, including Lebanese society, in the face of future crises.

REFERENCES

•Abd Al-Sattar. Rasha. (2015). Adaptation Resilience Scale (Psychological Resilience). Cairo: Anglo-Egyptian Library.

•Al-Hamdani. Rabi'a. (2018). Psychological Resilience and Emotional Dryness: A Contemporary Educational-Psychological Perspective. Amman: Al-Warraq Publishing and Distribution.

Al Jalalah, B. (2025). *Psychological Resilience: Ideas for Coping with Stress and Pressure to Achieve Success and Live Happily*. Dar Tashkeel for Publishing and Distribution.

Al-Sharika, M. (2025). *Social and Emotional Learning*. Arab Center for Educational Research for Gulf States.

Brown, T. (2021). *Resilience in times of crisis: A psychosocial perspective*. Routledge.

Caton, C. L.. (2021). Culture, religion and resilience in crisis contexts. *The Journal of Social Issues*, 356–372.

•Council on Foreign Relations. (2025). Lebanon's political crisis: Timeline and analysis. https://www.cfr.org

Creswell, J. W. (2014). *Research design: Qualitative, quantitative, and mixed methods approaches* (4th ed.). Sage.

Deeb, L., & Harb, M. (2022). Politics of sectarianism in Lebanon. Journal of Ethnographic Studies, 211–229.

Helou, J. (2024). Psychological impact of Beirut port explosion on Lebanese health workers. Lebanese Journal of Mental Health, 45–60.

Hussein, R. (2020). Social support networks in Arab societies. Arab Social Science Review, 77–95.

Keogh, J., Garvis, S., Pendergast, D., & Diamond, P. (2012). *Early childhood resilience: Processes, contexts, and outcomes*. Springer.

Laila, A. (2015). *Social Theory and Societal Issues: Mechanisms of Social Cohesion*. Anglo-Egyptian Library.

Luthar, S. S., & Cicchetti, D. (2021). Resilience in development: Progress and future directions. *Development and Psychopathology*, 341–355.

Luthar, S. S., & Crossman, E. J. (2014). Resilience: A multidimensional construct. In Cooper, M. J. (Ed.), *Handbook of developmental psychopathology* (pp. 125–140). Springer. DOI: 10.1007/978-1-4614-9608-3_7

Maalouf, F. T.. (2020). Youth mental health in Lebanon: Challenges and opportunities. *Frontiers in Psychiatry*, *11*, 987.

Masten, A. S. (2014). *Ordinary magic: Resilience in development*. Guilford Press.

Masten, A. S. (2018). Resilience theory and research on children and families: Past, present, and promise. *Journal of Family Theory & Review*, *10*(1), 12–31. DOI: 10.1111/jftr.12255

Melhem, S. (2014). *Developmental Psychology: Human Life Cycle*. Dar Al-Fikr for Publishing and Distribution.

Mikdashi, M. (2022). Sectarianism and migration in Lebanon. *The Middle East Journal*, •••, 543–561.

Odeh, J. (2021). *Psychological Resilience*. Guardian.

Rajab, A. (2025). *Self-Management Skills and Psychological Resilience: How to Overcome Stress and Challenges?* Azha Publishing.

Rutter, M. (2012). Resilience as a dynamic concept. *Development and Psychopathology*, *24*(2), 335–344. DOI: 10.1017/S0954579412000028 PMID: 22559117

Ungar, M. (2008). Resilience across cultures. *British Journal of Social Work*, •••, 218–235.

UNICEF. (2022). Youth in Lebanon: Aspirations and challenges. UNICEF Regional Report.

UNICEF. (2025). Mental health and resilience among Lebanese youth. UNICEF Country Office Report.

World Bank. (2025). *Lebanon economic monitor: Collapse and recovery*. World Bank Group.

Younes, L.. (2021). COVID-19, trust, and mental health in Lebanon. *Global Health Journal (Amsterdam, Netherlands)*, 301–310.

Zein, L.. (2021). Sectarian divisions and youth identity in Lebanon. *Social Identities*, •••, 811–828.

Chapter 8
Developing Entrepreneurial Skills among Lebanese University Students:
A Business Management Perspective

Hani Mohammad Haidoura
https://orcid.org/0000-0001-9164-2196
Université de Technologie et de Sciences Appliquées Libano-Française, Lebanon

ABSTRACT

The elements that affect entrepreneurial skill development in Lebanese university students are evaluated in a research study. Four important variables are used to study in this research, that are Entrepreneurship Education Quality (EEQ), Institutional Support for Entrepreneurship (ISE), Socioeconomic Factors (SEF), and Personal Characteristics and Mindset (PC). The major obstacle for Lebanese students is caused by the economic instability, though, these students believe diverse concepts will be developed in the end. For universities, programs should expand their mentorship programs and students should be allowed resource access points as well as educational approaches that build on technical knowledge alongside training of entrepreneurial personality.

DOI: 10.4018/979-8-3373-5651-8.ch008

INTRODUCTION

The Study Background

Entrepreneurship is now an essential element of the modern business world, and we acknowledge that it works as an important economic expansion mechanism, which is currently growing innovation in the present. The current economic situation in Lebanon works as a catalyst that reveals students must acquire entrepreneurial competencies at universities. The Lebanese economy faces a serious unemployment challenge that affects young people, which prompts the need to establish entrepreneurship as an occupational option. These vital functions of entrepreneurial abilities development help people gain personal empowerment and advance both the economic development and social progress of the country.

Technology-based economies use entrepreneurship as an important driving force recognized globally by the entire population. The creation of entrepreneurship drives economic employment growth because it allows businesses to maximize their resources while fostering innovative operational changes. According to Jia et al. (2021), entrepreneurial Education enables students to secure improved job prospects and increase the odds of launching their businesses. The importance of teaching entrepreneurship skills to students in every academic institution has become universally acknowledged because developing nations like Lebanon require inventive answers to economic disturbances.

Colleges in Lebanon should become fundamental institutions for crafting a worldwide method of entrepreneurship. Global business success requires Education from these establishments that teach students practical and basic knowledge. Technology infusion across higher Education supports entrepreneurial teaching by providing students with digital success skills, as reported by Hamzeh, Mershad and Vetokhin (2019).

The Study Problem

Increased entrepreneurship importance creates challenges for Lebanese universities in instructing entrepreneurial abilities to their student body. Students encounter obstacles that stem from the internal factors of educational institutions, together with economic surroundings found outside the institutions. International business education in Lebanon fails to establish competencies for developing both entrepreneurial thinking skills and business competencies, according to Bahhouth et al. (2012). The majority of university graduates fail to gain essential business leadership abilities that will enable them to start profitable ventures.

Multiple obstructions stand in the way of Lebanese university students who want to become entrepreneurs. Students face three main obstacles when pursuing entrepreneurship: they have restricted funding options along with insufficient mentorship for new business ventures together with scarce resources to cultivate their business concepts. The unstable and corrupted socio-political environment in Lebanon works as a strong deterrent for individuals looking to start a business. The business sector encounters limited access to opportunities because of economic development obstacles that stem from corruption, according to Fahed-Sreih (2023).

The present educational techniques fall short because they do not provide enough practical learning opportunities for entrepreneurship education. Theoretical Education remains essential, yet it fails to provide sufficient practical skills that business founders need to start and maintain firms. According to Sarouphim (2015), educational practices in Lebanon have changed slowly, which leads to a general decline in entrepreneurial skills among the population. Educational institutions should make a fundamental change from their current teaching methods to adopt experiential learning models.

Research Aim

This research examines the essential variables that affect Lebanese university students' acquisition of entrepreneurial skills through analysis of both educational experiences and institutional backing and sociological along with individual characteristics. This study examines how different factors affect student entrepreneurial capabilities while surveying possible methods to boost entrepreneurial Education for Lebanese higher education institutions.

Research Objectives

- To assess the quality of entrepreneurship education provided by Lebanese universities and its impact on developing entrepreneurial skills among students.
- To evaluate the role of institutional support (such as mentorship, networking, and resources) in fostering entrepreneurial skills in Lebanese university students.
- To analyze the influence of socioeconomic factors, including economic instability and access to capital, on Lebanese university students' entrepreneurial intentions and skill development.

- To explore how personal characteristics and entrepreneurial mindset (e.g., creativity, risk-taking, and proactivity) contribute to developing entrepreneurial skills in Lebanese students.

Research Questions

- How does the quality of entrepreneurship education in Lebanese universities impact the development of entrepreneurial skills among students?
- What role does institutional support (mentorship, resources, networking) play in developing entrepreneurial skills in Lebanese university students?
- How do socioeconomic factors, such as economic instability and limited access to capital, affect Lebanese university students' entrepreneurial intentions and skill development?
- In what ways do personal characteristics and mindset (e.g., creativity, risk tolerance, proactivity) influence the entrepreneurial skill development of Lebanese students?
- What improvements can be made to entrepreneurship education in Lebanese universities to better equip students with the skills needed for entrepreneurial success?

Significance of Research

The study demonstrates the importance to stakeholders comprising academic institutions and policymakers along with students. The research offers universities in Lebanon essential knowledge about their existing entrepreneurship education programs, allowing them to develop better educational approaches. Pruett and Şeşen (2017) establish that the quality of entrepreneurship education heavily depends on how faculty and students perceive each other. Research of these student perceptions enables universities to develop academic programs that meet the students' educational requirements and professional goals. The research informs state officials about the complex aspects of entrepreneurship development and available opportunities in the nation. The study offers data that policy developers can utilize to build programs endorsing entrepreneurial business formation and promoting better business conditions while lowering obstacles for entrepreneur entry. According to Zheng (2022), "Internet + maker education" represents a modern educational approach for enhancing student entrepreneurial abilities through a new method of business education.

The research benefits students by showing them which abilities and competencies should be mastered to be successful in entrepreneurial business activities. The study delivers information about how students can face their obstacles while identifying what resources they need to get past them. The improvement of students' entrepreneurial competencies will aid them in pursuing their business goals effectively and boost economic growth in Lebanon. This study adds detailed knowledge to what is already known about Lebanon's entrepreneurship education framework. The research provides comprehensive data on the keys that shape entrepreneurial skill development for university students, together with applicable solutions for system improvement toward promoting entrepreneurship in the country.

LITERATURE REVIEW

Theoretical Literature

Theory of Entrepreneurial Intentions

The Theory of Planned Behavior (TPB) delivers important information about entrepreneurial intentions after its application in Lebanese student groups. Entrepreneurial business initiation requires three core elements because of studies conducted by Kautonen et al. (2015) and Širola (2020) and subsequent investigations. The psychological components presented by Ali et al. provide important elements to build an extensive predictive model for student entrepreneurial intentions (Ali et al., 2019; Self-belief in entrepreneurial capabilities combined with self-confidence represents the core characteristics of personal attitudes that promote entrepreneurial intention, according to Miralles et al. According to Miralles et al., self-knowledge regarding entrepreneurship enables students to establish necessary behavioural control systems for developing their intentions (Miralles et al., 2015). The element of personal attitude plays a vital role in entrepreneurial success, and Education should use experiential learning alongside practical entrepreneurial activities to enhance these attitudes.

Entrepreneurial intentions receive their formation from both internal viewpoints and essential external factors that span socioeconomic realms to institutional backing. In their view, Biraglia and Kadile demonstrate that entrepreneurial passion stands as an essential internal force for motivating, but external environmental elements create a substantial impact on entrepreneurial intentions (Biraglia & Kadile, 2016). Political and economic instabilities in Lebanon pose significant barriers to Lebanese students who wish to become entrepreneurs, according to Bahhouth et al. (2012). Numerous external obstacles in higher education can be addressed by a strategic entrepreneurial curriculum that teaches real-life entrepreneurial scenarios and helps students believe that entrepreneurship is feasible (Mejjah et al., 2021).

Lebanese universities should develop their curriculum with lessons and skills teaching which follow the TPB framework. The findings from Rauch and Hulsink reveal that entrepreneurship education shapes genuine entrepreneurial conduct because students gain new business venture initiation intentions directly through their educational encounters (Rauch & Hulsink, 2015). Programming that let students solve practical issues while receiving mentoring support should help students become more aware and motivated about entrepreneurial activities.

Human Capital Theory and Skill Development

The Human Capital Theory establishes that education investments, as well as training, develop individual professional capabilities which generate better employment results for entrepreneurs (Liñán & Fayolle, 2015). Education focused on human capital development remains crucial for Lebanese university students because it helps create a link between classroom learning and real-world entrepreneurship practice. The theory demonstrates how entrepreneurs need to develop both technical and soft skills to reach business success. A variety of studies regarding entrepreneurial success demonstrate creativity with leadership alongside resilience forms fundamental skills (Ali et al., 2019; Esfandiar et al., 2019).

Schools need to develop spaces which allow students to learn essential entrepreneurial skills. Schools must combine their educational activities with external learning opportunities which give students direct practical experiences. Entrepreneurship courses need to incorporate simulations and applied projects, which force students to exhibit their abilities to solve problems and innovate, according to Mejjah et al. (2021). According to Ebewo and Shambare, entrepreneurial leadership shows potential to boost student intentions after COVID-19, so universities should modify their programs to address contemporary worldwide issues (Ebewo & Shambare, 2023).

Students' human capital development receives its greatest value from entrepreneurial Education. According to Zampetakis et al., student learning outcomes experience transformation through educational programs, which leads them to move from being passive learners towards entrepreneur status (Zampetakis et al., 2015). Curriculum adjustments focusing on entrepreneurial mindset education alongside practical learning experiences will empower Lebanese university students to handle entrepreneurial starting challenges.

Entrepreneurship Education Quality (EEQ) in Lebanese Universities

Higher-quality entrepreneur teaching at Lebanese universities functions as a core element that helps students gain entrepreneurial success requirements. A curriculum that combines theoretical concepts with real-world applications must be properly designed to achieve success. According to Liu, there should exist an efficient evaluation framework for entrepreneurship education because it leads to improved innovation levels combined with stronger student involvement in the field Liu (2015). The assessment framework of entrepreneurship education in Lebanon faces two main obstacles because it lacks standard evaluation systems alongside consistent educational methods. Saeed et al. establish that students; entrepreneurial intentions are affected by their perceptions of university support. Students demonstrate higher entrepreneurial intentions when their institutions provide resources and practical Education in combination with mentoring programs (Saeed et al., 2018). The data shows that both theoretical Education about entrepreneurship and practical business experience need emphasis to develop entrepreneurial competencies in students. According to Tlaiss students' diverse gender backgrounds changes their understanding of entrepreneurship along with their hopes to become entrepreneurs since quality education must recognize each group's unique Lebanese challenges (Tlaiss, 2022).

Institutional Support for Entrepreneurship (ISE) in Lebanon

The basis of Lebanese entrepreneurship growth derives from institutional support structures. Student entrepreneurial intention together with their outcomes depend strongly upon the operational environment surrounding them. University and government institutions need to create support for the development of entrepreneurial culture when economic and political environments remain unstable as Hill et al. (2018) explain. Such support includes provision of funding as well as mentorship services with network opportunities. Incubators of business play a necessary role by fulfilling this purpose. The platform created by university-business incubators at Hassan allows students and entrepreneurs to connect for resource sharing with mentorship service that enhances their entrepreneurial development capabilities (Hassan, 2020). Under the support system of incubator facilities in Lebanon students can build their business ideas in an enabling environment which counteracts limited traditional employment sector opportunities. Institutions must implement extended-term backing programs which will help achieve these targets. According to Nauffal and Skulte-Ouaiss (2018) studies reveal that enhanced higher education standards directly enhance performance in student employment markets and entrepreneurial ventures in Lebanon. Educational institutions need to build framework systems that adjust to be more effective for helping new entrepreneurs achieve better results.

Socioeconomic Factors (SEF) and Entrepreneurship among Lebanese Students

Socioeconomic conditions determine the entrepreneurial atmosphere which Lebanese students experience. The economic instability together with high joblessness and political disturbances simultaneously create entry barriers and motivational elements for business ventures among Lebanese students. Lebanese female entrepreneurs face severe conditions according to Kuran and Khabbaz (2024) as these challenges result from socioeconomic elements needing customized support systems. These elements generate varying entrepreneurial outcomes when personal characteristics combine with market situation elements. The research by Chakra and Jardali demonstrates how familial networking and communal support influences entrepreneurial targets in Lebanon since social and cultural aspects significantly influence business expansion in that country (Chakra & Jardali, 2022). Learning success depends heavily on educational institutions creating curricula which understand and address student socioeconomic realities together with their facing obstacles. Reliable access to funding resources remains out of reach for numerous students who want to start their businesses. Several young entrepreneurs in Lebanon encounter substantial challenges when trying to obtain both start-up funding and financial services as per Hill et al. (2018). Better financial training programs that merge with microfinance services will enhance the entrepreneurial environment by overcoming these economic barriers.

Personal Characteristics and Mindset (PC) in Entrepreneurial Skill Development

Individual personality types together with mental outlook serve as fundamental factors that affect student development of entrepreneurial competencies. The development of entrepreneurial intentions among students is fundamentally guided by three traits including self-efficacy and proactive personality along with resilience. Numerous studies reveal that students' entrepreneurial self-efficacy acts as a connecting factor which explains why entrepreneurship education shapes their business startup intentions. According to Hou et al. students' entrepreneurial intentions are influenced by Education which enhances their self-efficacy through essential entrepreneurial knowledge and skills Hou et al. (2019). Results verify that educational institutions must establish programs to increase student confidence in their entrepreneurial skills.

The formation of entrepreneurial intentions by students heavily depends on their proactive behavioral traits. Luo et al. (2022) studied the relationship between proactive personality and social and human capital which affects entrepreneurial intentions showing that proactive individuals tend toward entrepreneurial pursuits. Educational programs need to actively develop proactive mentalities among students based on this observed relationship between personal traits and entrepreneurial intentions. Students within the Lebanese economy will benefit from proactive personality development because it helps them create innovative solutions and start ventures even when facing economic limitations.

Global Perspectives on Entrepreneurial Education

The global educational models for entrepreneurial development create essential knowledge which Lebanon should apply to its educational system. The demand for international prestige within entrepreneurial programs keeps rising according to Zadravec who suggests diverse business and cultural environments enable students to build innovative skills and adaptability (Zadravec, 2021). The view supports the notion that world-oriented entrepreneurship education leads to increased student participation and superior market-readiness for interconnected environments.

A global approach to entrepreneurship education enables Lebanese students to study highly successful entrepreneurial environments from worldwide locations. Students need to learn multiple entrepreneurial approaches and understand how cultural differences affect an individual's entrepreneurial goals. The inclusion of international perspectives within Lebanese university curriculums enables better training for students who wish to establish multinational entrepreneurial operations and builds an expansive innovative environment.

The development of technological systems in entrepreneurship education represents an international educational direction that Lebanese academic institutions should utilize. The research by Santika et al. demonstrates how institutions should use digital platforms to develop student entrepreneurial capabilities which promotes network building for entrepreneurial business success. Santika et al. fail to explicitly state the precise connection between their research and entrepreneurial intention enhancement (Santika et al., 2022). A technology-based educational structure that bases its foundation on technology enables the development of adaptability and creativity assets which modern entrepreneurial environments require.

Conceptual Framework

Figure 1. Conceptual framework

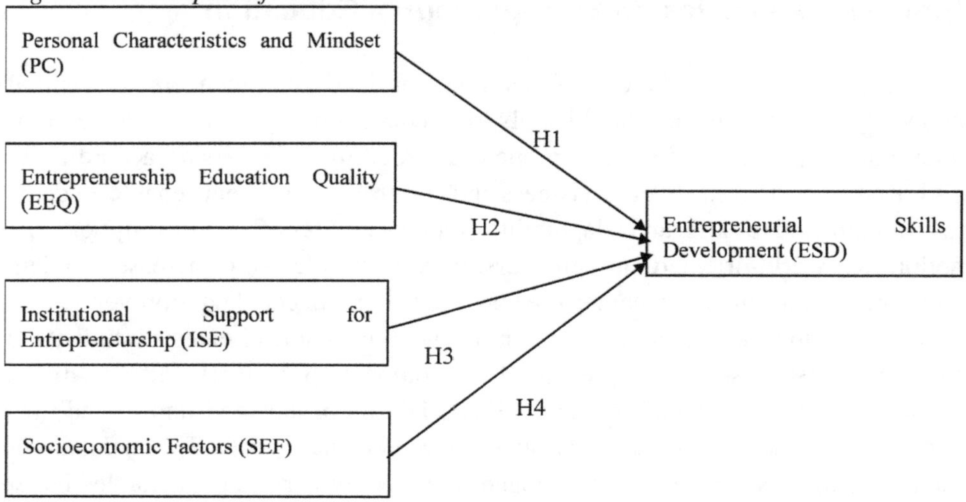

H1=Entrepreneurship Education Quality (EEQ) has a positive significant relationship with Entrepreneurial Skills Development (ESD)

H2=Institutional Support for Entrepreneurship (ISE) has a positive significant relationship with Entrepreneurial Skills Development (ESD)

H3=Socio-Economic Factors (SEF) has a positive significant relationship with Entrepreneurial Skills Development (ESD)

H4=Personal Characteristics and Mindset (PC) has a positive significant relationship with Entrepreneurial Skills Development (ESD)

Research Gap

The exploration of factors affecting entrepreneurial intentions through existing research fails to address specific obstacles which Lebanese university students experience in their development of entrepreneurial skills. Research focusing on Lebanese populations is scarce even though Rauch and Hulsink (2015) and Saeed et al. (2018) have noted the significance of entrepreneurial Education for skill development. Bahhouth et al. (2012) and Tlaiss (2022) produced research about Lebanese socio-political challenges, but their findings lack connections between these circumstances and university entrepreneurship education effectiveness.

Research about applying the Theory of Planned Behavior (TPB) to analyze entrepreneurial intentions among Lebanese university students remains scarce despite its widespread use in entrepreneurial intention research (Kautonen et al., 2015; Miralles et al., 2015). Entrepreneurial intentions in Lebanon receive recognition concerning socioeconomic impacts including economic disruption and funding restrictions (Hill et al., 2018) but still need further evaluation at the tertiary education level.

The global research about self-efficacy and proactive personality affecting entrepreneurial intentions (Wang et al., 2015; Luo et al., 2022) fails to appear sufficiently in Lebanese educational environments. The examination of these factors needs deeper investigation within Lebanon's distinct environment because it represents a substantial absence in existing scholarly works.

RESEARCH METHODOLOGY

The research methodology adopts mixed-methods to assess how entrepreneurial skills develop in Lebanese university students. Through this study design quantitative survey results merge with interview-generated qualitative information (Arab, 2018). The research employs various investigation techniques to obtain complete knowledge of the elements driving entrepreneurial skill growth in this particular environment.

Research Paradigm

This research is based on interpretivist research paradigm for the reason that it studies participant meanings and social contextual experiences and interpretations. Because of the ability to interpret complex dynamic elements, the interpretivism becomes ideal to measure entrepreneurial competences in Lebanese university students (Fahed-Sreih et al., 2009). This research methodology enables researchers to gain knowledge on how students view their entrepreneurial works and their experiences in academic settings and economic environment.

Initially, the research acknowledges the role of positivism for the collection of numerical data by quantitative methods of detecting the linkage between the analytical variables (Ghannage, 2018). When they use a mixed methods approach, researchers are given complete understanding since they combine numerical data with personal interpretation that allows them to validate their findings using this two-way.

Research Approach

The research framework in this investigation is the mixing of the qualitative and quantitative methods since they permit techniques of data collection and analysis to be merged into a unified approach. Mixed methods research generates optimal research conditions for the research of complex phenomena through combination of qualitative data analysis with quantitative investigation on important elements like educational quality, institutional support, economic factors and participant quality (Haj Youssef et al., 2025).

As regards the quantitative part of the study, they use survey methodology to measure a very large amount of Lebanese university students, while the qualitative part of the study utilizes the use of semi structured interviews. Qualitative interview data may be used with the mixed methods research design to help detect major relationship patterns and understand reasons underlying, L. Alexandre (2025). The combination method is a means of obtaining comprehensive knowledge about the impact of multiple related elements in entrepreneurial skills formation.

Research Design

The study uses a descriptive correlational methodology with case study elements to perform its quantitative research alongside qualitative research segments through case study methods. A descriptive correlational study traces relationships that exist between the variables entrepreneurship education quality, institutional support, socioeconomic factors, personal characteristics and entrepreneurial skill development (Lichy et al., 2020). The design system establishes measurement methods to determine the connection level between the study variables.

The qualitative study phase uses a case study methodology to conduct detailed assessments that supplement the findings obtained from statistical measurements (Saleh, 2025). A specific study method uses in-depth conversations with a limited number of participants to obtain detailed explanatory evidence about individual and organizational elements that impact entrepreneurial skill training.

Data Collection

The study depends on survey questionnaires and semi-structured interviews for data collection. The research survey distributed to 100 Lebanese university students seeks information about entrepreneurship education quality along with institutional support and socioeconomic background and personal characteristics (Toufaily and Bou Zakhem, 2024). The survey contains Likert scale and demographic sections. The qualitative section consists of five detailed interviews which grant researchers the ability to investigate participant-specific challenges and experiences. Together these methods obtain survey-based generalizable data and interview-based deep insights for explaining the research outcomes.

Data Analysis

The research analysis combines two sequential steps based on mixed-methods methodologies. Survey data gets summarized using descriptive statistics while regression analysis with correlation tests check the linkage between independent variables (EEQ, ISE, SEF, PC) and dependent variable (ESD) through inferential statistics. The data analysis method for qualitative information entails thematic analysis of semi-structured interview transcripts where identified patterns and themes receive emphasis (Youssef et al., 2024). Research objectives obtain deeper insights about entrepreneurial skills development factors using NVivo software to code and link themes detected in the data.

Sampling and Population

Lebanese undergraduate and graduate students enrolled in colleges make up the target sample for this research. A combination of stratified random sampling will ensure that the studied students come from different academic departments across multiple Lebanese universities while representing diverse economic and social backgrounds of the entire student demographic. students majoring in various academic disciplines at open and private institutions throughout Lebanon will make up the sampling frame to present research results that accurately reflect the entire student group (YOUSSEF, 2021). The quantitative analysis includes 100 participants as a research sample because this number enables us to obtain reliable statistical correlations between variables. Research data collection will proceed through personal distribution combined with online systems depending on participant place of residence. An exploration with 5 qualified students who possess entrepreneurial interests will serve the qualitative component of the research. The participants will be chosen according to their responses to the screening question in the survey after validating their alignment with entrepreneurial intent (Saleh, 2025). The limited number of participants in the interview section enables researchers to deeply study each participant's personal experiences.

Ethical Consideration

The rights and welfare of research participants depend on these ethical considerations, which need to be present throughout every step of the research process. The study foundation on informed consent provides essential elements about the research goal along with data acquisition procedures and participant rights. Participants need to review and agree to a written consent form to demonstrate their free consent to research participation. All participant information remains secret under a system that conceals their names and all identifying details. Each interview participant uses a false name to hide their identity (Ghannage, 2018). The right to withdraw from studying remains accessible to participants at all times without fear of adverse outcomes. In contrast, the study establishes complete disclosure at its beginning and reconfirms this right during data collection. Secure data protection occurs through password-protected storage of all collected information which includes both survey answers and interview recordings. All research data access lies with team members who utilize it only for their research activities (Arab, 2018). The authors obtain necessary review board approval to conduct data collection while keeping the review board standards in mind. The researcher adopts these standards to preserve both ethical standards and participant protection during all phases of research.

RESULTS AND DISCUSSION

Quantitative Results

Coding and Missing Values

The researchers implemented proper coding procedures to guarantee precise analysis of the dataset. The research team used appropriate techniques to handle missing data by applying mean imputation for continuous variables. By using these techniques researchers successfully protected the validity of their data collection while controlling missing data effects throughout their research. All necessary data points remained present in the results set.

Outliers

A box plot examination of Entrepreneurial Skills Development (ESD) shows no data points deviated from the norm in the database. All values display uniform distribution within the expected data range which confirms that extreme values are absent in the dataset. These data points support the observation that the dataset remains intact.

Figure 2. Outliers

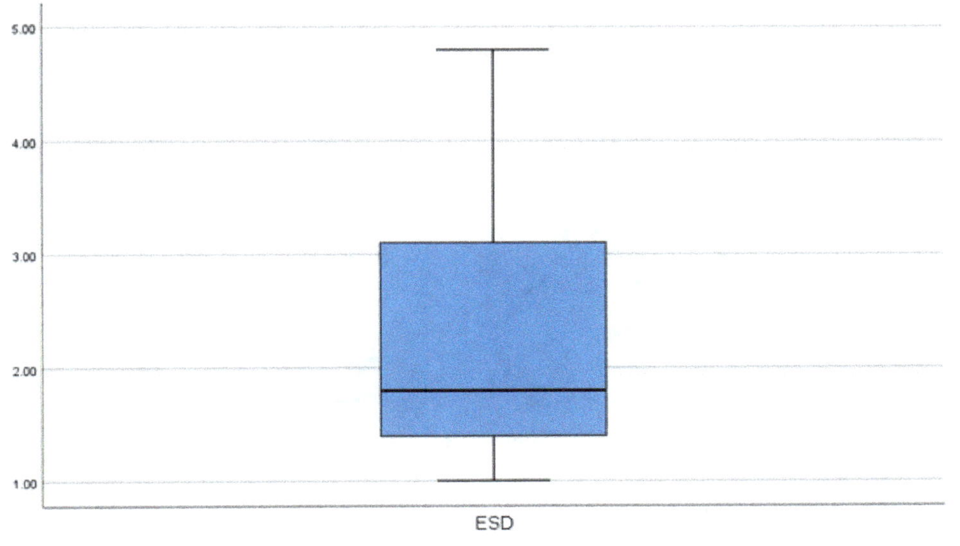

Normality Plot

The normality plot shows that Entrepreneurial Skills Development (ESD) regression standardized residuals comply with near-normal distribution patterns. The bell-shaped histogram for the data is symmetric and the P-P plot maintains an alignment with the diagonal line which indicates that the data follows a normal distribution. The regression results remain reliable since the analysis meets its assumptions regarding normality distribution.

Figure 3. Normality plot

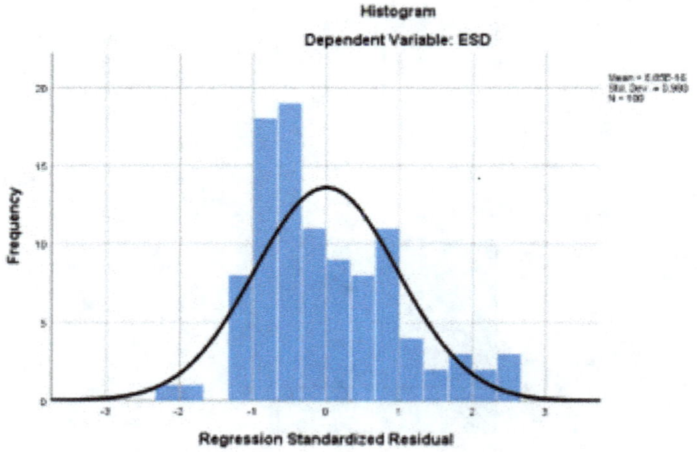

Figure 4. Normality Plot

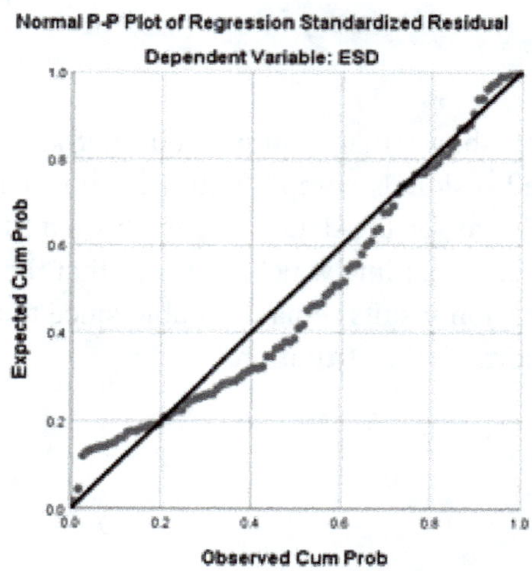

Demographic Analysis

The study participants follow an even distribution between male and female students making up 60% and 37% of the total respondents respectively. The survey participants primarily belong to the postgraduate level and make up 59% of the sample while students in their undergraduate years comprise 27% of the respondents and graduate students represent 14%. The participant sample consists mostly of students who study Business/Management at a rate of 49%. Most students have selected majors in Arts and Humanities (29%) or Science (15%) apart from Engineering (7%) among their other fields of study. The initial sample data shows strong representation among academic disciplines while Business/Management constitutes the dominant group likely to deliver vital information about entrepreneurship education. The research benefits from gender and academic profile diversity, which improves study generalization and reflects the complete student demographic of Lebanon.

Table 1. Demographic analysis

Demographics		N	%
Gender	Female	37	37
	Male	60	60
	Prefer not to say	3	3
What is your current level of study	Graduate	14	14
	Postgraduate	59	59
	Undergraduate	27	27
What is your field of study?	Arts and Humanities	29	29
	Business/Management	49	49
	Engineering	7	7
	Science	15	15

Correlation Analysis

Data from this research shows that strong, significant connections exist between the independent factors affecting Entrepreneurial Skills Development (ESD). The independent variables EEQ and SEF show the strongest direct influence on ESD based on Pearson correlation coefficients of 0.489 and 0.528 ($p < 0.01$). Institutional Support for Entrepreneurship (ISE) and Personal Characteristics and Mindset (PC) also show significant positive correlations with ESD ($r = 0.466$ and $r = 0.606$, respectively). Students develop their entrepreneurial skills most strongly from their education quality combined with their social economic conditions and institutional backing while being shaped by their identity. The relationship between EEQ and ESD holds according to Saeed et al. (2018) since quality educational systems directly develop entrepreneurial intentions and skills thus affirming educational frameworks' essential contribution to skill emergence.

Table 2. Correlation analysis

Correlations						
		EEQ	ISE	SEF	PC	ESD
EEQ	Pearson Correlation	1	.942**	.948**	.834**	.489**
	Sig. (2-tailed)		.000	.000	.000	.000
	N	100	100	100	100	100
ISE	Pearson Correlation	.942**	1	.916**	.809**	.466**
	Sig. (2-tailed)	.000		.000	.000	.000
	N	100	100	100	100	100
SEF	Pearson Correlation	.948**	.916**	1	.838**	.528**
	Sig. (2-tailed)	.000	.000		.000	.000
	N	100	100	100	100	100
PC	Pearson Correlation	.834**	.809**	.838**	1	.606**
	Sig. (2-tailed)	.000	.000	.000		.000
	N	100	100	100	100	100
ESD	Pearson Correlation	.489**	.466**	.528**	.606**	1
	Sig. (2-tailed)	.000	.000	.000	.000	
	N	100	100	100	100	100

**. Correlation is significant at the 0.01 level (2-tailed).

Descriptive Statistics

Survey data analysis demonstrates that the measurements follow a symmetric distribution since most of the variables maintain values within recommended skewness and kurtosis values. The measurements of EEQ, SEF and PC display modest skewness together with kurtosis effects that demonstrate students respond to a wide range of the rating scale regarding their experiences and perspectives. The students' responses to EEQ1 (where questions gauge entrepreneurial education quality) show 1.440 skewness with 1.878 kurtosis which implies that students across multiple viewpoints assess the educational quality. The data matches previous studies by Rauch and Hulsink (2015) regarding the inconsistent student satisfaction levels with their entrepreneurship education. The descriptive statistics indicate that participants hold different perspectives about how Education and socioeconomic variables influence their entrepreneurial capabilities thus enhancing the analysis and comprehension of collected data.

Table 3. Descriptive statistics

Descriptive Statistics					
	Skewness	Kurtosis		Skewness	Kurtosis
EEQ1	1.440	1.878	SEF4	1.290	2.150
EEQ2	1.466	1.726	SEF5	1.281	1.462
EEQ3	1.156	1.381	PC1	1.428	1.709
EEQ4	1.481	1.483	PC2	1.217	0.898
EEQ5	1.041	0.026	PC3	1.755	2.921
ISE1	1.268	1.236	PC4	1.545	2.739
ISE2	1.290	2.150	PC5	1.340	2.058
ISE3	1.440	1.878	ESD1	1.413	1.756
ISE4	1.466	1.726	ESD2	0.934	-0.383
ISE5	1.156	1.381	ESD3	0.605	-0.884
SEF1	1.481	1.483	ESD4	0.916	-0.199
SEF2	1.041	0.026	ESD5	0.558	-1.261
SEF3	1.268	1.236	Valid N (listwise)		

Regression Analysis

The analysis data confirms that all proposed hypotheses lead to meaningful correlations between the independent variables and their measurement of Entrepreneurial Skills Development (ESD). Research results using H1 (Entrepreneurship Education Quality) validate its positive connection to ESD with Beta = 0.489 and $R^2 = 0.239$ and p = 0.000. H2 (Institutional Support for Entrepreneurship, Beta = 0.466, $R^2 = 0.217$, p = 0.000) reveals the importance of institutional support. Socioeconomic Factors have a significant impact on entrepreneurial skills based on the findings presented in H3 where Beta = 0.528 with $R^2 = 0.279$ and p = 0.000. Data indicates that personal characteristics along with mindset (H4) demonstrate the maximum impact on ESD with Beta = 0.606 and $R^2 = 0.367$ at p = 0.000. The study reveals that all relationships contribute to positive development of entrepreneurial skills thereby demonstrating the significance of these defining factors.

Table 4. Regression analysis

Hypotheses	Beta	R square	Sig	Result
H1 = Entrepreneurship Education Quality (EEQ) has a positive and significant relationship with Entrepreneurial Skills Development (ESD)	0.489	0.239	0.000	Supported
H2 = Institutional Support for Entrepreneurship (ISE) has a positive and significant relationship with Entrepreneurial Skills Development (ESD)	0.466	0.217	0.000	Supported
H3 = Socioeconomic Factors (SEF) have a positive significant relationship with Entrepreneurial Skills Development (ESD)	0.528	0.279	0.000	Supported
H4 = Personal Characteristics and Mindset (PC) has a positive significant relationship with Entrepreneurial Skills Development (ESD)	0.606	0.367	0.000	Supported

Reliability Analysis

All variables in the reliability analysis showed strong internal consistency according to the results. The reliability statistics indicate EEQ with 0.848 and ESD at 0.854 and SEF with 0.827 PC at 0.831 and ISE with 0.819. The survey items display reliability through values that exceed the standard threshold of 0.7 for construct measurement. The high consistency values from the reliability analysis demonstrate that collected data shows trustworthy patterns which establishes the study results as valid for future analytical purposes.

Table 5. Reliability analysis

Variable	Items	Cronbach's Alpha
EEQ	5	0.848
ESD	5	0.854
SEF	5	0.827
PC	5	0.831
ISE	5	0.819

Qualitative Results

Theme 1: Entrepreneurial Mindset

The educational courses about entrepreneurship played a substantial role in shaping how these students address business problems. A respondent explained that course learning frameworks proved useful in solving actual business problems, as per Rauch and Hulsink (2015). Their theory highlights how entrepreneurship education creates critical thinking alongside problem-solving competencies. The students demonstrate real-world business strategy application via critical thinking education as presented in this student's response: "The courses focused on critical thinking which made it easier to solve complex business problems."

Students want the curriculum to devote increased attention to practical theory implementation. According to a student survey participants academic knowledge about business strategy development fell short because they needed more training on concrete implementation methods. Research by Miralles et al. (2015) supports the necessity of practical implementation methods in entrepreneurship education in order to improve student capabilities in business execution. Student responses demonstrate that both case-based learning examples and real examples of business scenarios have enhanced their entrepreneurial mindset because practical educational approaches matter for developing correct entrepreneurial behaviour (Saeed et al., 2018).

Theme 2: Barriers to Entrepreneurial Success

Owing to persistent political instability as well as economic turbulence, student entrepreneurship initiatives encounter high barriers in Lebanon. Economic instability also requires the respondent to choose both caution and inspiration to solve the creative entrepreneurial needs and problems. It shows how economic instability has the contradictory effect on entrepreneurial drive. Fahed-Sreih (2023) states that political instability in Lebanon brings about uncertainty that is prevalent in Lebanon which usually stifles new entrepreneurial development. A cause for discouragement during that startup phase was that one student explained that there are hurdles that businesses go through when looking for funding because of economic challenges. Research studies that involve such developing, challenging financial and economic limited countries which are examples of Lebanon as they are unable to generate entrepreneurial intentions (Hill et al., 2018).

Yet economic challenges are not enough to prevent students from becoming entrepreneurs as it is the economic adversity that gives them creative possibility to counter it. The entrepreneurs who search for distinctive business offers during periods of uncertainty are said to have problem solving capabilities and creative thinking abilities during periods of environmental challenges as stated by Kuran and Khabbaz (2024). Economic disruption forms entrepreneurial innovation, but the instability also creates a lot number of development challenges, rendering business startup support and resource acquisition cumbersome.

University mentorship programs became fundamental components that assist students in eliminating business startup obstacles. This participant stated that mentorship helps, but the program lacks enough individual interactions to meet all start-up business needs. According to Pruett and Şeşen (2017), mentorship availability coupled with quality enhances the development of entrepreneurial skills. The study participants demanded personalized mentorship from established business owners who could provide practical guidance for their needs. The findings establish an enduring necessity to deliver mentoring programs containing genuine professional tools for student development.

Theme 3: Institutional Support and Resources

The development of startup activities needs institutional backing, yet different institutions exhibit diverse resource availability and operational efficiency. One respondent mentioned, "The university provides enough resources, but they lack the structure to be fully effective," highlighting a mismatch between available resources and their actual utility. Hassan (2020) supports the perspective that Lebanese universities present resources, although their inadequate infrastructure and strategic alignment hinder their ability to support student entrepreneurship development. Similarly, another respondent observed, "There are a good number of resources, but the funding options are limited," which points to a major challenge in the Lebanese entrepreneurial ecosystem: the lack of adequate financial support for young entrepreneurs. The restricted opportunities for funding stand as a major impediment because they prevent students from making their business concepts become operational businesses (Kuran & Khabbaz, 2024).

In terms of mentorship, the responses indicate a generally positive view, with one student stating, "The mentorship programs are supportive, and the mentors provide valuable insights." The resources shared regarding mentorship programs show that students want structured mentorship opportunities along with individual attention in those programs. According to Saeed et al. (2018), mentorship provides value to students, though this value grows when the mentor and student need to align specifically. Students indicated that better practical learning avenues, such as internships combined with real-world projects, should be introduced to their Education. One respondent stated, "The practical learning opportunities help bridge the gap between theory and practice," indicating that experiential learning is an essential component of entrepreneurship education. The research findings by Liu (2015) show that hands-on learning must appear in entrepreneurship curricula because this method trains students to succeed in real business environments.

The survey evidence indicates institutional support proves essential for entrepreneurship development, yet resources, mentoring programs, and first-hand business training lack adequate methods of delivery. The development of vital aspects within Lebanese universities would produce substantial improvements to their entrepreneurial ecosystem.

Discussion

An analysis is provided of what entrepreneurship education, plus institutional headquarters, economic circumstances and personal attributes play in the acquisition of eloping entrepreneurial competency among Lebanese university students. Data analysis shows that high quality of Entrepreneurship Education Quality (EEQ) will have a positive and meaningful effect of Entrepreneurial Skills Development (ESD) that will make the students equip attributes of critical thinking, as well as the business strategy formulation skills and practical problem solving competencies (Rauch & Hulsink, 2015). As indicated by Saeed et al. (2018) who looked at how mentoring helps entrepreneurial intentions, institutional support activities for entrepreneurship (ISE) through mentoring and access to resources benefits students greatly. An entrepreneurial problem framed from the account of economic turbulence in Lebanon, and in the same time, it has its own possibilities becoming entrepreneurial. Initially, the students saw economic challenges as obstacles that prevented them from following the path they had chosen, however, these challenges pushed several students toward coming up with innovative solutions. The people, just like Kuran and Khabbaz (2024) found during their research, usually solve creative and entrepreneurial problems in times of challenge. Researchers Wang et al. (2015) have confirmed among various factors which have an impact on the entrepreneurial success, personal characteristics such as self-efficacy and creativity as well as resilience do have the strongest impact. The economic adversity brought students drove them towards entrepreneurial activities and the proactivity of students. Educational institutions must then come up with means of developing both technical competencies and fundamental traits to achieving entrepreneurial success for comprehensive student learning outcome.

CONCLUSION

Summary of Results

Research has investigated how EEQ, ISE, SEF, and PC contribute to Lebanese university students' development of entrepreneurial skills. Seventeen university students contributed quantitative data to explain how all four independent variables create positive effects on Entrepreneurial Skills Development (ESD). The analysis established that EEQ, ISE, SEF, and PC possess influential relationships but PC produced the highest effect size (Beta = 0.606, p = 0.000). The findings demonstrate how education standards combined with institutional backing from the socioeconomic elements alongside individual student characteristics jointly develop their entrepreneurial competencies. The reliability analysis showed positive results since Cronbach's alpha exceeded 0.7, indicating that the survey data was reliable and consistent. The descriptive statistics provided additional details about student experiences in entrepreneurship education by showing that numerous students wanted to experience more hands-on learning activities. The regression data analysis passed validation by normality plots because the plots showed that the data point distribution was normal. The correlation analysis showed a high dependency between EEQ and ESD ($r = 0.489$) and also between PC and ESD ($r = 0.606$), which strengthens the importance of formal Education and personal traits in the development of entrepreneurial capabilities.

Practical Implications

Various practical advantages from the study help educational groups as well as Lebanon's government sector and upcoming entrepreneurs through their development process. University institutions must make educational quality improvements for entrepreneurship training. According to the assessment, students need to experience genuine business education by implementing internship partnerships, realistic business assignments, and case studies to develop their practical entrepreneurial skills (Saeed et al., 2018). Students highlighted the need for practical business direction that helps them manage businesses effectively under Lebanon's intricate socioeconomic system since traditional Education does not provide this essential capability. The study demonstrates the importance of institutional support by providing mentoring services in addition to funding options and entrepreneurial relationships that help develop student entrepreneurship. Pruett and Şeşen (2017) recommend universities should create improved mentoring services alongside increased resource availability to improve their entrepreneurial ecosystem. The economic difficulties of Lebanon, along with funding limitations, need sustained substantial backing from outside parties according to the input of multiple research subjects.

According to this research study, individual personality traits combined with an entrepreneurial way of thinking help people reach their entrepreneurial success goals. Students possessing high self-efficacy and creativity combined with resilience factors will develop enhanced entrepreneurial incentives that lead to success (Wang et al., 2015). Academic institutions need to establish educational initiatives that train essential personality traits for student development of self-assurance and determination when converting academic concepts into practical business operations.

Limitations and Recommendations for Future Research

Additional research in this field must overcome multiple constraints that emerged from this study. The study faces a crucial limitation due to its small student sample of 100, which might make it difficult to apply the research results beyond the studied group. The development of a broader population, including pupils enrolled at multiple schools in different regions or nations of Lebanon beyond the current sample, would establish a more thorough knowledge of what shapes entrepreneurial abilities. The analysis depends on self-reported information because social desirability effects and problems with recall accuracy might influence these types of data. Future investigators need to integrate performance assessments along with extended tracking methods which monitor student entrepreneurial advancement throughout periods.

More extensive studies should explore how personal traits match external elements, including family aspects and previous entrepreneurial activities. These factors have remained unstudied in the present research project. Research should focus on evaluating particular programs dedicated to boosting entrepreneurship education within Lebanese socioeconomic environments which face additional difficulties. This study shows that the entrepreneurial skill development of Lebanese university students depends on multiple critical influencing factors. The research shows that students' entrepreneurial capabilities develop through the synergistic effects of excellent educational programs and institutional backing, as well as their economic background and individual characteristics. Future universities with policymakers need to unite their resources to solve these contributing elements and create spaces that help students achieve innovative entrepreneurial achievement.

REFERENCES

Alexandre, L. (2025). *Page Restricted.* [online] Elgaronline.com. Available at: https://www.elgaronline.com/abstract/edcoll/9781784717414/9781784717414.00019.xml [Accessed 28 Mar. 2025].

Ali, İ., Ali, M., & Badghish, S. (2019). Symmetric and asymmetric modelling of the entrepreneurial ecosystem in developing entrepreneurial intentions among female university students in Saudi Arabia. *International Journal of Gender and Entrepreneurship, 11*(4), 435–458. DOI: 10.1108/IJGE-02-2019-0039

Arab, L. A. (2018). Entrepreneurial Intentions of Undergraduate Students. (c2018). *Lau.edu.lb.* [online] doi:http://hdl.handle.net/10725/10471

Bahhouth, V., Spillane, J., Bahhouth, J., & Khoueiri, R. (2012). Students' perception of international business curriculum: An explorative study—case of Lebanon. *Journal of Teaching in International Business, 23*(3), 236–250. DOI: 10.1080/08975930.2012.744635

Biraglia, A., & Kadile, V. (2016). The role of entrepreneurial passion and creativity in developing entrepreneurial intentions: Insights from American homebrewers. *Journal of Small Business Management, 55*(1), 170–188. DOI: 10.1111/jsbm.12242

Chakra, S., & Jardali, H. (2022). Drivers of social entrepreneurial intentions of Palestinian refugees in Lebanon. *Journal of Research in Marketing and Entrepreneurship, 24*(2), 242–269. DOI: 10.1108/JRME-02-2021-0014

Ebewo, P., & Shambare, R. (2023). The effects of entrepreneurship leadership on youth entrepreneurial intentions post-COVID-19: The case of Gauteng. *Businesses, 3*(4), 569–584. DOI: 10.3390/businesses3040035

Esfandiar, K., Sharifi-Tehrani, M., Pratt, S., & Altınay, L. (2019). Understanding entrepreneurial intentions: A developed integrated structural model approach. *Journal of Business Research, 94*, 172–182. DOI: 10.1016/j.jbusres.2017.10.045

Fahed-Sreih, J., 2023. Corruption and new insights in Lebanon. https://doi.org/.DOI: 10.5772/intechopen.111903

Fahed-Sreih, J., Pistrui, D., Huang, W., & Welsch, H. (2009). Family contributions to entrepreneurial development in Lebanon. *The International Journal of Organizational Analysis, 17*(3), 248–261. DOI: 10.1108/19348830910974941

Ghannage, R. (2018). The Relevance of Entrepreneurial Education in Lebanese Higher Institutions: A Context Other than Business and Economic Studies. *KnE Social Sciences*, *3*(10), 68. DOI: 10.18502/kss.v3i10.2884

Haj Youssef, M., Harakeh, M., Hussein, H., & Sayour, N. (2025). Policy Brief: Addressing Barriers and Empowering Entrepreneurs in Lebanon. *F1000 Research*, *14*, 13. DOI: 10.12688/f1000research.156589.1

Hamzeh, W., Mershad, K., & Vetokhin, S. (2019). Integrating technology into higher Education: A case study in Lebanon. *Journal of Technology and Science Education*, *9*(3), 442. DOI: 10.3926/jotse.651

Hassan, N. (2020). University business incubators as a tool for accelerating entrepreneurship: A theoretical perspective. *Review of Economics and Political Science*, *9*(5), 434–453. DOI: 10.1108/REPS-10-2019-0142

Hill, S., Akhrass, E., & Barakat, H. (2018). Changing patterns of entrepreneurship in Lebanon. *Kne Social Sciences*, *3*(10), 10. DOI: 10.18502/kss.v3i10.2879

Hou, F., Su, Y., Lu, M., & Qi, M. (2019). Model of the entrepreneurial intention of university students in the Pearl River Delta of China. *Frontiers in Psychology*, *10*, 916. Advance online publication. DOI: 10.3389/fpsyg.2019.00916 PMID: 31114522

Jia, C., Zuo, J., & Lü, W. (2021). Influence of entrepreneurship education on employment quality and employment willingness. *International Journal of Emerging Technologies in Learning*, *16*(16), 65. DOI: 10.3991/ijet.v16i16.24897

Kautonen, T., Gelderen, M., & Fink, M. (2015). Robustness of the theory of planned behaviour in predicting entrepreneurial intentions and actions. *Entrepreneurship Theory and Practice*, *39*(3), 655–674. DOI: 10.1111/etap.12056

Kuran, O., & Khabbaz, L. (2024). Stakeholder dynamics in rural Lebanese women's entrepreneurship. *Journal of Enterprising Communities People and Places in the Global Economy*, *18*(5), 1098–1125. DOI: 10.1108/JEC-01-2024-0006

Lichy, J., Farquhar, J.D. and Kachour, M. (2020). Entrepreneurship via social networks – 'connected woman' in Lebanon. *Qualitative Market Research: An International Journal*, ahead-of-print(ahead-of-print). Doi:.DOI: 10.1108/QMR-01-2020-0004

Liñán, F., & Fayolle, A. (2015). A systematic literature review on entrepreneurial intentions: Citation, thematic analyses, and research agenda. *The International Entrepreneurship and Management Journal*, *11*(4), 907–933. DOI: 10.1007/s11365-015-0356-5

Liu, X. (2015). Research on the quality evaluation system of entrepreneurship education based on efficiency. *The Open Cybernetics & Systemics Journal*, *9*(1), 2969–2974. DOI: 10.2174/1874110X01509012969

Luo, Y., Huang, J., & Gao, S. (2022). Relationship between proactive personality and entrepreneurial intentions in college students: Mediation effects of social capital and human capital. *Frontiers in Psychology*, *13*, 861447. Advance online publication. DOI: 10.3389/fpsyg.2022.861447 PMID: 35783804

Mejjah, O., Ngulyavyangu, H., Peter, T., & Mwita, S. (2021). Assessment of entrepreneurial traits and intention among undergraduate students at Catholic University of Health and Allied Sciences - Mwanza, Tanzania. *Asian Journal of Business and Management*, *9*(5). Advance online publication. DOI: 10.24203/ajbm.v9i5.6808

Miralles, F., Giones, F., & Riverola, C. (2015). Evaluating the impact of prior experience in entrepreneurial intention. *The International Entrepreneurship and Management Journal*, *12*(3), 791–813. DOI: 10.1007/s11365-015-0365-4

Nauffal, D., & Skulte-Ouaiss, J. (2018). Quality higher Education drives employability in the Middle East. *Education + Training*, *60*(9), 1057–1069. DOI: 10.1108/ET-05-2017-0072

Pruett, M., & Şeşen, H. (2017). Faculty-student perceptions about entrepreneurship in six countries. *Education + Training*, *59*(1), 105–120. DOI: 10.1108/ET-03-2013-0028

Rauch, A., & Hulsink, W. (2015). Putting entrepreneurship education where the intention to act lies: An investigation into the impact of entrepreneurship education on entrepreneurial behaviour. *Academy of Management Learning & Education*, *14*(2), 187–204. DOI: 10.5465/amle.2012.0293

Saeed, S., Yousafzai, S., Yani-de-Soriano, M., & Muffatto, M. 2018. The role of perceived university support in the formation of students' entrepreneurial intention. pp.3-23. https://doi.org/DOI: 10.4324/9781315611495-1

Saleh, H. (2025). The perceptions of the Lebanese students of choosing their career in entrepreneurship. *Jordan Journal of Business Administration*, [online] 10(2). Available at: https://archives.ju.edu.jo/index.php/JJBA/article/view/6630 [Accessed 28 Mar. 2025].

Santika, I., Wardana, I., Setiawan, P., & Widagda, I. (2022). Entrepreneurship education and green entrepreneurial intention. *Linguistics and Culture Review*, *6*, 797–810. DOI: 10.21744/lingcure.v6nS1.2159

Sarouphim, K. (2015). Slowly but surely. *Journal for the Education of the Gifted*, *38*(2), 196–211. DOI: 10.1177/0162353215578278

Širola, D., 2020. Determinants of entrepreneurial intentions. *Zbornik Veleučilišta U Rijeci*, 8(1), pp.169-187. https://doi.org/.DOI: 10.31784/zvr.8.1.17

Tlaiss, H. (2022). Entrepreneurship, gender and success in Lebanon. *International Journal of Emerging Markets*, *18*(11), 4754–4771. DOI: 10.1108/IJOEM-01-2021-0145

Toufaily, B. and Bou Zakhem, N. (2024). Drivers of Student Social Entrepreneurial Intention Amid the Economic Crisis in Lebanon: A Mediation Model. *Sustainability*, [online] 16(7), p.2807. .DOI: 10.3390/su16072807

Wang, J., Chang, C., Yao, S., & Liang, C. (2015). The contribution of self-efficacy to the relationship between personality traits and entrepreneurial intention. *Higher Education*, *72*(2), 209–224. DOI: 10.1007/s10734-015-9946-y

YOUSSEF. S.A. (2021). Factors Affecting Entrepreneurial Intention of University Students: A Lebanese Case Study - STORE - University of Staffordshire Online Repository. *Staffs.ac.uk*. [online] doi:https://eprints.staffs.ac.uk/6892/2/Samia%20 Youssef-EThOS%20Deposit%20Agreement_tcm44-75119.doc

Youssef, H., Harakeh, M., Hussein, H., & Sayour, N. (2024). Brief on Entrepreneurship in Lebanon Unlocking Potentials: Addressing Barriers and Empowering Entrepreneurs in Lebanon | LJMU Research Online. *Ljmu.ac.uk*. [online] doi:https://researchonline .ljmu.ac.uk/id/eprint/24116/1/Policy%20Brief%20on%20Entrepreneurship%20in %20Lebanon.pdf

Zadravec, K. 2021. Developing intercultural and entrepreneurship competencies among students – the case of business, IT, and services disciplines. *SHS Web of Conferences*, 111, 01002. https://doi.org/DOI: 10.1051/shsconf/202111101002

Zampetakis, L., Kafetsios, K., Lerakis, M., & Moustakis, V. (2015). Investigating the role of self-construal in the formation of entrepreneurial intentions. *Frontiers in Psychology*, *6*. Advance online publication. DOI: 10.3389/fpsyg.2015.01085 PMID: 26284009

Zheng, T. 2022. Research on the current situation of university students' innovation and entrepreneurship under the background of "Internet + maker education". pp.668-676. https://doi.org/DOI: 10.2991/978-2-494069-89-3_79

Chapter 9
Entrepreneurship Training and Human Development:
Evidence From Rural Women in Lebanon

Hazem Harb
https://orcid.org/0000-0002-2488-1581
Saint Joseph University of Beirut, Lebanon

ABSTRACT

This chapter examines how entrepreneurship training functions as a multidimensional development tool for rural women in Lebanon, a fragile state marked by political instability and gendered socio-economic exclusion. Based on field research with 284 women across five governorates, it analyzes the impact of NGO-led programs on technical capacities, self-efficacy, economic participation, and civic engagement. Rooted in the Capability Approach and Human Capital Theory, the study finds that while training improved business skills and income generation, its deeper contributions lie in internal and social transformations. Participants reported increased confidence, leadership, and decision-making power. Yet, structural barriers such as limited finance access and patriarchal norms, limit the sustainability of these gains. The chapter critiques linear, skills-based models and proposes a gender-responsive, ecosystemic approach to entrepreneurship, offering theoretical, practical, and policy insights from a rarely studied Middle Eastern, post-crisis rural context.

DOI: 10.4018/979-8-3373-5651-8.ch009

INTRODUCTION

In recent decades, entrepreneurship has been promoted as a powerful vehicle for individual advancement and economic development. Its appeal lies in its promise of self-reliance, innovation, and bottom-up growth, particularly in contexts where formal employment is inaccessible or inadequate. Nowhere has this appeal been more pronounced than in fragile or conflict-affected countries, where state capacity is weak, market systems are volatile, and social services are often outsourced to NGOs and international agencies. Among the most visible beneficiaries of this trend are women, especially those residing in rural or marginalized areas who are frequently positioned as key agents of change within development discourse.

Lebanon presents a particularly urgent and complex case for examining this phenomenon. Once a regional commercial hub, the country has in recent years faced a cascading series of crises: financial collapse, widespread corruption, political paralysis, the COVID-19 pandemic, and the Beirut port explosion of 2020. These interlocking crises have severely undermined state institutions, deepened poverty, and intensified social inequalities. Women in rural Lebanon are among the hardest hit. They face limited access to education, high rates of informal labor, mobility restrictions, and entrenched patriarchal norms that limit their voice in public life. In the absence of meaningful state support, civil society actors have stepped in, and entrepreneurship training has emerged as one of the most visible development interventions targeting women.

This chapter engages critically with the question: What does entrepreneurship training actually achieve for rural women in fragile contexts like Lebanon? It investigates how such training influences not only technical skill acquisition and income generation, but also self-efficacy, civic participation, and broader dimensions of human development. Drawing from a mixed-methods field study involving 284 women across five rural governorates, it assesses the outcomes of NGO-led entrepreneurship training programs and explores the barriers that continue to shape women's post-training trajectories.

The chapter's argument is that entrepreneurship training, while often framed in economic terms, holds underutilized potential as a tool for expanding women's capabilities and reshaping their social positioning. However, this potential is frequently curtailed by programmatic limitations and structural inequalities that are rarely addressed in full. In this regard, entrepreneurship must be viewed not simply as a set of technical competencies, but as a relational and political process that unfolds within specific social, cultural, and institutional contexts.

To guide this inquiry, the chapter draws on two interrelated theoretical frameworks. First, Human Capital Theory (Becker, 1964) provides insight into how training enhances individual productivity and employability. It helps explain how women develop and apply business skills to generate income or improve household security. Yet Human Capital Theory, in its traditional form, is insufficient for capturing the full range of developmental changes women experience, particularly those that fall outside economic metrics.

Second, the chapter mobilizes Amartya Sen's Capability Approach (Sen, 1999), which shifts the analytical focus from resources or outputs to real freedoms: what individuals are actually able to do and be. This framework is especially relevant in fragile and gender-unequal settings where access to skills does not automatically translate into agency or opportunity. Capabilities such as voice, mobility, and decision-making are central to women's empowerment but are often excluded from program evaluations and policy assessments.

By applying these frameworks, the chapter offers a conceptual model that situates entrepreneurship training as a potential driver of multidimensional human development. It does so by analyzing not only whether women earn more or start businesses, but also whether they begin to speak in public, influence decisions in their households, organize with other women, and envision futures previously foreclosed by gendered limitations.

The contribution of this chapter is both empirical and theoretical. Empirically, it offers one of the few systematic, field-based evaluations of entrepreneurship training in the rural Middle East, a region often overlooked in gender and development scholarship. It builds on original data from five governorates in Lebanon and draws attention to regional and demographic variations in outcomes. Theoretically, it challenges linear models of empowerment that equate knowledge acquisition with autonomy, arguing instead for a situated understanding of how women convert skills into meaningful capabilities amidst ongoing fragility.

The chapter is structured as follows: Section 2 lays out the conceptual foundations and development policy context for entrepreneurship training in fragile environments. Section 3 presents the theoretical framework, integrating Human Capital Theory and the Capability Approach. Section 4 provides an expanded review of the literature on entrepreneurship, gender, and development outcomes. Section 5 outlines the methodology of the study. Section 6 presents the socio-political context of rural Lebanon. Section 7 details the study's findings across four dimensions: skill acquisition, self-efficacy, economic participation, and civic engagement. Section 8 offers a critical discussion of these results, followed by a reflection on theoretical contributions in Section 9. Section 10 provides concrete policy and programming recommendations. Section 11 addresses limitations and proposes directions for future research. The final section concludes with a synthesis of the study's implications for development practice and scholarship.

CONCEPTUAL FOUNDATIONS

Entrepreneurship is often promoted as a transformative force within development policy, one that empowers individuals to become self-reliant, resilient, and economically active. This framing is particularly dominant in post-crisis and fragile state contexts, where state institutions are either absent or dysfunctional. In such environments, entrepreneurship is positioned not only as a solution to unemployment but as a broader mechanism for rebuilding livelihoods and civic trust. However, the conceptual underpinnings of entrepreneurship are far from neutral especially when applied to marginalized women. They are entangled with global discourses of empowerment, neoliberal assumptions about individual responsibility, and complex socio-political dynamics.

This section lays the conceptual groundwork for the chapter by unpacking three core themes: the evolving role of entrepreneurship in development discourse, the intersection of gender and agency in fragile settings, and the central role played by non-state actors – particularly NGOs – in delivering training programs aimed at "empowering" women.

Entrepreneurship as Development Strategy

The rise of entrepreneurship as a favored development intervention can be traced to broader shifts in global economic policy since the 1980s. With the decline of Keynesian models and the rise of neoliberal frameworks, development strategies increasingly emphasized privatization, deregulation, and minimization of state roles. Within this climate, entrepreneurship was rebranded from a private sector pursuit into a tool for poverty alleviation and inclusion.

This reconceptualization framed individuals – especially women – as entrepreneurial subjects capable of driving economic change from the bottom up. As Dacin, Dacin, and Matear (2010) note, development organizations began to design programs that treat entrepreneurship not just as an economic activity but as a normative good. The "entrepreneurial poor" became central figures in policy discourse, valorized for their creativity, resilience, and potential to escape poverty through self-employment.

However, this framing has been critiqued for obscuring structural constraints. Critics argue that positioning entrepreneurship as a substitute for social protection or employment policy risks shifting the burden of development onto individuals. For rural women in fragile states, such a shift may result in greater exposure to risk, unpaid labor, and social backlash especially when training is offered without addressing the broader ecosystem needed to support entrepreneurship.

Gender, Agency, and Fragility

Entrepreneurship training for women is frequently framed in the language of empowerment. This framing reflects growing awareness within development policy that gender equity is not only a human rights imperative but also an economic efficiency strategy. The assumption is that empowered women are more likely to invest in their families, contribute to community stability, and support democratic governance (Cornwall & Rivas, 2015).

Yet empowerment itself is a contested concept. Kabeer (1999) defines empowerment as the process by which people gain the ability to make strategic life choices in contexts where this ability was previously denied to them. This definition moves beyond instrumentalist or economic metrics and invites a more expansive analysis of power, autonomy, and social transformation.

In fragile contexts, however, the path to empowerment is rarely linear. Structural fragility is characterized by weak institutions, chronic insecurity, and lack of state responsiveness, intersects with gendered constraints to produce highly constrained environments. Women may gain technical skills without being able to act on them. They may increase income but face backlash within their families. They may be encouraged to lead but excluded from decision-making spaces.

Indeed, research from fragile states suggests that empowerment must be analyzed relationally and contextually. It is not enough to ask whether women "benefited" from a program; rather, we must ask how these benefits were shaped, negotiated, and potentially constrained by family roles, religious norms, local politics, and intersecting forms of disadvantage (Molyneux, 1985; Van Eerdewijk et al., 2017). In this context, entrepreneurship training can be both enabling and problematic. It can offer new tools for self-assertion while simultaneously reinforcing the individualization of risk and responsibility.

The Role of NGOs in Delivering Entrepreneurship Training

In fragile states, the retreat or failure of state institutions has created a vacuum often filled by non-state actors. NGOs have emerged as key players in service provision, capacity building, and social mobilization. In Lebanon, the NGO sector has become a parallel infrastructure—providing education, healthcare, and economic programming to populations underserved by public institutions.

Entrepreneurship training has become one of the most prominent interventions led by NGOs, particularly those focusing on women. These programs vary in scope and approach but generally include modules on business planning, marketing, budgeting, and financial literacy. Many also incorporate components on leadership, communication, and confidence-building.

While NGO-led training programs are often lauded for their flexibility, local embeddedness, and participatory methods, they are also subject to critique. Development scholars point to three key challenges:

1. Projectization and short-termism: Many training programs are funded through short project cycles, limiting their ability to provide follow-up support, mentorship, or access to capital. This undermines sustainability and reduces the likelihood of meaningful transformation (Lewis & Opoku-Mensah, 2006).
2. Donor-driven priorities: Programs are often shaped more by donor reporting requirements than by community needs. This can result in overemphasis on quantifiable outputs—such as number of women trained or businesses launched—rather than qualitative outcomes like changes in agency or social status (Mosse, 2005).
3. Cultural misalignment: When imported models of entrepreneurship are applied without local adaptation, they may fail to resonate with participants' lived realities. For example, training programs that emphasize individual competition may conflict with communal or cooperative modes of livelihood common in rural areas.

Despite these challenges, NGOs remain vital conduits for entrepreneurship programming in Lebanon and elsewhere. They possess local networks, community legitimacy, and experience in working with marginalized groups. The key question, then, is how to harness these strengths while designing programs that are holistic, inclusive, and capable of supporting long-term development outcomes.

THEORETICAL FRAMEWORK

The study draws upon two complementary theoretical frameworks; Human Capital Theory (HCT) and the Capability Approach (CA), to analyze the effects of entrepreneurship training on rural women in Lebanon. These frameworks provide distinct yet intersecting lenses that help explain not only how women acquire skills but also whether and how those skills translate into agency, empowerment, and developmental change in fragile socio-political contexts.

Human Capital Theory: Individual Skill as Investment

Human Capital Theory, originally developed by Gary Becker (1964), posits that individuals can increase their productivity and economic value through investments in education, training, and health. In this framework, human capital is understood as a stock of skills and knowledge that enhances individual labor market outcomes and, by extension, contributes to economic growth at the societal level.

Entrepreneurship training programs are often designed with this theory in mind. They are structured to build specific competencies such as financial literacy, business planning, marketing, and risk management that increase a participant's ability to generate income or establish a sustainable enterprise. Under this model, successful training leads to measurable outputs: business creation, profit generation, job provision, or improved household income. The more skills an individual accumulates, the greater their potential to be economically productive.

This framework aligns with policy thinking that entrepreneurship views as a pragmatic solution to structural labor market failures, especially in fragile states. Where public employment is limited and private sectors are weak, encouraging self-employment becomes a viable alternative. In Lebanon, where rural economies suffer from underinvestment and a lack of formal employment opportunities, NGOs and donors have promoted entrepreneurship training as a means of plugging these institutional gaps.

However, critiques of HCT in development contexts have pointed out its limitations. The theory often assumes linear progression from skill acquisition to improved outcomes, abstracting away from the social and structural barriers that affect this process. For instance, Lazear's (2005) "jack-of-all-trades" theory argues that successful entrepreneurs require a diverse mix of generalist skills rather than deep specialization. Yet many training programs focus narrowly on technical aspects, overlooking the need for adaptability, negotiation, and resilience in volatile environments.

Moreover, HCT tends to individualize responsibility. It assumes that once equipped with knowledge, it is the individual's job to navigate obstacles and pursue economic success. In fragile and patriarchal societies, however, this assumption breaks down. Structural issues such as gender norms, lack of capital access, household labor burdens, and mobility restrictions significantly limit what even the most skilled women can achieve (Brush et al., 2009; Roomi & Parrott, 2008).

Thus, while HCT provides a useful foundation for understanding how training enhances women's productive potential, it does not sufficiently capture the relational and contextual elements of empowerment. For that, a broader framework is required, one that includes freedom, opportunity, and agency as core dimensions of development.

The Capability Approach: Freedom, Functioning, and Agency

The Capability Approach, developed by Amartya Sen (1999), shifts the focus of development from resources or utilities to the actual freedoms individuals have to lead lives they value. According to Sen, development should be assessed in terms of capabilities, defined as the real opportunities available to people, not just in terms of income or formal rights.

This perspective is particularly valuable for analyzing women's empowerment through entrepreneurship training. It emphasizes not just whether women have access to training or start businesses, but whether they can exercise agency, participate in public life, and make decisions that affect their futures. The Capability Approach recognizes that the same set of resources, such as knowledge or credit, can lead to very different outcomes depending on a person's social position, cultural context, and institutional environment.

Robeyns (2006) expands on this by introducing the concept of conversion factors, which determine how effectively individuals can translate resources into capabilities. These include personal factors (e.g., literacy, health), social factors (e.g., gender norms, stigma), and environmental factors (e.g., infrastructure, legal systems). In the context of rural Lebanon, even when women acquire business knowledge, they may be unable to apply it due to restricted mobility, lack of access to markets, or opposition from family members.

Applied to entrepreneurship training, the Capability Approach suggests that empowerment occurs not just when a woman starts a business, but when she can choose whether or not to do so freely, and when that choice is supported by structural conditions that make it sustainable. Empowerment, in this sense, is about expanding the range of real options available to individuals, not just providing technical skills.

The Capability Approach also aligns with feminist critiques of development that emphasize process over outcome. As Cornwall and Rivas (2015) argue, genuine empowerment involves transforming power relations, not just achieving isolated outcomes. For example, a woman who speaks in a village council meeting for the first time after completing a training program may not have launched a business, but she has expanded her capability to participate in civic life, a core dimension of development.

Integrating the Frameworks

The integration of Human Capital Theory and the Capability Approach provide a comprehensive framework for evaluating entrepreneurship training programs in fragile settings. While HCT helps identify and measure skill acquisition and economic output, CA allows for a deeper understanding of how these skills are converted into meaningful action.

This integrated framework is operationalized in the study by measuring both tangible outcomes (e.g., income-generating activity, enterprise creation) and intangible shifts (e.g., self-confidence, civic participation, leadership behavior). It also guides the analysis of barriers that affect outcome sustainability, such as access to finance, social support, and institutional infrastructure.

The benefit of this combined framework is that it avoids both economistic reductionism and abstract idealism. It acknowledges that skills matter and that training can be transformative, but it also insists that empowerment is relational, contingent, and context-dependent. In this way, the framework positions entrepreneurship training not as a stand-alone solution, but as one component within a broader ecosystem of human development.

This dual perspective is particularly critical in the case of Lebanon. As a fragile state with highly uneven development and deep gender inequality, Lebanon offers a real-world laboratory for understanding how development interventions operate under constraint. The integration of HCT and CA allows the study to explore both the potential and the limits of entrepreneurship training and to propose more context-sensitive, durable approaches to women's empowerment.

LITERATURE REVIEW

Entrepreneurship Training and Human Development: Conceptual Trends

This dual Over the past two decades, entrepreneurship training has gained traction as a development strategy aimed at addressing poverty, labor exclusion, and gender inequality. Rooted in microeconomic theory, entrepreneurship has been framed as a pathway toward individual economic independence and a mechanism for social transformation, particularly in settings where formal employment is scarce or inaccessible (Lazear, 2005; Martin et al., 2013). Training programs, typically delivered by NGOs or international agencies, often promise to empower marginalized populations by equipping them with the technical and managerial skills needed to launch and sustain microenterprises. While the economic rationale for these programs is well-established, questions persist about their actual developmental value, particularly in fragile and gender-unequal contexts.

Human development, as conceptualized by Sen (1999), extends beyond income and employment to include capabilities, freedoms, and the agency to pursue lives that individuals value. When seen through this lens, entrepreneurship training should be evaluated not solely on whether participants start businesses, but on whether it expands their ability to make meaningful life choices, exercise voice, and influence their environments. This reconceptualization has significant implications for how training programs are designed, delivered, and assessed.

In practice, however, most evaluations remain narrowly focused. Metrics such as post-training income, business registration, and job creation dominate impact assessments, leaving intangible outcomes like self-confidence, leadership, and civic participation underexplored (McKenzie & Woodruff, 2014; Brixiová et al., 2020). This tendency is even more problematic in fragile states, where institutional instability, gender norms, and social fragmentation fundamentally shape how skills are acquired and used.

Gender, Entrepreneurship, and Training Outcomes

Women's entrepreneurship has received increasing attention in both academic and policy discourse, yet it continues to be treated as a technical intervention rather than a social process. Scholars have long critiqued the assumption that women simply need skills to succeed. Brush et al. (2009) argue that entrepreneurship for women is deeply shaped by access to networks, family responsibilities, and gendered expectations, which often go unaddressed in training curricula.

Recent studies suggest that entrepreneurship training can lead to enhanced agency and self-efficacy, but only when interventions are contextualized and holistic. For example, Maritz and Brown (2013) show that programs emphasizing leadership, decision-making, and public speaking often have greater long-term impact than those focused solely on business planning. Similarly, Roomi and Parrott (2008), in their study of Pakistani women, found that participants' ability to apply skills post-training was strongly influenced by male approval, mobility restrictions, and household dynamics, factors that rarely appear in formal program evaluations.

In the MENA region, the gendered nature of entrepreneurship is particularly pronounced. Cultural norms, religious frameworks, and legal restrictions often limit women's access to capital, markets, and even physical spaces outside the home. In Lebanon, women's labor force participation remains among the lowest in the region, and rural women face even steeper challenges due to weak infrastructure, conservative gender roles, and limited institutional support (World Bank, 2022).

Despite these constraints, evidence suggests that entrepreneurship training can foster meaningful shifts in self-perception and community engagement. For instance, a study by Brixiová, Kangoye, and Said (2020) in Ethiopia found that women who completed training programs were more likely to participate in cooperatives, join advocacy efforts, and express higher levels of political interest. Such outcomes suggest that training may operate not only as an economic intervention but as a form of civic and psychological empowerment.

The Capability Approach and Women's Empowerment

The Capability Approach offers valuable theoretical lens through which to evaluate the impact of entrepreneurship training. Introduced by Amartya Sen (1999), the approach emphasizes what individuals are able to do and to be, rather than what they have. This shift places focus on the real freedoms people enjoy in converting resources into valuable functioning, such as participating in public life, making economic decisions, or living with dignity.

Robeyns (2006) argues that the Capability Approach is especially relevant to gender analysis because it forces us to interrogate structural constraints – such as patriarchal norms, time poverty, and domestic labor – that shape women's ability to act. In rural areas of fragile states, even when women possess entrepreneurial knowledge, they may lack the conversion factors (e.g., mobility, confidence, familial support) needed to turn knowledge into action. Alkire (2005) further contends that empowerment must be seen as an expansion of agency across multiple spheres, including social, political, and economic dimensions.

Applied to entrepreneurship training, the Capability Approach demands a broader evaluation framework. Instead of merely tracking whether women launch businesses, researchers must examine whether training enhances women's control over their time, increases their decision-making power, or enables them to participate meaningfully in community affairs. This approach aligns with findings from development studies indicating that social capital, peer networks, and civic involvement are often more predictive of sustained empowerment than technical proficiency alone (Cornwall & Rivas, 2015; Kabeer, 1999).

Human Capital Theory and Its Limits

Human Capital Theory (Becker, 1964) has long underpinned training interventions in development contexts. The theory posits that investments in education, health, and skill development increase productivity and, by extension, income. This framework provides a clear justification for entrepreneurship training: by improving individual capabilities, particularly among marginalized populations, such programs are assumed to generate returns both for participants and the broader economy.

While HCT remains influential, it has been criticized for its individualistic and economistic assumptions. It tends to abstract individuals from their social environments, assuming that knowledge acquisition naturally leads to upward mobility. In practice, this is often not the case, particularly for women in fragile and patriarchal societies. For example, Brush et al. (2009) argue that HCT fails to account for gendered constraints that mediate access to markets, clients, and capital. Moreover, Lazear's (2005) theory of "jack-of-all-trades" entrepreneurship suggests that generalists are more successful than specialists, highlighting the need for broad-based training, yet many programs focus narrowly on technical or financial content.

In fragile contexts, the assumption that knowledge leads to productivity is particularly problematic. Political instability, weak legal systems, and informal labor markets create barriers that no individual skill can overcome (Narayan et al., 2000). Thus, entrepreneurship training must be embedded within an ecosystem that includes follow-up mentoring, access to microfinance, legal literacy, and collective organization.

Fragility, Gender, and the Limits of Linear Models

Literature on development in fragile contexts warns against simplistic assumptions of linear progress. Gender-focused programs often assume that empowerment follows a predictable path from knowledge to confidence to action. However, research shows that this path is often nonlinear, interrupted by backlash, social fatigue, and structural inertia. Molyneux (1985) famously distinguished between practical and strategic gender interests, noting that many development programs address short-term needs (e.g., income) without altering the power structures that sustain gender inequality.

In fragile states like Lebanon, this dynamic is exacerbated by the collapse of institutional trust and the weakening of public infrastructure. As a result, NGOs often assume the dual role of service provider and civic educator. This places significant pressure on entrepreneurship training to deliver results not only in income generation, but in building capacity for leadership, negotiation, and public participation. Research suggests that programs that include components such as group work, peer mentoring, and community engagement are more likely to yield transformative outcomes (Van Eerdewijk et al., 2017).

Moreover, intersectional perspectives are crucial for understanding how age, education, marital status, and region interact to shape women's experiences with training. Rural women, for example, may face additional hurdles compared to their urban counterparts due to geographic isolation, limited transportation, and lower literacy rates. Programs that ignore these nuances risk reinforcing existing inequalities.

METHODOLOGY

This chapter draws on a cross-sectional, field-based research design to assess the developmental outcomes of entrepreneurship training among rural women in Lebanon. The methodology was carefully structured to align with both the theoretical framework – anchored in Human Capital Theory and the Capability Approach – and the socio-political realities of working in a fragile state context. This section outlines research design, sampling strategy, data collection instruments, analytical methods, and ethical considerations.

Research Design

The study adopts a quantitative, descriptive-analytical approach. A structured survey instrument was used to collect data from a representative sample of women who had participated in entrepreneurship training programs delivered by local NGOs across five rural governorates in Lebanon. The design aimed to capture both tangible (economic) and intangible (capability-related) outcomes resulting from training, as well as the contextual factors that shaped participants' experiences.

While qualitative approaches offer depth in understanding lived experiences, the use of a quantitative instrument allows comparability across diverse regions and demographic groups. The intention was to generate a broad-based empirical foundation that could support policy recommendations, while still incorporating agency-centered indicators reflective of the Capability Approach.

Population and Sampling

The target population consisted of women who completed entrepreneurship training programs between 2020 and 2022 in five rural governorates: Akkar, Baalbek-Hermel, Bekaa, Nabatieh, and South Lebanon. These regions were selected due to their geographic diversity, rural characteristics, and high levels of economic marginalization. They also reflect different sectarian compositions and political affiliations, adding richness to the analysis.

A stratified purposive sampling technique was used to ensure representation across regions and to account for variations in training providers and curriculum content. The final sample included 284 respondents, with the following regional breakdown:

- Bekaa: 76
- South Lebanon: 63
- Akkar: 59
- Nabatieh: 47
- Baalbek-Hermel: 39

Participants were identified through program completion lists provided by NGO partners. Inclusion criteria required that participants had completed the full training curriculum and were over the age of 18.

Data Collection Instrument

The survey instrument was developed based on a synthesis of entrepreneurship evaluation tools, empowerment scales (e.g., CARE's Women's Empowerment Index), and items aligned with the Capability Approach. The questionnaire was designed to assess five key domains:

1. Demographics and background characteristics: Age, marital status, education, household size, income level.
2. Training characteristics: Type, duration, delivery method, satisfaction.
3. Entrepreneurial skills acquisition: Business planning, marketing, budgeting, resource mobilization.
4. Self-efficacy and confidence: Public speaking, leadership roles, decision-making.
5. Post-training outcomes:
   - Economic participation (e.g., income generation, enterprise creation)
   - Civic engagement (e.g., group membership, advocacy participation)
   - Structural barriers (e.g., capital access, social constraints)

The instrument was developed in English and translated into Arabic, with back-translation used to ensure accuracy. It was pilot-tested with a sample of 20 participants from a neighboring region not included in the main study. Feedback from the pilot informed revisions to question phrasing and sequence.

Data Collection Process

Data collection was carried out between February and April 2023. Due to COVID-19 precautions and regional access challenges, surveys were administered in person by trained female field researchers with prior experience in community development. The use of local female enumerators was intentional to build trust and encourage open responses, particularly on sensitive topics such as household dynamics and decision-making power.

Field teams underwent a two-day training session covering ethical research conduct, survey protocol, and gender-sensitive interviewing techniques. Enumerators were supervised by regional coordinators who monitored data quality and provided on-site support.

In total, 284 completed surveys were collected, resulting in a high response rate (over 90%) due to the pre-established relationship between participants and local NGO partners.

Data Analysis

Quantitative data were entered into SPSS and cleaned prior to analysis. Descriptive statistics were used to profile participants and summarize responses across the five domains. Cross-tabulations and Chi-square tests of association were employed to examine relationships between independent variables (e.g., region, age group, education level) and key outcomes (e.g., confidence increase, income generation, civic engagement).

Where appropriate, open-ended responses were grouped thematically to provide additional context and explanation for quantitative trends. These qualitative inputs, while not systematically coded, helped illustrate participant voice and contextual nuance.

Ethical Considerations

Ethical approval for the study was obtained from the relevant institutional review board at Saint Joseph University of Beirut. All participants provided informed consent, and confidentiality was guaranteed throughout the process. Participation was voluntary, and no personal identifiers were collected

Special attention was given to safeguarding protocols, particularly given the gender-sensitive nature of the research and the fragile context in which it was conducted. Enumerators were instructed to terminate any interview if respondents appeared distressed or uncomfortable.

Additionally, the study adhered to Do No Harm principles and was guided by ethical standards for research with vulnerable populations, including the right to withdraw and the avoidance of coercive incentives.

EMPIRICAL CONTEXT: RURAL LEBANON

Understanding the outcomes of entrepreneurship training in Lebanon requires a close reading of the socioeconomic, political, and institutional context in which rural women live and operate. This section presents a contextual analysis that situates the research findings within the country's ongoing fragility, rural development challenges, and NGO-driven service provision model.

Lebanon's Fragile Statehood and Economic Collapse

Lebanon has experienced multiple crises over the past decades, but the period following 2019 marks an especially severe phase of economic and institutional breakdown. Triggered by years of fiscal mismanagement, corruption, and political gridlock, the country's financial system collapsed in late 2019. The national currency lost over 90% of its value, banks froze deposits, and inflation skyrocketed, pushing more than 70% of the population into poverty (World Bank, 2022).

This collapse was compounded by the COVID-19 pandemic and the Beirut port explosion in August 2020, which further strained the already fragile healthcare, education, and social welfare systems. Public sector institutions have struggled to provide even basic services, and trust in government is at a historic low. Amid this vacuum, the development landscape has become increasingly shaped by non-state actors, particularly international and local NGOs, who have stepped in to deliver essential services.

In this context of state fragility and humanitarian dependency, entrepreneurship training has emerged as a substitute for more systemic employment or welfare policies. It is framed as a tool for resilience and self-reliance, with a focus on helping individuals generate income independently. However, this approach assumes that individuals—particularly rural women—possess the structural support needed to apply their training in practice. As this chapter argues, such assumptions often obscure the realities of rural underdevelopment and gender inequality.

Rural Lebanon: Geography, Infrastructure, and Marginalization

While Lebanon is often portrayed through the lens of urban centers such as Beirut, rural areas make up a significant proportion of the country's geography and population. Regions such as Bekaa, Akkar, Baalbek-Hermel, South Lebanon, and Nabatieh are characterized by lower levels of infrastructure, high unemployment, poor access to healthcare and education, and limited economic diversification.

In many cases, rural communities have been historically neglected in national development plans. Public investment in these regions is minimal, and residents face significant disparities in service provision compared to urban areas. Rural women, in particular, encounter a layered set of constraints:

- Limited mobility due to lack of public transportation or cultural restrictions
- Gendered expectations that prioritize caregiving and domestic roles
- Low access to capital due to informal lending practices and male-dominated financial institutions
- Weak digital infrastructure that limits access to markets, training, and online platforms

These factors contribute to persistent economic exclusion and social invisibility. Although agriculture remains a primary economic activity in some regions, the sector has faced declining productivity and investment, leaving few formal employment options for women. As a result, many rural women turn to informal economic activities, such as home-based production, food processing, or seasonal labor – none of which offer social protection or long-term sustainability.

Sociopolitical Dimensions of Gender and Development

Rural women's marginalization in Lebanon is shaped not only by geography but also by deeply rooted gender norms and political dynamics. The country's confessional system distributes power along sectarian lines, and in many rural areas, political allegiances intersect with family structures, tribal affiliations, and religious institutions. These dynamics often reinforce patriarchal control over women's mobility, education, and labor force participation.

Studies have shown that women in rural Lebanon face greater restrictions on voice and visibility than their urban counterparts. For example, participation in municipal councils, cooperatives, or community associations remains low, and women who seek public roles may face backlash or social isolation. At the same time, familial and social networks serve as both constraints and support systems, offering protection and shared labor, but also enforcing conformity and limiting autonomy (Joseph, 1996).

In this landscape, any intervention aimed at empowering women must grapple with the realities of social surveillance, community politics, and informal power structures. Entrepreneurship training programs may offer tools for self-employment, but their long-term impact depends on whether women can navigate and negotiate these power dynamics.

The Role of NGOs and the Entrepreneurship Training Ecosystem

In the absence of effective state intervention, NGOs have emerged as primary agents of development in rural Lebanon. With support from bilateral donors, multilateral agencies, and diaspora philanthropy, NGOs have developed a wide array of training programs focused on women's economic empowerment. These programs vary in scope and methodology, but generally include:

- Business development modules: Covering planning, finance, marketing, and operations
- Soft skills training: Emphasizing leadership, communication, and confidence
- Sector-specific workshops: Focused on agriculture, food processing, crafts, and services

- Incubation and grants: In some cases, participants receive seed funding or mentorship

Despite these efforts, the NGO ecosystem also reflects broader challenges:

- Duplication and fragmentation: Lack of coordination among NGOs often leads to overlapping or inconsistent programming. Participants may attend multiple training courses with varying content quality, with no integrated follow-up strategy.
- Donor accountability pressures: Programs are often designed to meet funder demands for quick, quantifiable results (e.g., number of beneficiaries trained), leading to short-termism and underinvestment in sustainability mechanisms such as mentorship or market linkages (Mosse, 2005).
- Training fatigue: In some regions, women have attended multiple training programs without tangible outcomes, resulting in skepticism and reduced motivation. This phenomenon highlights the need for better evaluation, alignment with local needs, and clear pathways from training to implementation.

Nevertheless, NGOs remain trusted actors in many rural communities. Their ability to work through local networks, use culturally sensitive approaches, and provide direct service delivery positions them as critical intermediaries between women and broader economic opportunities.

Regional Profiles of Study Areas

The research spans five rural governorates, each with distinct demographic, economic, and political characteristics. Below is a brief profile of each region:

- Bekaa: A diverse agricultural region with a high refugee population and long-standing development challenges. Women here often engage in food processing, farming, and seasonal trade. NGO presence is strong but fragmented.
- South Lebanon: Characterized by its post-conflict reconstruction context. Economic life is shaped by agriculture, remittances, and informal labor. Women face traditional gender roles, but some community-based organizations support cooperative models.
- Akkar: Among the poorest regions in Lebanon, with limited infrastructure and high youth unemployment. Women here often lack access to training and markets. Mobility restrictions are severe, and early marriage remains prevalent.

- Nabatieh: A conservative region with limited women's civic participation. While some women engage in small-scale home businesses, access to credit and public space is minimal.
- Baalbek-Hermel: Marked by deep underdevelopment and tribal influence. Women's economic activities are often invisible or unpaid. NGO interventions face logistical and cultural challenges but remain crucial sources of support.

These regional contexts are not merely background, they are active determinants of how entrepreneurship training is received, applied, and sustained. Program outcomes must be interpreted with sensitivity to these diverse structural and cultural dynamics.

FINDINGS

This section presents the empirical findings from the field study involving 284 women across five governorates in rural Lebanon. The results are organized into four primary domains aligned with the study's theoretical framework: entrepreneurial skill acquisition, self-efficacy and confidence, economic participation, and civic engagement. Each domain is examined through descriptive and comparative analysis, highlighting both quantitative outcomes and structural limitations reported by participants.

Entrepreneurial Skill Acquisition

Entrepreneurship training programs delivered by NGOs aimed to equip participants with a suite of technical and managerial competencies. These included business planning, budgeting, marketing, inventory control, customer relations, and digital tools for enterprise management.

Among respondents:

- 92% reported gaining new business skills through the training, with the most frequently cited competencies budgeting (81%), marketing (76%), and business planning (72%).
- 88% stated that the training content was relevant to their daily lives, and 85% reported satisfaction with the delivery method and trainer engagement.

These figures suggest that, from a Human Capital Theory perspective, the programs effectively improved individual capacity. However, the acquisition of skills did not automatically translate into enterprise creation or income generation for all participants.

Notably, regional disparities emerged:

- Women in Bekaa and South Lebanon reported higher engagement with digital tools, likely reflecting better internet infrastructure and NGO support in those regions.
- Participants from Baalbek-Hermel and Akkar were less likely to report mastery of budgeting or digital literacy, potentially due to lower baseline education levels and limited follow-up support.

While the majority of participants felt better prepared to manage small businesses, some expressed concerns about information overload and the absence of context-specific examples in the training materials. A number of respondents described sessions as "too abstract" or "not linked to real market conditions," indicating a need for localized curriculum design.

Self-Efficacy and Confidence

A key dimension of the Capability Approach is whether individuals feel empowered to act, whether or not they ultimately do. In this domain, the training had a measurable impact.

- 88% of respondents reported feeling more confident in expressing opinions during group discussions or public gatherings.
- 79% indicated improved ability to make decisions about financial matters at home or in their businesses.
- 65% stated they had spoken publicly for the first time during or after the training program.

This boost in confidence was particularly significant among younger women (ages 20–35), who were more likely to report shifts in self-perception and ambition. Several participants described the training as "transformative," emphasizing that it gave them the language and legitimacy to speak up in both private and public settings.

However, qualitative responses revealed the limits of this self-efficacy. Many women noted that while they had gained confidence, they were still expected to defer to male relatives in financial or public decisions. Some respondents spoke of husbands or fathers "allowing" them to attend trainings but discouraging them from acting on the knowledge gained. In this sense, self-efficacy increased, but the social conditions to exercise agencies remained constrained.

These findings underscore the need to distinguish between internal transformation and external permission structures—an analytical strength of the Capability Approach.

Economic Participation and Enterprise Creation

From an HCT standpoint, the ultimate goal of entrepreneurship training is to increase participants' economic productivity and independence. In this domain, outcomes were mixed.

- **69% of respondents reported engaging in some form of income-generating activity** following the training.
- Of these, **38% launched new micro-enterprises**, often home-based and in sectors such as food processing, tailoring, or beauty services.
- The remaining 31% engaged in informal trade, seasonal labor, or expanded existing family businesses using the knowledge gained.

However, only 12% of those who launched businesses reported consistent monthly profit, indicating that enterprise sustainability remains elusive.

The main barriers cited by participants included:

- Lack of start-up capital (62%)
- No access to credit or microfinance (49%)
- Saturated local markets (45%)
- Household responsibilities limiting time availability (41%)

These findings challenge the linear assumption embedded in HCT—that skills lead directly to economic outcomes. They highlight the need for integrated post-training support mechanisms, including access to finance, mentoring, and market linkages. Several women reported attending multiple trainings across different NGOs without any program offering tangible next steps such as incubation, seed funding, or cooperative building.

The entrepreneurial activities that were launched tended to be low-capital, home-based, and labor-intensive. This reflects both the limitations of local markets and the constraints imposed by conservative gender norms that restrict women's movement and visibility.

Civic Participation and Leadership

One of the most significant findings relates to civic engagement, a domain rarely measured in training program evaluations. The study found that training programs often served as an entry point for women to join community groups, cooperatives, or advocacy initiatives.

- 72% of respondents joined at least one local group or association following the training. These included agricultural cooperatives, savings groups, and informal support networks.
- 46% assumed leadership roles within those groups or in community events.
- 34% participated in advocacy efforts, such as local development planning meetings or NGO consultation sessions.

These outcomes suggest that entrepreneurship training had unexpected spillover effects into civic life. The group-based format of many programs created spaces for solidarity and collective learning, allowing women to form peer networks that extended beyond the training itself.

However, civic engagement also varies regionally:

- In Bekaa and South Lebanon, NGO-facilitated cooperatives provided structured platforms for women's leadership.
- In Baalbek-Hermel and Akkar, engagement was lower, partly due to weaker NGO presence and more restrictive social norms.

Importantly, several women framed their participation in civic activities as an extension of their entrepreneurial identity. They described "helping other women start" or "speaking up for our needs" as part of their development, blurring the line between economic and political agency. This reinforces the Capability Approach's emphasis on multidimensional empowerment.

Demographic and Regional Trends

Cross-tabulation analyses revealed several noteworthy trends:

- Education level was positively associated with both confidence gains and likelihood of launching a business.
- Married women with children were less likely to engage in sustained economic activity, primarily due to household labor burdens and time poverty.
- Participants from Bekaa had the highest rates of business creation and group membership, suggesting a more supportive training and post-training environment.

While these associations are not causal, they point to important intersectional dynamics that must be considered in program design and policy formulation.

Summary of Findings

The study reveals that entrepreneurship training programs in rural Lebanon produced multidimensional impacts, with outcomes spanning beyond economic productivity to include voice, self-confidence, and community participation. However, these gains were often partial, uneven, and dependent on enabling conditions, such as access to capital, supportive family structures, and NGO follow-up mechanisms.

The findings affirm the value of integrating Human Capital Theory and the Capability Approach. HCT helps identify what skills were acquired, while CA allows us to assess whether those skills were convertible into meaningful opportunities. The analysis also underscores the importance of moving beyond simplistic outcome metrics and embracing relational, context-sensitive frameworks that reflect the lived realities of women in fragile states.

Relationship Analysis: Linking Training to Developmental Outcomes

To examine whether observed development outcomes were statistically associated with participation in entrepreneurship training, chi-square (χ^2) tests of association were conducted. These analyses explored expected relationships based on prior literature and the chapter's theoretical framework, particularly the intersections between capability expansion, skill acquisition, and civic or economic activation.

Self-Efficacy and Training Participation

The first analytical expectation was that participation in entrepreneurship training – especially modules focused on leadership, communication, and confidence – would be positively associated with increased self-efficacy among rural women. This was measured through reported comfort in public speaking, decision-making, and willingness to assume leadership roles.

χ^2 (2, N = 284) = 24.78, p < .001

The relationship was statistically significant. Women who attended confidence-building modules were more likely to report increased self-belief, autonomy in household decision-making, and willingness to engage publicly. This supports the notion that training serves not only technical but also psychosocial developmental functions.

Civic Engagement and Skill Acquisition

The second expectation concerned the relationship between entrepreneurial and leadership training and civic or community engagement. Indicators included participation in local associations, speaking at public forums, and initiating community-based projects.

χ^2 (2, N = 284) = 18.13, p = .002

This result supports the hypothesis that leadership training is significantly associated with increased civic involvement. Women with limited prior public experience appeared especially likely to engage in cooperative or group-based structures after training.

Economic Activation and Business Knowledge

A third analytical expectation was that women who acquired business management skills would be more likely to engage in income-generating activity post-training, either through home-based enterprise, freelance work, or informal product sales.

χ^2 (2, N = 284) = 16.42, p = .005

Findings affirm a positive association between technical knowledge and post-training economic participation. Women who reported proficiency in pricing, marketing, and budgeting were more likely to report earning income independently, although this did not always translate into formal business registration or financial sustainability.

DISCUSSION

This section interprets the study's findings through the theoretical lenses of Human Capital Theory and the Capability Approach, while situating them within broader debates on gender, development, and fragility. The discussion aims to move beyond descriptive results and examine what the observed outcomes reveal about the nature of empowerment, the limits of training-centric interventions, and the relational dynamics that shape women's post-training trajectories in fragile rural contexts.

Beyond Income: Rethinking Empowerment in Fragile Contexts

Traditional assessments of entrepreneurship programs tend to prioritize economic indicators: how many businesses were launched, how much income was generated, how many jobs were created. While these metrics are important, they offer a limited view of what training achieves, particularly in fragile settings marked by instability, exclusion, and gender inequality.

The study shows that entrepreneurship training had its most profound effects in non-economic domains: self-efficacy, civic participation, public speaking, and community engagement. These findings challenge the assumption that economic productivity is the most important or even the most realistic outcome of training. For many women, the ability to voice opinions publicly, join a community group, or influence household decisions represented a radical departure from previous life possibilities, even in the absence of formal business success.

This reorientation aligns with the Capability Approach's emphasis on expanding real freedoms and valuable functioning. From this perspective, empowerment is not just about doing (e.g., starting a business) but about being able to choose – with dignity, agency, and support – whether to pursue that path or another. It also reflects what Kabeer (1999) calls "strategic life choices," which may be incremental but still transformative in their social meaning.

Capability Conversion and Structural Constraints

The gap between skill acquisition and sustainable outcomes highlights a core insight of the Capability Approach: resources alone are insufficient unless individuals can convert them into action. This study identified several conversion factors – both enabling and constraining – that shaped women's ability to act on their training:

- Positive conversion factors included peer networks formed during training, NGO follow-up support, and access to cooperative structures in some regions.
- Negative conversion factors included unpaid care burdens, male-dominated household decision-making, lack of mobility, saturated markets, and absence of microfinance.

This underscores that capability development is not simply additive or individualistic. It is relational, dependent on supportive ecosystems that include family, community, institutions, and infrastructure. For example, a woman with excellent marketing skills may still fail to launch a business if she cannot travel to urban centers or access startup capital.

Moreover, these conversion barriers vary by region, education level, and marital status, reinforcing the need for intersectional analysis. Programs that do not adapt to these differences risk reproducing the very inequalities they seek to challenge.

Training as Civic Activation

One of the most striking findings was the extent to which training served as a platform for civic engagement. Participants reported joining groups, assuming leadership roles, and participating in advocacy efforts, none of which were explicit goals of the training programs. These outcomes suggest that training created social spaces for interaction, solidarity, and mutual recognition, allowing women to see themselves as legitimate public actors.

This challenges the narrow framing of entrepreneurship as a private economic pursuit and affirms its potential as a civic and political intervention. It also resonates with feminist critiques that argue for a more expansive understanding of agency, one that includes collective action, voice, and contestation of social norms (Cornwall & Rivas, 2015).

In fragile contexts, where trust in institutions is low and civic space is limited, these outcomes are particularly significant. Women's increased presence in public or semi-public spheres may contribute not only to personal empowerment but to broader processes of community resilience, accountability, and social cohesion.

Fragility and the Problem of Short-Termism

The Lebanese context adds a layer of complexity to these findings. As a fragile state undergoing economic collapse, Lebanon lacks the institutional infrastructure to sustain development gains. The reliance on NGOs as primary service providers creates both opportunity and vulnerability. While NGOs can reach marginalized groups and tailor programs to local contexts, they are often constrained by short-term funding cycles, donor-driven metrics, and limited ability to follow through on long-term outcomes.

This study found that women often completed multiple trainings across different NGOs without coordinated follow-up, access to credit, or business incubation. This "training fatigue" points to a deeper issue: entrepreneurship programs are frequently treated as isolated technical fixes rather than components of an integrated development strategy. Without systems for financial inclusion, legal support, mentorship, and infrastructure, the long-term impact of training remains uncertain.

These patterns mirror global critiques of project-based development in fragile states, where interventions are fragmented, reactive, and rarely systemic (Mosse,

2005). To move beyond this, there is a need for ecosystem thinking, an approach that treats empowerment as a layered process requiring structural and relational support.

Implications for Theory and Practice

The findings call for a reframing of entrepreneurship training along two key lines:

1. **Theoretically**, the study supports a hybrid framework that integrates Human Capital Theory and the Capability Approach. HCT helps explain how skill acquisition enhances productivity, while CA captures the broader social and relational conditions necessary for empowerment. Together, they provide a more comprehensive understanding of what training can and cannot achieve in fragile, gendered contexts.
2. **Practically**, the findings highlight the need for program designs that are context-sensitive, long-term, and multidimensional. Training should be coupled with mentorship, group training, access to finance, and engagement with family and community power structures. Programs must also be evaluated not just on economic metrics but on indicators of voice, participation, and psychological empowerment.

Finally, the study affirms the importance of empirical research in under-studied contexts. Much of the literature on women's entrepreneurship comes from stable or urban environments. By focusing on rural Lebanon – a fragile, politically complex, and socially conservative setting – this chapter contributes to a more nuanced understanding of how development interventions unfold in difficult terrain.

THEORETICAL CONTRIBUTIONS

This chapter offers several contributions to the theoretical understanding of women's empowerment, entrepreneurship, and development interventions in fragile and gender-constrained contexts. By integrating Human Capital Theory and the Capability Approach, the study challenges narrow formulations of entrepreneurial success and proposes a more nuanced, multidimensional framework for assessing training outcomes.

Advancing a Dual-Framework Model for Development Interventions

The combined use of Human Capital Theory (HCT) and the Capability Approach (CA) offers a theoretical advance for analyzing training interventions in marginalized settings. HCT has traditionally informed program design and evaluation by emphasizing skill acquisition and individual productivity. It allows for the measurement of technical gains and income-related outcomes—metrics that are often favored by donors and development practitioners.

However, this chapter demonstrates that HCT alone is insufficient for capturing the range of impacts experienced by rural women. Without attending to structural constraints, relational dynamics, and socio-cultural conversion factors, HCT can obscure the real-life obstacles that limit post-training agencies. For example, acquiring financial literacy may not lead to entrepreneurial activity if a woman cannot leave her home without permission or lacks access to startup capital.

By incorporating the Capability Approach, the analysis is expanded to include freedom, voice, and choice—outcomes that matter deeply to women but are rarely prioritized in entrepreneurship program logic models. CA also allows for recognition of partial or intangible progress: a woman who has not launched a business but has joined a local group, spoken in public, or influenced household decisions has nonetheless experienced empowerment in meaningful terms. This dual-framework model can serve as a template for future studies seeking to balance outcome measurement with contextual depth.

Reframing Empowerment as Relational and Context-Dependent

Another key theoretical contribution is the reframing of empowerment as a relational and context-dependent process. Much of the entrepreneurship literature continues to treat empowerment as an individual-level transformation, measured through metrics such as increased income or business registration. Yet this study shows that empowerment unfolds within a matrix of social expectations, household dynamics, political norms, and gendered institutions.

The findings emphasize that rural women's ability to act is shaped not just by internal confidence or external resources, but by how these are negotiated within families and communities. Some women described gaining the confidence to speak out but also experiencing backlash from male relatives. Others found empowerment through collective action – joining cooperatives or initiating local advocacy – not through isolated entrepreneurial success.

These insights align with feminist development scholars (e.g., Kabeer, 1999; Cornwall & Rivas, 2015) who argue that empowerment cannot be divorced from power relations. It is not a state to be achieved, but a process to be negotiated. By foregrounding this relational perspective, the chapter contributes to a growing body of literature that critiques the individualization of development outcomes and calls for a more ecosystemic understanding of agency.

Contextualizing Women's Entrepreneurship in Fragile States

A third contribution lies in the study's regional and political specificity. Much of the global literature on women's entrepreneurship draws from urban, middle-income, or relatively stable environments. In contrast, this chapter is grounded in a fragile, rural, and politically fragmented setting, adding empirical diversity and theoretical nuance to the field.

Lebanon's compounded crises – economic collapse, institutional failure, and social fragmentation – have created an environment where the meanings and mechanisms of empowerment differ dramatically from mainstream assumptions. In such a context, launching a business may be less feasible or meaningful than achieving voice in a village council, securing informal income, or building a mutual aid network.

This study therefore contributes to a context-sensitive theorization of empowerment, showing how fragile environments require tailored frameworks that account for instability, mistrust in institutions, and reliance on non-state actors. It also demonstrates that in contexts of structural constraints, even small changes like increased self-efficacy or participation in a cooperative, can be transformative.

Bridging Development Studies and Human Resource Management (HRM)

Finally, the chapter speaks of an important interdisciplinary gap: the link between development studies and human resource management in non-corporate contexts. While HRM traditionally focuses on organizational performance and employee productivity, there is increasing recognition of the need to understand capacity-building and training in humanitarian, community, and informal sector settings.

By applying HRM concepts (e.g., skills development, leadership training, performance outcomes) to a development context, the study shows how women's training is both a human resource intervention and a form of grassroots development. It calls for HRM scholars to engage with the social and political dimensions of training, especially when it occurs outside formal employment structures and is delivered by NGOs rather than corporations.

This contribution opens the door for more cross-sectoral dialogue on how human development can be supported in informal and fragile contexts, where traditional

HRM frameworks are not directly applicable but where principles of capacity, support, and empowerment remain relevant.

POLICY AND PROGRAMMATIC RECOMMENDATIONS

Based on the findings of this study and the multidimensional outcomes observed across rural regions of Lebanon, this section outlines a set of policy and programmatic recommendations for NGOs, donors, and state actors involved in designing and implementing entrepreneurship training. These recommendations aim to move beyond short-term skill-building and toward more integrated, sustainable models of women's empowerment.

Reframe Training Objectives Around Capability Expansion

Most training programs remain oriented around predefined economic goals: launching microenterprises, generating income, or securing employment. While these are valid aims, they often fall short in fragile contexts where structural constraints limit women's ability to act on acquired skills.

Recommendation

Reframe entrepreneurship training not solely as an economic intervention, but as a **capability-building process**. This would include metrics and objectives related to:

- Confidence and voice in public spaces
- Leadership and decision-making within families or cooperatives
- Participation in civic forums and advocacy
- Strengthened peer networks and social capital

Such a reframing would align programming goals more closely with women's lived realities and allow for more diverse, context-sensitive indicators of success.

Extend Program Duration and Deepen Follow-Up Support

A consistent finding in this study was that women lacked sustained support after training. Many reported being "left alone" to navigate challenges such as capital access, market entry, or product development. Without ongoing mentorship or incubation, skill acquisition rarely translates into long-term outcomes.

Recommendation

Funders and NGOs should prioritize **multi-phase programming**, in which training is followed by:

- Access to tailored coaching and mentorship
- Market assessments and business incubation
- Peer-led cooperative structures for continued collaboration
- Grant or microcredit access with capacity-building safeguards

This approach requires longer funding cycles and flexible program design, but it significantly enhances the likelihood of meaningful, lasting empowerment.

Address Structural Constraints Through Ecosystem Engagement

Women in the study frequently cited barriers beyond their control: social norms, household responsibilities, transport limitations, lack of capital, and absence of supportive institutions. Tackling these requires more than training content; it demands systemic thinking.

Recommendation

Programs should **embed structural diagnostics** into their design and advocate for change at the ecosystem level. This may include:

- Partnering with financial institutions to create **gender-sensitive loan products**
- Engaging with male family members to reduce household resistance
- Advocating for transport subsidies or digital access in rural areas
- Coordinating with ministries to align training with **local economic development plans**

NGOs must not only equip individuals but also shape environments in which those individuals can thrive.

Build on Collective Models and Women's Networks

The findings showed that many women flourished when they worked in groups or cooperatives. These structures provide mutual support, reduce risk, and increase bargaining power. They also offer an alternative to the hyper-individualized entrepreneurship model, which may not be realistic or appealing for many rural women.

Recommendation

Expand support for **collective entrepreneurship** models, such as:

- Women-led cooperatives in food processing, textiles, or agriculture
- Rotating savings and credit associations (ROSCAs) linked to enterprise
- Advocacy networks focused on women's rights and economic inclusion

These collective formats reinforce social cohesion and often perform better in fragile, high-risk environments than solo businesses.

Monitor Non-Economic Empowerment Outcomes

Current monitoring and evaluation (M&E) frameworks are heavily weighted toward quantitative outputs: number of businesses launched, income increases, products sold. While valuable, these metrics underrepresent important transformations in identity, agency, and leadership.

Recommendation

Develop hybrid M&E tools that combine quantitative and qualitative indicators across empowerment dimensions, including:

- Voice and leadership in community structures
- Changes in household decision-making dynamics
- Shifts in self-perception and aspiration
- Degree of participation in local governance

Integrating these into project evaluations can offer a fuller picture of impact and support donor learning beyond narrow success/failure metrics.

Strengthen Coordination Among NGOs and Donors

Participants reported attending multiple training courses from different organizations—sometimes with conflicting methodologies or duplicate content. This reflects the lack of coordination and strategic planning in Lebanon's NGO landscape.

Recommendation

Promote **inter-NGO collaboration** through:

- Shared databases of training beneficiaries.
- Joint development of standardized training modules.
- Coordination platforms by region or sector.
- Pooling of follow-up and mentorship resources.

Donors can encourage this by requiring coordination mechanisms as part of funding agreements and by supporting **capacity-building coalitions** among local actors.

Contextualize Training to Regional Realities

The study revealed significant regional differences in outcomes, shaped by infrastructure, mobility, cultural norms, and NGO presence. One-size-fits-all programs risk exacerbating inequalities or missing opportunities.

Recommendation

Customize training content, delivery methods, and post-training support to **local realities**. For example:

- In conservative regions, use women-only spaces and female trainers.
- In areas with poor digital access, prioritize analog tools or local fairs.
- In high-migration areas, link training to diaspora markets or remittances.

Tailoring interventions in this way ensures relevance, acceptance, and impact.

Entrepreneurship training for rural women in fragile contexts must evolve beyond technical workshops and embrace holistic, context-aware strategies that attend to both skills and systems. These recommendations provide a roadmap for reimagining what it means to empower women – not as isolated actors, but as agents navigating complex social, economic, and political terrains.

LIMITATIONS AND DIRECTIONS FOR FUTURE RESEARCH

While this study provides valuable insights into the outcomes of entrepreneurship training for rural women in Lebanon, it is important to recognize its methodological and conceptual limitations. These caveats should guide the interpretation of findings and inform future research agendas that seek to deepen understanding of empowerment dynamics in fragile contexts.

Limitations of Scope and Generalizability

The study focused on five rural governorates in Lebanon, offering a strong regional representation. However, the findings cannot be generalized to all rural women across Lebanon, nor to other fragile states without caution. Differences in political structures, cultural norms, economic infrastructure, and NGO ecosystems may yield very different results in other national settings.

Furthermore, the sample was drawn from women who had completed NGO-run training programs. As such, the study does not capture the perspectives of women who were excluded from or who declined participation, groups that may experience even greater marginalization. This limits the generalizability of the empowerment outcomes reported here, as participants were already engaged in some form of development-oriented programming.

Measurement Constraints

While the study employed a multi-dimensional survey grounded in the Capability Approach and Human Capital Theory, several empowerment indicators are difficult to capture through structured questionnaires, particularly those related to voice, leadership, and civic participation. Nuances such as subtle shifts in household power dynamics, internal motivation, or changes in social perception are better suited to qualitative or ethnographic methods.

The use of self-reported data also introduces potential response bias, especially in environments where beneficiaries may perceive NGOs as sources of future support. Respondents may have overstated program benefits due to gratitude, loyalty, or social desirability effects. While efforts were made to ensure confidentiality and trust, this remains an inherent limitation in field-based research.

Absence of Longitudinal Data

This study employed a cross-sectional design, collecting data at a single point in time after training completion. As a result, it cannot capture long-term sustainability of outcomes. Are businesses still operating a year later? Do women retain or grow their confidence and civic engagement? Are initial gains in income, leadership, or mobility sustained? or do they fade without continued support?

Longitudinal studies would offer deeper insight into trajectories of empowerment, identifying when, how, and why changes persist or reverse. They would also help capture life events such as marriage, childbirth, or crisis shocks that influence empowerment dynamics over time.

Complexity of Attribution

The study design does not allow for strict causal attribution between training and observed outcomes. While statistical associations were found between program participation and various development indicators, other confounding variables may be at play. For example, women who joined cooperatives or launched businesses may have had pre-existing motivation, supportive family structures, or access to additional resources that were not fully accounted for.

Although purposive sampling and hypothesis testing provided some analytical rigor, future studies could benefit from quasi-experimental designs, such as matched samples or control groups, to better isolate program effects.

Contextual Fluidity and External Shocks

Lebanon's ongoing crises – economic, political, and institutional – introduce a degree of volatility that may impact program outcomes in unpredictable ways. A training intervention may appear effective one month and become irrelevant the next due to currency collapse, electricity shortages, or conflict flare-ups. These shifting conditions make it difficult to assess what portion of observed outcomes stem from training and what are shaped by external macro-level events.

As such, development programming and research in fragile contexts must build in adaptive mechanisms and contingency planning. Static program models and rigid research instruments are often outpaced by real-world changes.

Directions for Future Research

To build on this study and address its limitations, several future research directions are recommended:

- Longitudinal studies track women's empowerment over time.
- Mixed-methods research combining surveys with in-depth interviews or participatory approaches.
- Comparative studies across rural and urban areas, or across different fragile states
- Institutional ethnographies examine how NGOs shape empowerment narratives and delivery.

- Intersectional analyses of how age, sect, disability, or migration status mediate training outcomes.
- Policy-oriented evaluations exploring how to integrate entrepreneurship training into national development strategies.

By pursuing these lines of inquiry, researchers can contribute to more grounded, responsive, and effective empowerment interventions that reflect the lived complexities of women in under-resourced settings.

CONCLUSION

This chapter has examined the developmental potential of entrepreneurship training for rural women in Lebanon, offering both empirical evidence and conceptual analysis rooted in the frameworks of Human Capital Theory and the Capability Approach. Drawing on survey data from 284 participants across five rural governorates, the study explored how training interventions shape women's economic, social, and civic lives in the context of a fragile state.

Summary of Key Findings

The findings reveal that entrepreneurship training programs produce a range of multidimensional outcomes, extending well beyond conventional economic indicators. Women reported significant gains in entrepreneurial skills, self-confidence, decision-making, and public participation. Many launched or expanded small income-generating activities, while others joined cooperatives, assumed leadership roles, or engaged in local advocacy efforts.

These outcomes suggest that, when properly designed and delivered, training programs can serve as catalysts for broader empowerment, offering women not only tools for economic activity, but also platforms for voice, visibility, and agency. However, the impact of training is highly contingent on social and structural factors, including access to finance, family support, mobility, and post-training mentorship.

Theoretical and Practical Contributions

The chapter contributes to development theory by proposing a dual-framework model that combines Human Capital Theory's focus on skill acquisition with the Capability Approach's emphasis on freedom, agency, and contextual conversion factors. This model offers a more comprehensive lens for evaluating training outcomes in fragile, gender-constrained environments.

From a practical standpoint, the study offers clear recommendations for policy and program design. These include extending program duration, embedding follow-up support, aligning training with local economic realities, supporting collective models of entrepreneurship, and measuring non-economic indicators of empowerment. Taken together, these recommendations offer a roadmap for more holistic and sustainable empowerment strategies in fragile settings.

Reframing Empowerment in Fragile Contexts

Perhaps the most important insight from this study is that empowerment in fragile contexts is rarely linear, often partial, and always negotiated. It does not necessarily take the form of business success or formal employment. For many women in rural Lebanon, empowerment meant being heard, being connected, or being able to choose small but meaningful shifts that challenge structural inequality and expand the space of possibility.

This calls for a rethinking of how success is defined and measured in development programs. It invites practitioners and scholars alike to move beyond instrumental, output-based models and toward frameworks that honor women's diverse aspirations, capabilities, and constraints. It also challenges us to reimagine entrepreneurship not only as an economic activity, but as a form of social transformation.

Looking Ahead

In a world marked by increasing fragility, displacement, and crisis, the insights from this chapter are more relevant than ever. Women in rural areas must be placed at the center of development strategies, not only as beneficiaries but as agents of change. Training interventions are a start, but true empowerment requires ecosystems of support, political will, and sustained commitment to justice and inclusion.

The journey of empowerment is ongoing, collective, and deeply contextual. By recognizing its complexity and supporting it holistically, policymakers, donors, and NGOs can move closer to the vision of inclusive, transformative development, one that enables all women, regardless of geography or background, to live lives they have reason to value.

REFERENCES

Alkire, S. (2005). Why the capability approach? *Journal of Human Development*, *6*(1), 115–135. DOI: 10.1080/146498805200034275

Becker, G. S. (1964). *Human capital: A theoretical and empirical analysis, with special reference to education* (1st ed.). University of Chicago Press.

Brixiová, Z., Kangoye, T., & Said, M. (2020). Training and support programs for youth entrepreneurs in Africa: A mixed methods evaluation from Ethiopia. *Journal of Development Effectiveness*, *12*(4), 299–319. DOI: 10.1080/19439342.2020.1834876

Brush, C. G., de Bruin, A., & Welter, F. (2009). A gender-aware framework for women's entrepreneurship. *International Journal of Gender and Entrepreneurship*, *1*(1), 8–24. DOI: 10.1108/17566260910942318

Cornwall, A., & Rivas, A. M. (2015). From 'gender equality and 'women's empowerment' to global justice: Reclaiming a transformative agenda for gender and development. *Third World Quarterly*, *36*(2), 396–415. DOI: 10.1080/01436597.2015.1013341

Dacin, P. A., Dacin, M. T., & Matear, M. (2010). Social entrepreneurship: Why we don't need a new theory and how we move forward from here. *The Academy of Management Perspectives*, *24*(3), 37–57.

Joseph, S. (1996). Patriarchy and development in the Arab world. *Gender and Development*, *4*(2), 14–19. DOI: 10.1080/741922010 PMID: 12291310

Kabeer, N. (1999). Resources, agency, achievements: Reflections on the measurement of women's empowerment. *Development and Change*, *30*(3), 435–464. DOI: 10.1111/1467-7660.00125

Lazear, E. P. (2005). Entrepreneurship. *Journal of Labor Economics*, *23*(4), 649–680. DOI: 10.1086/491605

Lewis, D., & Opoku-Mensah, P. (2006). Moving forward research agendas on international NGOs: Theory, agency and context. *Journal of International Development*, *18*(5), 665–675. DOI: 10.1002/jid.1306

Maritz, A., & Brown, C. (2013). Illuminating the black box of entrepreneurship education programs. *Education + Training*, *55*(3), 234–252. DOI: 10.1108/00400911311309305

Martin, B. C., McNally, J. J., & Kay, M. J. (2013). Examining the formation of human capital in entrepreneurship: A meta-analysis of entrepreneurship education outcomes. *Journal of Business Venturing*, *28*(2), 211–224. DOI: 10.1016/j.jbusvent.2012.03.002

McKenzie, D., & Woodruff, C. (2014). What are we learning from business training and entrepreneurship evaluations around the developing world? *The World Bank Research Observer*, *29*(1), 48–82. DOI: 10.1093/wbro/lkt007

Molyneux, M. (1985). Family reform in socialist states: The hidden agenda. *Feminist Review*, *21*(1), 47–64. DOI: 10.1057/fr.1985.28

Mosse, D. (2005). *Cultivating development: An ethnography of aid policy and practice*. Pluto Press.

Narayan, D., Patel, R., Schafft, K., Rademacher, A., & Koch-Schulte, S. (2000). Voices of the poor. Can anyone hear us? (pp. xi+-343). DOI: 10.1596/0-1952-1601-6

Rae, D. (2006). Entrepreneurial learning: A conceptual framework for technology-based enterprise. *Technology Analysis and Strategic Management*, *18*(1), 39–56. DOI: 10.1080/09537320500520494

Robeyns, I. (2006). The capability approach in practice. *Journal of Political Philosophy*, *14*(3), 351–376. DOI: 10.1111/j.1467-9760.2006.00263.x

Roomi, M. A., & Parrott, G. (2008). Barriers to development and progression of women entrepreneurs in Pakistan. *The Journal of Entrepreneurship*, *17*(1), 59–72. DOI: 10.1177/097135570701700105

Sen, A. (1999). *Development as freedom*. Oxford University Press.

Van Eerdewijk, A. H. J. M., Wong, F., Vaast, C., Newton, J., Tyszler, M., & Pennington, A. (2017). White paper: A conceptual model on women and girls' empowerment.

World Bank. (2022). *Labor force participation rate, female (% of female population ages 15+)* [World Bank Open Data Portal]. https://data.worldbank.org/indicator/SL.TLF.CACT.FE.ZS

Chapter 10
Reinforcing Trust Through CSR:
A Study of Lebanese Banking Sector

Mireille Wafaa Gereij
https://orcid.org/0009-0007-9797-8193
Jinan University, Lebanon

Elie Abdallah Basbous
Jinan University, Lebanon

ABSTRACT

Corporate Social Responsibility (CSR) is gaining relevance in Lebanon's banking sector amid severe economic and social crises. This study explores how banks implement CSR, the challenges they face, and stakeholder roles. Though not legally required, CSR is seen as a strategic necessity driven by competition, global trends, and citizen awareness. Banks like Bank Audi, SGBL, and BLF lead efforts in education, health, environment, and ethics. Using mixed methods, the research combines interviews with banking professionals and quantitative analysis of CSR's effect on staff and client engagement. Results show that despite weak regulations, banks use CSR to build trust, improve reputation, and boost internal morale. Yet, instability and inconsistent standards hinder progress. The study recommends creating a standard CSR framework, fostering collaboration, and aligning with global sustainability models to enhance resilience in the Lebanese banking sector.

DOI: 10.4018/979-8-3373-5651-8.ch010

INTRODUCTION

In the past two decades Lebanon has embarked upon an escalating process of market liberalization, unbundling its finance sector and exposing major industries to international competition. Such reforms have been accompanied by recurring political crises, currency crises and banking crises, and have provided the context against which the social license of business has had to been tested. On this background, Corporate Social Responsibility (CSR) has also become a tool with the help of which banks in Lebanon could demonstrate their credibility, take care of stakeholders to solve their issues and promote national development objectives (Chedrawi, Osta & Osta 2020). The practice in Lebanon has not caught up with most jurisdictions in the West where CSR is anchored in the law or compulsory reporting, thus the banking sector should prove to be the right example through which the mobilization of voluntary CSR during the period of institutional weakness is analyzed (Yehia & Szabados 2018).

CSR has ceased to be a peripheral activity of philanthropic nature to Lebanese banks who have found it as a mindful tool of reputation, risk management, and competitive positioning. Studies conducted on emerging economies have established that in a case where formal institutions are weak in a region or country, socially responsible initiatives transmit a market signal that replaces hard quality and governance information (Su, Peng, Tan & Cheung 2014). This signaling role is further populated by the fact that customers in the Middle East consider the role played by banks in social stability in addition to their financial performance (Haddad, Bachkirov & Grishina 2021). Research in the Gulf and Levantine banks validates the claims that CSR reports decrease the information defectiveness and boost stakeholder trustworthiness even in areas with little statutory surveillance (Semenescu & Curmei 2015; Gali, Mohammad & Jamali 2016).

The Lebanese sector illustrates both the opportunities and pitfalls of this voluntary model. Leading institutions such as Bank Audi, Société Générale de Banque au Liban (SGBL) and Banque Libano-Française (BLF) publicize extensive programs in education, healthcare, environmental stewardship and ethical finance, often aligning their narratives with the UN Sustainable Development Goals. Yet analysts continue to question the depth and consistency of these commitments (Adikaram & Holcomb 2023). CSR reports are rarely standardized, external assurance is limited, and macro-financial stress has forced many banks to retrench community spending. A neo-institutional study by Chedrawi, Osta, and Osta (2020) finds that Lebanese banks deploy CSR primarily to secure stakeholder legitimacy rather than to comply with any coherent national policy framework, a conclusion echoed in wider surveys of Arab firms' CSR motivators (Chehimi, Hejase & Hejase 2019).

These contradictions raise two broad research puzzles: (i) Which CSR domains attract the greatest investment and public visibility in Lebanese banking, and why? (ii) How do partnerships with regulators, NGOs, universities and the media shape the effectiveness of bank-led social interventions? Addressing these questions is vital because recent work links targeted CSR to higher employee engagement (Bizri, Wahbi & Al 2021), superior customer satisfaction (Quoc 2022) and, under certain conditions, improved financial returns (Gangi, Mustilli & Varrone 2019; Saeed, Mudliar & Kumari 2023). Conversely, poorly governed CSR can backfire, increasing stakeholder skepticism and eroding brand equity (Usmani & Nayak 2022).

To investigate these issues the present study adopts a dual-method design. First, an inductive qualitative phase uses semi-structured interviews with senior managers in the three largest domestic banks to map how CSR priorities are set, resourced and communicated. Qualitative enquiry is particularly suited to Lebanon's fluid institutional environment, allowing researchers to surface informal routines and power dynamics that survey instruments may miss (Hao, Farooq & Sun 2018). Second, a confirmatory quantitative phase surveys employees and clients to test whether the diffusion factors identified by managers—such as competitive pressure, analyst scrutiny or social-media visibility—actually predict higher stakeholder engagement. Mixed-method CSR studies in other emerging markets have proved valuable for reconciling top-management narratives with grassroots perceptions (Adapa 2015; Waheed et al. 2021).

The Lebanese policy landscape provides only limited scaffolding for such initiatives. Attempts to codify CSR through the Lebanese Standards Institution (Libnor) have been sporadic and largely donor-driven, echoing regional findings that CSR policy often advances faster through market imitation than through government mandate (Hao et al. 2018). In the absence of strong regulation, competitive dynamics become decisive. CSR allows banks to pursue optimal distinctiveness—being similar enough to meet industry norms yet different enough to attract stakeholder attention (Al-Shammari, Banerjee & Dadanlar 2024). For institutions battered by currency devaluation and capital-control controversies, visible CSR programs can also repair damaged legitimacy by demonstrating moral commitment beyond profit (Gómez, et al., 2016).

Yet voluntary systems suffer from informational fragmentation. Separate surveys of Lebanese executives, regulators and journalists reveal uneven awareness of CSR objectives and very limited data sharing across sectors (Chedrawi & Osta 2017). Such fragmentation mirrors global evidence that CSR outcomes improve when disclosure frameworks are standardized and analyst coverage is dense (Chakroun, Matoussi & Mbirki 2017; Adikaram & Holcomb 2023). Without shared metrics it is difficult for Lebanese stakeholders to benchmark progress or to identify green-washing.

Worsening macro-economic conditions further complicate implementation. Since 2019 hyper-inflation and liquidity shortages have curtailed banks' ability to fund long-term projects, while severe public distrust has reduced the reputational dividend of CSR (Chedrawi, Osta, & Osta, 2020). International evidence suggests that during crises firms tend to prioritize CSR activities that directly protect core stakeholders—employees, depositors and local communities—because these groups generate the most immediate legitimacy return (Rhee, Park & Petersen 2018; Muchiri & Muigai 2019). In Lebanon employee-centered initiatives, such as mental-health support and upskilling programs, appear particularly salient, consistent with studies showing that internal CSR can buffer job insecurity and lower turnover intentions (Sayad & Diab 2022; Mengjiao, Arshad & Yating 2024).

Stakeholder engagement therefore becomes the linchpin of credible CSR. Banks that co-create projects with NGOs and academic partners report stronger workforce commitment and clearer development impacts (Bizri et al. 2021). Social-media research likewise indicates that multi-channel dialogue enhances transparency and mitigates the risk of stakeholder cynicism (Hao et al. 2018). Nonetheless, engagement must be matched by robust governance. Cross-country analyses show that board diversity and independent oversight significantly raise the quality of CSR disclosure, especially in finance (Yousfi & Béji 2022; Zaid, Wang & Abuhijleh 2019). Where Lebanese boards remain highly concentrated, advancing CSR may thus require parallel reforms in corporate governance structures.

Adopting regionally compatible reporting templates would allow analysts and civil society to benchmark performance, encouraging a race to the top (Chakroun et al. 2017). Evidence from Asian and European banks shows that CSR delivers sustained value only when embedded in risk management, innovation and product design (Poussing 2018; Singh, Misra & Yadav 2021). Collaborative platforms involving NGOs, municipalities and academic units generate shared ownership and reduce duplication, in line with findings that socially grounded CSR accelerates financial recovery and cultural resilience (M.K. Hastjarjo & Slamet 2020; Wang et al. 2024).

By analyzing how these levers operate under Lebanon's unique constraints, the study contributes to the growing literature on CSR in fragile and emerging banking systems. More broadly, it offers practitioners a roadmap for transforming voluntary CSR from a discretionary marketing tool into a core pillar of sustainable banking strategy, capable of restoring stakeholder trust and reinforcing the sector's role in national recovery.

RESEARCH METHODOLOGY

In fulfilling the exploratory phase of this research, a qualitative design of research was used to find out how CSR practices are viewed and carried out in the banking industry in Lebanon. Because of the changing and multifaceted nature of CSR in Lebanon, semi-structured interviews were chosen as the key approach to primary data, due to the rich potential to explore the realities on the ground and the bank leaders' perspectives on the best approach. The strategy is consistent with the contemporary scholarship of CSR in the banking environment, which holds that in such settings, open-ended qualitative interviews permit permission to capture the subtle perceptions and emerging themes (Adikaram & Holcomb, 2023; Bizri, Wahbi & Al, 2021). Purposive selection of three big identified Lebanese banks namely, Bank Audi, Soci ete Generale de Banque au Liban (SGBL) and Banque Libano-Francaise (BLF) was made because they actively involve in CSR initiatives. They were selected because they are representative, accessible, and diverse in CSR strategies per the request to study weak economies through location-specific CSR orientations (Chedrawi, Osta & Osta, 2020).

A total of 15 in-depth interviews on a one-by-one basis were conducted in a span of six months and each interview duration was about 90 minutes. In order to ensure their consistency, all the interviews were carried out by the same researcher with the help of specific reasoning that applied an iterative approach of moving between empirical data and theory, which is in accordance with current standards of the modern CSR inquiry (Waheed et al., 2021; Haddad, Bachkirov & Grishina, 2021). Within the interview guide, five main areas were targeted to discuss the background of the participants in terms of professional background, knowledge of CSR, history of the institution, implementation of CSR program, and strategic partnerships. Such a loose format allowed the researcher to create specific follow-up questions depending on the experience of the respondents as it proved to increase the reliability of the qualitative research on CSR (Sayad & Diab, 2022). Some interviews happened (because of the COVID-19 restrictions) through Zoom, which recent CSR study successfully proved to be a valid method of fulfilling the purpose of collecting authentic answers in distanced settings (Mengjiao, Arshad & Yating, 2024). Transcribed data of the complete interviews were thematically analyzed to acquire information on how CSR is being diffused within the banks and between banks.

FINDINGS OF CASE STUDY ANALYSIS

Case Study: Bank Audi

Bank Audi traces its origins back to the 19th century and has since evolved into one of Lebanon's largest universal banks. Its transition from a family-controlled institution to a publicly listed joint-stock company has heightened external scrutiny and encouraged a more formalized approach to corporate governance and disclosure. Recent studies argue that analyst coverage intensifies pressure on banks to articulate distinctive, reputation-enhancing CSR agendas (Al-Shammari, Banerjee & Dadanlar 2024). For Audi, that scrutiny coincides with Lebanon's protracted financial crisis, prompting the bank to use CSR both as a legitimacy shield and as a strategic differentiator, a pattern consistent with broader evidence from Middle-Eastern banking (Chedrawi, Osta & Osta 2020).

Interviews reveal that Audi frames CSR as integral to its value-creation model rather than as discretionary philanthropy. Executives describe a portfolio that channels resources into education, primary healthcare and climate-related projects—domains highlighted by Bizri, Wahbi and Al (2021) as most likely to strengthen employee engagement in crisis economies. Internally, the bank emphasizes transparent governance, enterprise-wide risk management and digital process automation. The latter frees operational capacity and finances that can be redeployed to social projects, reflecting findings that technology-enabled efficiency frequently underwrites credible CSR in emerging markets (Waheed et al. 2021). Externally, Audi collaborates with universities and NGOs to co-design scholarship schemes and community recycling campaigns. Such multi-stakeholder alliances correspond to evidence that collaborative CSR platforms generate stronger legitimacy than firm-only initiatives (Haddad, Bachkirov & Grishina 2021). Nevertheless, executives concede that hyper-inflation and deposit flight have constrained budgets, echoing Adikaram and Holcomb's (2023) observation that market turbulence forces banks to prioritize CSR projects with immediate reputational returns.

Case Study: Banque Libano-Française (BLF)

BLF occupies a leading position in retail and corporate finance and has cultivated a reputation for stringent compliance. In the absence of robust national CSR regulation, the bank leverages international frameworks—particularly ISO standards and UN Sustainable Development Goals—to signal responsible conduct, mirroring the signaling logic highlighted by Su, Peng, Tan and Cheung (2014).

Senior managers describe CSR as "embedded in risk culture," a stance aligning with research that links integrated CSR–ERM systems to higher reputational capital (Usmani & Nayak 2022). BLF's internal agenda centers on ethics training, equal-opportunity hiring and employee-led volunteering. Recent work by Sayad and Diab (2022) shows that staff who perceive authentic CSR are more likely to champion organizational change—an outcome BLF seeks to harness as it digitalizes retail channels. Externally, the bank concentrates on public-health campaigns and reforestation drives, often co-funded with municipal authorities. These programs aim to build community resilience, which Rhee, Park and Petersen (2018) identify as critical where local stakeholder pressures are intense. Interviewees note that partnerships with academic institutions have boosted the credibility of impact assessments, answering Chakroun, Matoussi and Mbirki's (2017) call for multi-support disclosure to curb stakeholder skepticism. Despite such progress, BLF managers acknowledge that limited national data-sharing platforms hamper benchmarking, a concern echoed in regional CSR scholarship (Gangi, Mustilli & Varrone 2019).

Case Study: Société Générale de Banque au Liban (SGBL)

SGBL is part of an international banking group and therefore faces dual accountability to Lebanese regulators and global headquarters. Comparative research suggests that multinational affiliation can raise the salience of CSR norms and accelerate adoption of global best practice (Gali, Mohammad & Jamali 2016).

Internally, SGBL frames CSR as a pillar of its "responsible banking" charter. It has expanded financial-inclusion lines for women-led micro-enterprises and refugees, reflecting findings that CSR coupled with inclusive finance widens developmental impact (Singh, Misra & Yadav 2021). The bank's governance team ties executive bonuses to ESG targets, in step with emerging evidence that incentive alignment strengthens CSR authenticity (Valentine 2024). Externally, SGBL funds vocational-training hubs and mobile health clinics in under-served regions. These projects respond to local expectations of socially restorative banking and mirror initiatives documented in other crisis-ridden economies where banks act as quasi-public welfare agents (Muchiri & Muigai 2019). Interview data indicate that SGBL's international parent exerts upward pressure for transparent carbon reporting, leading the Lebanese subsidiary to pilot renewable-energy lending products. Such product innovation supports Poussing's (2018) thesis that CSR can catalyze sustainable innovation adoption. Nonetheless, managers admit that the liquidity squeeze has curtailed new project rollouts, underscoring Waheed et al.'s (2021) insight that external CSR is highly vulnerable to macro-economic shocks.

COMPARATIVE INSIGHTS AND CONCLUSION

The cross-case analysis highlights both convergence and divergence in Lebanese banks' CSR strategies. All three institutions locate CSR at the intersection of legitimacy repair and competitive differentiation, supporting Al-Shammari, Banerjee and Dadanlar's (2024) argument that firms pursue "optimal distinctiveness" under heavy analyst gaze. Each bank embeds CSR in governance rhetoric—transparency, anti-money laundering, diversity—yet the scope and funding intensity vary with ownership structure and international exposure. Banks with multinational ties (SGBL) or strong compliance reputations (BLF) exhibit more formalized ESG metrics, whereas locally rooted Audi emphasizes community-embedded projects, resonating with Chedrawi, Osta, and Osta's (2020) neo-institutional reading of stakeholder legitimacy.

The findings also reaffirm the contingency of CSR under financial distress. While interviewees across banks champion education, healthcare and environmental stewardship, budget reallocations towards core solvency imperatives have narrowed programme breadth. This mirrors global evidence that crisis contexts push banks to select CSR activities delivering rapid reputational dividends and internal cohesion (Adikaram & Holcomb 2023). Despite these constraints, the cases reveal an emerging consensus that CSR must shift from ad-hoc philanthropy to integrated strategy, intertwining with digital transformation, risk management and talent retention. Such integration promises to reinforce stakeholder trust and to embed Lebanese banks within international sustainability circuits, yet realization depends on transparent impact measurement and collaborative platforms to counteract the country's chronic data fragmentation.

Future quantitative analysis—outlined in the following chapter—will test whether the diffusion factors identified qualitatively, such as analyst scrutiny, international affiliation and employee engagement, statistically predict higher stakeholder satisfaction and financial resilience. In doing so, it will extend regional CSR scholarship and offer actionable insights to practitioners seeking to balance survival imperatives with the long-run socio-economic responsibilities of Lebanese banking.

ANALYSIS OF RESULTS

Referring to the interview questions previously noted, this section provides a conclusion that incorporates the quantitative results. Each question is listed and followed by relevant conclusions, highlighting key points that support the existing body of knowledge in the available literature on the subject. It is important to note that these questions are derived based on the interview themes, structured by the researcher to ensure relevance, the theoretical framework of the study, sample specification, dialogue quality, and the strategy for data reporting future perspectives. The questions and corresponding conclusions are outlined below.

Questions and Relevant Conclusions

Table 1.

Sections	Questions
Contact	Mr. Director, thank you for taking the time to discuss the perception of CSR among clients and employees of the bank.
Theme 1: Employees' Perception of CSR	Are social responsibility and sustainable development topics regularly discussed in your professional environment at the bank? Are employees satisfied? Are they engaged in this bank? To what extent do they experience workplace anxiety? Does your organization frequently communicate its environmental and social commitments? What specific responsibilities are assigned to you regarding sustainable development?
Theme 2: Clients' Perception of CSR	Do your clients understand CSR? Are they satisfied? What image do they have of your bank? Do you play an active role in communicating CSR initiatives to a clientele that may or may not be aware of the bank's sustainability actions?
Theme 3: Future Perspectives	Based on your expertise, what do you consider the most pressing ethical commitments for banks regarding sustainable development and the current financial crisis? As a direct representative of the bank's clients, do you believe CSR communication benefits your bank? Do you find CSR communication to be authentic, reliable, and transparent? Can we now consider CSR as a marketing positioning strategy in the sustainable development market?
Conclusion	Thank you again, Mr. Director, for your time.

The following table presents the various themes covered in our interview guide.

Themes of the Interview Guide

- Personal introduction and purpose of the interviews.
- Profile of the bank executive (age, education, values, principles, and objectives) and bank profile (history, size, etc.).
- Employees' perception of CSR.
- Clients' perception of CSR.
- Future perspectives (the real ethical challenges for banks regarding sustainable development, marketing positioning in the sustainable development market).

The interviews were conducted with executives face-to-face at bank branches between May and September 2020, using the interview guide. Each interview lasted between one and two hours. All interviews were recorded with the respondents' consent and fully transcribed. The sample size was determined based on semantic saturation.

Overview of Bank Executives

Table 2.

Bank Name	Year Founded	Region	Position	Gender	Age	Education
Audi	1962	Beirut	CSR Manager	Female	50	Master's Degree
		Mina	Regional Director	Male	59	Bachelor's Degree
		Azmi	Director	Female	46	Bachelor's Degree
		Boulevard	Director	Male	44	Master's Degree
		Amioun	Director	Male	48	Bachelor's Degree
BLF	1967	Beirut	Quality Manager	Male	43	Master's Degree
		Kousba	Director	Male	43	Master's Degree
		Mina	Director	Female	40	Bachelor's Degree
		Tripoli	Director	Male	46	Master's Degree
		Batroun	Director	Female	50	Bachelor's Degree
SGBL	1953	Beirut	CSR Manager	Female	42	Master's Degree
		Tripoli	Director	Female	60	Master's Degree
		Kfaraaka	Director	Male	46	Bachelor's Degree
		Boulevard	Director	Male	43	Bachelor's Degree
		Halba	Director	Male	54	Bachelor's Degree

This section of our research aims to illustrate and explain the CSR phenomenon in Lebanese banks from the perspective of executives and directors. We rely on theoretical knowledge to contextualize our empirical observations, conducting a back-and-forth analysis between collected data and existing theories (Charreire-Petit & Durieux, 2014).

By their nature, banks especially in Lebanon exhibit a high degree of homogeneity as organizations. Consequently, CSR strategies tend to be similar across institutions, driven by shared training, values, and business environments. This section presents the results of our empirical study conducted on three Lebanese banks that adhere to CSR standards. Our findings reveal key elements that shape the strategic vision of fifteen executives, including CSR and quality managers, as well as general directors. We also highlight how these executives perceive CSR certification and its role in enhancing corporate social responsibility both within the banks and among their clients.

Interviews with Bank Audi Executives

Audi Bank's CSR approach follows a strategic framework based on five fundamental pillars: corporate governance, economic development, human development, community development, and environmental protection. The CSR Manager at the Beirut headquarters stated: "I have always been passionate about ethics and CSR theories… everything related to sustainable development and the well-being of employees and clients." She emphasized: "Audi Bank's practices are at the core of its business model, ensuring the responsibility and transparency that the bank aspires to maintain in order to remain the most reliable financial institution for both internal and external stakeholders."

Four executives at Audi insisted that the bank's goal is to amplify its social impact by leveraging its corporate strength to collaborate with stakeholders and contribute to a more resilient and stable economy. Mr. Khodr from the Azmi branch added: "Audi collaborates with various organizations to build a strong societal ecosystem and resilient communities while addressing humanitarian, health, social, educational, and cultural aspects."

Meanwhile, Mrs. Asad, director of the Amioun branch, explained that the bank aligns its objectives with international standards such as ISO 26000, the UN Global Compact, and the Sustainable Development Goals (SDGs): "At Audi Bank, employees are generally satisfied and engaged. There is constant communication regarding environmental and social commitments, including high-quality education, gender equality, and infrastructure development. However, the current economic situation has increased workplace anxiety and daily stress across the banking sector."

Mr. Dabliz, director of the Boulevard branch, emphasized that customer loyalty and trust in the bank have declined significantly since the onset of the financial crisis, altering the once impeccable image of the Lebanese banking sector.

Regarding the real ethical challenges banks face in sustainable development, all Audi executives agreed that overcoming current difficulties requires leveraging the bank's longstanding reputation and investing in innovation. The regional director stated: "Our competitive advantage stems from Audi Bank's historically strong image. Banks have long been perceived as market leaders with solid positioning. Innovation has allowed us to expand and explore new markets."

Executives believe that CSR can only be achieved through strong leadership and a deeply ingrained corporate culture, following the philosophy: 'We innovate internally… and enjoy the process.'

Interviews with SGBL Bank

SGBL asserts its adherence to the bank's ethical code and corporate social responsibility (CSR) principles. The CSR officer stated: "Our ambition is to become a leader in sustainable finance and development, leveraging innovation and expertise to establish new standards. However, the current economic situation has altered our future projects, leading to heightened workplace anxiety among employees and decreased satisfaction. Meanwhile, the bank's reputation has suffered, causing a loss of customer trust."

She further added:

"Despite these challenges, the core pillars of our CSR ambition remain interwoven. Two of them define our responsible banking framework: fostering a culture of accountability and being a responsible employer. The other two serve as catalysts for positive transformation: facilitating ecological transition and supporting regional development."

The branch director of Boulevard commented: "Our CSR goals before 2019 were closely linked to our commitments €300 billion allocated to sustainable finance from 2022 to 2025. We also introduced a new remuneration policy aimed at retaining employees through attractive and fair salaries while aligning compensation with the group's social and environmental responsibility objectives." Meanwhile, the branch director of Mina emphasized SGBL's commitment to CSR and the importance of sustainable development, stating: "The bank has always been responsible towards its employees and customers, who appreciate the commitment of SGBL bankers and the positive image our bank maintains among its clientele."

She also noted: "We voluntarily contribute to sustainable development, both within our operations and in our interactions with stakeholders, particularly employees and clients." The branch director of Halba highlighted the role of customer communication in promoting sustainable development initiatives, explaining SGBL's efforts and achievements in this field: "Due to the ongoing crisis, the bank's image has deteriorated among both clients and employees, resulting in dissatisfaction and a loss of trust." He added: "As a direct liaison with bank customers, I find CSR communication to be an advantage for our group. It contributes to a better image of the bank among the staff and the customers, which results in the development of a specific approach to marketing and newly enhances the positioning of the bank and its reputation (SGBL).

In relation to the first theme, the existing body of knowledge implies that diverse ways in which corporate social responsibility (CSR) can contribute to creating a sustainable environment at the workplace should be considered. When established in an organization, CSR can be used as a strategic tool to improve engagement between the employees and also to be used to make our environment sustainable.

When employees adopt concept of CSR, they will be more satisfied about their jobs which in turn, may help the company in building a better reputation and develop good working climate. The mentioned notion is rather well-documented in the existing literature, which highlights the effects of CSR-driven activities on the employee satisfaction levels and the overall organizational perception.

Besides employee engagement, a direct effect on organization performance and business outcome is the workplace commitment. In the present research, employees in banks have a stronger level of job satisfaction when they can recognize CSR as being part of the business strategy in the company they work. Glimmers of the same have been found in the studies done before thus supporting the notion that CSR instills feelings of belonging and purpose among the employees. The work is also useful to complement the literature by defining the context of CSR in banking firms and by using a mixed-method to examine the perceptions of the employees.

EMPLOYEE WELL-BEING AND WORKPLACE ANXIETY

Interestingly, the analysis also showed that the employee experience is not universal in the context of workplace anxiety, which is why it is necessary and practical to establish such support mechanisms as counseling, mentorship, and mental health programs. When managing stress and anxiety in the work place, this can have a positive effect in the attitude that employees are having regarding the CSR activities and also it can make the employees have a good overall job satisfaction. Scholars have stressed the importance of the strategy that helps to eliminate burnout and incidence of employee retention using well-being approaches. Through the convergence of the CSR efforts with employee well-being programs, the banks would be able to generate a long-term positive impact capable of improving work efficiency and productivity. Also, great leadership is capable of building the culture of sustainability in the organization, which will pressure the employees to engage in environmentally friendly practices not only within the professional field but also in personal life.

CUSTOMER PERCEPTION AND CSR COMMUNICATION

About the second theme, the results have shown that the level of workers to be aware of CSR has drastically increased especially following the COVID-19 outbreak. It is believed that companies with high importance of the sustainability initiatives are more likely to have a better customer loyalty and brand perception. Also, customers are likely to record improved levels of satisfaction where banks are pursuing CSR initiatives, which are in line with sustainable objectives. There is a direct relationship between perception and customer perception and the CSR business strategy. This idea can also be viewed in line with the available literature that indicates that transparent CSR communication plays a crucial role in boosting corporate reputation.

Corporate Social Responsibility (CSO) has become an important aspect in the realms of Lebanese banking industry which dictates its direction, operations and its interaction with various stakeholders. This paper has covered the role of CSR in financial institutions in Lebanon, its major challenges, practices and the stakeholders in it. Examination of the CSR practices in the large Lebanese banks including the Bank Audi, Societe Generale De Banque Au Liban (SGBL) and Banque Libano-Francaise (BLF) has revealed the way banks have incorporated the social responsibility aspect in their lines of businesses, notwithstanding the economic and the financial uncertainty in the country.

DISCUSSION: CSR PROGRESS, CONSTRAINTS, AND STAKEHOLDER DYNAMICS IN LEBANESE BANKING

Lebanese banks have in the last 10 years amassed CSR portfolios into expanding beyond discretionary philanthropy into policies that sound much like international standards. Indeed, empirical research on finance in the Middle East demonstrates the growing tendency of sustainability pledges to be formed with reference to global schemes, such as the ISO 26000, the UN Global Compact, and industry-tailored AML instructions, in order to safeguard the legitimacy and cover ethical investors (Chedrawi, Osta, & Osta, 2020). This has been the trend in other emerging markets whose adoption of CSR is driven by the market and not the state (Adapa, 2015). The primary-health Units, education grants, as well as carbon-reduction initiatives, focus on leading Lebanese institutions based on the strategic logic that CSR can enhance corporate ethical identity and brand differentiation (Dutta et al., 2021). Scholar-practitioner research also demonstrates that socially responsible initiatives integrated with a risk management policy may enhance the valuation of analysts and minimize the capital-cost penalties (Adikaram & Holcomb, 2023; Al-Shammari, Banerjee & Dadanlar, 2024). Concisely, the industry has registered progress in institutionalizing CSR as one of the pillars of competitive strategy.

Yet these achievements are tempered by structural and macro-economic constraints. First, Lebanon lacks a binding CSR statute or uniform disclosure code; compliance remains voluntary and patchy, mirroring broader Levantine trends (Chehimi, Hejase & Hejase, 2019). Comparative evidence from Tunisia and Morocco suggests that the absence of enforceable rules amplifies information asymmetry and weakens stakeholder confidence (Chakroun, Matoussi & Mbirki, 2017). Second, the country's prolonged financial crisis—marked by currency collapse and liquidity shortages—has forced banks to divert resources from long-term social projects to balance-sheet survival, echoing findings that economic stress stifles CSR outlays in frontier economies (Saeed, Mudliar & Kumari, 2023). Directors interviewed for this study confirm that budget cuts have delayed new community programs and curtailed impact-assessment efforts, a scenario consistent with Haddad, Bachkirov and Grishina's (2021) crisis-driven CSR retrenchment model. Finally, inadequate governance oversight limits accountability: board diversity and independent CSR committees—both linked to superior disclosure quality—remain underdeveloped in Lebanese banking (Yousfi & Béji, 2022).

Notwithstanding these barriers, stakeholder engagement emerges as the decisive mediator between CSR intent and social impact. Consistent with Rhee, Park and Petersen's (2018) work on local pressures, Lebanese banks rely on coalitions with NGOs, universities, and municipalities to deliver health clinics, reforestation drives, and financial-literacy workshops. Such partnerships not only pool resources but also signal authenticity to external watchdogs (Chedrawi & Osta, 2017). Internally, CSR is associated with stronger affective commitment and lower turnover intentions; survey research in Egypt and China demonstrates that employees who perceive their employer as socially responsible report higher engagement and resilience during crises (Sayad & Diab, 2022; Waheed et al., 2021). Our interviews echo these findings: staff view CSR volunteering days and scholarship funds as tangible proof of the bank's long-run commitment, even under austerity. Where programs falter, workplace anxiety escalates—a dynamic flagged in recent meta-reviews on CSR and employee well-being (Mengjiao, Arshad & Yating, 2024).

Customer-facing outcomes are equally salient. Studies across Asia and Europe show that visible CSR boosts satisfaction and, via reputational gains, financial performance (Quoc, 2022; Gangi, Mustilli & Varrone, 2019). In Lebanon, however, widespread distrust of financial institutions has diluted this premium. Marketing research indicates that transparent story-telling and third-party audits are essential to re-establish consumer faith after scandals (Tinker, Barnes & Wilson, 2017). Lebanese banks have responded by heightening disclosure on social media and embedding ESG metrics in annual reports—tactics proven to moderate skepticism and differentiate brands in saturated markets (Hao, Farooq & Sun, 2018).

Taken together, the findings suggest that CSR in Lebanon's banking sector occupies a liminal space: sufficiently developed to influence stakeholder attitudes, yet vulnerable to economic shocks and regulatory gaps. Addressing these limitations requires a three-pronged agenda. First, policymakers and the Banque du Liban could introduce a baseline reporting framework, drawing on regional precedents that link mandatory disclosure to reduced information asymmetry (Semenescu & Curmei,

2015). Second, banks should integrate CSR KPIs into enterprise-risk dashboards, ensuring that cutbacks do not erode long-term reputational capital—a recommendation supported by enterprise-wide studies on CSR and financial stability (Usmani & Nayak, 2022). Third, deeper collaboration with civil society and academia can enhance monitoring, foster social innovation and amplify legitimacy, consistent with community-centered CSR paradigms observed in post-pandemic contexts (M.K, Hastjarjo & Slamet, 2020).

In conclusion, while Lebanese banks have advanced appreciably in aligning CSR with international norms, a fragile economic setting and weak regulatory scaffolding continue to constrain systemic impact. Sustained stakeholder engagement, rigorous governance and strategic integration will be critical if CSR is to transcend symbolic gestures and become a durable lever of trust and resilience in Lebanon's financial landscape.

CHALLENGES HINDERING EFFECTIVE CSR IMPLEMENTATION

This empirical evidence proves that Lebanese banks have been adopting the CSR rhetoric, but there still exists a series of inter-locking constraints that are making it blunt. First of all, there is a lack of binding regulatory framework: CSR disclosure is not compulsory, enforcement is inefficient and the reporting template differs across institutions, and the same thing applies in other Levantine markets where laissez-faire regulation is the driving force behind inconsistency (Chedrawi, Osta & Osta 2020; Chakroun, Matoussi & Mbirki 2017). This weakness is further aggravated by economic turbulence. Chronic liquidity crunches, devalued currencies and evaporating provisioning expenses have influenced the focus of management regarding the long-term sustainability programs to quick fix solvency in case as much so in crisis economies where financial stress decreases CSR outlay (Saeed, Mudliar & Kumari 2023; Haddad, Bachkirov & Grishina 2021).

The third impediment is the participation of the population. Research indicates that the problem of low visibility and community participation, which decreases legitimacy and behavioral change, can be addressed through the use of digital channels that boost CSR narratives, but most Lebanese initiatives continue to impede scope and functionality because of the limited use of digital channels (Hao, Farooq & Sun 2018; M.K, Hastjarjo & Slamet 2020). The fourth obstacle is fragmentation-related: since there is no common industry platform, projects tend to either become redundant or closely grouped by location, multiplying value and hindering system-wide benefit and stakeholder cynicism (Chedrawi & Osta 2017; Gali, Mohammad & Jamali 2016). Last but not least, inconsistent standards of governance allow to transfer a certain level of green-window dressing; without diverse boards or CSR committees, some banks consider social responsibility as a marketing facade but not as a management process (Yousfi & Beji 2022; Zaid, Wang & Abuhijleh 2019).

RECOMMENDATIONS FOR ENHANCING CSR IN LEBANESE BANKS

Addressing these constraints calls for a multilayered reform agenda. At the policy level, legislators and the Banque du Liban could introduce a light-touch but compulsory disclosure code aligned with global benchmarks, thereby narrowing information asymmetry and helping analysts to reward genuine performers (Al-Shammari, Banerjee & Dadanlar 2024; Semenescu & Curmei 2015). Within the banks, building financial buffers for sustainability spending is vital; ring-fenced CSR budgets linked to balanced-scorecard targets can insulate projects from cyclical shocks and have been shown to correlate with stronger reputational pay-offs (Valentine 2024; Waheed et al. 2021). Because legitimacy is socially co-constructed, deeper collaboration with NGOs, universities and municipal bodies should be made routine, a step proven to enhance programme reach and employee engagement (Rhee, Park & Petersen 2018; Bizri, Wahbi & Al 2021). Awareness deficits can be tackled through data-driven storytelling on social media and community roadshows, strategies that international research links to improved stakeholder trust and participation (Hao et al. 2018). Critically, CSR must migrate from a standalone department into the core of competitive strategy: embedding social metrics in enterprise dashboards ensures that executives weigh societal impact alongside return on equity, a linkage shown to strengthen the CSR–performance relationship in Asian banking (Aprillianto & Sayekti 2017; Dutta et al. 2021). Last, transparency must extend beyond glossy reports; real-time dashboards on AML compliance, climate exposure and social-impact KPIs would answer analyst demands for verifiable evidence and reinforce the positive signaling effect documented by Adikaram and Holcomb (2023) and Usmani and Nayak (2022).

THE FUTURE OF CSR IN LEBANON'S BANKING SECTOR

Looking forward, the trajectory of CSR will hinge on the sector's ability to formalize these reforms while capitalizing on technology. Fintech applications, open-banking APIs and blockchain-based traceability tools can streamline giving, widen financial inclusion and generate auditable data flows, thereby lowering the cost of trust (Poussing 2018; Wang et al. 2024). Simultaneously, integrating environmental stress tests and social-impact scenarios into risk models will align CSR with prudential oversight, an approach that has lifted performance resilience in European institutions (Gangi, Mustilli & Varrone 2019). Strategic change theory also suggests that banks able to balance conformity to global ESG norms with distinctive local initiatives will attract socially conscious investors without losing cultural relevance (Al-Shammari et al. 2024). Customer studies reinforce this outlook: retail clients increasingly reward transparent, impact-linked products such as green loans and social bonds, signaling a market space for innovation-driven CSR (Quoc 2022).

FINAL THOUGHTS

In sum, CSR has become an indispensable—if still imperfect—lever of legitimacy and competitive renewal for Lebanese banks. Progress is visible in the alignment with international standards and the proliferation of community programs, yet regulatory lacunae, macro-economic stress and governance asymmetries continue to curb full impact. Elevating CSR from episodic philanthropy to strategic imperative will require clearer rules, sturdier financial underpinnings, richer stakeholder coalitions and data-centric transparency. Implemented cohesively, these measures can convert CSR into a durable asset that safeguards reputational capital, nurtures social inclusion and reinforces the banking sector's capacity to drive Lebanon's economic recovery (Chedrawi, Osta, &Osta, 2020; Gangi et al. 2019).

REFERENCES

Adapa, S. (2015). The extent of corporate social responsibility engagement in malaysian banks offering islamic banking services. *Journal of Internet Banking and Commerce, 20*(2). Advance online publication. DOI: 10.4172/1204-5357.1000113

Adikaram, R., & Holcomb, A. (2023). Exploring the role of analysts in identifying and communicating the value of bank CSR activity. *International Journal of Bank Marketing*. Advance online publication. DOI: 10.1108/IJBM-12-2022-0557

Al-Shammari, M., Banerjee, S. N., & Dadanlar, H. H. (2024). Strategic change and firms' optimal distinctiveness: The moderating effect of analyst coverage. *British Journal of Management*. Advance online publication. DOI: 10.1111/1467-8551.12873

Aprillianto, B., & Sayekti, Y. (2017). *The role of firm strategy to intervene the influence of corporate social performance on corporate financial performance*. Gatr Journal of Finance and Banking Review., DOI: 10.35609/jfbr.2017.2.3(4)

Bizri, R.M., Wahbi, M. and Al, J.H. (2021). The impact of CSR best practices on job performance: The mediating roles of affective commitment and work engagement. *Journal of Organizational Effectiveness People and Performance*. .DOI: 10.1108/JOEPP-01-2020-0015

Brizek, M.G. (2018). An examination regarding the role of corporate social responsibility (CSR) within financial institutions: A literature review. *Journal of Business Management & Economics*. .DOI: 10.15520/jbme.v6i04.2117

Chakroun, R., Matoussi, H., & Mbirki, S. (2017). Determinants of CSR disclosure of tunisian listed banks: A multi-support analysis. *Social Responsibility Journal, 13*(3), 552–584. Advance online publication. DOI: 10.1108/SRJ-04-2016-0055

Charreire-Petit, S., & Durieux, F. (2014). Chapitre 3. Explorer et tester: les deux voies de la recherche. In *Méthodes de recherche en management* (Vol. 4, pp. 76-104). Dunod.

Chedrawi, C., Osta, A., & Osta, S. (2020). CSR in the lebanese banking sector: A neo-institutional approach to stakeholders' legitimacy. *Journal of Asia Business Studies, 14*(2), 143–157. Advance online publication. DOI: 10.1108/JABS-03-2018-0093

Chedrawi, C., & Osta, S. (2017). *CSR in the banking sector: A legitimacy approach to the shareholders' and stakeholders' debate*. Marketing and Branding Research., DOI: 10.33844/mbr.2017.60272

Chehimi, Ghada M, Hejase, Ale J and Hejase, Nour H (2019). An assessment of lebanese companies' motivators to adopt CSR strategies. *Open Journal of Business and Management*. .DOI: 10.4236/ojbm.2019.74130

Corporate social responsibility in international business. (2024). *Qalaai Zanist Scientific Journal*. .DOI: 10.25212/lfu.qzj.8.5.50

Dutta, S., Katti, S., Phani, B. V., & Zhu, P. (2021). Corporate social responsibility spending as a building block for sustainable corporate ethical identity: Lessons from indian business groups. *Managerial and Decision Economics*. Advance online publication. DOI: 10.1002/mde.3412

Gali, N., Hajjar, D., & Jamali, I. (2016, June 06). Mohammad and Jamali, I. (2016). The corporate governance and social responsibility nexus in the lebanese banking industry. *Corporate Governance (Bradford)*, *16*(3), 609–638. Advance online publication. DOI: 10.1108/CG-08-2015-0109

Gangi, F., Mustilli, M., & Varrone, N. (2019). The impact of corporate social responsibility (CSR) knowledge on corporate financial performance: Evidence from the european banking industry. *Journal of Knowledge Management*, *23*(1), 110–134. Advance online publication. DOI: 10.1108/JKM-04-2018-0267

Gómez, S.G., Eroğul, Murat Şakir and Barragan, S. (2016). Similarities and differences in teaching corporate social responsibility: Evidence from mexico and canada. *Journal on Efficiency and Responsibility in Education and Science*. .DOI: 10.7160/eriesj.2016.090303

Haddad, P.E., Bachkirov, Alexandre Anatolievich and Grishina, O.A. (2021). Comparative CSR decision-making in the middle east: An exploratory study. *International Journal of Islamic and Middle Eastern Finance and Management*. .DOI: 10.1108/IMEFM-01-2020-0017

Hao, Y., Farooq, Q., & Sun, Y. (2018). Development of theoretical framework and measures for the role of social media in realizing corporate social responsibility through native and Non-native communication modes: Moderating effects of Cross-cultural management. *Corporate Social Responsibility and Environmental Management*, *25*(4), 704–711. Advance online publication. DOI: 10.1002/csr.1523

Jamali, D., Abdallah, H., & Matar, F. (2016). Opportunities and challenges for CSR mainstreaming in business schools. *International Journal of Technology and Educational Marketing*, *6*(2), 1–29. Advance online publication. DOI: 10.4018/IJTEM.2016070101

Jia, X. (2019). Corporate social responsibility activities and firm performance: The moderating role of strategic emphasis and industry competition. *Corporate Social Responsibility and Environmental Management*. Advance online publication. DOI: 10.1002/csr.1774

Mengjiao, Z., Arshad, M. A., & Yating, L. (2024). Examining the relationship between corporate social responsibility and employee turnover intentions: A comprehensive review. *International Journal of Academic Research in Business & Social Sciences, 14*(1). Advance online publication. DOI: 10.6007/IJARBSS/v14-i1/20003

M.K, D., Hastjarjo, S. and Slamet, Y. (2020). CSR in new normal: Realizing good citizenship and corporate social performance for community development. .DOI: 10.4108/eai.4-8-2020.2302429

Muchiri, M., & Muigai, R. G. (2019). *Effect of environmental CSR activities on the financial performance of financial institutions in kirinyaga county*. Rjfa., DOI: 10.7176/RJFA/10-5-07

Poussing, N. (2018). Does corporate social responsibility encourage sustainable innovation Adoption?¡scp¿E¡/Scp¿mpirical evidence From¡scp¿L¡/Scp¿uxembourg. *Corporate Social Responsibility and Environmental Management*. Advance online publication. DOI: 10.1002/csr.1712

Quoc, K. (2022). Corporate social responsibility and bank's performance under the mediating role of customer satisfaction and bank reputation. *Emerging Science Journal, 6*(6), 1409–1429. Advance online publication. DOI: 10.28991/ESJ-2022-06-06-012

Rhee, Y. P., Park, C., & Petersen, B. K. (2018). The effect of local stakeholder pressures on responsive and strategic CSR activities. *Business & Society*. Advance online publication. DOI: 10.1177/0007650318816454

Saeed, M. M., Mudliar, M., & Kumari, M. (2023). Corporate social responsibility and financial performance nexus: Empirical evidence from ghana. *Corporate Social Responsibility and Environmental Management, 30*(6), 2799–2815. Advance online publication. DOI: 10.1002/csr.2516

Sayad, S. E., & Diab, A. (2022). Bank employee perceptions of corporate social responsibility practices: Evidence from egypt. *Sustainability (Basel), 14*(3), 1862. Advance online publication. DOI: 10.3390/su14031862

Semenescu, A. and Curmei, Cătălin Valeriu (2015). Using CSR to mitigate information asymmetry in the banking sector. *Management & Marketing*. .DOI: 10.1515/mmcks-2015-0021

Singh, K., Misra, M., & Yadav, J. (2021). Corporate social responsibility and financial inclusion: Evaluating the moderating effect of income. *MDE. Managerial and Decision Economics*, *42*(5), 1263–1274. Advance online publication. DOI: 10.1002/mde.3306

Su, W., Peng, M. W., Tan, W., & Cheung, Y. (2014). The signaling effect of corporate social responsibility in emerging economies. *Journal of Business Ethics*. Advance online publication. DOI: 10.1007/s10551-014-2404-4

Tinker, M., Barnes, L. and Wilson, D.T. (2017). Corporate social responsibility strategy and its influence over consumer purchasing decisions in financial institutions. *Us-China Education Review B*. .DOI: 10.17265/2161-6248/2017.05.003

Ullah, M.H. and Rahman, M.A. (2015). Corporate social responsibility reporting practices in banking companies in bangladesh. *Journal of Financial Reporting and Accounting*. .DOI: 10.1108/JFRA-05-2013-0038

Usmani, A. and Nayak, P. (2022). Effect of CSR and enterprise risk management system on corporate reputation: An empirical study. *International Journal of Social Sciences & Economic Environment*. .DOI: 10.53882/IJSSEE.2022.0702004

Valentine, I. (2024). Corporate social responsibility (CSR) practices and financial performance of firms. *American Journal of Finance*. .DOI: 10.47672/ajf.1811

Waheed, A., Zhang, Q., Zafar, A. U., Zameer, H., Ashfaq, M., & Nusrat, A. (2021). Impact of internal and external CSR on organizational performance with moderating role of culture: Empirical evidence from chinese banking sector. *International Journal of Bank Marketing*, *39*(4), 499–515. Advance online publication. DOI: 10.1108/IJBM-04-2020-0215

Wang, H., Zhang, T., Wang, X., Zheng, J., Zhao, Y., Cai, R., Liu, X., Jia, Q., Zhu, Z., & Jiang, X. (2024). How contingency adjusts corporate social responsibility (CSR) in the tourism industry: A quasi-experiment in china. *Systems*, *12*(3), 83. Advance online publication. DOI: 10.3390/systems12030083

Yehia, A.-D. and Szabados, G. (2018). The middle east perspective of corporate social responsibility. *International Journal of Engineering and Management Sciences*. doi:https://doi.org/..DOI: 10.21791/ijems.2018.4.24

Yousfi, O. and Béji, R. (2022). CSR: What does board diversity bring to the table? .DOI: 10.5772/intechopen.94342

Zaid, M. A. A., Wang, M., & Abuhijleh, S. T. F. (2019). The effect of corporate governance practices on corporate social responsibility disclosure. *Journal of Global Responsibility*, *10*(2), 134–160. Advance online publication. DOI: 10.1108/JGR-10-2018-0053

Chapter 11
From Philanthropy to Policy:
The Role of CSR in Strengthening Societal and Financial Resilience

Mireille Wafaa Gereij
https://orcid.org/0009-0007-9797-8193
Jinan University, Lebanon

Elie abdallah Basbous
Jinan University, Lebanon

Radwan Nazif Choughari
Jinan University, Lebanon

ABSTRACT

Corporate and Banking Social Responsibility (CBSR) has come a long way as a philanthropic program to a serious model of corporate governance, sustainability, and stakeholders. This study focuses on historical evolution, theory, as well as practice of CSR in corporate and banking spheres with the accent on its role in corporate governance, environmental sustainability, and ethical decision making. This paper similarly looks at how banking institutions can facilitate financial inclusion, responsible investment, and sustainable development. y. The results highlight the importance of ethical business conducts in promoting economic robustness, regulatory conformance, and business legitimacy, especially in financial establishments where governance is very crucial in risk-management and assurance to stakeholders.

DOI: 10.4018/979-8-3373-5651-8.ch011

INTRODUCTION

Corporate Social Responsibility (CSR) has risen to be a key pillar in corporate governance in contemporary business environment, and this has been attributed to switch of focus on value creation to stockholders, instead of profit maximization (Carroll, 1999; Freeman, 1984). CSR has historically been conceptualized as voluntary philanthropic activities (Bowen, 1953), but it is currently viewed as a strategic way of influencing competitiveness and sustainability of firms (Kramer and Porter, 2006; Aggeni and Godard, 2006). Due to the growing demands of stakeholders in the context of globalization and environmental deterioration, companies have no option but to integrate CSR in their governance systems in order to remain legitimate and sustainable (Acquier, 2007; Gond, 2006; Gond and Igalens, 2008).

Being a systemic element of financial and economic stability, the banking sector has a special ethical burden (Branco, 2006; Marsiglia and Falautano, 2005). Regulatory and institutional responses to the 2008 global financial meltdown emphasized the absence of ethical behaviors thus requiring banks to incorporate CSR in their mainstream business plans (Adams and Mehran, 2004; Cartapanis, 2003). Financial institutions, in turn, are moving towards green lending, ethical investment, and social inclusion practices in ever increasing numbers in an attempt to reconcile profitability with environmental and social governance (ESG) objectives (Esposito, Chen and Skouloudis, 2023; Brundtland, 1987).

The paper is a critical discussion of the relationship between corporate governance (CG) and CSR banking and finance sectors. It also examines how CG mechanisms: board composition, transparency, and stakeholder engagement, aid in the implementation of CSR (Carroll, 1979; Chauveau and Rosé, 2003; Chau, Nguyen and Bui, 2022). Basing on the stakeholder theory, legitimacy theory, and sustainable development models, the paper examines the ability of CSR to create a competitive advantage and firm performance (Chen, Wang and Zhang, 2023; Ratnawati et al., 2018; Mahmood and Khan, 2022). Through the synthesis of both academic and industry literature, the study sheds light on the interdependence between good governance and responsible social behavior and forms part of a better comprehension of how ethical strategies can build resilience and stakeholder confidence in financial institutions.

THE CONCEPT OF CORPORATE AND BANKING SOCIAL RESPONSIBILITY

Corporate Social Responsibility (CSR) has passed the stage of voluntary and charity idea to the stage of strategic development that determines the organizational legitimacy and its relevance to the market (Carroll, 1999; Bowen, 1953; Acquier, 2007). Nowadays, in the context of the globalized economy, CSR represents the manner in which the businesses address the ethical, environmental and socio-economic concerns within the framework of their governance, going beyond the regulations (Aggeni and Godard, 2006; Gond and Igalens, 2008). Such a change is especially marked in the banking industry where financial institutions are becoming increasingly responsible due to their actions in the wider society (Branco, 2006; Adams and Mehran, 2004).

The notion that companies ought to relate with stakeholders; those impacted by or who can impact the organization has been central to CSR systems (Freeman, 1984). The method also makes banks and corporations address strategic goals in line with customer, investor, civil society and regulatory anticipations (Chang, Kim and Lee, 2022; Chau, Nguyen and Bui, 2022). This sort of alignment becomes crucial not merely to have ethical legitimacy but to be able to maintain a competitive advantage in a place where transparency and accountability are expected by an ever more educated populace (Adamik and Nowicki, 2019; Hadj et al., 2020).

THE ROOTS OF CSR: FROM SUSTAINABLE DEVELOPMENT TO CSR

Ensuring the well-being of both the environment and humanity, while maintaining financial viability, is essential for global sustainability. However, the industrial revolution of the past century often prioritized profit maximization over other critical considerations, frequently at the expense of both the natural world and human welfare (Costa, 2008, p. 5).

Sustainable development, often cited as the philosophical underpinning of CSR, emerged as a response to the adverse consequences of industrial growth and profit-centric capitalism (Costa, 2008; Brundtland, 1987). The widely accepted Brundtland definition posits that development should meet current needs without jeopardizing the ability of future generations to meet theirs (Brundtland, 1987). This three-bottom-line model (economic performance, social equity and environmental protection) has since come to dominate CSR approaches in all sectors including the financial sector (Ferone, Morselli and Ricci, 2001; Chauveau and Rosé, 2003). European model, particularly after the Laeken Declaration and the EU guidelines, institutionalized CSR by incorporating principles of community participation, ethical governance and solidarity into corporate codes (Aggeni and Godard, 2006; Ardiansyah and Alnoor, 2024). The CSR therefore was no longer a series of optional activities but a calculated reaction to socio-political demands, especially dealing with globalization issues (Chaudhry, Raza and Khurshid, 2021).

Within banking, CSR practices encompass financial inclusion, responsible lending, and green investment portfolios—all oriented toward mitigating reputational and systemic risk (Branco, 2006). Effective CSR in banking is underpinned by governance mechanisms such as board structure, risk management protocols, and stakeholder dialogue, which together enhance institutional trust and operational resilience (Adams and Mehran, 2004; Cartapanis, 2003).

The integration of CSR into core business strategies also signifies a move toward sustainable management, whereby internal functions—marketing, human resources, compliance—adopt practices that reinforce ethical and ecological goals (Gallardo-Vázquez, Sánchez-Hernández and Barcik, 2019; Chau, Nguyen and Bui, 2022). Firms increasingly use CSR communication and performance metrics to improve brand image, investor appeal, and customer loyalty (Chen, Wang and Zhang, 2023). This reflects a dual motive: fulfilling moral obligations while securing a competitive edge in dynamic markets (Cegliński and Wiśniewska, 2017).

Moreover, CSR implementation is now influenced by the digitalization and stakeholder co-creation trends of Industry 4.0, making transparency, innovation, and agile governance more critical than ever (Ratnawati et al., 2018). Scholars also emphasize that CSR must move beyond symbolic gestures toward substantial integration into strategic decision-making to yield tangible benefits in firm performance and public trust (Ho, Ahmed and Sultana, 2021). Thus, CSR has matured into a governance and strategy paradigm that transcends traditional compliance. It demands proactive engagement with societal challenges while embedding ethical considerations into financial and operational systems—especially within banking, where trust and responsibility are fundamental to long-term sustainability (Green, 1989; Epstein and Cornelius, 2003).

SUSTAINABLE MANAGEMENT AS THE INTERSECTION OF THREE AXES

Ferone et al. (2001, p. 5) define sustainable management as the ability to continuously reconcile conflicting interests by implementing administrative strategies that address the diverse expectations of corporate stakeholders. Companies develop CSR-based strategies in response to legislative changes, sectoral economic growth, and technological advancements. The emphasis remains on establishing credibility and fostering trust in their operational environment.

Sustainable development, as a concept, ensures that present needs are met without endangering future generations' ability to meet their own. The term originates from the 1987 report by the World Commission on Environment and Development, led by Brundtland (Brundtland Report, p. 40). Since the Rio Conference in 1992, sustainable development has gained significant traction within the business sector (Chauveau & Rosé, 2003). Initially, the social aspect of sustainability was overshadowed by environmental concerns (Fraissé & Guerfel-Henda, 2005, p. 5). However, Aggeni and Godard (2006) highlight the Global Compact as a major milestone in merging environmental considerations with human rights, as outlined in corporate conduct agreements introduced in 2001. Today, the equilibrium between environmental, social, and economic factors remains a defining characteristic of sustainable development.

Within corporate frameworks, sustainable development is often synonymous with CSR, as CSR represents the practical application of sustainability principles at the organizational level. According to ISO 26000, CSR plays a crucial role in advancing sustainable development by integrating ethical, social, and environmental considerations into business operations.

THE DEVELOPMENT OF THE CSR CONCEPT

The Context of the Emergence of a CSR Doctrine

Corporate Social Responsibility (CSR) has evolved over a century from ethical reflection to a strategic business function. In tracing its development, scholars emphasize the importance of understanding the socio-historical context in which CSR emerged as a managerial concern (Acquier, 2007). The initial seeds of CSR were sown in the United States during the rise of industrial capitalism, where early debates centered on the ethical responsibilities of business leaders in addressing corruption and misrepresentation (Ardiansyah and Alnoor, 2024). By the late 19th century, American industrialists began articulating their obligations to society, a sentiment which grew stronger in the 1920s as corporate legitimacy was increasingly scrutinized (Bartók, 2018). The publication of Social Responsibilities of the Businessman by Bowen in 1953 is often regarded as a foundational milestone in formalizing CSR theory (Bowen, 1953). Historical analyses such as those by Heald and others helped establish CSR's intellectual lineage (Costa, 2008).

The History of Corporate Social Responsibility

The Anglo-American context played a central role in institutionalizing CSR, particularly through philanthropic traditions exemplified by figures like George Pullman, who created "Pullman City" to accommodate factory workers (Branco, 2006). By 1920, paternalistic employer practices were visible in nearly 15% of U.S. firms, showcasing early social considerations in management (Adams and Mehran, 2004). This phase of paternalism was not only driven by competition for skilled labor but also reflected shifting social expectations around capitalism and corporate obligations (Cantrell, Kyriazis and Noble, 2014). As corporations expanded, their societal roles were questioned, prompting management to seek broader legitimacy through social responsiveness (Aggeni and Godard, 2006).

Starting in the 1880s, philanthropy became the dominant form of CSR, underpinned by Protestant values that promoted the ethical stewardship of wealth (Gallardo-Vázquez, Sánchez-Hernández and Barcik, 2019). Andrew Carnegie's The Gospel of Wealth is emblematic of this era, promoting the moral duty of the rich to reinvest in society (Carroll, 1999).

CSR later crystallized into a doctrine during the early 20th century, aligning with reforms such as antitrust legislation and public relations campaigns aimed at preserving corporate image and societal trust (Chauveau and Rosé, 2003). The case of Standard Oil under Rockefeller, which pioneered public relations departments, illustrates early efforts to manage corporate accountability through communication (Chaudhry, Raza and Khurshid, 2021). The Great Depression in 1929 disrupted this evolution, diminishing public confidence in business leaders and prompting a pause in CSR initiatives. In response, Roosevelt's New Deal reasserted state oversight of corporations to shield citizens from corporate misconduct (Cartapanis, 2003). After World War II, voluntary CSR practices resumed, often to rehabilitate corporate reputations through social engagement (Alawamleh and Giacaman, 2020).

By the mid-20th century, CSR had transitioned from business discourse to academic inquiry, becoming a recognized field of study (Locket, Moon and Visser, 2006). Bowen's work in particular is considered pivotal in this shift, as it attempted to define why business leaders were embracing social responsibilities and how they viewed their roles in society (Carroll, 1979).

Bowen argued that modern managers were adopting social obligations either out of necessity, moral conviction, or as a result of the changing nature of corporate ownership and control (Flannery, 1998). His broader insight suggested that large corporations had become central to economic and social governance, influencing countless lives and warranting public scrutiny (Donaldson and Dunfee, 1999).

Moreover, CSR's origins are closely tied to religious traditions. Bowen's study was commissioned by the Federal Council of the Churches of Christ in America, highlighting the intersection of Protestant ethics and corporate purpose (Adamik and Nowicki, 2019). His work bridged theological ideals with emerging business ethics. In the 1960s, figures like Keith Davis redefined CSR to include executive actions that considered the ethical consequences of corporate decisions (Davis, 1967). Scholars such as Davis, and subsequently Levitt and Friedman, argued about the possibility of whether CSR was a moral obligation or a diversionary tactic to the profit-making motive (Friedman, 1962).

William Frederick added by putting forward the CSR as a societal tool of corporate and societal resource management in perspective of the common good, and not the strict personal benefit (Frederick, 1960). His insistence on the role of financial and human capital to fulfill larger social ends was still another innovation outside the close circle of shareholder-oriented governance (Gémar and Espinar, 2015). However, opponents of CSR believed that profit should be the primary goal of the business with ethical considerations and cautioned that venturing too far into the social realm could create a situation where corporate focus and responsibility were diffused (Chang, Kim and Lee, 2022).

The Emergence of CSR as a Managerial Doctrine

Corporate Social Responsibility (CSR) has substantially turned out to be a moral issue to a managerial philosophy. Its development could be retraced to the ethical discussion of corporate behavior in the initial stages of industrial capitalism in the United States (Acquier, 2007). The earlier complaints centered on business conducts such as false advertising and monopoly, as society was beginning to desire that businesses ought to be ethical (Chauveau and Rosé, 2003). Towards the end of the 19th century, the debate on CSR started gathering steam amongst the industrialists in America, which would lead to the more structured approaches of the 20th century (Heald, 1970). Social Responsibilities of the Businessman (1953) by Howard Bowen is generally recognized as the initial effort to scientificize and managerialize the daylight savings time as an academic and management concept (Bowen, 1953). His views were supported later by historical analyses demonstrating how the transformations of society have caused business leaders to reconsider their roles (Gond, 2006).

The Evolution of Corporate Social Responsibility

Early antecedents of CSR can be found in philanthropic activities of American industrialists, especially the building of model towns like Pullman City, Illinois, in the 1880s (Costa, 2008). These initiatives were characteristic of a paternalistic corporate welfare strategy whereby companies provided in the welfare of their employees in terms of housing and healthcare (Ferone, Morselli and Ricci, 2001).

CSR during this time was driven by social shifts and the rise of large corporations, prompting society to demand more responsible corporate behavior (Aggeni and Godard, 2006). Scholars argue that this phase marked a legitimization process for capitalism, where CSR became a tool for enhancing corporate image and societal acceptance (Freeman, 1984). Protestant ethics and stewardship also played a crucial role in shaping CSR, emphasizing a moral obligation among the wealthy to support social welfare (Gond and Igalens, 2008). Andrew Carnegie's The Gospel of Wealth (1889) encapsulated this ideology, calling on industrialists to redistribute their wealth for societal good (Brundtland, 1987).

Between 1900 and 1920, reform movements and antitrust legislation gave rise to the idea of trusteeship, where corporations were expected to act as stewards of public interest (Igalens and Joras, 2002). Companies like Standard Oil developed public relations departments to manage reputational risks, reflecting a shift toward structured social responsibility (Adams and Mehran, 2004). The Great Depression (1929) led to a decline in voluntary CSR practices but also spurred regulatory frameworks such as Roosevelt's New Deal, which reinforced corporate accountability to society (Cartapanis, 2003). Post-WWII, businesses resumed CSR initiatives to restore public trust and repair reputational damage (Chaudhry, Raza and Khurshid, 2021).

From the 1950s onwards, CSR became a legitimate academic discipline. Bowen's definition of CSR emphasized that corporate leaders should make decisions aligned with societal values, laws, and ethical norms (Bowen, 1953). He argued that motivations for CSR included regulatory pressures, ethical beliefs, and the separation of ownership from management, which required public accountability (Acquier, 2007). Bowen's research also highlighted the religious foundations of CSR, as his book was commissioned by religious institutions keen to promote ethical business leadership (Donaldson and Dunfee, 1999). The discussion was further developed later by scholars such as Keith Davis who established a connection between CSR and managerial ethics, as well as effects on stakeholders (Davis, 1960).

William Frederick also developed the CSR in his mind beyond the economic objective to comprise the wider social concerns (Freeman, 1984). Critics such as Milton Friedman, however maintained that the only duty of a firm is to make as much profits as possible within the legal framework (Friedman, 1962), others such as Levitt sounded warnings that an over-emphasis on the social aspect would derail firms of their primary economic role (Levitt, 1958).

THE ENVIRONMENTAL DIMENSION

Corporate Social Responsibility (CSR) environmental dimension highlights the need of businesses to review the environmental impacts of their operations. Corporations have a huge responsibility in ensuring that they reduce environmental degradation due to industrial activities as highlighted by Fraissé and Guerfel-Henda (2005). The responsibility also includes the attempts to make the business strategies consistent with ecological sustainability through minimizing carbon emissions, utilizing resources efficiently and making long-term commitments to environmental protection.

The current CSR initiatives promote the methodical decrease in energy use and reduction of greenhouse gaseous emissions by means of reforms in operations. As an example, Chauveau and Rosé (2003) claim that organization actively needs to implement environmentally responsible behaviors like increased recycling, reduction of waste, and investment on cleaner technologies. Besides helping in building ecological integrity, such initiatives help in strengthening the image of a firm in the mind of consumers who are increasingly becoming environmentally conscious. What is more, the programs prolonging lifecycles of the products and focusing on their energy efficiency are the important elements of sustainable development (Costa, 2008).

The incorporation of renewable resources in the supply chains is another important aspect in the environmental CSR. Brundtland (1987) argued that sustainability should be attained through fulfilling the current needs without jeopardizing the future generation capabilities of fulfilling their own needs. This principle requires the corporations to adopt a sourcing and production policy in preference to renewable or biodegradable or recyclable materials. By prioritizing suppliers that reach sustainability standards, responsible procurement policies, as proposed by Kopel (2021), can ensure accountability across the supply chain.

Corporate environmental responsibility also relies on protection of biodiversity. Companies that are involved in land-based or extraction-oriented businesses, i.e. agriculture or mining, are particularly supposed to employ conservation measures. Hadj et al. (2020) argue that biodiversity actions, including forest restoration, habitat conservation, and pollution prevention, are critical in the reduction of environmental damage, without sacrificing stakeholder legality.

In an effort to show their commitments towards the environment, many companies are aiming at being in line with the set global standards. ISO 14001 and other certifications that act as references in environmental management system are deemed as important next indicators of CSR integration (Flannery, 1998). The frameworks help companies formalize their sustainability practices and enhance transparency that ultimately helps them to engage better with their stakeholders and gain competitiveness in the long run. In the same line of thought, eco-labels can help consumers to generate ethical decisions and motivate firms to develop green innovations (Zhao, Xu and Wu, 2022). In such a way, the environmental aspect of the CSR outline summarizes the holistic approach in which compliance, innovation, and ethics overlap. It represents a wider paradigm shift in which companies are supposed to do more than what is required by law and incorporate the ideas of sustainability into the very ethos of their operation (Gallardo-Vazquez, Sanchez-Hernandez and Barcik, 2019).

THE SOCIAL DIMENSION

The social aspect of CSR deals with the internal and external aspects and they cover different aspects of societal responsibilities of an organization. In the internal context, CSR deals with the employees and their rights, welfare, and development, whereas the external context relates to how a company affects the communities in which it operates, in general.

The social dimension internally focuses on human resource practices which are ethical, fair work conditions and the wellbeing of the employees. Saini and Ram (2025) claim that socially responsible firms also spend on equal recruitment, inclusive selection, and continuous education of workers so that they could create a motivated and faithful labor force. Internal CSR is also focused on health and safety, equitable remuneration and career development. Companies integrating these concepts into their HR policies can achieve greater staff involvement and elasticity (Do and Sum, 2021).

In addition, organizations are supposed to assist employees who are victims of industrial shifts. It involves providing retraining or upskilling to those workers who are affected by technological change or downsizing. As noted by Wahyuni, Firmansyah and Setiawan (2024), these types of proactive actions not only help avoid this social disruption but also build a reputation of corporate empathy and ethical leadership.

Another important factor is equality at the workplace especially in terms of gender, race and ability diversity. Chau, Nguyen and Bui (2022) argue that besides being an ethical requirement, diversity and inclusivity promote innovation and improve decision-making by exposing decision-makers to different points of view. At that, CSR turns into a strategic instrument that connects ethical aspects and competitive advantages. On the outside, the social aspect mandates that the corporations add value to their society. It involves funding education activities, infrastructure development, or health initiative. As an illustration, Ardiansyah and Alnoor (2024) mention the benefits of CSR activities like sponsoring youth entrepreneurship or investing in the local education project, strengthening the social license to operate and creating long-term goodwill.

Besides philanthropic giving, companies also need to produce goods and services in an ethical manner, consumable and sustainable to the environment. According to Lee-Wong and More (2016), providing consumers with products of high quality, which adhere to social values, increases consumer trust and brand loyalty. Such expectations apply particularly to industries like food, healthcare and consumer goods, where consumer trust in social impact is great.

The social involvement of corporations also involves the observance of international labor and human rights conventions. The concerns like the abolition of child labor, disallowance of forced labor and inclusion of rights of the workers are critical to practice ethical integrity. According to Ekawarti, Sugiyanto and Sutaryo (2023), multinational corporations, in particular, are under pressure regarding adherence to international ethical norms, in particular, supply chain management in developing nations.

The social CSR practices also require the businesses to promote free conversation with both their employees and external stakeholders. Opening up collective bargaining, feedback, and consultation channels promotes the culture of inclusion and responsibility (Cantrell, Kyriazis and Noble, 2014). Through such engagement mechanisms, corporate transparency is enhanced, and organizational policies are streamlined towards the expectations of the stakeholders.

However, such social initiatives need both financial and strategic capability that small and medium enterprises (SMEs) may not afford. SMEs are less likely to have the budgetary limits and departments devoted to CSR, which means it is more difficult to make such practice institutionalized (Nadányiová, 2021). Nevertheless, SMEs are becoming more innovative in the integration of scalable social initiatives that are within their limits and in accordance with the local demands (Yasrawan, Sukarya and Tamba, 2023). Overall, the social aspect of CSR translates ethical duty of companies towards the internal stakeholders and society at large. Conscious efforts must be made to inculcate principles of equity, inclusiveness and community participation in business paradigms. By doing this, businesses not only satisfy their ethical duties, but also help to achieve long-term viability and legitimacy.

THEORETICAL FOUNDATIONS OF CSR

The theoretical foundations of Corporate Social Responsibility (CSR) have been shifting throughout the decades, mirroring the active interaction of economic, ethical and social demands. The conceptual model of CSR is influenced by philosophical discourse and managerial practice, which implies conflicting views on the role of corporations and responsibilities of stakeholders. Historically, as Igalens and Gond (2008) assume, the theory of CSR has developed along two main lines: the first one is the efforts to specify the meaning, scope, and moral content of CSR; the second one is the efforts to define the stakeholders to whom companies are responsible (duty of care).

The first academic efforts at defining CSR date back to the middle of the 20th century, when a landmark piece of research, Social Responsibilities of the Businessman, was produced by Howard R. Bowen in 1953 and is cited as the birth of contemporary CSR thinking. Bowen expressed the idea that business leaders should work out strategies which consider the expectations of society, ethical norms, and submitted a model of voluntary responsibility, which aimed to anticipate the undue governmental regulation (Bowen, 1953; Gond and Igalens, 2008). This initial formulation was strongly tinted by the Protestant moral philosophy, especially the spirit of stewardship, trusteeship and charity (Acquier, Gond and IgALENS, 2005).

During the 1960s, theorists such as Keith Davis expanded the scope of CSR. Davis (1960) proposed that corporate decisions should go beyond technological and financial concerns to include social considerations. In a refined 1967 definition, Davis emphasized the ethical ramifications of business decisions and their impact on stakeholders, arguing for the integration of morality into corporate governance structures (Davis, 1967). This perspective aligned with the broader shift towards stakeholder-oriented management theories that gained traction in subsequent decades (Freeman, 1984).

One of the most influential models of CSR was introduced by Archie B. Carroll in 1979, who conceptualized CSR as a multi-layered responsibility encompassing economic, legal, ethical, and philanthropic dimensions. According to Carroll, a company's primary obligation is to remain profitable and comply with laws, but it must also operate ethically and contribute voluntarily to societal well-being (Carroll, 1979; Carroll, 1999). Carroll's CSR pyramid became a foundational framework, widely cited in management literature and integrated into corporate strategy discussions globally (Chang, Kim and Lee, 2022).

Building upon this foundation, Jonas (1979) added a philosophical dimension by proposing that corporate responsibility must involve a forward-locking, moral justification of business practices. In The Imperative of Responsibility, Jonas applied Kantian ethics to corporate conduct, emphasizing the need for businesses to anticipate and mitigate long-term consequences of their actions. According to him, this moral vision is crucial in sustainable development and intergenerational fairness (Acquier, 2007; Saeidi et al., 2018).

Another advancement in the stakeholder approach was made by Jones (1980), who opined that CSR ought not to be limited to the shareholders, but rather to all groups affected by the business operations, even in situations where there is no express contractual requirement. The views expressed by Jones are a part of a wider trend in the development of the CSR theory to leave shareholder primacy behind and move towards stakeholder inclusivity (Donaldson and Dunfee, 1999; Cegliński and Wi dawska, 2017). This development reached a point when Wood (1991) wrote that business and society are inseparable and society has the right to expect ethical corporate conduct.

THE RELIGIOUS FOUNDATIONS OF CSR

One can also trace the origins of CSR to religious theories, especially the Christian one, which was instrumental in the development of initial theories regarding corporate ethics. An example is the Catholic social theology, which established the basis of the institutional discourse on justice and economic fairness. One of the milestones towards this formation was the issuance of Rerum Novarum in 1891 by Pope Leo XIII. This papal encyclical was a reaction straight forward to the social-economic injustices of industrial capitalism as well as the growing power of the Marxist ideology (Acquier, 2007). It denounced the unjust wage, human dignity, protection of individual property, and solidarity between labor and capital (Acquier, Gond and Igalens, 2005).

Theological ideas (in particular, stewardship and social trusteeship) shaped CSR in the Protestant tradition. The Protestant ethics urged business leaders to regard wealth, not only as their personal asset, but as an entrustment that they managed on behalf of the society and in the presence of God (Acquier, Gond and Igalens, 2005). Bowen was raised as a Protestant, and his background was heavily involved in his take on the meaning of CSR thus Bowen believed in self-regulation of the business as a way of preserving moral legitimacy and social balance. His contribution still graces the debate on ethics in CSR especially in the Anglo-American environment where the Protestant tradition still influences the institutional conduct (Bowen, 1953).

The formulation by Bowen concurs with three main concepts of protestant business ethics. First, stewardship calls on corporate leaders to be responsible in the management of resources by having a long-term perspective. Second, trusteeship implies that holders of corporate power cannot be held accountable merely to fellow shareholders but to the whole society. Third, the principle of charity requires companies to make voluntary contributions to the welfare of the population, which remains a significant basis of modern CSR practice in the areas of philanthropy and community support (Chaudhry, Raza and Khurshid, 2021; Ardiansyah and Alnoor, 2024).

The modern CSR rhetoric has since developed to embrace secular and pluralistic orientation. Whereas, religious traditions gave birth to moral basis of CSR during its initial development, contemporary CSR theory alludes economic, legal, and environmental systems to meet the challenge of globalization, stakeholder jugglery, and sustainability (Brundtland, 1987; Gond, 2006). However, effects of religious ethics are still evident especially in the areas of guiding corporate values, purpose-driven leadership, and voluntary social action in much of the world. To sum up, the theoretical bases of CSR have been formed as a result of a cross of management experience, philosophical reflections and religious ethics. The historical development of the initial ethical theories into the modern stakeholder and sustainability frameworks depicts the multidimensionality aspect of CSR. With corporations increasingly under pressure to behave ethically and in a transparent manner, knowledge of these theoretical origins is likely to continue playing a central role in the development of responsible and sustainable business practices throughout the 21st century.

GOVERNANCE AS THE CORNERSTONE OF CSR IN BANKS

Corporate governance has offered a structure on which Corporate Social Responsibility (CSR) can be practiced especially in the banking industry, whereby financial institutions are required to pursue economic goals alongside ethical, social and environmental issues. Since banks have extensive range of stakeholders such as customers, employees, shareholders, regulators and communities, they have to embrace governance structures that embrace accountability, transparency and sustainable development. Good governance system, as Adams and Mehran, (2004) reckon, helps the bank to develop competency in all its risk management efforts, foster trust of the stakeholders and integrate the objectives of CSR with the general strategic objectives. Sound governance processes and systems (board supervision and ethical leadership) give an institution the ability to integrate the principles of CSR into regular banking activities.

In the financial services sector, CSR and governance are interlinked through the concept of fiduciary duty and systemic stability. Marsiglia and Falautano (2005) argue that sustainability and ethical responsibility in financial institutions are not only desirable but necessary for maintaining public confidence and long-term performance. Indeed, weak governance practices have historically contributed to financial crises, prompting calls for more rigorous CSR standards in banking (Cartapanis, 2003). Green (1989) further asserts that trust, a fundamental component of banking, is preserved when institutions demonstrate social accountability and environmental consciousness.

CHALLENGES OF CSR IMPLEMENTATION

Despite the growing prominence of CSR, businesses—especially in the banking sector—face numerous challenges in implementing these initiatives. The classical economic argument raised by scholars such as Levitt (1958) and Friedman (1962) questions the legitimacy of CSR by positing that a corporation's sole purpose is to maximize profits. This position suggests that diverting corporate resources towards social causes compromises competitiveness. However, modern scholars increasingly counter this by emphasizing that CSR can drive long-term profitability through brand loyalty, employee satisfaction, and risk mitigation (Hadj et al., 2020; Chaudhry, Raza and Khurshid, 2021).

Igalens and Joras (2002) identify the inherent complexity of aligning diverse CSR goals—social equity, environmental preservation, and economic performance—without undermining market competition. Indeed, businesses must ensure that their commitment to CSR does not place them at a disadvantage relative to competitors who avoid such obligations. Crifo and Ponssard (2008) add that credibility in CSR strategies is closely tied to regulatory compliance and voluntary certifications such as ISO 26000 and ISO 14001, which help to operationalize CSR in a way that is transparent and externally verifiable.

Locket, Moon and Visser (2006) categorize CSR challenges into three key domains: strategic decision-making, operational execution, and stakeholder communication. Each of these areas requires institutional adjustments and cultural shifts within the organization to ensure that CSR is embedded, not superficial.

Strategic Challenges: Balancing Short-Term and Long-Term Objectives

Strategically, one of the most pressing dilemmas in CSR is reconciling short-term financial goals with long-term sustainability. As observed by Crifo and Ponssard (2008), investment in CSR is often influenced by market trends and competitive positioning, particularly in sectors such as green energy, healthcare, and ethical finance. In banking, CSR must be integrated into a broader strategic framework rather than treated as a peripheral activity. Wahyuni, Firmansyah and Setiawan (2024) suggest that CSR-driven strategic planning enables firms to capitalize on emerging sustainability opportunities while enhancing reputational capital.

The role of corporate communication teams becomes central in framing CSR as part of the company's long-term narrative. If a bank, for instance, decides to restructure operations or shut down a branch, providing timely notice and support to affected employees demonstrates ethical foresight and social sensitivity (Ardiansyah and Alnoor, 2024). Strategic CSR thus not only supports corporate risk management but also contributes to long-term resilience by fostering goodwill among stakeholders.

Operational Challenges: Aligning CSR with Business Efficiency

Operationalizing CSR within day-to-day activities introduces further challenges. According to Kramer and Porter (2006), CSR must address issues such as ethical labor practices, workplace safety, and environmental compliance. These responsibilities fall largely on middle management, who must balance cost-efficiency with social accountability. Epstein and Cornelius (2003) recommend tools like the

Sustainable Balanced Scorecard to embed CSR metrics into business performance systems, thereby allowing companies to track non-financial impacts alongside financial outcomes.

However, operational alignment often comes at a cost. CSR investments may increase expenditure in the short term—through higher wages, eco-friendly technologies, or social programs—but can yield long-term benefits in employee retention and customer loyalty (Mahmood et al., 2020). Executives must thus shift from a cost-centric view of CSR to one that recognizes its strategic value, even when tools for evaluation remain underdeveloped.

Managerial Challenges: Balancing Economic and Social Commitments

The most important issue in management level is how to balance profitability and ethical behaviors. Corporate leadership is deemed instrumental towards shaping the corporate culture and inculcating the values of CSR within various departments. The financial returns of CSR might not come into reality soon as pointed out by Déjean and Gond (2003), which means that executives must think long term despite the quarterly pressures. The leaders have to tread a fine line between various pressures, the demands of the shareholders, the requirements of the regulators and the lobbying of the stakeholders.

The banking industry, specifically, has to deal with CSR under a banner of fiduciary duty and financial control. According to Branco (2006), the banking industry is faced with a situation where they have to show ethical behaviors in their risk assessment and capital allocation exercise. Managers are increasingly being forced to integrate CSR with the corporate governance system, as opposed to considering it as a nice-to-have, because of external pressures in the form of media attention, consumer activism, and ESG investment criteria.

In short, although the idea of CSR holds promise of competitive advantage and ethical leadership, it poses in itself significant strategic, operational and managerial difficulties. Banking is a business of financial intermediation and due to public confidence in this sector, the banking sector should consider governance as the foundation of CSR in order to achieve sustainability, legitimacy and stakeholder trust.

CSR IN BANKING INSTITUTIONS: CHALLENGES AND STRATEGIC INITIATIVES

Corporate social responsibility (CSR) is an essential aspect through which the banking sector is providing leadership in the financial systems. Being one of the key economic intermediaries, banks have a significant impact on sustainable development because of their ability to deploy financial capital in industry. This centrality however presents them with a different set of CSR issues that are quite distinct to those faced in the manufacturing or retail industries (Marsiglia and Falautano, 2005; Branco, 2006). The most CSR-sensitive issues in banking are sustainable finance, ethical lending, social risk management and transparent governance.

Due to the growing stakeholder scrutiny, changing regulatory requirements and the augmented societal understanding of the implications of financial procedures, banks are beneath pressure to incorporate CSR in their corporate governance framework (Adams and Mehran, 2004). Because financial institutions can decide to support or freeze environmentally or socially questionable operations, they are in a special position to promote sustainability through the advancement of ethical investment policies and green financial products (Branco, 2006; Marsiglia and Falautano, 2005). However, the story of CSR practice in this industry is not challenge-free. Most banks are ill at ease in trying to balance profit making with social responsibility particularly in unstable or deregulated sectors where the bottom line is likely to take precedence over the welfare of stakeholders (Green, 1989).

Regulation regimes like the Basel Accords have started pointing out CSR as part of risk management and financial oversight. The guidelines have promoted the setting up of internal frameworks by banks to track social, environmental risks in addition to the conventional financial risks (Cartapanis, 2003; Demirguc-Kunt, Kane and Laeven, 2008). The 2008 global financial crisis was a bruising reminder of the costs of failing to recognize these extended duties and led to a sector-wide review of governance expectations and requirements of transparency (Epstein and Cornelius, 2003).

Current CSR approaches to banking are no longer about legislation or acts of charity; they constitute part of business practices. Worker's welfare is also cited as a pillar of ethical internal governance and banks have enacted systems to ascertain that there is fair hiring practice, equitable pay, and career advancement prospects (Marsiglia and Falautano, 2005). Internally, banks recognize their workforce as a critical stakeholder group, the satisfaction and engagement of which directly impact service quality and ability to innovate. To make these commitments official, most banks establish them in either a collective bargaining agreement or a CSR charter which describes the institutional responsibility towards employees (Freeman, 1984; Igalens and Joras, 2002).

On the surface, CSR can be seen through the act of promoting environmentally friendly banking services, i.e. carbon footprint reduction, moving towards a paperless banking system and venture capital in renewable energy sources (Chauveau and Rosé, 2003; Fraiss, and Guerfel-Henda, 2005). More importantly, banks are designing financial products that serve sustainable development goals (SDGs) such as launching green loans, microfinance programs, or socially responsible investment portfolios. Not only are these instruments favorable to the environment and society but also increase long-term profitability by reducing reputational and regulatory risks (Ardiansyah and Alnoor, 2024; Bartók, 2018).

Transparency as one of the major pillars of CSR in banking sphere relates both to the establishment of the proper reporting practices within the company and the disclosure practices of the company. The banks are also moving towards using sustainability reporting standards like the Global Reporting Initiative (GRI) or the Corporate Sustainability Reporting Directive (CSRD) in the EU to report their social and environmental performance to their stakeholders (Esposito, Chen and Skouloudis, 2023). Transparency of CSR reporting, according to Clare (2019), creates stakeholder trust, which subsequently translates to a source of intangible capital and brand equity in competitive financial markets.

Banks also engage in extensive stakeholder partnerships with civil society organizations, NGOs, and regulatory bodies to co-create social value. These partnerships facilitate the development of inclusive financial products, community development programs, and educational initiatives promoting financial literacy (Cantrell, Kyriazis and Noble, 2014; Chaudhry, Raza and Khurshid, 2021). From a strategic perspective, these CSR activities are aligned with the broader shift towards stakeholder capitalism, in which the interests of customers, employees, suppliers, and society at large are balanced with shareholder value (Carroll, 1999; Kramer and Porter, 2006).

Nevertheless, effective CSR governance within banks requires robust institutional mechanisms. Risk management departments have been tasked not only with identifying financial vulnerabilities but also with overseeing social and environmental impacts. According to Marsiglia and Falautano (2005), banking professionals are increasingly trained in sustainability assessments, ensuring that lending decisions factors in externalities such as pollution, labor conditions, or community displacement. These practices, when embedded within broader governance structures, reduce the risk of financial mismanagement and foster sustainable economic growth.

Furthermore, diversity and inclusion have emerged as critical CSR themes in banking. Governance reforms have mandated more gender-balanced and culturally diverse boards, reflecting the growing recognition that inclusive leadership strengthens ethical decision-making and enhances strategic foresight (Adams and Mehran, 2004). Customers too are increasingly valuing banks that prioritize ethical practices, inclusive policies, and social innovation, contributing to long-term customer loyalty and reputational resilience (Branco, 2006; Liu, Wong and Tseng, 2014).

Ultimately, CSR in the banking sector is no longer an optional or symbolic endeavor. It is a strategic imperative that contributes to market differentiation, risk mitigation, and societal value creation. As Epstein and Cornelius (2003) argue, CSR performance must be measured and managed with the same rigor as financial performance. Banks that embed sustainability into their core business strategies not only enhance stakeholder confidence but also contribute to building resilient and ethical financial ecosystems in an increasingly interconnected global economy.

KEY ASPECTS OF BANKING GOVERNANCE

Before 2011, the term banking governance did not have a uniform definition. Nevertheless, the critical nature of banks in the stable functioning of the world economy has provoked increasing academic concern in the governance frameworks in financial institutions. Menkhoff and Suwanaporn (2007) noted that ineffective governance mechanisms usually beckon the basis of financial crisis especially in the economies that have financial markets that are deregulated. In the same line of thought, Minsky (1996) propounded that weaknesses in governance lead to the amplification of systemic instability, especially when the regulatory structures are unable to restrain speculative tendencies within the banking systems.

Risk exposure also has direct relationship with governance. Pathan (2009) establishes a relation between ineffective governance and excessive risk-taking and Adams and Mehran (2004) observe that strong board structures are related to financial strength. Cartapanis (2003) highlighted that prudential regulation cannot simply be relied upon to stop the financial disruption; governance should be more active in systemic risks identification and deployment of internal control measures that go beyond the simple compliance expectations. Macey and O'hara (2003) cautioned that excessive rigid regulatory structures could impair the flexibility of governance, generating agency issues that diminish effectiveness. In their place, researchers are supporting a combination of regulation control and market-based governance practices to instill ethical conduct and financial accountability (Prowse, 1997; Polo, 2007).

Banking Regulations and Compliance

Due to their role in credit intermediation, capital allocation, and deposit protection, banks are subject to rigorous regulatory oversight. Demirgüç-Kunt, Kane, and Laeven (2005) highlighted deposit insurance schemes as vital safeguards for consumer confidence, while Flannery (1998) advocated for the use of market information in prudential supervision to complement formal regulation.

Rochet (2003) differentiated banking oversight into micro-prudential regulation, which protects individual depositors, and macro-prudential regulation, which stabilizes the entire financial system. Merton (1977) recommended central bank supervision and capital adequacy enforcement to mitigate systemic risks. However, overly prescriptive regulations may limit banks' flexibility to innovate and manage stakeholder interests effectively. This has led to calls for hybrid governance models that combine regulatory compliance with ethical leadership and transparency (Polo, 2007). Such models enable banks to pursue both financial performance and responsible conduct, aligning with the goals of corporate social responsibility.

THE LINK BETWEEN GOVERNANCE AND CSR

CSR has evolved into a core governance concern, especially in banking, where financial institutions face scrutiny from regulators, investors, and the public. This section explores four theoretical models that explain how governance and CSR intersect.

Stakeholder Theory

Stakeholder theory maintains that firms should be accountable to all parties affected by their operations, not just shareholders. Donaldson and Preston (1995) emphasized the ethical and social obligations of firms towards diverse stakeholder groups. Governance mechanisms, under this framework, ensure that stakeholder conflicts are managed and that information asymmetry does not disadvantage weaker parties (Chang, Kim and Lee, 2022). CSR turns into a strategic reaction to the demands of the stakeholders; profitability is intertwined with ethical responsibilities. It changes corporate governance into a more encompassing process that conveys the financial and non-financial goals equally.

Corporate-Society Contract Theory

According to this theory, businesses acquire their legitimacy through delivering on a larger social contract than economic benefits (Donaldson and Dunfee, 1999). Gendron (2000) advanced that, since businesses are powerful agents in society, they should ensure that their practices conform to the norms and expectations of the community members in order to remain legitimate. This conception of CSR is not a voluntary act of philanthropy, but a responsibility which firms have due to the implicit contract between society and the firms. This theoretical position supports the line of argument that CSR is a governance task that is motivated by outer accountability as opposed to inner discretion.

Neo-Institutional Theory

Rojot (1999) argues that no business operates in a vacuum but is situated within institutional settings that provide a context within which behavior is constrained. According to Meyer and Rowan (1977), the governance structures and CSR initiatives are commonly institutionalized answers to legitimacy pressures. Organizations also abide by the expectation of societies as a way of gaining a foothold in the competitive markets. Following the impression management theory, CSR can also be pursued shallowly to create an ethical image without any meaningful attachment (Gémar and Espinar, 2015).

Psychological Contract Theory

The theory brings into spotlight the unspoken rules between the employer and the employees. According to Ring and Van de Ven (1994), stakeholders, particularly the internal stakeholders, enter into psychological contract informed by ethical governance, fair treatment, and social responsibility. When this unwritten agreement is broken, employee trust and dedication are destroyed. CSR, then, becomes a critical aspect of the interior governance, which makes sure that the organizational culture reinforces ethical conduct and stakeholder relationship (Alawamleh and Giacaman, 2020).

CSR DURING ECONOMIC CRISES

CSR gains heightened relevance during financial crises. Mercier (2000) proposed CSR as an alternative governance tool to mitigate reputational risk and reinforce stakeholder trust when conventional financial strategies falter. Yet, the gap between CSR rhetoric and actual practice remains significant (Lindgreen and Swaen, 2010). Martinet and Reynaud (2004) argued that sustainability reporting often serves as a communication strategy rather than an accurate reflection of corporate behavior. Lordon (2003) cautioned that firms may highlight green initiatives while concealing governance failures.

Tonkiss (2009) observed that most corporations have been selective in implementing CSR when crisis hit so that they can become responsible in the eyes of the reform without actually doing it. The same symbolic compliance was also cited by Mauléon and Saulquin (2008) as one of the key issues in implementation of CSR. Even so, CSR may provide protection against both financial and reputational loss (Vanhamme and Grobben, 2009), since it is part of a wider governance and resilience plan. Yoon, Gürhan-Canli, and Schwarz (2006) have shown that CSR can be utilized in maintaining corporate image and customer loyalty in bad economic times. Descolonges and Saincy (2004), however, cautioned that when companies are hard hit by the economic condition, then they tend to relegate CSR to implement cost cuts. Thus, it is crucial to incorporate CSR in the long-term governance frameworks to create resilience and trusted relationships that go beyond the financial quarterly or annual reporting.

CONCLUSION

The combination of corporate governance and Corporate Social Responsibility (CSR) in banking industry is the main change of the balance between the economic goals and ethical, social and environmental responsibilities of financial institutions. This development of CSR as a voluntary philanthropic activity to a formal corporate governance system highlights the importance of CSR in achieving financial sustainability, compliance with laws and stakeholder confidence. The banks, being the significant stakeholders in the economic stability and growth, can specifically exhibit the CSR throughout their business processes not only in the internal management aspect, but also in the overall financial system.

As outlined in the findings of this research, effective CSR implementation requires a well-established corporate governance structure, more so in an industry where risk management, transparency and ethical responsibility directly affect the levels of public confidence. CSR initiatives are anchored on governance structures that guide banks through the web of regulations, stakeholder demands, and the UN sustainable development objectives. Incorporating the CSR in their strategic decision-making processes, banks can ensure the reduction of financial and reputational risks, increase in customer trust, and sustainable economic growth.

Also, the research findings demonstrate CSR as a strategic asset rather than a compliance burden, which enabled the establishment of competitive advantage and long-term sustainability. By focusing on sustainability reporting, ethical lending practices, and responsible investment strategies, banks can establish themselves as leaders in the new financial environment. The rise of the importance of Environmental, Social and Governance (ESG) factors in the investment process is another reason why banks should consider CSR in their governance framework.

Nonetheless, there are still issues on how to achieve a balance between short-term profitability and long-term CSR commitments. The conflict between economic performance and social responsibility is something that is still defining the choices of corporations, especially when financial crises come into play. The study highlights that even though CSR improves corporate image and stakeholder interactions, it must be noted that it has to be done with sincerity and not a form of tokenism.

Banking institutions should not just focus on the superficial CSR activities but need to integrate responsible business practices as part and parcel of their operational models. Going forward, with the ever-changing global markets, banks have to take the initiatives to adjust to the regulatory trends, social demands and risks that are developing. Embracing a culture of governance, which emphasizes transparency, accountability, and sustainability, the financial institutions can encourage beneficial societal change without compromising their economic sustainability. Further studies ought to be conducted on how digital transformation, fintech developments, and regulatory changes across the world are influencing CSR in banking, as a way of ensuring the governance structures are dynamic and flexible to accommodate varying market conditions.

In conclusion, effective implementation of CSR in banking governance will eventually shape responsibility finance since it will affirm the responsibility of bank in stimulating economic stability, responsible investment and sustainable growth in a more interdependent world.

REFERENCES

Acquier, A. (2007). *Corporate Social Responsibility: A Historical Perspective.* Springer.

Acquier, A., Gond, J. P., & Igalens, J. (2005). Des fondements religieux de la responsabilité sociale de l'entreprise à la responsabilité sociale de l'entreprise comme religion. *Cahier de recherche, 166*, 1-31.

Adamik, A., & Nowicki, M. (2019). Pathologies and Paradoxes of Co-Creation: A Contribution to the Discussion about Corporate Social Responsibility in Building a Competitive Advantage in the Age of Industry 4.0. *Sustainability (Basel), 11*(18), 4954. DOI: 10.3390/su11184954

Adams, R. B., & Mehran, H. (2004). Corporate Performance, Board Structure, and Their Determinants in the Banking Industry. *Journal of Financial Economics, 74*(2), 189–234.

Aggeni, F., & Godard, O. (2006). *Corporate Responsibility and Sustainable Development: Beyond Compliance.* Palgrave Macmillan.

Alawamleh, M., & Giacaman, R. (2020). Corporate social responsibility impacts on Palestinian and Jordanian consumer purchasing. *The International Journal of Organizational Analysis, 28*(4), 879–894. DOI: 10.1108/IJOA-03-2020-2109

Ardiansyah, A., & Alnoor, A. (2024). Integrating Corporate Social Responsibility into Business Strategy: Creating Sustainable Value. *International Insight Journal of Business, 1*(1), 12–23. DOI: 10.62569/iijb.v1i1.5

Bartók, O. (2018). The Use of CSR in E-Commerce as a Way to Compete. *Journal of Competitiveness, 10*(4), 5–20. DOI: 10.7441/joc.2018.04.01

Bowen, H. R. (1953). *Social Responsibilities of the Businessman.* Harper & Row.

Branco, M. C. (2006). Corporate Social Responsibility and Banking Sector Performance. *Journal of Business Ethics, 69*(1), 111–132. DOI: 10.1007/s10551-006-9071-z

Brundtland, G. H. (1987). *Our Common Future: The World Commission on Environment and Development.* Oxford University Press.

Cantrell, J., Kyriazis, E., & Noble, G. (2014). Developing CSR Giving as a Dynamic Capability for Salient Stakeholder Management. *Journal of Business Ethics, 119*(2), 123–136. DOI: 10.1007/s10551-014-2229-1

Carroll, A. B. (1979). A Three-Dimensional Conceptual Model of Corporate Performance. *Academy of Management Review*, *4*(4), 497–505. DOI: 10.2307/257850

Carroll, A. B. (1999). Corporate Social Responsibility: Evolution of a Definitional Construct. *Business & Society*, *38*(3), 268–295. DOI: 10.1177/000765039903800303

Cartapanis, A. (2003). The Governance of Global Banking Crises. *Review of International Political Economy*, *10*(1), 1–25.

Cegliński, P., & Wiśniewska, A. (2017). CSR as a Source of Competitive Advantage: The Case Study of Polpharma Group. *Journal of Corporate Responsibility and Leadership*, *3*(1), 29–47. DOI: 10.12775/JCRL.2016.020

Chang, H., Kim, H., & Lee, Y. (2022). Corporate Social Responsibility and Product Market Power. *Economies*, *10*(6), 151. DOI: 10.3390/economies10060151

Chau, T. H. H., Nguyen, P. T. T., & Bui, T. H. (2022). The Impact of Corporate Social Responsibility on Strategic Objectives: Evidence from Vietnam. *VNU University of Economics and Business Journal*, *2*(6), 114–125. DOI: 10.57110/vnujeb.v2i6.136

Chaudhry, N. I., Raza, S. A., & Khurshid, M. A. (2021). Enhancing Shareholder Value through Corporate Social Responsibility Disclosure in a Developing Country. *Estudios de Economía Aplicada*, *39*(4), 1–13. DOI: 10.25115/eea.v39i4.4315

Chauveau, A., & Rosé, J. J. (2003). *Sustainable Development and Corporate Responsibility in Practice*. Routledge.

Chen, Y., Wang, S., & Zhang, W. (2023). Paradox of CSR distinctiveness: The tension between stakeholder legitimacy evaluation and stakeholder integration efficiency. *Corporate Social Responsibility and Environmental Management*, *30*(5), 2053–2065. DOI: 10.1002/csr.2523

Clare, R. (2019). Corporate Image and Competitive Advantage for Apparel Companies. *Trends in Textile Engineering & Fashion Technology*, *5*(3), 345–351. DOI: 10.31031/TTEFT.2019.05.000618

Costa, J. (2008). *The New Frontiers of Sustainable Development in Business*. Springer.

Crifo, P., & Ponssard, J. P. (2008). RSE et/ou performance financière: points de repère et pistes de recherche.

Davis, K. (1960). Can Business Afford to Ignore Social Responsibilities? *California Management Review*, *2*(3), 70–76. DOI: 10.2307/41166246

Davis, K. (1967). Understanding the Social Responsibility Puzzle. *Business Horizons*, *10*(4), 45–50. DOI: 10.1016/0007-6813(67)90007-9

Demirgüç-Kunt, A., Kane, E. J., & Laeven, L. (2008). Determinants of deposit-insurance adoption and design. *Journal of Financial Intermediation, 17*(3), 407–438. DOI: 10.1016/j.jfi.2007.03.009

Descolonges, M., & Saincy, B. (2004). *Les entreprises seront-elles un jour responsables?*. La dispute.

Do, H. T., & Sum, V. (2021). Business student attitudes toward corporate social responsibility: A Vietnamese study. *Journal of Management & Organization, 27*(4), 737–754. DOI: 10.1017/jmo.2021.28

Donaldson, T., & Dunfee, T. W. (1999). *Ties That Bind: A Social Contracts Approach to Business Ethics*. Harvard Business Press.

Donaldson, T., & Preston, L. E. (1995). The stakeholder theory of the corporation: Concepts, evidence, and implications. *Academy of Management Review, 20*(1), 65–91. DOI: 10.2307/258887

Ekawarti, L., Sugiyanto, T. and Sutaryo, S. (2023) 'Corporate Social Responsibility (CSR): A New Paradigm of Accounting for Environmentally Based Companies', *JEMASI: Jurnal Ekonomi Manajemen dan Akuntansi*, 19(1), pp. 1–12. .DOI: 10.35449/jemasi.v19i1.617

Epstein, M. J., & Cornelius, B. (2003). *Measuring Corporate Social Responsibility Performance: Accounting for Stakeholder Value*. Wiley.

Esposito, D. E., Chen, S., & Skouloudis, A. (2023). Beyond the business case? Retracing the walk of Corporate Social Responsibility and financial performance relationship in the Oil and Gas sector. *Corporate Social Responsibility and Environmental Management, 30*(5), 1953–1966. DOI: 10.1002/csr.2676

Ferone, S., Morselli, G., & Ricci, P. (2001). *Managing Sustainable Enterprises: Strategic Approaches*. Emerald Publishing.

Flannery, M. J. (1998). Using Market Information in Prudential Bank Supervision: A Review of the U.S. Empirical Evidence. *Journal of Money, Credit and Banking, 30*(3), 273–305. DOI: 10.2307/2601102

Fraissé, P., & Guerfel-Henda, S. (2005). The Environmental Dimension of CSR. *Journal of Business Ethics, 60*(2), 153–170.

Frederick, W. C. (1960). The growing concern over business responsibility. *California Management Review, 2*(4), 54–61. DOI: 10.2307/41165405

Freeman, R. E. (1984). *Strategic Management: A Stakeholder Approach*. Cambridge University Press.

Friedman, M. (1962). *Capitalism and Freedom*. University of Chicago Press.

Gallardo-Vázquez, D., Sánchez-Hernández, M. I., & Barcik, A. (2019). Corporate Social Responsibility as an Antecedent of Innovation, Reputation, Performance, and Competitive Success: A Multiple Mediation Analysis. *Sustainability (Basel), 11*(20), 5614. DOI: 10.3390/su11205614

Gémar, G., & Espinar, R. (2015). Communication about Corporate Social Responsibility practices and Return on Equity. *European Journal of Family Business, 5*(1), 15–22. DOI: 10.24310/ejfbejfb.v5i1.5060

Gond, J. P. (2006). *The Institutionalization of CSR in Business Strategy*. Palgrave Macmillan.

Gond, J. P., & Igalens, J. (2008). *Corporate Governance and Social Responsibility*. Pearson.

Green, T. (1989). *Financial Institutions and Corporate Responsibility: The Role of Trust in Banking*. Oxford University Press.

Hadj, T. B., Albashabsheh, A., Jamal, A., & Cherif, M. (2020). Mediation role of responsible innovation between CSR strategy and competitive advantage: Empirical evidence for the case of Saudi Arabia enterprises. *Management Science Letters, 10*(3), 727–738. DOI: 10.5267/j.msl.2019.10.022

Heald, M. (1970). *The Social Responsibilities of Business: Company and Community, 1900–1960*. Case Western Reserve University Press.

Ho, K. H., Ahmed, M., & Sultana, N. (2021). Does engaging in corporate social responsibility activities influence firm performance? The moderating effects of risk preferences and performance measurement systems. *Management Decision, 59*(12), 3020–3041. DOI: 10.1108/MD-07-2020-0925

Igalens, J. and Joras, M. (2002) *Corporate Sustainability and Business Ethics: A Global Perspective*.

Kanojia, P., Pradhan, R., & Patel, A. (2016). A Study on the Impact of Corporate Social Responsibility on Brand Image and Customer Loyalty. *Journal of Global Business Insights, 1*(2), 73–84. DOI: 10.5038/2640-6489.1.2

Kopel, M. (2021). CSR leadership, spillovers, and first-mover advantage. *Decisions in Economics and Finance, 44*(3), 857–872. DOI: 10.1007/s10203-021-00328-9

Kramer, M., & Porter, M. (2006). Strategy and Society: The Link Between Competitive Advantage and Corporate Social Responsibility. *Harvard Business Review, 84*(12), 78–92.

Le, T.T. (2022) 'Corporate social responsibility and SMEs' performance: mediating role of corporate image, corporate reputation and customer loyalty', *International Journal of Emerging Markets*, ahead-of-print. .DOI: 10.1108/IJOEM-07-2021-1164

Lee-Wong, M., & More, E. (2016). Management of corporate social responsibility in Hong Kong small and medium enterprises. *Journal of Global Responsibility*, *7*(2), 304–322. DOI: 10.1108/JGR-07-2016-0018

Levitt, T. (1958). The Dangers of Social Responsibility. *Harvard Business Review*, *36*(5), 41–50.

Lindgreen, A., & Swaen, V. (2010). Corporate social responsibility. *International Journal of Management Reviews*, *12*(1), 1–7. DOI: 10.1111/j.1468-2370.2009.00277.x

Liu, M. T., Wong, I. A., & Tseng, T. H. (2014). The impact of corporate social responsibility (CSR) performance and perceived brand quality on customer-based brand preference. *Journal of Services Marketing*, *28*(3), 181–194. DOI: 10.1108/JSM-09-2012-0171

Locket, J., Moon, J., & Visser, W. (2006). Corporate Social Responsibility in Management Research. *Journal of Business Ethics*, *69*(1), 1–27.

Lordon, F. (2003). Conatus et institutions: pour un structuralisme énergétique. In *L'Année de la régulation n° 7 (2003-2004)* (pp. 111–146). Presses de Sciences Po. DOI: 10.3917/scpo.colle.2003.01.0111

Macey, J. R., & O'hara, M. (2003). The corporate governance of banks. *Economic Policy Review*, *9*(1).

Mahmood, A., & Khan, M. T. (2022). CSR and customer orientation: A sustainable pathway towards competitive advantage? A customer's perspective. *Social Responsibility Journal*, *18*(5), 1002–1018. DOI: 10.1108/SRJ-04-2020-0139

Mahmood, M., Zaman, M., Ahmed, M., & Khan, S. (2020). 'Corporate Social Responsibility and Firms' Financial Performance: A New Insight'. *Sustainability (Basel)*, *12*(10), 4211. DOI: 10.3390/su12104211

Marsiglia, E., & Falautano, I. (2005). Corporate Social Responsibility and Sustainability in Financial Services. *Journal of Financial Services Marketing*, *9*(4), 331–345.

Martinet, A. C., & Reynaud, E. (2004). Entreprise durable, finance et stratégie. *Revue française de gestion, 152*(5), 121-136.

Mauléon, F., & Saulquin, J. Y. (2008). L'affichage éthique dans les rapports annuels des sociétés du CAC 40: Déterminants et implications pratiques. *Management & Avenir*, *20*(6), 129–146. DOI: 10.3917/mav.020.0129

Menkhoff, L., & Suwanaporn, C. (2007). On the rationale of bank lending in pre-crisis Thailand. *Applied Economics, 39*(9), 1077–1089. DOI: 10.1080/00036840500447815

Mercier, A. (2000, October). L'observation du travail des élèves, quels en sont les objets élémentaires et comment peut-on les produire? In *Journées Franco-Québécoises de Didactique*.

Merton, R. C. (1977). An analytic derivation of the cost of deposit insurance and loan guarantees an application of modern option pricing theory. *Journal of Banking & Finance, 1*(1), 3–11. DOI: 10.1016/0378-4266(77)90015-2

Meyer, J. W., & Rowan, B. (1977). Institutionalized organizations: Formal structure as myth and ceremony. *American Journal of Sociology, 83*(2), 340–363. DOI: 10.1086/226550

Minsky, H. P. (1996). Uncertainty and the institutional structure of capitalist economies: Remarks upon receiving the Veblen-Commons award. *Journal of Economic Issues, 30*(2), 357–368. DOI: 10.1080/00213624.1996.11505800

Nadányiová, M. (2021) 'The perception of corporate social responsibility and its impact on consumer buying behaviour in the process of globalization', *SHS Web of Conferences*, 92, p. 06024. DOI: 10.1051/shsconf/20219206024

Pathan, S. (2009). Strong boards, CEO power and bank risk-taking. *Journal of Banking & Finance, 33*(7), 1340–1350. DOI: 10.1016/j.jbankfin.2009.02.001

Polo, A. (2007). Corporate governance of banks: the current state of the debate.

Prowse, S. (1997). Corporate control in commercial banks. *Journal of Financial Research, 20*(4), 509–527. DOI: 10.1111/j.1475-6803.1997.tb00263.x

Ratnawati, K., & Rahayu, S. M. (2018). 'The Role of SMEs' Innovation and Learning Orientation in Mediating the Effect of CSR Programme on SMEs' Performance and Competitive Advantage'. *Global Business Review, 19*(3, suppl), S23–S41. DOI: 10.1177/0972150918757842

Ring, P. S., & Van de Ven, A. H. (1994). Developmental processes of cooperative interorganizational relationships. *Academy of Management Review, 19*(1), 90–118. DOI: 10.2307/258836

Saeidi, S. P., Sofian, S., Saeidi, P., Saeidi, S. P., & Saaeidi, S. A. (2018). 'The Utilitarian Aspect of the Philosophy of Ecology: The Case of Corporate Social Responsibility', *Filosofija. Sociologija, 29*(1), 49–57. DOI: 10.6001/fil-soc.v29i1.3630

Saini, R. and Ram, K. (2025) 'Unlocking Sustainable Growth: Social Innovation as a Link Between CSR and Competitive Advantage in Haryana's MSMEs', *SEDME (Small Enterprises Development, Management & Extension Journal)*, 52(1), pp. 34–47. .DOI: 10.1177/09708464251326387

Thành, N.T., Phuong, H.L. and Hieu, T. (2023) 'Corporate social responsibility, corporate reputation and intention to apply for a job: evidence from students in an emerging economy', *International Journal of Organizational Analysis*, ahead-of-print. .DOI: 10.1108/IJOA-11-2022-3484

Vanhamme, J., & Grobben, B. (2009). "Too good to be true!". The effectiveness of CSR history in countering negative publicity. *Journal of Business Ethics*, *85*(S2, Suppl 2), 273–283. DOI: 10.1007/s10551-008-9731-2

Wahyuni, D., Firmansyah, A., & Setiawan, H. (2024). Sustainability in Business: Integrating CSR into Long-term Strategy. *Asian Business and Innovation Management*, *2*(3), 45–59. DOI: 10.60079/abim.v2i3.354

Yasrawan, Y., Sukarya, R., & Tamba, J. (2023). Competitive Advantage as A Link Between the Influence of Intellectual Capital and CSR on Financial Performance. *Jurnal Ilmiah Akuntansi*, *8*(1), 41–53. DOI: 10.23887/jia.v8i1.46981

Yoon, Y., Gürhan-Canli, Z., & Schwarz, N. (2006). The effect of corporate social responsibility (CSR) activities on companies with bad reputations. *Journal of Consumer Psychology*, *16*(4), 377–390. DOI: 10.1207/s15327663jcp1604_9

Zhao, H., Xu, S., & Wu, L. (2022). The Trade-off between Corporate Social Responsibility and Competitive Advantage: A Biform Game Model. *Technological and Economic Development of Economy*, *28*(3), 638–662. DOI: 10.3846/tede.2022.16483

Chapter 12
Strategic Leadership and Crisis Adaptation:
Navigating Business Uncertainty Amid Global Upheaval

Zahraa Hussein Mokdad
https://orcid.org/0009-0006-0933-528X
Intelligentsia Center for Research and Studies, Lebanon

ABSTRACT

In an era marked by persistent political, environmental, and cultural upheavals, the resilience of business organizations has become increasingly dependent on the quality and agility of their strategic leadership. This chapter explores how strategic leadership functions as a critical factor in crisis adaptation, focusing on decision-making processes, organizational culture, and leadership communication during times of uncertainty. Drawing on contemporary case studies and theoretical frameworks in business management, the study highlights how effective leaders not only respond to external shocks but also anticipate future disruptions through proactive planning, risk management, and innovation-driven strategies. The chapter begins by mapping the global context of recent crises including the COVID-19 pandemic, geopolitical conflicts, economic instability, and climate-related disruptions and their collective impact on business continuity. It then examines the role of transformational and adaptive leadership styles in fostering resilience at both organizational and community levels. By integrating qualitative and quantitative insights, the chapter identifies key success factors and common pitfalls in leadership response strategies. Special attention is given to small and medium-sized enterprises (SMEs), which often lack institutional buffers yet demonstrate significant innovation and adaptability. The study concludes with a strategic resilience framework designed to guide leaders in building long-term organizational sustainability, emphasizing agility, stakeholder

DOI: 10.4018/979-8-3373-5651-8.ch012

engagement, and ethical governance. This chapter contributes to the broader discourse on global societal resilience by positioning business leadership not only as a driver of economic recovery but also as a pillar of social stability in the face of multidimensional crises.

1. INTRODUCTION

The contemporary business landscape is increasingly characterized by an intricate mixture of volatility, uncertainty, complexity, and ambiguity popularly recognized as VUCA. Although the concept originates from military doctrine, it has become essential for understanding modern organizational environments (Sucharita, 2016; Ungureanu et al., 2018) and is reinforced by frameworks from Deloitte (Arkenberg, 2019). These forces combine to create what many now refer to as a "polycrisis" a condition of overlapping systemic crises, including ecological breakdown, geopolitical upheaval, digital transformation, democratic erosion, and public health threats (Morin, 1993; Lawrence et al., 2024). This polycrisis context isn't merely a succession of discrete events, but a simultaneous convergence where challenges reinforce and intensify each other. For example, the COVID-19 pandemic disrupted global supply networks and accelerated digital adoption, while geopolitical tensions and climate events further amplified cost pressures and resource scarcity. Such interplay mirrors the concept of "permacrisis" a quasi-permanent state of instability that demands continuous strategic vigilance (Morin, 1993; Sucharita, 2016).

Yet VUCA alone does not fully capture this era's complexity. Recent studies argue leaders now face polycrisis dynamics interwoven crises that defy traditional cause-and-effect logic (Lawrence et al., 2024; Tooze, 2022). As interdependencies across sectors intensify, decision environments become less predictable, necessitating heightened sensing, scenario planning, and adaptive feedback loops. McKinsey (2024) emphasizes that firms with built-in capabilities for anticipation and rapid response outperformed their peers during the 2007–2011 financial crisis, underscoring operational resilience as a strategic imperative. Such forces demand more than reactive crisis protocols—they require leadership that can engage complexity directly. Managers often underestimate the intricate nature of modern disruptions; linear fixes often fail, and oversimplification can worsen outcomes (Henning Reschke et al., 2022). Agile and adaptive approaches, originally from software development, have been repurposed for organizational leadership: enabling iterative learning, decentralizing decision-making, and tolerating controlled failure (Appelo, 2011; Hayward, 2018; Solga, 2021).

Moreover, polycrisis demands that leaders manage both technological and human dynamics. Technological resilience such as flexible IT architectures and digitized operations is vital. Yet, social cohesion, trust, and emotional stability among teams are equally critical. Without trust, strategic agility may fail to translate into coherent action. Recent empirical work on organizational resilience characterizes it as the capacity to anticipate, absorb, adapt, and learn from crisis (Duchek, 2020; Bundy et al., 2017). Leadership is central: leaders' structure distributed sensemaking processes; they build infrastructures for exploration and routine execution (Jansen et al., 2009), and gauge when to direct and when to empower. Ambidextrous leaders intentionally shift between enabling experimentation and enforcing stability, crucial in complex, shifting environments.

Against this backdrop, strategic leadership can no longer be perceived as a long-range planning exercise. Instead, it must operate as a real-time orchestration of sensing, framing, and responding. Leaders must engage in what IMD describes as "constructive ambiguity": embracing uncertainty, exploring alternate futures, and inviting diverse perspectives before deciding (IMD, 2025). The stakes are global. Mishandling polycrisis can destroy value, erode trust, and jeopardize socio-economic stability. But effective strategic leadership can transform turbulence into opportunity by initiating faster innovation, rallying stakeholders, and embedding resilience into culture. This introduction establishes the central challenge: global instability is not episodic but systemic, permanent, and accelerating. Firms need leaders equipped with frameworks and behavioral agility to steer through these dynamics. The following sections will dissect strategic leadership practices, illustrate them with case studies, and propose a framework for resilience that aligns with contemporary realities.

Relevance of Strategic Leadership

Strategic leadership serves as the pivotal mechanism transforming a reactive organization into one capable of navigating systemic turbulence. Scholars argue that traditional command-and-control models falter under layered unpredictability characteristic of today's polycrisis environment (Sunoto & Daryanto, 2024). Instead, strategic leadership equips organizations with the capacity to align rapidly shifting priorities with long-term sustainability, balancing exploration with exploitation from incremental refinement to disruptive innovation (Jansen, Vera, & Crossan, 2009; Wikipedia, 2025).

A recent study in Egypt's hospitality sector revealed how leaders with strategic foresight enhanced operational resilience by leveraging digital readiness and crisis-responsive strategies (Elbanna et al., 2023). Strategic leadership proved to be the catalyst linking digital preparedness to real-time crisis response, thus ensuring business continuity when standard protocols failed (Elbanna et al., 2023). The distinction between strategic and crisis leadership lies in proactivity: strategic leaders embed crisis-readiness into organizational DNA. They anticipate emerging threats, cultivate scenario-planning routines, and invest in sensing capabilities. This enables them to act on early warning signals rather than reacting to fully formed threats (James & Wooten, 2022). Research supports this: firms that integrated scouting mechanisms and diverse leadership teams were significantly more prepared for COVID-19 driven disruptions (James & Wooten, 2022).

Furthermore, strategic leadership also weaves ethical decision-making and stakeholder inclusion into resilience frameworks. In the United Arab Emirates, SMEs that combined ethical leadership with sustainability and adaptive practices navigated COVID-19 more effectively maintaining workforce morale and stakeholder trust (Al-Suwaidi & Tlemsani, 2025). Finally, strategic leadership fosters organizational ambidexterity shifting dynamically between stability and innovation. Empirical evidence from psychological capital research shows that leaders who balance transformational inspiration with directive clarity enhance team resilience (PMC, 2024). This duality encouraging innovation while ensuring clarity enables teams to adapt quickly and recover stronger. In sum, strategic leadership is no longer optional. It functions as the linchpin that integrates sense-making, digital readiness, ethical governance, and organizational ambidexterity. Under global upheaval, such leadership transforms external threats into strategic inflection points driving sustained resilience and competitive advantage.

This chapter aims to explore the role of strategic leadership in enhancing organizational resilience under conditions of sustained global upheaval. Specifically, it addresses how leaders can shape decision-making environments, culture, and structures to enable rapid adaptation, innovation, and continuity in the face of crisis. Unlike reactive crisis management approaches, strategic leadership integrates long-term vision with real-time flexibility positioning resilience not as a survival tactic but as a strategic capability embedded in governance, operations, and culture.

The chapter is structured around four core objectives:

1. To conceptualize strategic leadership as a multi-level function that aligns foresight, adaptability, and organizational values in turbulent contexts.
2. To analyze how strategic leadership fosters organizational resilience, particularly by building systems of agility, learning, and psychological safety.
3. To assess empirical cases across different organizational types, especially small and medium-sized enterprises (SMEs), which are often the most exposed yet most innovative during crisis scenarios.
4. To propose a strategic resilience framework that synthesizes leadership behaviors, digital integration, ethical governance, and ambidextrous decision-making.

THEORETICAL FRAMEWORK

A solid understanding of strategic leadership begins with clarifying its foundational concepts: leadership, resilience, and crisis adaptation.

Leadership

Leadership encompasses the capacity to influence organizational direction and foster collective achievement. Theories vary from contingency views suggesting effectiveness depends on the fit between leader behavior and situation (Fiedler & Garcia, 1987) to models that emphasize relational flexibility; for instance, situational leadership posits leaders should shift between directing and supporting based on team readiness (Hersey & Blanchard, 1969). Transformational leadership transcends transactional exchanges by inspiring vision, intellectual stimulation, and individualized support—particularly effective in turbulent contexts (Bass & Riggio, 2006). Under pressure, leaders draw upon cognitive resource theory: intelligence aids in low-stress contexts, while experience becomes critical during high-stress situations (Fiedler & García, 1987).

Resilience

Resilience refers to the capacity of an organization to absorb, adapt, and learn from adversity, maintaining or restoring its core functions (United Nations, 2009; Duchek, 2020). It is not merely bouncing back but includes coping accepting reality and executing adaptive responses and learning, which transforms failure into future preparedness (Wildavsky, 1991; Kayes, 2018). Resilient organizations combine technical systems with human dynamics, fostering psychological safety, collective sensemaking, and emergent improvisation to respond in real time (Weick et al., 2005; Duchek, 2020).

Crisis Adaptation

Crisis adaptation is the dynamic capability enabling an organization to adjust structures, mindsets, and resources as crises unfold. It implicates both strategic foresight anticipating scenarios and operational pivoting realigning actions under duress (Heifetz et al., 2009). Adaptive leadership prioritizes learning, collaboration, and decentralized decision-making to transform crisis into opportunity (Heifetz & Linsky, 2002; Wu et al., 2021). Empirical studies during COVID-19 identify transformational leadership and directive clarity (a blend of vision and decisiveness) as jointly enhancing organizational resilience, mediated by employees' psychological capital (Hope, efficacy, resilience, optimism) (Frontiers in Psychology, 2023).

MODELS OF STRATEGIC LEADERSHIP

Strategic leadership draws on a variety of models, each offering a unique lens through which leaders navigate uncertainty and drive resilience. The most relevant frameworks in turbulent environments include transformational leadership, adaptive leadership, ambidextrous leadership, and strategic foresight leadership.

Transformational leadership is anchored in vision, inspiration, and individualized consideration. It moves beyond transactional exchanges by motivating followers to transcend self-interest for organizational goals (Bass & Avolio, 1994). Especially during crises, such leaders foster trust, emotional commitment, and shared purpose critical factors in sustaining employee morale and engagement (Tucker & Russell, 2004). Transformational leadership has been positively correlated with innovation and resilience in high-stakes industries such as healthcare and education, where rapid adaptation and team cohesion are essential (Garcia-Sánchez et al., 2021).

Heifetz and Linsky (2002) introduced adaptive leadership as a framework for leading through change when problems lack clear solutions. Unlike technical problems with known fixes, adaptive challenges require experimentation, learning, and mobilizing diverse stakeholders. Adaptive leaders engage in sensemaking, distribute responsibility, and challenge status quo norms. In complex environments, this model proves essential for facilitating continuous adjustment while maintaining legitimacy and direction (Northouse, 2022).

Ambidextrous leadership refers to a leader's ability to simultaneously support exploration (innovation) and exploitation (efficiency). This model becomes especially relevant in volatile markets where organizations must pivot between stability and transformation. Leader's balance competing demands empowering innovation units while reinforcing performance structures (Rosing et al., 2011). Empirical research shows that firms with ambidextrous leadership report higher adaptive capacity and faster crisis recovery (Mom et al., 2015).

Strategic foresight refers to the practice of anticipating emerging trends and preparing for multiple plausible futures. Leaders applying this model utilize tools like scenario planning, environmental scanning, and trend impact analysis to reduce uncertainty. Foresight is not prediction it is preparation. Effective foresight leaders use weak signals to build adaptive strategies, strengthening long-term resilience (Rohrbeck & Kum, 2018).

LINK TO BUSINESS MANAGEMENT THEORIES

Strategic leadership is not an isolated concept it is deeply intertwined with foundational theories in business management. These theoretical intersections clarify how leadership drives organizational behavior, resource allocation, and adaptive performance in turbulent environments.

The Resource-Based View (Barney, 1991) positions internal capabilities as key to sustainable competitive advantage. Leadership is central to this framework because leaders shape, deploy, and renew valuable, rare, inimitable, and non-substitutable (VRIN) resources. Strategic leaders act as "resource orchestrators," aligning intangible assets such as culture, knowledge, and innovation potential with firm strategy. Especially during crises, leadership becomes a dynamic capability that determines whether firms can reconfigure resources to address shifting demands (Teece, 2014).

Contingency theory suggests there is no single best way to lead; instead, effectiveness depends on aligning leadership style with contextual factors such as organizational size, structure, and environmental uncertainty (Fiedler, 1967). Strategic leadership aligns with this theory by emphasizing the need for situational awareness and adaptability. During periods of upheaval, leaders who tailor their decision-making processes to fit internal constraints and external volatility are more likely to maintain operational continuity.

Strategic leadership also resonates with stakeholder theory, which extends managerial responsibility beyond shareholders to include employees, suppliers, customers, and the broader community (Freeman, 1984). Resilient leaders adopt inclusive approaches that build trust, particularly during crises where stakeholder confidence is fragile. By embedding transparency and ethical responsiveness, strategic leaders enhance reputational capital and social legitimacy critical assets in post-crisis recovery.

Lastly, systems theory reinforces the interconnectedness of leadership decisions across organizational functions. Strategic leaders recognize the organization as a living system, where disruptions in one area (e.g., supply chain) have ripple effects elsewhere (e.g., customer service or HR). Systems thinking enables leaders to identify leverage points, coordinate cross-functional responses, and develop feedback loops to support learning and continuous adaptation (Senge, 2006).

GLOBAL CRISIS CONTEXT

Overview of Recent Global Crises

The contemporary global environment is widely described as a polycrisis—a condition in which multiple, simultaneous crises intersect, reinforcing each other and creating cascading systemic instability (Tooze, 2022; Morin, 1993). These crises are no longer isolated events but are interdependent, amplifying risk across social, economic, and political domains.

The COVID-19 pandemic marked the onset of this turbulence. Beginning in early 2020, it disrupted global supply chains, caused widespread job losses, and triggered one of the deepest recessions in recent history (Baldwin & Weder di Mauro, 2020). Beyond health effects, the pandemic exposed organizational vulnerabilities and forced rapid digital adaptation, remote operations, and crisis management innovation.

Following this, the global supply chain crisis between 2021 and 2023 compounded recovery challenges. Port congestions, shipping container shortages, and semiconductor scarcities paralyzed production in key industries such as automotive, electronics, and pharmaceuticals (Gupta et al., 2022). The situation revealed critical flaws in just-in-time inventory models and prompted a shift toward more resilient "just-in-case" strategies (Sheffi, 2022).

Geopolitical instability has intensified global uncertainty. The Russia–Ukraine war disrupted global energy and food markets, while rising U.S.–China tensions have led to trade fragmentation and strategic decoupling in sectors like technology and clean energy (Goldstein, 2023). These developments have accelerated a trend toward economic nationalism and regional supply chain realignment.

Concurrently, the world is facing energy and food crises. Energy price volatility, aggravated by geopolitical conflict and fossil fuel dependence, has driven inflation and fiscal stress in many countries (IEA, 2023). Food insecurity has also surged due to export restrictions, climate-related crop failures, and fertilizer shortages, disproportionately impacting developing economies (UNDP, 2022).

The intensifying impact of climate change adds another layer of risk. Extreme weather events—wildfires, floods, and prolonged droughts—have become more frequent, with direct consequences for agriculture, infrastructure, and urban resilience (IPCC, 2023). Regulatory requirements for carbon disclosure, ESG compliance, and climate adaptation are placing new burdens on firms across industries.

Finally, a growing financial vulnerability is emerging. High inflation, aggressive interest rate hikes, and unsustainable sovereign debt levels have increased credit risk and reduced fiscal buffers in many economies (BIS, 2024). These pressures are particularly acute in emerging markets, where capital flight and exchange rate volatility further constrain policy space.

Compounding and Converging Effects

These crises are not isolated but interconnected. For example, pandemic-driven supply disruptions exacerbated inflation, which was then further intensified by war-related energy shocks and climate-induced food scarcity. Scholars are now referring to this escalation as a hypercrisis—a dynamic where convergence, speed, and unpredictability outpace institutional response capacity (Tooze, 2022).

Implications for Organizations

This volatile context presents several challenges for firms:

- Operational complexity: Simultaneous disruptions in logistics, labor, and capital flows increase risk exposure.
- Strategic ambiguity: Leaders face shortened planning cycles and frequent strategy reassessments.
- Stakeholder pressure: Customers, investors, and regulators increasingly demand ethical, adaptive, and sustainable behavior.

Impact on Business Systems and Leadership Structures

The convergence of global crises has placed unprecedented pressure on organizational systems and challenged the viability of traditional leadership structures. In this volatile context, linear planning models, rigid hierarchies, and siloed functions are proving increasingly ineffective. Instead, organizations are being pushed toward decentralized, adaptive, and networked leadership systems that can respond to dynamic threats in real time (Uhl-Bien & Arena, 2018).

From a systems perspective, global disruptions have revealed the fragility of extended supply chains, lean inventories, and over-centralized operations. Companies heavily reliant on global sourcing and just-in-time models faced severe operational paralysis during the pandemic and subsequent supply chain crises (Ivanov & Dolgui, 2020). These vulnerabilities prompted firms to re-engineer their systems around principles of redundancy, agility, and digital visibility, supported by leadership committed to proactive scenario planning and cross-functional coordination.

Leadership structures are also undergoing a transformation. Traditional top-down models—characterized by command-and-control decision-making—have struggled to cope with the speed and complexity of overlapping crises (Ancona et al., 2007). In contrast, distributed leadership frameworks, which empower decision-making at multiple levels, have been more effective in enhancing organizational responsiveness and resilience. This shift necessitates a cultural transformation in leadership development, emphasizing emotional intelligence, systems thinking, and collaborative learning (Raelin, 2016).

Crises have also reshaped strategic priorities, pushing resilience, digital transformation, and stakeholder trust to the forefront of business agendas. Leaders are increasingly expected to act as sensemakers—interpreting complex, ambiguous signals to guide collective action under uncertainty (Weick, Sutcliffe, & Obstfeld, 2005). As such, leadership roles are expanding beyond performance and profitability to include risk anticipation, ethical governance, and sustainability integration.

Furthermore, the evolving crisis landscape has highlighted the importance of organizational learning loops, where feedback from failure is rapidly incorporated into strategy and operations. Strategic leaders now need to build adaptive cultures that normalize learning from disruptions, enabling continuous improvement and resilience building (Schoemaker et al., 2018).

STRATEGIC LEADERSHIP IN ACTION

Successful Crisis Leadership

Philips – Rapid Ecosystem Adaptation

During the COVID-19 crisis, Philips demonstrated extraordinary organizational agility. Within days, senior leaders reconfigured supply chains, delegated authority to frontline teams, and accelerated production of life-saving medical technologies (Time, 2025). This act of ecosystem leadership shifted control into adaptive networks, combining top-down support with collaborative partnerships—resulting in swift innovation and responsiveness (Time, 2025).

Johnson & Johnson – The Tylenol Case

In 1982, Johnson & Johnson faced a lethal crisis when cyanide-laced Tylenol capsules resulted in consumer deaths. CEO James Burke immediately launched a national recall of 31 million bottles, publicly accepted responsibility, and introduced tamper-proof packaging to restore public trust (Wikipedia, 2025). The company's transparent, stakeholder-centered strategy became a benchmark in ethical, reputation-saving leadership.

Failed or Weak Crisis Leadership

Boeing 737 MAX Scandal

Following two deadly crashes linked to Boeing's 737 MAX design, critics accused company leadership of downplaying safety concerns and delaying decisive action. Major airline customers publicly lost confidence and pushed for executive accountability (Reuters, 2024). The event underscored how defensive culture, lack of urgency, and failure to prioritize stakeholder trust can severely damage organizational legitimacy.

Lehman Brothers Collapse (2008)

Lehman Brothers' leadership exemplifies failure in anticipating financial and systemic risk. Heavily leveraged and deeply invested in subprime mortgages, their decision-makers inflated risk through short-term profitability focus. When the housing bubble burst, the unchecked exposure and delayed response with no credible rescue plan triggered the largest U.S. bankruptcy at the time (Investopedia, 2009).

Comparative Analysis

Leadership Element	Success Cases	Failure Cases
Decision Speed	Philips and J&J acted decisively and early	Boeing and Lehman delayed critical responses
Stakeholder Trust	Proactive, transparent communication	Attempts to protect reputation backfired
Organizational Learning	Adopted structural change (tamper-proof packaging, adaptive supply chains)	Repeated mistakes, failure to adapt crisis culture
Leadership Network	Distributed authority, ecosystem partnerships	Centralized hierarchy, lack of external input

Decision-Making Under Pressure

Effective decision-making under pressure is a cornerstone of strategic leadership during crisis. High-stakes environments—characterized by incomplete information, time scarcity, and emotional intensity—require leaders to shift from routine analysis to intuitive judgment, often relying on accumulated experience and rapid sensemaking (Klein, 1998). Strategic leaders who perform well in such conditions are those who combine speed with disciplined flexibility, accepting ambiguity while continuously reassessing options as new data emerges (Eisenhardt & Zbaracki, 1992).

Crisis decisions are rarely about finding the optimal choice—they're about identifying the least damaging viable path quickly and clearly. Research has shown that organizations led by individuals who embrace bounded rationality and iterative decision cycles recover more effectively (Christensen et al., 2016). Notably, success depends less on initial accuracy than on leaders' willingness to adjust course decisively as feedback unfolds.

Moreover, distributed decision-making—empowering mid-level and frontline teams to act independently—has emerged as a best practice in fast-moving crises. Such decentralization increases the speed and quality of response by allowing situationally informed decisions to be made without bureaucratic delay (Uhl-Bien & Arena, 2018).

Communication and Stakeholder Management

Transparent and timely communication is equally essential to crisis leadership. Stakeholders—employees, customers, investors, and regulators—expect clarity, reassurance, and accountability. Leaders who fail to communicate proactively risk losing trust, increasing panic, and worsening reputational harm (Coombs, 2007).

Strategic communication during crises follows a few key principles: speed, honesty, empathy, and consistency. Johnson & Johnson's Tylenol response in 1982 remains a textbook example. CEO James Burke's early acknowledgment of the crisis, clear public messaging, and direct engagement with regulators and media enabled the company to preserve consumer trust and brand equity (Fink, 2013).

Furthermore, communication must be two-way. Stakeholder engagement is not limited to broadcasting information—it requires listening, feedback loops, and adapting actions to stakeholder concerns (Heath, 2010). This is particularly vital in complex crises where stakeholder expectations evolve rapidly.

Internally, leaders must ensure that employees are informed, safe, and heard. Externally, they must manage investor relations, regulatory scrutiny, and public image—all while delivering consistent messaging across platforms and geographies. Failures in this area, as in Boeing's 737 MAX case, demonstrate how misalignment in communication strategy can significantly amplify operational crises into long-term trust deficits (Gillespie & Dietz, 2009).

FOCUS ON SMES

Unique Challenges and Advantages in Crisis Scenarios

Small and medium-sized enterprises (SMEs) are vital to global economies, accounting for over 90% of all firms and a majority of employment worldwide (OECD, 2021). Yet in times of crisis, they face disproportionate vulnerabilities and resource constraints compared to larger corporations. Despite this, SMEs often demonstrate surprising agility and adaptability—making their experience both cautionary and instructive for strategic leadership.

The most pressing challenge for SMEs during crises is their limited financial and operational buffer. Many operate on thin margins with restricted access to credit, making them acutely sensitive to cash flow disruptions and supply chain delays (Cowling et al., 2020). Their lack of diversification—in revenue streams, suppliers, or markets—amplifies the risk of localized shocks. A single disrupted contract or vendor issue can have cascading effects across the firm's operations.

Additionally, SMEs often lack formalized risk management systems. Crisis plans, contingency strategies, and business continuity protocols are less likely to be in place, especially in micro-enterprises (Herbane, 2013). As a result, decision-making under pressure is often reactive rather than strategic, heavily influenced by immediate operational concerns and emotional leadership responses.

Leadership concentration presents another constraint. Many SMEs are led by single-owner-managers or family heads whose decision-making authority is centralized. While this allows speed, it can limit input diversity and delay structural adaptation, especially when leaders resist change or lack crisis experience (Doern, 2016).

Paradoxically, the same lean structures that expose SMEs to crisis also provide agility and responsiveness. Without layers of bureaucracy, SMEs can pivot rapidly in response to new market signals. During the COVID-19 pandemic, many small manufacturers quickly retooled production for medical supplies or hygiene products—an example of adaptive innovation at speed (Bourletidis & Triantafyllopoulos, 2014).

Moreover, SMEs tend to be deeply embedded in their local communities and customer networks, allowing for more personalized service and faster feedback loops. These tight stakeholder relationships foster loyalty and enable leaders to co-create solutions with clients, suppliers, and even competitors.

Leadership proximity also strengthens internal trust and cohesion. With flatter hierarchies and closer interpersonal ties, SME leaders can communicate decisions transparently, reinforce shared purpose, and mobilize employees quickly. This informal resilience, based on relational capital rather than formal strategy, plays a crucial role in withstanding external shocks (Pal et al., 2014).

In sum, while SMEs are structurally vulnerable to crisis, they also possess distinct strengths—flexibility, trust-based networks, and responsiveness—that can be leveraged through intentional leadership. The next section explores strategic approaches SMEs can adopt to institutionalize resilience beyond improvised survival.

Leadership Innovation and Grassroots Resilience

In the face of recurrent global disruptions, many SMEs have responded not with scale, but with leadership innovation—creative approaches rooted in community networks, frugal resourcefulness, and cultural proximity. Unlike larger firms that often rely on formalized corporate structures and institutional strategies, SMEs frequently survive and recover through grassroots resilience, underpinned by adaptive leadership, informal systems, and embedded trust.

Leadership innovation in SMEs is typically characterized by non-traditional problem-solving, often driven by necessity rather than resources. During crises, SME leaders tend to wear multiple hats—strategist, communicator, and executor—allowing them to integrate sensing and acting without bureaucratic delay (Linnenluecke, 2017). This proximity to frontline operations creates a unique feedback-rich environment, enabling rapid iteration and adjustment in strategy.

Moreover, leadership in SMEs often draws strength from community alignment. Especially in family-run or regionally embedded firms, resilience is not just economic—it is social and moral. Leaders often prioritize staff welfare, community continuity, and local supplier survival, reinforcing a mutual support system in times of stress (De Vries & Shields, 2006). These networks become vital sources of shared intelligence, collective action, and emotional support.

SME resilience is also visible in micro-level innovations. During the COVID-19 pandemic, local restaurants in Lebanon and Jordan restructured their menus, shifted to delivery-only operations, and used social media to crowdsource funding or barter services—demonstrating a form of informal entrepreneurialism underpinned by leadership improvisation (Kraus et al., 2020). These responses, while often lacking formality, proved highly effective and replicable.

A core enabler of grassroots resilience is relational capital—long-standing trust among employees, suppliers, and customers. SME leaders who have cultivated open communication, shared norms, and transparency are better positioned to mobilize these relationships under duress. Trust-based leadership compensates for the lack of formal risk protocols by creating a culture of mutual commitment and shared sacrifice (Sullivan-Taylor & Branicki, 2011).

Innovation in SME leadership also involves learning-by-doing. Crises push small firms to develop adaptive competencies—such as digitalization, remote work systems, or new product offerings—not as part of long-term strategies, but as real-time survival tactics that become permanent innovations (Battisti & Deakins, 2017). This emergent learning process strengthens organizational capacity over time.

In conclusion, the resilience of SMEs is not accidental. It emerges from leadership that is locally grounded, relationally rich, and creatively adaptive. While lacking the scale and systems of large corporations, SMEs often outperform them in navigating uncertainty—not through formal planning, but through innovative leadership embedded in community logic.

STRATEGIC RESILIENCE FRAMEWORK

Overview

In a world grappling with polycrisis and hypercrisis dynamics, strategic resilience has emerged as a cornerstone of business continuity. Strategic resilience extends beyond risk mitigation—it centers on the intentional design of systems, practices, and policies that absorb, adapt, and transform amid systemic shocks. This section outlines a robust, multidimensional framework comprising four interrelated components:

1. Environmental Sensing & Anticipatory Governance
2. Adaptive Operational Architecture
3. Relational & Ethical Stakeholder Alignment
4. Learning, Innovation & Transformational Culture

Each component includes actionable guidelines and policy levers, enabling both firms and ecosystems to navigate uncertainty proactively.

Proposed Model for Business Continuity and Adaptation

ENVIRONMENTAL SENSING & ANTICIPATORY GOVERNANCE

Definition & Rationale

Environmental sensing refers to an organization's ability to continuously monitor macro and micro signals—geopolitical moves, technology shifts, social sentiment, regulatory trends—to identify emerging threats and opportunities. Anticipatory governance integrates forecasting, scenario development, and governance mechanisms to translate foresight into strategic action.

Key Practices

- Dynamic Scenario Planning

Drawn from strategic foresight theory (Rohrbeck & Kum, 2018), firms should conduct annual scenario exercises incorporating low-probability, high-impact events—such as pandemics, cyber disruptions, or political upheavals.

- Red Teaming & Weak-Signal Detection

Cross-functional "red teams" should challenge organizational blind spots. Systems should be in place to capture weak signals—such as customer sentiment shifts or supply volatility—with analytical capability to assess potential futures.

- Board-Level Resilience Oversight

Boards must include risk and resilience committees responsible for periodic stress testing and linkage between forward-looking insights and corporate strategy. This ensures governance becomes anticipatory rather than reactive.

Organizational Guidelines & Policy Implications

- Embed foresight functions within C-suite roles (e.g., Chief Resilience Officer) with dedicated budgets.
- Mandate scenario planning and stress-testing in annual filings.
- Incentivize environmental scanning through both reward systems and policy recognition.

Definition & Rationale

Benefits of redundant flexibility: adaptive operational architecture combines digital infrastructure, modular physical assets, and organizational agility. It ensures operations remain functional during unpredictable disruptions by enabling quick reconfiguration or substitution.

Key Practices

- Modular Supply Chains

Leverage nearshoring, multi-sourcing, and buffer inventories to reduce dependency on single geographies. Digital twins and Blockchain-based traceability enhance supply visibility and verification.

- Digital Operational Platforms

Use cloud-based ERP systems that allow dynamic resource reallocation—whether workforce, manufacturing, or logistics—depending on real-time demand or disruption.

- Emergency Activation Protocols

Predefined "playbooks"—for remote work, alternative suppliers, financial stewardship—should be ready for rapid execution with clear command channels.

- Cross-Training & Workforce Flexibility

Equip employees to rotate through similar roles, reducing critical single-point dependencies. This builds operational elasticity with minimal cost.

Organizational Guidelines & Policy Implications

- Offer government-subsidized programs for digitization and workforce cross-training.
- Provide tax credits for modular operations and dual-sourcing procurement.
- Encourage industry coalitions to share contingency infrastructure (e.g. shared warehousing, transport flexibility).

RELATIONAL & ETHICAL STAKEHOLDER ALIGNMENT

Definition & Rationale

This component emphasizes building trust and legitimacy across employees, suppliers, communities, investors, and regulators—anchored in ethical transparency and constant engagement.

Key Practices

- Empathetic Internal Communication

Leaders must foster open communication during crises—explaining challenges, acknowledging uncertainty, and soliciting feedback. Psychological safety is key to collaboration and innovation (Heifetz et al., 2009).

- Transparent External Reporting

Crisis disclosures should include both operational and reputational metrics—such as employee welfare, emissions targets, and supply chain resilience.

- Collaborative Ecosystem Leadership

SMEs and large firms alike benefit from collective problem-solving. Cross-sector hubs or roundtables help them share insights, risk, and innovation.

- Ethical Governance Charter

A living framework that commits the firm to equitable decisions in layoffs, supplier payments, and community support.

Organizational Guidelines & Policy Implications

- Regulators should require crisis disclosures beyond financials—e.g. human capital and operational risk.
- Recognize stakeholder engagement and ecosystem leadership in ESG ratings.
- Provide policy support for cross-sector resilience networks (e.g. Small Business Administration grants for coalition-building).

Definition & Rationale

Long-term resilience depends on the institutionalization of learning loops, innovation mindsets, and cultural openness. This subsection combines crisis-driven improvisation with ongoing capability enhancement.

Key Practices

- After-Action Reviews (AARs)

Post-crisis evaluations should be structured and obligatory. Document root causes, decision pathways, and knowledge integration into strategy.

- Experimentation Sandbox

Firms should allocate resources for testing emergent ideas in low-risk environments—for example, piloting remote work tools or supply redundancies before emergencies.

- Ambidextrous Leadership Training

Leaders should be trained to switch between exploration and exploitation modes. Digital dashboards help track both operational KPIs and innovation metrics.

- Resilience Champions & Communities of Practice

Internal networks promote knowledge sharing across departments and with external partners.

Organizational Guidelines & Policy Implications

- Policy could mandate AARs (similar to aviation and defense practices).
- Provide R&D credits not just for technology but for resilience-driven innovation.
- Incorporate resilience competencies in leadership certification programs (e.g. public-private partnerships in business schools).

Framework Integration: Dynamic Resilience Loop

To help integrate these practices, the framework operates as a continuous resilience loop:

1. Sense & Govern: Environmental scanning and board review.
2. Activate: Engage operational protocols and stakeholder communication.
3. Learn: Conduct reviews and derive lessons.
4. Re-Engineer: Update governance, structures, and culture.

These components form a reinforcing cycle that embeds resilience into organizational DNA.

Strategic Recommendations

For leaders:

- Appoint a designated resilience lead with cross-functional mandate.
- Run quarterly scenario planning with external expert input.
- Include resilience metrics in performance evaluation.
- Cultivate cross-functional networks and reflection forums for adaptability.

For policymakers and industry bodies:

- Offer vouchers and tax incentives for SME digitalization and resilience planning.
- Incorporate resilience content in public reporting standards.
- Support industry consortia and public-private innovation sandboxes.

Policy Implications

Financial incentives – resilience grants and tax expense deductibility accelerate organizational adoption.

Regulatory alignment – resilience commitments should be institutionalized in governance standards with soft enforcement (e.g. ESG reporting).

Capacity building – public-private partnerships in leadership development, cross-functional training, and digitalization platforms.

Ecosystem activation – interfirm resilience hubs promote knowledge sharing and shared infrastructure investments.

CONCLUSION

Key Findings

This chapter has systematically demonstrated that in today's polycrisis environment—characterized by intersecting and compounding disruptions—strategic resilience is no longer optional for businesses; it is essential. Several key findings emerge:

1. Global Crisis Landscape is Systemic and Persistent

The brief overview in Section 3 makes clear that the recent global crises—pandemics, geopolitical turbulence, supply chain breakdowns, climate shocks, and financial instability—are not isolated incidents. Instead, they form an intertwined system with reinforcing feedback loops, producing a persistent baseline of uncertainty. The concept of hypercrisis confirms that instability has become the new normal.

2. Strategic Leadership is Central to Resilience

Strategic leadership—encompassing foresight, adaptability, ambidexterity, and ethical stakeholder engagement—is the central mechanism through which organizations build and sustain resilience. Leaders who embrace iterative scenario planning, empower distributed decision-making, and operate with transparency transform adversity into organizational strength.

3. Diverse Leadership Models Complement One Another

Transformational, adaptive, ambidextrous, and foresight-based leadership models each provide critical competencies. Table-based analysis in Section 2 shows how transformational leaders inspire collective purpose; adaptive leaders leverage uncertainty; ambidextrous leaders balance innovation with efficiency; and foresight leaders map possible futures. These models, when integrated, create leadership that is both visionary and practical in crisis.

4. Business Theories Underpin Strategic Action

Connections to business management theories such as the Resource-Based View, Contingency Theory, Stakeholder Theory, and Systems Theory provide theoretical support for strategic resilience. Leaders orchestrate internal capabilities, align style with context, engage diverse stakeholders, and navigate interdependent systems.

5. SMEs Possess Unique Strengths and Constraints

SMEs face constraints—limited liquidity, governance shortcomings, and leadership centrality—but also exhibit advantages, including dexterity, local embedding, and relational capital. Leadership innovation at the grassroots level enables many SMEs to rapidly pivot, improvise solutions, and co-create survival strategies with stakeholders.

6. The Strategic Resilience Framework is Integrative and Actionable

The four-component model—Environmental Sensing, Adaptive Architecture, Ethical Stakeholder Alignment, and Transformational Learning—formulates a continuous resilience loop. This framework integrates systems, processes, and culture, providing a coherent pathway to embedding resilience into organizational DNA.

Recommendations for Future Leaders

Based on the above findings, here are ten targeted recommendations for leaders aiming to navigate future unknowns effectively:

1. Institute Resilience Leadership Roles

Create a dedicated role or committee (e.g., Chief Resilience Officer) with visibility into strategy, operations, and stakeholder relations. Ensure board-level oversight of resilience planning.

2. Embed Foresight into Strategy Cycles

Integrate dynamic scenario exercises into annual planning. Use cross-functional teams to scan weak signals and assess plausible futures. Allocate resources to foreseen but low-probability events.

3. Empower Distributed Decision Rights

Push appropriate decision authority to local teams, providing guidelines and guardrails, but trusting judgment and reducing bottlenecks. Recognize this as a central feature of responsive operations.

4. Design Flexible Systems of Work

Adopt modular supply chains and digital platforms. Build buffer capabilities (financial, inventory, labor rotation) to offset disruptions. Invest in digital twins or advanced analytics to monitor system health.

5. Establish Crisis-Ready Protocols

Develop and maintain playbooks for crisis modes—covering remote work, emergency communication, supplier switches, and incident command. Conduct drills and exercises regularly.

6. Foster Transparent Stakeholder Engagement

Maintain ongoing, empathetic communication with employees, investors, regulators, and communities. Position crises as joint challenges rather than firm-centric issues. Gather and act on feedback.

7. Institutionalize Feedback Loops

After every disruption—tabletop exercise or live crisis—conduct structured After-Action Reviews. Deposit insights into centralized knowledge systems and adapt governance accordingly.

8. Cultivate Ambidextrous Leadership Teams

Promote executives who can pivot between pragmatic stabilization and exploratory innovation. Balance performance metrics (e.g., cost targets) with resilience-oriented indicators (e.g., supply chain robustness).

9. Amplify Community-Based Resilience in SMEs

- Encourage SMEs to join resilience coalitions that pool resources and share intelligence.
- Provide professional development in sensemaking, participatory leadership, and community engagement.

10. Measure Resilience Progress Over Time

Track resilience KPIs such as scenario-readiness scores, supply chain flexibility metrics, crisis response speed, and stakeholder confidence surveys. Publicly report progress in ESG and integrated reporting frameworks.

Contribution to Societal Resilience

This chapter's impact extends beyond individual firms, supporting broader societal resilience through three mechanisms:

A. Strengthening Economic Stability

Resilient firms prevent economic collapse during crises, safeguarding jobs, infrastructure, and supply continuity. They reduce pressure on governments and support stable business ecosystems. SMEs, in particular, anchor communities by sustaining local production, employment, and social cohesion.

B. Reinforcing Social Trust and Governance

Transparency and ethical leadership rebuild public trust in institutions. Crisis empathy, stakeholder inclusion, and equitable practices protect reputational capital and social legitimacy—essential in preventing political and social fragmentation.

C. Enhancing Systemic Adaptability

The proposed framework encourages resource sharing, collaborative forecasting, and cross-institutional learning. When firms enact resilience collectively—through coalitions and interorganizational networks—they foster robust response systems capable of accommodating fluid change.

Taken together, resilient firms form critical nodes in resilient societies. Organizational practices scale upward supporting emergency response, sustaining essential services during disruptions, and enabling quicker recovery.

Final Thoughts

The global context has shifted from episodic turbulence to dynamic continuity of uncertainty. In this environment, only organizations that integrate strategic resilience into their DNA—via proactive leadership, flexible systems, stakeholder trust, and continuous innovation—will survive and thrive.

The proposed model guides leaders from reactive, tactical responses to proactive, transformative practices, reshaping the role of business from mere economic actors to partners in systemic resilience. As the climate crisis intensifies, geopolitical landscapes fragment, and technological disruption accelerates, the capacity to sense, adapt, and learn will define organizational relevance.

This chapter calls on future leaders to rethink their mandate. Leadership must be reimagined as stewardship—balancing stability with change, performance with purpose, and corporate ambition with civic responsibility. Through this lens, businesses not only weather storms—they become catalysts for resilient societies.

REFERENCES

Al-Suwaidi, T., & Tlemsani, I. (2025). Sustainable leadership styles adopted by small businesses in the UAE during and post COVID-19. *Discover Sustainability*, •••, 5–6.

Ancona, D., Malone, T. W., Orlikowski, W. J., & Senge, P. (2007). In praise of the incomplete leader. *Harvard Business Review*, 85(2), 92–100. pp. 94–96.

Appelo, J. (2011). *Management 3.0: Leading agile developers, developing agile leaders*. Addison-Wesley.

Arkenberg, C. (2019, October 23). Leading in a VUCA world: How 21st-century leaders can find success amid volatility, uncertainty, complexity, and ambiguity. *Deloitte Insights*.

Barney, J. B. (1991). Firm resources and sustained competitive advantage. *Journal of Management, 17*(1), 99–120. pp. 105–107.

Bass, B. M., & Avolio, B. J. (1994). *Improving organizational effectiveness through transformational leadership*. Sage Publications.

Bass, B. M., & Riggio, R. E. (2006). *Transformational leadership* (2nd ed.). Lawrence Erlbaum Associates. DOI: 10.4324/9781410617095

Battisti, M., & Deakins, D. (2017). The relationship between dynamic capabilities, the firm's resource base and performance in a post-disaster environment. *International Small Business Journal, 35*(1), 78–98. pp. 80–82.

Bourletidis, K., & Triantafyllopoulos, Y. (2014). SMEs survival in time of crisis: Strategies, tactics and commercial success stories. *Procedia - Social and Behavioral Sciences, 148*, 639–644. pp. 641–642.

Bundy, J., Pfarrer, M. D., Short, C. E., & Coombs, W. T. (2017). Crises and crisis management: Integration, interpretation, and research development. *Journal of Management, 43*(6), 1661–1692. p. 1662.

Cowling, M., Brown, R., & Rocha, A. (2020). Did you save some cash for a rainy COVID-19 day? The crisis and SMEs. *International Small Business Journal, 38*(7), 593–604. pp. 595–596.

De Vries, H. P., & Shields, M. (2006). Leadership and small business resilience. *New Zealand Journal of Applied Business Research, 5*(1), 55–66. pp. 58–60.

Doern, R. (2016). Entrepreneurship and crisis management: The experiences of small businesses during the London 2011 riots. *International Small Business Journal, 34*(3), 276–302. pp. 278–281.

Duchek, S. (2020). Organizational resilience: A capability-based conceptualization. *Business Research, 13*, 215–246. pp. 220–221.

Elbanna, S., Callum, M., & Aly, R. (2023). Leveraging strategic leadership for boosting operational resilience in hotels: The role of crisis response strategies and e-readiness. *International Journal of Contemporary Hospitality Management.*

Fiedler, F. E. (1967). *A theory of leadership effectiveness*. McGraw-Hill.

Fiedler, F. E., & García, J. E. (1987). *New approaches to leadership: Cognitive resources and organizational performance*. Wiley.

Freeman, R. E. (1984). *Strategic management: A stakeholder approach*. Pitman.

Frontiers in Psychology. (2023). Surviving a crisis: A multilevel model of leadership styles, psychological capital, and organizational resilience. pp. 1–2.

Hayward, S. (2018). *The agile leader: How to create an agile business in the digital age*. Kogan Page.

Heifetz, R., & Linsky, M. (2002). *Leadership on the Line*. Harvard Business Review Press.

Heifetz, R. A., Grashow, A., & Linsky, M. (2009). *The practice of adaptive leadership*. Harvard Business Press.

Herbane, B. (2013). Exploring crisis management in UK small- and medium-sized enterprises. *Journal of Contingencies and Crisis Management, 21*(2), 82–95. pp. 85–86.

Hersey, P., & Blanchard, K. H. (1969). Life cycle theory of leadership. *Training and Development Journal, 23*, 26–34. pp. 28–29.

IMD. (2025). Shift 5: From decisive closure to constructive ambiguity. *IMD – Leadership in a turbulent world.*

Investopedia. (2009). *The Collapse of Lehman Brothers: A case study*. pp. 1–2.

Ivanov, D., & Dolgui, A. (2020). Viability of intertwined supply networks: Extending the supply chain resilience angles towards survivability. *International Journal of Production Research, 58*(10), 2904–2915. pp. 2907–2909.

James, E. H., & Wooten, L. P. (2022). Why business leaders should always be thinking about the next crisis. *Time*.

Jansen, J. J. P., Vera, D., & Crossan, M. (2009). Strategic leadership for exploration and exploitation: The moderating role of environmental dynamism. *The Leadership Quarterly*, *20*(1), 5–18. DOI: 10.1016/j.leaqua.2008.11.008

Jansen, J. J. P., Vera, D., & Crossan, M. (2009). Strategic leadership for exploration and exploitation: The moderating role of environmental dynamism. *The Leadership Quarterly*, *20*(1), 5–18. DOI: 10.1016/j.leaqua.2008.11.008

Kayes, D. C. (2018). *Organizational resilience: How learning sustains organizations in crisis*. The Learning Organization, 25(2), 143–146. p. 145.

Kraus, S., Clauss, T., Breier, M., Gast, J., Zardini, A., & Tiberius, V. (2020). The economics of COVID-19: Initial empirical evidence on how family firms in five European countries cope with the corona crisis. *International Journal of Entrepreneurial Behavior & Research, 26*(5), 1067–1092. pp. 1073–1075.

Lawrence, M. M., Homer-Dixon, T., Janzwood, S., Rockström, J., & Renn, O. (2024). *Global polycrisis: The causal mechanisms of crisis entanglement*. Global Sustainability.

Linnenluecke, M. K. (2017). Resilience in business and management research: A review of influential publications and a research agenda. *International Journal of Management Reviews, 19*(1), 4–30. pp. 9–11.

McKinsey & Company. (2024). The resilience imperative: Succeeding in uncertain times.

Mom, T. J. M., Fourné, S. P. L., & Jansen, J. J. P. (2015). Managers' work experience, ambidexterity, and performance. *Human Resource Management, 54*(S1), s133–s153. pp. s139–s140.

Morin, E. (1993). *Terre-Patrie*. Éditions du Seuil.

Northouse, P. G. (2022). *Leadership: Theory and practice* (9th ed.). Sage.

OECD. (2021). *The Digital Transformation of SMEs*. OECD Publishing.

Pal, R., Torstensson, H., & Mattila, H. (2014). Antecedents of organizational resilience in economic crises—An empirical study of Swedish textile and clothing SMEs. *International Journal of Production Economics, 147*, 410–428. pp. 412–414.

PMC. (2024). Surviving a crisis: A multilevel model of leadership styles, employees' psychological capital and organizational resilience. *Frontiers in Psychology*.

Raelin, J. A. (2016). *Leadership-as-practice: Theory and application*. Routledge. DOI: 10.4324/9781315684123

Reschke, H.. (2022). *Management in times of crises: reflections on characteristics, avoiding pitfalls, and pathways out*. PMC.

Reuters. (2024, March 27). *How Boeing's leadership was 'fired' by its own customers*.

Rohrbeck, R., & Kum, M. E. (2018). Corporate foresight and its impact on firm performance. *Technological Forecasting and Social Change, 129*, 105–116. pp. 108–109.

Rosing, K., Frese, M., & Bausch, A. (2011). Explaining the effectiveness of leadership styles: The role of innovation. *Journal of Organizational Behavior, 32*(1), 104–128. pp. 106–107.

Schoemaker, P. J. H., Heaton, S., & Teece, D. (2018). Innovation, dynamic capabilities, and leadership. *California Management Review, 61*(1), 15–42. pp. 20–22.

Senge, P. M. (2006). *The fifth discipline: The art & practice of the learning organization* (Revised ed.). Doubleday.

Sucharita. (2016). Clarifying the conceptual map of VUCA: A systematic review. *International Journal of Organizational Analysis*, 24(3), 484–502.

Sullivan-Taylor, B., & Branicki, L. (2011). Creating resilient SMEs: Why one size might not fit all. *International Journal of Production Research, 49*(18), 5565–5579. pp. 5568–5570.

Sunoto, P., & Daryanto, E. (2024). The role of strategic leadership in building organizational resilience. *International Journal of Integrative Sciences*, *4*(3), 89–105.

Teece, D. J. (2014). The foundations of enterprise performance: Dynamic and ordinary capabilities in an economic theory of firms. *Academy of Management Perspectives, 28*(4), 328–352. pp. 335

Time. (2025, July 10). *The Only Way That Companies Can Build Lasting Transformation*.

Tucker, B. A., & Russell, R. F. (2004). The influence of the transformational leader. *Journal of Leadership & Organizational Studies, 10*(4), 103–111. pp. 105–106.

Uhl-Bien, M., & Arena, M. (2018). Leadership for organizational adaptability: A theoretical synthesis and integrative framework. *The Leadership Quarterly, 29*(1), 89–104. pp. 90–92.

Ungureanu, M.. (2018). Clarifying the conceptual map of VUCA: A systematic review. *The International Journal of Organizational Analysis*, *26*(4), 563–585.

Weick, K. E., Sutcliffe, K. M., & Obstfeld, D. (2005). Organizing and the process of sensemaking. *Organization Science, 16*(4), 409–421. pp. 417–418.

Wildavsky, A. (1991). *Searching for safety*. Transaction Publishers.

Wu, H., Hu, M. Y., & Smith, K. (2021). Turning crisis into opportunity: Adaptive leadership for organizational resilience. *Leadership & Organization Development Journal, 42*(2), 158–171. p. 162.

Chapter 13
Governing Disruption:
A New Architecture for Global Leadership in the Age of Intelligent Systems

Ali Al Mokdad
https://orcid.org/0009-0000-2825-5855
Operations Copilot L.L.C., UAE

ABSTRACT

The 2020s have normalized disruption as a governance baseline. Interacting shocks—climate extremes, conflicts, pandemics, cyber incidents, and algorithmic harms—propagate across energy, food, finance, information, and security systems, producing a polycrisis that overwhelms linear, siloed policy models. This chapter reframes the polycrisis as structural turbulence and advances an architecture for governing it in the age of intelligent systems. Artificial intelligence is treated not only as a tool but as a sociotechnical governance actor whose affordances and failure modes jointly reshape legitimacy and effectiveness. The chapter proposes design imperatives for resilient governance, translates them into institutional innovations, and outlines competencies for disruption-ready leadership, offering a blueprint for orchestration under uncertainty.

INTRODUCTION: THE POLYCRISIS AS A NEW GOVERNANCE REGIME

The third decade of the twenty-first century is not merely marked by an accumulation of crises but by their interaction. A pandemic has destabilized global health systems, disrupted trade flows, and accelerated digital dependency (Kickbusch &

DOI: 10.4018/979-8-3373-5651-8.ch013

Agrawal, 2022). Climate extremes continue to trigger displacement, food insecurity, and conflict (IPCC, 2023). Geopolitical conflicts, such as the war in Ukraine, reverberate across energy and food markets, producing secondary shocks in regions far removed from the battlefield (FAO, 2022). Artificial intelligence (AI) and digital platforms—once framed as instruments of resilience—are simultaneously enablers of efficiency and vectors of disruption, amplifying bias, misinformation, and surveillance (Cath, 2023). These dynamics constitute what the World Economic Forum (2023) and other scholars have described as a *polycrisis*: a condition in which multiple crises are causally entangled, producing outcomes more severe and unpredictable than the sum of their parts.

The polycrisis is best understood not as a temporary spike in volatility but as a new governance regime of structural turbulence (Homer-Dixon et al., 2021). Complex adaptive systems theory explains why tightly coupled global networks—supply chains, financial systems, cyber infrastructures, and food-energy-water nexuses—are prone to cascading failures, feedback loops, and tipping points (Meadows, 2008; Helbing, 2013). The 2021 blockage of the Suez Canal, which interrupted 12 percent of global trade within days, illustrated how localized disruptions propagate globally through interdependent systems (UNCTAD, 2021). Similarly, climate-driven crop failures interact with market shocks and conflict dynamics, producing emergent humanitarian crises that exceed the explanatory power of traditional models (Benton & Froggatt, 2021).

Traditional architectures of global governance—the United Nations system, Bretton Woods institutions, and regional organizations—were designed for a mid-twentieth-century world characterized by state-centrism, episodic crises, and slower communication cycles (Weiss & Wilkinson, 2021). These institutions emphasize stability, restoration, and hierarchical decision-making. Yet in conditions of permanent turbulence, this architecture systematically underperforms. As Adam Day (2023) argues, governance designed for "episodic disruption" lacks the foresight, agility, and modularity required for continuous, systemic turbulence.

This mismatch is not primarily a failure of political will but a failure of method. Policymaking remains dominated by descriptive and correlational approaches, which treat crises as discrete events rather than interdependent phenomena. Without incorporating systems thinking—feedback loops, emergence, and nonlinearity—policy responses risk producing unintended consequences. The COVID-19 pandemic exemplified this: lockdown policies, necessary to slow viral transmission, also triggered cascading impacts in education, gender equity, and labor markets (UNDP, 2022). These outcomes reveal the inadequacy of linear models in governing complex realities.

Within this turbulent regime, AI has emerged as both a tool and an actor in governance. AI augments decision-making through speed, scale, and pattern recognition: computer vision models can compress post-disaster assessments from weeks to hours; natural language processing pipelines can detect weak signals of epidemic outbreaks across dozens of languages (WHO, 2024). However, AI also introduces novel governance risks: entrenched bias, opacity, automation over-reliance, and surveillance creep. Recent failures—the UK's 2020 A-levels grading algorithm, which disproportionately penalized disadvantaged students (Williamson, 2020), and the EU's iBorderCtrl project, widely criticized as pseudoscience—demonstrate how *claims that outpace evidence* can rapidly erode legitimacy.

Crucially, AI systems increasingly act in ways that shape institutional outcomes without continuous human supervision, raising what can be called the agential-alignment problem: how to ensure semi-autonomous systems align with human rights, public values, and democratic accountability (Bryson, 2019). The European Union's AI Act—entered into force in 2024—represents the most ambitious regulatory attempt to date, establishing bans on "unacceptable-risk" systems, obligations for high-risk deployments, and mechanisms for transparency and oversight (European Parliament, 2024). This signals a recognition that AI is not merely a technical tool but a governance actor whose design and deployment require regulation akin to other forms of political and economic power.

The convergence of polycrisis dynamics and AI's rise as a governance actor demands a new architecture of global leadership. Such an architecture must be:

- **Anticipatory**: embedding foresight, scenario planning, and horizon scanning into routine governance cycles (Guston, 2014).
- **Resilient-by-design**: engineering systems for graceful degradation, redundancy, and recovery under stress (Linkov & Trump, 2019).
- **Human-centric and rights-preserving**: prioritizing fairness, dignity, and contestability over efficiency alone (UNESCO, 2021).
- **Polycentric**: distributing leadership functions across states, cities, firms, civil society, standards bodies, and increasingly AI systems themselves, coordinated through accountability frameworks (Ostrom, 2010; Abbott & Snidal, 2020).

This chapter develops these arguments in five steps. Section 1 lays the conceptual foundations, situating the polycrisis within systems theory and positioning AI as a sociotechnical actor. Section 2 analyzes empirical cases from 2020–2025, including humanitarian damage assessment, epidemic intelligence, forecast-based financing, digital identity and payments, and algorithmic failures in education and border management. Section 3 distills design principles for resilient, tech-enabled governance and examines their institutionalization in frameworks such as the EU AI Act, the UN Global Digital Compact, and the G7 Hiroshima Process. Section 4 explores the transformation of diplomacy and leadership, highlighting hybrid diplomacy, regime complexes, and polycentric governance. Section 5 proposes institutional design moves and a competency model for disruption-ready leaders. The chapter concludes with a forward-looking agenda for research and practice, emphasizing the need for **orchestration under uncertainty**—the capacity to align diverse intelligence, both human and artificial, in the pursuit of resilient, ethical, and inclusive governance.

CONCEPTUAL FOUNDATIONS: POLYCRISIS, SYSTEMS, AND AI AS A SOCIOTECHNICAL ACTOR

Disruption, Polycrisis, and Systems Fragility

The concept of *disruption* has shifted from being episodic to structural. In earlier paradigms, crises were understood as discrete events—wars, economic downturns, or natural disasters—occurring within otherwise stable systems. Today, global risks are characterized less by their individual severity than by their entanglement. This condition, widely termed a *polycrisis*, describes the simultaneity and interaction of crises across multiple systems—ecological, technological, social, and geopolitical—that amplify one another's effects (Homer-Dixon et al., 2021; World Economic Forum, 2023).

A pandemic is no longer solely a health crisis; it cascades into food insecurity, education disruption, political instability, and digital dependency. A regional war, such as the conflict in Ukraine, ripples into global energy markets, accelerates food inflation, and triggers refugee flows across continents (FAO, 2022). Climate extremes, including prolonged droughts and floods, interact with these dynamics to exacerbate displacement, resource competition, and fragility (IPCC, 2023).

Systems theory helps explain this transformation. Global infrastructures are tightly coupled and highly interdependent. In such conditions, disturbances propagate rapidly, producing nonlinear effects, feedback loops, and tipping points (Helbing, 2013; Meadows, 2008). The 2021 blockage of the Suez Canal, which disrupted 12% of global trade in less than a week, demonstrated how a local event can cascade into global supply chain fragility (UNCTAD, 2021). The fragility is not accidental; it is a product of optimization for efficiency—"just-in-time" logistics, lean energy networks, and globalized food systems—at the expense of redundancy and resilience.

Traditional governance models, designed to restore equilibrium after singular shocks, struggle in this environment. As Adam Day (2023) argues, they are optimized for "episodic disruption" rather than *permanent turbulence*. The governance challenge, therefore, is no longer to restore a prior order but to build adaptive capacity: the ability to absorb disturbances, reconfigure institutional arrangements, and learn under uncertainty. This requires moving from linear, event-driven governance to complexity-aware, resilience-oriented governance.

The Rise of AI as a Governance Actor

Artificial intelligence (AI) has emerged as a defining feature of this turbulent landscape. Initially deployed as a tool for efficiency, AI is increasingly assuming the role of a governance actor, shaping decision cycles in crisis response, resource allocation, and policy implementation. Its comparative advantage lies in its ability to process vast, heterogeneous datasets at speed and scale, identifying weak signals and generating actionable insights beyond human analytic capacity.

In humanitarian contexts, computer vision models now compress disaster assessments from weeks to hours, enabling early triage and targeted interventions. Natural language processing systems like the WHO's Epidemic Intelligence from Open Sources (EIOS) detect outbreak signals across multiple languages and platforms, providing precious lead time for public health response (WHO, 2024). Predictive analytics in forecast-based financing trigger the pre-emptive release of humanitarian funds, shifting the logic of aid from reaction to prevention (IFRC, 2024).

Yet these benefits are entangled with profound risks. Bias remains a structural problem, particularly when proxy variables stand in for complex social realities. A model that infers infrastructural needs based on call-center activity may over-represent communities with better connectivity while under-representing marginalized groups. Automation bias compounds this risk, as human decision-makers often defer to algorithmic outputs even when contradicted by other evidence (Yeung, 2022). Opacity further undermines legitimacy: many AI models function as "black boxes," making it difficult to trace errors or contest outputs. Beyond technical concerns, AI systems can enable surveillance creep and the weaponization of disinformation, as seen in the use of synthetic media in conflict zones (Cath, 2023).

Most critically, AI is no longer confined to passive data processing. Increasingly, it acts agentially within institutions, dynamically shaping priorities, allocating resources, and adapting to feedback in semi-autonomous ways. In doing so, AI assumes characteristics traditionally reserved for bureaucrats, diplomats, and administrators. This transition—from tool to actor—creates new institutional design challenges: how to embed AI systems in governance frameworks such that their agency is aligned with human rights, public values, and democratic accountability.

The Agential-Alignment Problem: From Tools to Actors

The transition of AI from a technical instrument to a sociotechnical actor generates what can be described as the agential-alignment problem. Unlike traditional bureaucratic agents, AI systems lack intrinsic social values, moral reasoning, or political accountability. Their "objectives" are encoded through training data, optimization functions, and institutional embedding. Yet in practice, their actions increasingly shape governance outcomes.

High-profile failures demonstrate the risks of misalignment. The UK's 2020 A-levels algorithm systematically downgraded students from disadvantaged schools while favoring private school students, sparking public outrage and forcing government reversal (Williamson, 2020). The EU's iBorderCtrl project, premised on AI "lie detection" through unvalidated "biomarkers of deceit," was condemned as pseudoscience and abandoned, leaving a legacy of mistrust. In both cases, the failures were not merely technical but institutional: claims outpaced evidence, external validation was absent, and contestability mechanisms were weak.

These cases underscore that AI governance cannot rely solely on technical fixes. Alignment requires institutional design:

- **Hard accountability mechanisms**, including evaluation, audit, continuous monitoring, and accessible appeals.
- **Regulatory anchors**, such as the EU AI Act, which bans "unacceptable-risk" systems (e.g., social scoring, real-time biometric identification in public spaces) and imposes strict obligations on high-risk applications.
- **Polycentric oversight**, involving states, regulators, standards bodies, civil society, and affected communities.

The agential-alignment problem reframes AI as part of the governance ecosystem: neither a neutral tool nor an autonomous sovereign, but a **co-actor** whose behavior must be shaped, constrained, and aligned with public interest. Solving this challenge is central to building legitimacy and resilience in governance systems under conditions of permanent disruption.

AI IN CRISIS AND GLOBAL SYSTEMS: CASE ANALYSIS (2020–2025)

AI's contribution to crisis governance is most credible when it augments human expertise and is embedded in robust operating procedures, rather than attempting to replace human judgment in high-stakes contexts. The following cases illustrate both sides of this spectrum—demonstrating how AI can accelerate assessment, coordination, and anticipatory action, while also show the risks of overreach, bias, and legitimacy loss when deploy without sufficient safeguards.

Rapid Damage Assessment & Anticipatory Relief (WFP / UN)

System Overview

The World Food Program (WFP), in partnership with Google Research and UN Global Pulse, developed SKAI, an open-source computer-vision pipeline that classifies building damage from satellite imagery to generate near-real-time crisis maps. This has been complemented by DEEP, a drone-based imagery tool designed to provide high-resolution assessments in environments where satellite imagery is limited by cloud cover or angle of capture.

Data/Model Pipeline

SKAI processes multispectral satellite scenes, tiles and pre-processes images, and applies trained classifiers calibrated on curated damage datasets. Outputs are then aggregated into settlement-level maps for operational planning. DEEP extends this pipeline into the field, enabling offline drone-based analysis where internet access is unavailable, and incorporates a feedback loop in which field teams annotate and correct outputs to continuously improve model accuracy.

Deployment and Results (2023–2025)

In multiple emergency operations, SKAI delivered building-damage assessments within 24–48 hours, achieving results that were 13 times faster and 77 percent cheaper than traditional manual methods. During the 2023 Türkiye–Syria earthquakes, SKAI produced damage maps within the first day, enabling faster routing of supplies and deployment of first responders.

Governance and Ethics

- *Human-in-the-loop.* WFP and UN workflows require expert validation and escalation protocols for ambiguous results, with explainability overlays (e.g., confidence scores) guiding field verification.
- *Data provenance.* Imagery licensing and capture priorities are carefully documented to mitigate tasking bias, such as prioritizing urban cores over rural areas.
- *Equity risk.* Imbalances between damaged and intact buildings, along with differences in construction typologies, can degrade model performance. Standard operating procedures therefore mandate ground-truth sampling and local calibration before outputs inform resource allocation.

Transferability and Limits

Effectiveness varies with building type, hazard context, and imagery latency. Integration into anticipatory action frameworks is promising but requires pre-agreed financial triggers and budgetary readiness (explored further in §2.3).

Epidemic Intelligence from Open Sources (WHO EIOS)

System Overview

The World Health Organization's Epidemic Intelligence from Open Sources (EIOS) platform combines multilingual media, social media, and official open-source streams to detect early outbreak signals. EIOS is not only a technology platform but also a global community of practice that links analysts across over 100 countries and dozens of partner organizations.

Data/Model Pipeline

The system ingests data streams, filters duplicates, applies language detection and event classification, and clusters potential signals before presenting them to human analysts through dashboards and collaborative workflows. Importantly, machine-generated scores are treated as inputs, not final judgments—analysts validate signals before escalation.

Deployment and Results (2024–2025)

A 2025 evaluation across 28 African states found that EIOS generated signals for over 80 percent of outbreaks later formally reported to WHO, often days before official notification by national authorities. This time advantage is critical for activating surveillance and response systems. However, as adoption has expanded, the system has faced challenges of scalability, functionality, and resource disparities across member states.

Governance and Ethics

- *False positives.* Noise and rumor detection remain critical; without expert triage, false alarms risk reputational damage and wasted resources.
- *Equity in coverage.* Low-connectivity areas remain under-represented, introducing bias in global situational awareness. WHO has responded by expanding regional source coverage and analyst networks.

- *Data responsibility.* WHO guidelines ensure cross-border data use is ethically managed, with community practices designed to prevent amplification of harmful misinformation.

Transferability and Limits

EIOS functions best as a sociotechnical system—a hybrid of machine signal detection and coordinated human interpretation. Its success depends on both the density of the media ecosystem and the analytical capacity of participating states.

Forecast-Based Financing & Anticipatory Action (IFRC / Red Cross System)

System Overview

Forecast-based financing (FbF) links hydrometeorological forecasts with predefined "impact thresholds" to trigger the release of humanitarian funds before disasters strike. Managed by the International Federation of Red Cross and Red Crescent Societies (IFRC), FbF is institutionalized through the Disaster Response Emergency Fund (DREF).

Deployment and Results (2024)

In 2024 alone, IFRC allocated CHF 85.3 million across 167 operations, supporting 97 National Societies and reaching more than 24 million people. Cost–benefit analyses estimate that every dollar spent on anticipatory measures saves up to US$15 in recovery costs by reducing losses and accelerating recovery.

Governance and Ethics

- *Trigger governance.* Thresholds must be calibrated carefully: too conservative and aid arrives too late; too liberal and funds may be misallocated.
- *Inclusion.* The effectiveness of anticipatory cash hinges on last-mile delivery systems and safeguards to reach marginalized groups, such as undocumented or remote populations.
- *Accountability.* Activation protocols require after-action reviews, linking forecast performance directly to outcomes to refine future thresholds.

Transferability and Limits

FbF relies not only on forecast accuracy but also on institutional readiness: pre-positioned logistics, functional payment systems, and trained local partners. Without these, anticipatory finance risks becoming an empty mechanism.

Digital Identity & Payments in Humanitarian Contexts (WFP Building Blocks)

System Overview

The WFP's Building Blocks (BB) platform is the humanitarian sector's largest blockchain-based payment and coordination system. It combines iris authentication with pseudonymous identifiers to facilitate cash and voucher assistance while enabling cross-agency deduplication of aid recipients.

Deployment and Results (2017–2025)

- *Scale and efficiency.* Since 2017, BB has processed more than US$555 million across ~25 million transactions, serving over one million refugees in Jordan and Bangladesh, while saving US$3.5 million in banking fees.
- *Ukraine operations.* Between May and August 2022, BB coordinated US$337 million in aid across 18 organizations and prevented US$35 million in duplication (equivalent to ~185,000 avoided duplicate transfers). By 2024, BB had scaled to 60+ partners, preventing an estimated US$170–200 million in duplication cumulatively, including US$67 million in 2024 alone.

Governance and Ethics

- *Rights and consent.* While BB uses pseudonymous identifiers on a consortium ledger, biometric capture raises concerns over consent, privacy, and redress in humanitarian contexts where opting out may not be realistic.
- *Coordination legitimacy.* BB's effectiveness depends on neutral governance structures and inter-agency agreements. In Ukraine, the Cash Working Group formally embedded BB as the sector-wide deduplication tool, legitimizing its role in equitable resource distribution.

Transferability and Limits

BB scales effectively where partner adoption is high and common identifiers exist. In contexts lacking robust ID systems, deduplication relies on fragile matching rules and is prone to error—raising questions about accountability and remediation.

Algorithmic Overreach and Public Trust (EU iBorderCtrl; UK A-levels)

iBorderCtrl (EU)

The EU's Horizon 2020-funded iBorderCtrl project sought to automate deception detection at borders using AI-driven analysis of facial micro-expressions. Independent reviews dismissed the methodology as pseudoscientific and lacking validation. Amid legal challenges, rights-based critiques, and growing skepticism, the project was abandoned. The case underscores that in high-rights environments such as border control, *claims must be proportionate to evidence*, and unvalidated technologies are governance non-starters.

UK A-Levels Algorithm (2020)

In response to COVID-19 disruptions, the UK's Office of Qualifications and Examinations Regulation (Ofqual) deployed an algorithm to moderate teacher-predicted grades. By relying heavily on historical school performance, the system disproportionately downgraded students from state schools while advantaging those from private schools. Roughly 39–40 percent of grades were reduced. Public outcry forced the government to abandon the system and reinstate teacher predictions. The episode demonstrates the risks of substituting human judgment with opaque algorithmic proxies in high-stakes contexts.

Governance Lessons

Both cases highlight institutional, not merely technical, failure. They lacked transparency, independent validation, and contestability mechanisms—eroding public trust and exacerbating inequities.

Table 1. Comparative analysis of AI in crisis response (2020–2025)

Case	AI Function	Outcomes (selected)	Governance Strengths	Governance Gaps / Risks
WFP SKAI	Satellite CV for damage assessment	≤24–48h delivery; **13× faster, 77% cheaper**	Human-in-the-loop validation; escalation protocols; explainability	Class imbalance; domain shift; image latency; licensing dependencies
WFP DEEP	Drone-based ML for damage analysis	Hours-level assessments; analyst feedback loop	Field calibration; offline capability	Airspace/weather constraints; analyst workload
WHO EIOS	OSINT/NLP outbreak signal detection	Detected >80% of notified events; lead time advantage	Global analyst community; validation workflows	False positives; uneven capacity; scalability challenges
IFRC FbF / DREF	Forecast-triggered humanitarian finance	CHF 85.3m in 167 ops (2024); up to US$15 saved per $1	Pre-agreed triggers; after-action reviews	Threshold calibration; last-mile capacity; equity in access
WFP Building Blocks	Blockchain + biometrics for cash transfers	US$555m processed; US$3.5m fees saved; prevented >US$170m duplication	Neutral consortium governance; auditability; inter-agency rails	Consent and privacy risks; matching errors; governance of identifiers
EU iBorderCtrl	Automated deception detection	Abandoned as pseudoscience	—	Invalid construct; absence of validation; rights violations
UK A-levels	Algorithmic grading moderation	~40% downgrades; reversal after protests	—	Proxy bias; opacity; lack of appeals mechanism

Cross-Case Synthesis: What Travels—and What Does Not

1. **Augment, don't replace.** SKAI, EIOS, and FbF succeeded when AI accelerated human workflows with clear escalation and uncertainty management. High-stakes substitution (iBorderCtrl; A-levels) collapsed under scrutiny.
2. **Triggers beat intuition.** FbF shows how pre-agreed triggers convert predictive analytics into governable actions by linking forecasts directly to funding and mandates.
3. **Shared rails matter.** Building Blocks demonstrates that digital coordination infrastructure—identity, payments, deduplication—can save hundreds of millions if paired with strong governance and rights safeguards.
4. **Evidence before deployment.** Pseudoscientific claims or poorly designed proxies erode legitimacy quickly. Validation, transparency, and contestability are prerequisites for public trust.

PRINCIPLES FOR RESILIENT, TECH-ENABLED GOVERNANCE

Section 2 shows a clear pattern: AI succeeds when it accelerates expert judgment through disciplined procedures and fails when it substitutes for legitimacy-critical human decisions without validation or recourse. Translating those lessons into a durable operating doctrine yields four mutually reinforcing principles. Each principle below includes (a) a rationale anchored in the cases; (b) institutional design moves; (c) concrete implementation tools (policies, processes, procurement); and (d) common failure modes to avoid.

Anticipatory Governance and Foresight-to-Budget Linkages

Rationale

In a polycrisis regime, speed and preparedness determine outcomes. FbF worked because pre-agreed triggers linked forecasts to money and mandates; SKAI created immediate "time advantage" because mission playbooks were prepared in advance; EIOS added days of lead time because analysts and workflows already existed.

Design Moves

- **Institutionalize foresight.** Stand up a permanent cross-government foresight unit with statutory authority to (i) run horizon scans and multi-hazard scenarios; (ii) publish annual risk outlooks; and (iii) convene "whole-of-society" drills (cities, utilities, platforms, civil society).
- **Tie foresight to budgets.** Require every ministry/agency to translate top risks into If-This-Then-That (ITTT) budget lines (e.g., "If flood risk index > X for Y days, release Z funds to pre-position supplies via partners A/B").
- **Codify triggers.** For priority hazards, adopt trigger handbooks (thresholds, confidence intervals, actions, roles, and communications). Review post-activation and iterate.
- **Exercise often.** Quarterly table-tops stress-testing critical sociotechnical systems (identity, payments, communications, logistics, electricity, cloud dependencies), including AI failure scenarios (model drift, data poisoning, synthetic media surges).

Implementation Tools

- **Policy:** "Foresight-to-Budget Directive" mandating ITTT appropriations, cross-agency drills, and public after-action reports.

- **Process:** Red-teaming of models used in anticipatory actions; pre-approved partner rosters and framework agreements to avoid procurement delays.
- **Procurement:** Indefinite Delivery/Indefinite Quantity (IDIQ) contracts for surge analytics, imagery, and evaluation; pre-negotiated satellite/drone tasking.

Failure Modes to Avoid

• Over-fitting triggers to historical events; failing to include equity criteria (who benefits first); triggers that are unfunded or lack last-mile logistics.

Resilience-by-Design: Modularity, Redundancy, and Graceful Degradation

Rationale

Complex systems fail in unexpected ways. SKAI/DEEP worked because human checks and offline capability allowed continuity under connectivity constraints. Building Blocks functioned because a shared rail persisted across many actors. By contrast, iBorderCtrl and A-levels failed "hard" because there were no safe fallbacks or contestability channels.

Design Moves

- **Modular architectures.** Decompose critical services (e.g., benefits issuance, damage mapping, outbreak alerts) into interoperable components with versioned APIs so one component can fail without collapsing the whole.
- **Graceful degradation.** Define "minimum viable service" tiers (e.g., when imagery is delayed, use last-known conditions + local reports; when models are offline, revert to rules-based triage).
- **Redundancy.** Dual data sources (satellite + drone + local enumeration), dual comms (fiber + cellular + radio), dual payment rails (banking + mobile money), dual clouds/regions with tested failover.
- **Provenance and logging.** Immutable logs for data lineage and decision traces to enable audits, after-action learning, and redress.

Implementation Tools

- **Policy:** "Continuity of AI-Mediated Operations" standard requiring fallback modes, run-books, and periodic failover exercises.

- **Process:** Quarterly "chaos drills" that intentionally disable components to test recovery time objectives; service level objectives (SLOs) for latency, accuracy, and recovery.
- **Procurement:** Require open interfaces and data export; disallow single-vendor lock-ins for critical functions; mandate architecture diagrams and resilience test reports at acceptance.

Failure Modes to Avoid

- Single points of failure (a lone model or platform); hidden dependencies (unmapped reliance on a specific provider or region); resilience promised but never tested.

Human-Centric, Risk-Based AI: From Principles to Regulation

Rationale

The difference between success (SKAI, EIOS, FbF) and failure (iBorderCtrl, A-levels) was not simply technical capability but governance proportional to risk. High-stakes deployments require stricter obligations than low-risk tools.

Design Moves

- **Risk tiering.** Classify systems by potential harms:
    - o *Unacceptable risk:* e.g., AI "lie detection," social scoring—prohibit.
    - o *High risk:* identity, benefits allocation, border/security, health triage—permit only with obligations (evaluation, audit, monitoring, oversight, documentation).
    - o *Limited/minimal risk:* productivity aids, non-decisional analytics—light transparency.
- **Assurance lifecycle.** Shift from "once-and-done" approvals to continuous assurance: pre-deployment evaluation → staged rollout with guardrails → live monitoring and incident reporting → periodic re-evaluation.
- **System-level documentation.** Beyond model cards: System Cards describing the end-to-end pipeline, data inputs, controls, human roles, failure modes, and recourse.
- **Separation of powers.** Distinct roles for builders (vendors), deployers (agencies), independent evaluators/auditors, and an ombudsperson/redress function for affected people.

Implementation Tools

- **Policy:** AI Risk Management Policy aligning with a tiered regime; public AI registers listing purpose, data categories, evaluation summaries, and contact points for appeal.
- **Process:**
  - o *Before deployment:* external evaluation against harm taxonomies (bias, safety, robustness, privacy).
  - o *During deployment:* canary rollouts; human supervision thresholds; audit trails.
  - o *After deployment:* quarterly performance drift checks; incident disclosure within defined timelines.
- **Procurement: Assurance-by-contract** clauses—vendor must (i) provide evaluation artifacts and red-team reports; (ii) support independent audits; (iii) commit to duty-of-candor on material risks; (iv) enable data access for oversight with privacy safeguards.

Failure Modes to Avoid

- Treating "principles" as sufficient without binding obligations; deploying with unvalidated proxies; ignoring population heterogeneity; lack of contestability and human accountability for consequential decisions.

Safeguarding Rights and Information Integrity in the Digital Era

Rationale

Rights and legitimacy are core system properties, not afterthoughts. Building Blocks surfaces tensions between efficiency and consent; EIOS highlights responsibilities around rumor amplification; iBorderCtrl shows how rights violations destroy trust; the A-levels controversy shows how opaque proxies can harm equity at scale. Meanwhile, synthetic media and coordinated manipulation campaigns threaten the epistemic environment on which governance depends.

Design Moves

- **Rights-by-design.** Bake privacy, fairness, accessibility, and contestability into requirements and testing; require Data Protection Impact Assessments and Human Rights Impact Assessments for high-risk deployments.

- **Provenance & authenticity.** Adopt content provenance standards (e.g., cryptographic signing of official communications, tamper-evident metadata) and verification workflows during crises.
- **Low-friction redress.** Provide simple, multilingual channels for individuals to challenge automated decisions; require timely human review and documented outcomes.
- **Information resilience.** Establish a Civic Information Operations function that monitors, counters, and pre-bunks harmful disinformation during emergencies—coordinated with platforms and independent media.
- **Data minimization & role-based access.** Especially in humanitarian contexts, limit collection to necessity, enforce tight access controls, and define retention/deletion schedules.

Implementation Tools

- **Policy:** Rights-Preserving AI Standard; Crisis Communications Provenance Policy; Automated Decision Notice requirements (plain-language reasons + appeal path).
- **Process:** Community review boards for sensitive deployments; periodic equity audits; joint simulations with media platforms for misinformation surges.
- **Procurement:** Mandate privacy-enhancing technologies (e.g., differential privacy where appropriate), configurable consent/opt-out mechanisms, and exportable decision logs for oversight.

Failure Modes to Avoid

- Consent theater (formal consent with no meaningful alternative); "black-box" deployments in rights-sensitive settings; absence of provenance in official messaging during crises; redress that is slow, opaque, or inaccessible to vulnerable groups.

Bringing the Principles Together

These four principles function as a stack:

1. **Anticipate** risks and pre-wire budgets and doctrine (so decisions can be fast and legitimate).
2. **Engineer for resilience** (so services continue under stress and failures are bounded).

3. **Scale assurance with risk** (so high-stakes uses face rigorous evaluation, oversight, and recourse).
4. **Safeguard rights and information integrity** (so legitimacy is preserved when it is most fragile).

In practice, agencies should codify this stack in a Governance Playbook that includes trigger handbooks, resilience run-books, AI risk and assurance policies, rights-preserving design standards, and procurement templates. This playbook turns abstract principles into repeatable routines that withstand the turbulence of real-world crises—and it provides the bridge from the case evidence in Section 2 to the diplomatic and institutional architectures we develop in the subsequent sections.

THE TRANSFORMATION OF GLOBAL LEADERSHIP AND DIPLOMACY

The diplomatic landscape is undergoing profound transformation in response to technological disruption, systemic crises, and shifting power balances. Traditional, state-centric diplomacy is no longer sufficient to manage challenges that transcend borders and domains, from artificial intelligence (AI) governance to data sovereignty and climate resilience. What emerges instead is a hybrid, polycentric, and contested diplomatic order, where states, multilateral institutions, cities, corporations, and civil society share overlapping and sometimes competing roles. This section examines how diplomacy and leadership are being reshaped in this new environment.

Hybrid Diplomacy: From Virtual Meetings to Institutionalized Practice

The COVID-19 pandemic accelerated the digitization of diplomacy. Emergency conditions forced the United Nations Security Council and other high-level bodies to adopt virtual formats, once seen as temporary workarounds but now increasingly institutionalized. The hybrid model—combining face-to-face negotiations for trust-building with digital platforms for efficiency—has become a durable feature of international governance.

Benefits

Hybrid diplomacy lowers barriers to participation, allowing smaller states and civil society groups to contribute more consistently without prohibitive travel costs.

It can increase transparency, broaden inclusivity, and reduce the environmental footprint of diplomatic activity.

Risks

Virtual platforms introduce new vulnerabilities: cybersecurity threats, unequal access due to bandwidth disparities, and the erosion of informal diplomacy, which often relies on unstructured interactions outside formal meetings. Scholars note that while digital platforms expand access, they risk weakening the thick trust networks upon which durable agreements are built.

Institutionalization

Organizations are beginning to formalize hybrid processes. The EU External Action Service now maintains secure virtual platforms for multilateral negotiations, and regional bodies in Africa and Asia have integrated hybrid procedures into their charters. Hybrid diplomacy is thus no longer an exception but part of the routine machinery of governance.

The Emergence of Frontier-AI Regimes

The governance of advanced AI illustrates how diplomacy is adapting to disruptive technologies. Instead of a single global treaty, AI governance is evolving into a regime complex—a patchwork of overlapping, sometimes competing frameworks that differ in scope, legal authority, and inclusivity. Three key initiatives exemplify this landscape.

UN Global Digital Compact

Adopted at the Summit of the Future in 2024, the Global Digital Compact (GDC) reflects a multilateral, consensus-based approach. It emphasizes digital inclusion, human rights, and global cooperation. The Compact calls for a Global Dialogue on AI Governance and proposes an International Scientific Panel on AI, modeled loosely on the Intergovernmental Panel on Climate Change. While non-binding, it represents the most globally inclusive forum to date, with contributions from governments, civil society, academia, and industry.

G7 Hiroshima Process

In contrast, the G7 Hiroshima Process on AI Governance reflects a club-based, voluntary model. It focuses on frontier AI risks, particularly large language models and generative systems, and promotes shared principles such as "red-teaming," transparency reporting, and content provenance. The process produced a voluntary Code of Conduct for Advanced AI Developers, adopted by leading firms in 2023–2024. Its limited membership, however, raises concerns of exclusivity, with Global South states and smaller economies largely excluded from norm-setting.

EU AI Act and the "Brussels Effect"

The European Union's AI Act, finalized in 2024, represents the most ambitious and binding attempt to regulate AI to date. Built around a risk-based framework, it prohibits "unacceptable" uses (e.g., social scoring, real-time biometric surveillance in public spaces) and imposes strict obligations on high-risk systems. The Act has extraterritorial reach, requiring any company deploying AI in the EU market to comply. This regulatory spillover—known as the *"Brussels effect"*—means the Act will likely shape global practices, even beyond Europe.

Table 2. Comparative governance frameworks for frontier AI

Framework	Nature	Focus	Approach to Risk	Key Mechanisms
UN Global Digital Compact	Multilateral, consensus-based, non-binding	Digital rights, inclusion, AI dialogue	Broad, principle-based	Global Dialogue on AI; International Scientific Panel on AI
G7 Hiroshima Process	Voluntary, club-based, non-binding	Frontier AI safety (LLMs, generative models)	Safety principles, best practices	Code of Conduct for developers; voluntary red-teaming
EU AI Act	Binding, legal, extra-territorial	Comprehensive AI regulation	Four-tier risk framework (unacceptable, high, limited, minimal)	Prohibitions, mandatory audits, compliance obligations

Polycentric Leadership in Action: States, Cities, Standards, Civil Society

Global governance is increasingly polycentric, distributed across multiple centers of authority. Rather than a single hierarchy, governance emerges from the interaction of diverse actors.

- **States.** Still the central actors in formal law-making, states anchor frameworks such as the EU AI Act and contribute to multilateral processes like the GDC.
- **Cities and sub-national actors.** Networks such as C40 Cities exercise "city diplomacy," pushing ambitious climate and resilience policies even when national governments lag. Their collective action illustrates how sub-national actors are now diplomatic entrepreneurs.
- **Standards bodies.** Technical standards organizations, such as ISO/IEC and IEEE, translate broad legal principles into operational norms. For AI, the development of ISO/IEC 42001 (AI management systems) is a crucial bridge between regulation and practice.
- **Civil society and epistemic communities.** Groups such as the Algorithmic Justice League, the Partnership on AI, and academic consortia play a watchdog role, highlighting blind spots, advocating for fairness, and shaping public discourse.

Table 3. Dimensions of polycentric leadership in practice

Actor	Role in Governance	Example	Mechanism of Influence
States	Anchor formal treaties, regulation	EU AI Act; GDC negotiations	Legal frameworks; diplomatic compacts
Cities & networks	Sub-national diplomacy, implementation	C40 Cities climate diplomacy	Policy diffusion; peer-to-peer governance
Standards bodies	Technical translation of regulation	ISO/IEC 42001; IEEE standards	Market-driven standards; certification
Civil society & academia	Oversight, advocacy, epistemic input	Algorithmic Justice League; research consortia	Advocacy campaigns; expert evidence; norm entrepreneurship

Geopolitical Fragmentation and Data Sovereignty: Risks of a Splinternet

A defining risk of digital-era governance is the rise of divergent models of data sovereignty. The United States emphasizes market-driven innovation, the European Union promotes rights-based regulation, while China and Russia prioritize state control and data localization. These divergent paradigms fuel fears of a "Splinternet"—a fragmented global digital order characterized by incompatible infrastructures and governance logics.

Consequences

- **Interoperability breakdown.** Cross-border data sharing for health surveillance, AI safety research, and humanitarian coordination becomes constrained.
- **Scientific fragmentation.** International collaboration in frontier research is weakened by incompatible data regimes and political mistrust.
- **Diplomatic gridlock.** Norm-setting forums face stalemate as states protect their divergent models.

Governance Challenge

The task for global leadership is to navigate between fragmentation and convergence. A polycentric system can foster innovation and pluralism, but excessive divergence risks eroding the very foundations of collective problem-solving.

Synthesis: Toward Integrated Diplomacy for the Digital Age

Hybrid diplomacy, frontier-AI regimes, and polycentric governance illustrate a world where authority is dispersed, institutional designs are contested, and risks are transboundary. The challenge is not to restore a unitary order but to orchestrate these overlapping systems into a form of integrated diplomacy. Future governance will depend on aligning hybrid diplomatic practices with inclusive multilateral compacts, ensuring that technical standards complement legal frameworks, and embedding rights protections into digital infrastructures. This integrated approach provides the scaffolding for the next stage of analysis: the role of the United Nations, AI diplomacy, science diplomacy, and green diplomacy in building resilient global leadership.

THE UN'S ROLE AND FUTURE IN GOVERNING DISRUPTION

The United Nations (UN) remains the most universal platform for global governance, encompassing 193 member states, dozens of specialized agencies, and a mandate that spans peace, development, human rights, humanitarian assistance, and increasingly, digital governance. Yet, the defining question in the age of polycrisis is whether the UN's architecture, instruments, and legitimacy are fit for purpose in managing global disruption.

Scholars and policymakers diverge on this question. Some view the UN as indispensable, emphasizing its legitimacy, inclusivity, and unparalleled convening power. Others highlight persistent inefficiencies, structural imbalances, and political inertia, particularly in the Security Council. Pragmatists increasingly frame the UN not as a singular problem-solver, but as a hub within a polycentric governance system, coordinating but not monopolizing authority. This section examines the UN's evolving role across key dimensions, from its convening power to institutional reform and normative entrepreneurship.

The United Nations as a Convening Platform for Polycrisis Governance

Comparative Advantage

The UN's ability to convene diverse actors remains unmatched. In a governance landscape fragmented between regional blocs, private actors, and civil society coalitions, the UN's universality enables it to frame challenges as *global* rather than sectional.

- **Humanitarian coordination.** Through the Office for the Coordination of Humanitarian Affairs (OCHA) and the Inter-Agency Standing Committee (IASC), the UN orchestrates multi-actor responses. The Central Emergency Response Fund (CERF) disbursed more than US$700 million in 2023 alone, demonstrating the UN's role in mobilizing flexible, rapid financing across crises.
- **Development agenda-setting.** The Sustainable Development Goals (SDGs), adopted in 2015, continue to structure international cooperation, despite uneven progress. They serve as a lingua franca across governments, donors, and civil society, anchoring policy and resource allocation in measurable, time-bound targets.
- **Norm entrepreneurship.** The UN frames new domains as governance issues—cybersecurity (through the Open-Ended Working Group), responsible AI (via the Global Digital Compact), and climate security (debated in the Security Council). This role is especially important for amplifying smaller states and marginalized communities, ensuring that norm-setting is not monopolized by powerful blocs.

Limitations

Convening power is necessary but not sufficient. Without enforcement, many UN initiatives risk remaining symbolic, what scholars call "declaratory multilateralism."

The challenge is bridging legitimacy with implementation capacity—a tension visible in climate agreements, disarmament resolutions, and digital compacts.

Reforming Multilateral Institutions for the AI and Climate Age

The UN was designed in 1945, reflecting a post-war geopolitical order. Its architecture is state-centric, hierarchical, and concentrated in the hands of the five permanent members of the Security Council. Today's disruptions—AI, cyber threats, climate breakdown, pandemics—demand institutions that are agile, inclusive, and cross-sectoral.

Institutional Gaps

Current UN structures lack dedicated mechanisms for frontier technologies and systemic risks:

- WHO manages pandemics, WMO handles meteorology, and FAO oversees food systems. Yet no agency integrates cross-sectoral risks, such as AI-enabled disinformation fueling conflict during climate disasters.
- The Security Council continues to privilege kinetic conflict, with limited engagement in systemic issues such as digital sovereignty or climate-induced migration.

Reform Proposals

Scholars and practitioners have advanced several ideas:

- A **UN Agency for Digital and AI Governance**, modeled on the International Atomic Energy Agency (IAEA), to oversee high-stakes technologies.
- Elevating **climate security** formally into the Security Council's agenda or creating a standing Climate Security Council.
- Institutionalizing **anticipatory action**, linking AI forecasting tools to humanitarian financing, so the UN shifts from reactive to preventive governance.

Fit-for-Purpose Debate

- *Critics* argue the UN's consensus-based decision-making is too slow for fast-moving technological disruption, and that veto politics undermine legitimacy.

- *Proponents* counter that while flawed, no other institution combines legitimacy, scale, and operational reach across peace, development, and humanitarian domains.
- Increasingly, scholars emphasize adaptation: the question is less whether the UN is obsolete, and more how it can recalibrate its structures for 21st-century governance.

The Summit of the Future and the Global Digital Compact: From Declaration to Implementation

The Summit of the Future (2024) produced the *Pact for the Future* and its annex, the Global Digital Compact (GDC)—the UN's most ambitious attempt to govern digital technologies multilaterally.

Promise

- The GDC establishes principles for digital cooperation, data governance, and AI ethics.
- It calls for a Global Dialogue on AI Governance and the creation of an International Scientific Panel on AI, modeled after the IPCC, to provide evidence-based guidance.
- The process was notable for its inclusiveness, involving states, civil society, academia, and the private sector.

Challenges

- The GDC is non-binding. Without follow-up, it risks becoming symbolic.
- Implementation requires embedding the Compact into institutional practice, such as:
    - Mandating periodic reviews by the General Assembly or ECOSOC.
    - Linking GDC principles to financing mechanisms (e.g., digital development funds).
    - Ensuring interoperability with binding frameworks like the EU AI Act.

The GDC thus exemplifies both the UN's strength—as a platform for inclusive norm-setting—and its weakness: the gap between declaratory principles and enforceable action.

Rethinking the UN Security Council and Trusteeship Council for Digital & Climate Governance

The UN Security Council (UNSC) remains the most powerful organ, but its design reflects a mid-20th-century balance of power. The five permanent members (P5) wield vetoes that often paralyze responses to contemporary crises, as seen in debates over Ukraine, Gaza, or Myanmar. Meanwhile, systemic risks like AI misuse, cyber conflict, and climate security remain peripheral.

Security Council Reform

- Proposals include expanding permanent membership to reflect geopolitical realities (India, Brazil, South Africa, Nigeria, or Japan), and establishing standing committees on climate security, cyber governance, and AI.
- However, entrenched veto politics makes structural reform unlikely in the short term.

Trusteeship Council Repurposing

- Dormant since 1994, the Trusteeship Council has been proposed as a forum for global commons governance.
- Scholars suggest reactivating it as a Digital and Climate Trusteeship Council, responsible for stewardship of cyberspace, AI, and transboundary environmental challenges.
- While politically ambitious, this idea reflects recognition that new institutional homes are needed for global commons issues that transcend sovereignty.

Feasibility. Reform faces political resistance from major powers, but debate is shifting: adaptation is increasingly framed not as optional but as necessary for the UN's survival as a relevant institution.

Beyond Structures: The UN's Role in Humanitarian Action, Development, and Diplomacy

Beyond debates over structure, the UN exercises real-world authority in three interlinked domains:

- **Humanitarian action.** Agencies such as WFP, UNHCR, and UNICEF operate at scale, increasingly integrating AI-enabled systems (e.g., WFP's SKAI for damage assessment, Building Blocks for digital identity). These innovations reinforce the UN's operational legitimacy, especially in fragile states.

- **Development.** Through UNDP, UNESCO, and UNEP, the UN anchors global development norms while supporting local implementation. Its role in digital inclusion and climate adaptation is particularly critical for Global South states navigating AI adoption.
- **Diplomacy.** The UN General Assembly and specialized fora remain the only universal diplomatic platforms. While regional clubs like the G7 or G20 exert influence, the UN legitimizes the participation of smaller states and civil society, ensuring inclusivity in norm-setting.

This multidimensional role—operational, normative, and diplomatic—demonstrates why the UN remains indispensable, even as its structures are questioned.

Fit-for-Purpose? Competing Perspectives

The debate on the UN's adequacy can be grouped into three perspectives:

- **Optimists.** Emphasize the UN's unparalleled legitimacy, universality, and agenda-setting ability. They argue that, despite inefficiencies, the UN remains indispensable for convening actors across regions and sectors.
- **Realists.** Highlight institutional inertia, power asymmetries, and veto politics. For them, the UN is structurally incapable of addressing fast-moving, systemic risks. The stalled Security Council reform exemplifies systemic gridlock.
- **Pragmatists.** Propose a middle ground: the UN should function as a hub in a polycentric system. It need not enforce every standard but should coordinate agendas, build capacity, and provide legitimacy, while regional organizations, standards bodies, and state–corporate partnerships handle implementation.

Synthesis: The UN in the Age of Disruption

The UN is simultaneously indispensable and insufficient. Its universal membership, convening power, and operational presence make it a cornerstone of global governance. Yet its structures, designed for a different era, require adaptation to manage the speed, complexity, and interdependence of 21st-century crises.

The central challenge is not whether the UN should lead, but how it can evolve. Options include:

- Reforming legacy organs (Security Council, Trusteeship Council).

- Mainstreaming anticipatory action and digital governance across humanitarian and development programming.
- Acting as a hub within a polycentric system, orchestrating diverse actors while safeguarding legitimacy and rights.

Ultimately, the UN's future relevance will depend on whether it can transform from a 20th-century architecture—anchored in sovereignty and hierarchy—into a 21st-century institution capable of orchestrating collective intelligence, human and artificial alike, in the service of global resilience.

AI DIPLOMACY AND ALGORITHMIC GEOPOLITICS

Artificial intelligence (AI) has rapidly become both a strategic asset and a subject of diplomacy. States now negotiate over semiconductors, computational infrastructure, data flows, and technical standards with an intensity once reserved for energy resources or arms control. At the same time, AI itself is being used to sense crises, shape narratives, and support negotiations. This dual role—AI as both the object and medium of statecraft—has given rise to what scholars term *algorithmic geopolitics*. In this new environment, power is not only measured through treaties or tariffs but also embedded in model weights, safety evaluations, and technical standards. The resulting diplomatic terrain is fragmented and competitive: major power rivalries unfold alongside efforts to establish cooperative safety institutes, global standards bodies, and South-led initiatives aimed at reclaiming agency and equity.

AI as a Strategic Asset and Diplomatic Subject

By mid-2024, more than sixty countries had adopted formal AI strategies, reframing AI from a sector-specific technology into a core national capability that combines economic competitiveness, security imperatives, and normative influence. This shift is visible in the tangible foundations of AI power—advanced chips, hyperscale computing capacity, access to critical datasets, and the skilled talent required to develop frontier systems. States are competing not only over these resources but also over the policy instruments that determine access and control.

The United States illustrates this dynamic. Between 2022 and 2023, Washington introduced successive export controls that restricted China's access to high-end logic chips and semiconductor fabrication equipment, restructuring supply chains and generating lasting diplomatic tensions (Council on Foreign Relations, 2024). At the same time, domestic industrial policy has been reinforced through incentives to boost homegrown innovation. Executive Order 14179, issued in 2025, marked a new phase by establishing an AI Action Plan and centralized White House coordination—signaling a recalibration of AI policy toward sustained global competitiveness in both economic and security terms (White House, 2025).

AI is also a growing focus of multilateral governance. The United Nations has elevated AI to the level of cross-cutting global priority. In 2024, the General Assembly adopted resolutions on safe and trustworthy AI and on narrowing digital divides. That same year, the Pact for the Future, including its annexed *Global Digital Compact*, established an inclusive but non-binding framework for cooperation on digital and AI governance. Meanwhile, the Council of Europe launched the world's first legally binding treaty on AI—its Framework Convention on AI, Human Rights, Democracy, and the Rule of Law (September 2024)—which set a minimum legal baseline for state use of AI while involving negotiators from beyond Europe. Together, these developments reflect a layered governance architecture: universal declarations provide broad direction, while treaties codify enforceable obligations (UNGA, 2024a, 2024b; United Nations, 2024; Council of Europe, 2024).

Domestic regulation is equally significant in shaping global practice. The EU AI Act, which entered into force between 2024 and 2025, introduced a risk-based framework that prohibits "unacceptable-risk" uses and imposes strict obligations on "high-risk" systems. Because firms seeking access to the EU market must comply, the Act extends what has been called the "Brussels effect," influencing organizational behavior well beyond Europe's borders. Similarly, ISO/IEC 42001:2023—the first auditable management system standard for AI—is spreading through procurement rules and certification schemes. In combination, these frameworks embed safety, documentation, and accountability into the economic incentives of AI deployment (ISO, 2023; A-LIGN, 2025).

Algorithmic Negotiation and Peacebuilding

AI tools are not only topics of negotiation; they are increasingly instruments of diplomacy and peacebuilding. Natural-language processing and network analytics now help mediators filter vast volumes of documents, detect early signs of escalation, and identify overlapping positions between conflicting parties. Early experiments with negotiation support systems trained on past agreements suggest modest but measurable improvements in efficiency (approximately 15%), while also exposing limitations around cultural nuance, legitimacy, and the risks of excessive reliance (Bicchi & Bremberg, 2022).

In practical settings, initiatives by UN Global Pulse and the Department of Political and Peacebuilding Affairs have piloted language-monitoring systems to detect incitement and geospatial machine learning tools to verify ceasefire compliance. These applications augment rather than replace political judgment and local expertise (UN Global Pulse, 2023).

Yet these innovations raise questions of legitimacy and consent. If communities perceive external models as opaque, biased, or imposed, peacebuilding efforts can lose credibility. Best practice now emphasizes transparency about model limitations, human-in-the-loop decision-making, and mechanisms allowing affected populations to contest algorithmic outputs. These principles echo the safeguards imposed on high-risk systems under the EU AI Act. The underlying principle is clear: AI can support negotiation and conflict resolution, but authority must remain human, accountable, and contextually grounded.

AI Safety Institutes and Cross-Border Standards Cooperation

Between late 2023 and 2025, several governments established specialized AI safety institutes—such as the UK's AI Safety Institute, the U.S. NIST-anchored counterpart, and the EU's AI Office. Their shared mandates include red-teaming, evaluation, and characterization of model risks. Importantly, these institutes are increasingly collaborating across borders, exchanging test methods and risk data in what has been described as "epistemic diplomacy."

This form of cooperation was accelerated by the Bletchley Park AI Safety Summit (2023), which produced the Bletchley Declaration on frontier risks, and the AI Seoul Summit (2024), which broadened participation and advanced commitments to complementary testing and shared methodologies. These developments represent pragmatic, technical cooperation that does not resolve broader political conflicts but builds interoperability and shared understanding of risks (UK Government, 2023; UK Government, 2024; ROK MOFA, 2024).

Standards bodies such as ISO/IEC JTC 1/SC 42 are increasingly arenas of strategic contestation. Seemingly technical details—taxonomies, compliance requirements, audit evidence—carry major commercial and political implications. ISO/IEC 42001, as the first certifiable AI management standard, sits at the center of this contest, linking safety science to compliance obligations and procurement contracts. The politics of "who writes the test methods" or "what counts as sufficient assurance" now directly shapes which actors gain trust and market access (Dafoe, 2024; ISO, 2023; KPMG, 2024).

Risks of AI Nationalism and Technological Protectionism

Growing divergences in regulatory models and supply chain security have revived fears of digital fragmentation. Rights-oriented frameworks, such as the GDPR and EU AI Act, prioritize transparency and individual protection, while sovereignty-centered approaches emphasize localization and state control. Parallel to this, semiconductor export restrictions and access limitations have multiplied. Although often justified on national security grounds, these moves risk undermining cross-border scientific collaboration, global health surveillance, and humanitarian information flows (European Parliament, 2023; Council on Foreign Relations, 2024).

The global distribution of AI resources remains highly unequal. Concentrations of computer power, venture capital, and elite research talent in a handful of countries threaten to lock low- and middle-income states into dependency. Yet many of the greatest global challenges—climate change, pandemic preparedness, disaster risk reduction—require precisely the kind of cross-border data and compute sharing that protectionist policies undermine. Balancing security concerns with functional interdependence is therefore a central governance challenge (UNDP, 2024).

AI Diplomacy in the Global South: Strategy, Not Afterthought

AI diplomacy cannot achieve legitimacy if its agenda is dictated by a narrow group of advanced economies. Analyses of participation in initiatives such as the G7 Hiroshima Process or OECD committees reveal a consistent underrepresentation of Global South states. Exclusion is not merely symbolic; it risks creating structural dependencies, where norms, standards, and testing protocols are imported wholesale without local adaptation (DiploFoundation, 2025).

True AI sovereignty requires infrastructure and institutions. Regional compute hubs, sovereign cloud regions, data trusts governed under local law, shared accelerators, and fabrication partnerships are not only technical assets but also political tools that shape autonomy. Without these, policy choices are constrained by technical dependency. A *Compute and Data Compact*—financing regional GPU clusters, privacy-preserving environments, and cross-border research pipelines—could transform bottlenecks into diplomatic assets (UNESCO, 2025).

Despite narratives of marginality, many Global South actors are actively shaping the AI landscape. India promotes "AI for All" and exports digital public infrastructure models; Kenya builds regional hubs and contributes to African-led standards; Brazil advances rights-based legislation; Indonesia develops indigenous-language models; and Gulf states leverage compute investments to act as geopolitical brokers. Regional organizations such as the AU, ASEAN, GCC, and MERCOSUR are increasingly positioned to pool bargaining power. The constraint is not only technical capacity but strategic assertion: the ability to convene summits, draft proposals, and co-author global rules.

AI Diplomacy as the New Chessboard of Foreign Policy

The United States has explicitly framed artificial intelligence as a pillar of national strategy, elevating it beyond a technological issue into a system-level capability central to both domestic renewal and geopolitical positioning. Executive Order 14179, issued in 2025, institutionalized AI policy coordination at the White House level, underscoring its cross-cutting importance. The U.S. approach combines three strands: fostering frontier innovation through industrial incentives, constraining adversaries through export controls, and aligning allies through safety-testing cooperation and shared governance initiatives. In this sense, AI is treated not only as a driver of economic growth but also as a strategic multiplier that underpins military readiness, cyber defense, and diplomatic leverage. The consistent signal across administrations has been clear: leadership in AI equates to leadership in the twenty-first-century international order (White House, 2025).

China, by contrast, has pursued a strategy of digital statecraft that integrates AI into its broader Belt and Road Initiative. Through the *Digital Silk Road*, Beijing exports not only infrastructure but also governance templates and technical standards. This external engagement is reinforced by domestic capacity to deploy large-scale models and infrastructure. The 2025 release of DeepSeek-R1, which achieved competitive reasoning performance at significantly lower costs, was emblematic of China's positioning: it used technical performance as a diplomatic signal, projecting innovation capacity while raising debates about censorship, data integrity, and the sustainability of chip access under ongoing export controls (Nature, 2025; Reuters, 2025a, 2025b). China's strategy thus blends material capability with normative influence, positioning AI as a vehicle for alternative models of digital governance and as a bargaining chip in negotiations over trade, technology, and security.

The European Union has taken yet another route, leveraging regulation as its primary instrument of geopolitical influence. The EU AI Act, which introduces a risk-based framework banning "unacceptable-risk" uses and imposing strict requirements on high-risk systems, is intended not only to regulate internal markets but also to set global standards. This is complemented by industrial initiatives designed to scale indigenous compute capacity and strengthen Europe's ability to host frontier innovation. The EU's wager is strategic: that a rights-based regulatory framework can coexist with innovation, and that its credibility in enforcing "trustworthy AI" will establish Brussels as a global reference point for both firms and governments. In this approach, the EU transforms its comparative advantage in normative governance into a form of digital statecraft, hoping to anchor global AI governance in principles of transparency, accountability, and human dignity (Reuters, 2025b).

Taken together, these strategies highlight the emerging chessboard of AI diplomacy. The United States foregrounds industrial capacity, security alliances, and export control; China leverages infrastructure, cost-effective models, and standards diplomacy; and the European Union relies on regulatory power and normative legitimacy. While their approaches differ, they converge on the same understanding: AI is not a sectoral technology but a systemic capability that reconfigures economic competitiveness, national security, and international influence simultaneously.

Across these approaches, the decisive arenas of contestation and cooperation are becoming increasingly clear. They include the setting of standards and regulations, which define the legal and technical baselines for deployment; access to compute and semiconductors, which determine who can build and run frontier models; governance of data flows, balancing sovereignty with interdependence; the cultivation of talent and research ecosystems, which sustain long-term innovation; the application of AI to global public goods, such as climate and health; and the forging of security alliances that define permissible and impermissible uses of autonomy, cyber operations, and information manipulation. Influence will accrue not only to those actors who dominate any single arena but especially to those capable of aligning several dimensions simultaneously and convening coalitions that bridge geopolitical divides. The contest ahead is therefore not merely about technological capability—it is about the ability to integrate governance, infrastructure, and legitimacy into a coherent, scalable strategy for global leadership.

Rewiring Multilateralism: Governing AI in a Divided World

The global system of AI governance remains heavily weighted toward aspirational principles and political declarations yet lags in the creation of enforceable and interoperable mechanisms. Multilateral statements frequently affirm commitments to "trustworthy" or "human-centered" AI, but the translation of these values into binding obligations remains uneven. A notable exception is the Council of Europe's *Framework Convention on Artificial Intelligence, Human Rights, Democracy and the Rule of Law*, which establishes legally enforceable standards for state behavior and embeds human rights safeguards into the deployment of AI (Council of Europe, 2024). By contrast, the United Nations' *Global Digital Compact*, annexed to the Pact for the Future, is broad and inclusive in scope but lacks legal force, functioning primarily as a political signal of global convergence (United Nations, 2024). This imbalance highlights the core dilemma: risks associated with AI—ranging from synthetic media and deepfakes that undermine information integrity, to dual-use autonomy in weapons systems—are proliferating faster than national regulations or unilateral controls can adapt. Without mechanisms that are simultaneously binding, interoperable, and adaptive, the governance gap will continue to widen.

A feasible architecture for global AI governance must therefore blend multiple instruments: binding treaties where consensus is possible, political compacts where flexibility is required, and technical institutions where practical cooperation can scale. Several proposals illustrate this hybrid approach:

1. **A Global AI Framework Convention** under UN auspices could provide the backbone of the system, modeled procedurally on the Paris Agreement. Rather than imposing rigid uniformity, it would establish a core set of baseline duties—such as safety testing, transparency in system documentation, and accountability safeguards—while leaving states free to determine national implementation pathways. Peer-review mechanisms and reporting cycles would ensure accountability, allowing governance to evolve alongside technological advances.
2. **An AI Peace and Security Council**, embedded within or networked to the UN system, would address risks that threaten international stability. This body could monitor escalation triggers in areas such as autonomous targeting, AI-enabled cyber operations, or large-scale disinformation campaigns. Equipped with early-warning capacity and empowered to recommend moratoria or confidence-building measures, such a council would provide a dedicated forum for security-related deliberation, mitigating the risk of AI crises escalating unchecked. Crucially, its decision-making rules would need to be inclusive and designed to avoid the chronic veto paralysis that undermines existing UN structures.
3. **A Global Data and Compute Compact** would directly address inequities in access to resources that underpin AI development. By financing regional compute hubs, privacy-preserving research environments, and cross-border data pipelines, this compact would transform infrastructure from a bottleneck into a shared diplomatic asset. It would also provide a platform for collaborative science in areas where collective intelligence is indispensable, such as climate modeling, pandemic surveillance, and disaster-risk reduction.
4. **A Multilateral Fund for AI and Development**, co-governed with Global South constituencies, would ensure that AI applications for the Sustainable Development Goals (SDGs) are scaled equitably. Independent audits, red-teaming exercises, and human-rights impact assessments would embed accountability into its operations, preventing the fund from reproducing donor-driven dependency models. Such a mechanism would not only finance innovation but also rebalance voice and agency in AI's global trajectory.
5. **A Network of AI Safety Institutes** would formalize technical cooperation across borders, consolidating the epistemic diplomacy that began at the Bletchley and Seoul summits. By sharing evaluation methods, benchmark suites, and incident reports, this network would create a scientific commons for AI safety. Such cooperation would not erase geopolitical rivalry but would help standardize the technical underpinnings of governance, ensuring that safety science can be interoperable even in a divided world.

The guiding principle across these proposals is to align sovereignty with shared responsibility. States retain the right to pursue national strategies, yet access to global markets, compute resources, and financing would be conditional upon verifiable commitments to safety and human rights. This reciprocal design prevents governance from being framed as a zero-sum constraint on innovation. Instead, it positions governance as a co-evolutionary process: as AI capabilities expand, so too must the global frameworks that guarantee equity, resilience, and security.

Synthesis

AI diplomacy now lies at a critical crossroads where multiple domains—security, trade, standards, and human rights—intersect and increasingly collide. Unlike past technological eras, where governance evolved around discrete issues such as nuclear deterrence or trade liberalization, artificial intelligence permeates all sectors simultaneously. It affects global supply chains, alters the conduct of warfare, reshapes financial markets, and redefines basic questions of dignity and rights in digital societies. This multidimensionality means that the governance challenge is not confined to one institution or treaty; instead, it requires overarching architecture capable of holding together diverse and often contradictory imperatives.

The first challenge is institutionalizing interoperability between science and law. AI is evolving in real time, with frontier systems displaying capabilities that policymakers only dimly understood just a few years ago. Governance cannot rely solely on fixed treaties or political declarations; it must embed mechanisms that translate technical insights into enforceable norms and vice versa. For example, safety institutes in the UK, United States, and EU are developing evaluation protocols that will only be meaningful if regulators can incorporate them into binding requirements and if courts and oversight bodies can adjudicate them. Conversely, legal concepts such as accountability, proportionality, or due process must be translated into design requirements that engineers can operationalize. Without this mutual translation, governance risks a dangerous lag: law chasing technology or science operating in a normative vacuum.

A second challenge is building bridges between regional compacts and global baselines. The governance landscape is already characterized by diversity. The European Union has chosen a rights-centric and risk-based regulatory model; the United States emphasizes innovation, industrial policy, and national security; China integrates AI governance into its model of state-led digital control; and many Global South states seek space to establish sovereignty while accessing resources and expertise. These approaches are not easily reconcilable. Yet, in areas such as safety testing, compute-sharing for public-good science, or prohibition of the most harmful uses, minimum global baselines are indispensable. What is needed is a governance model that resembles a "variable geometry": a universal core of principles and obligations, complemented by regionally adapted compacts that reflect different social contracts. Without such a layered system, governance will either fragment into incompatible silos or collapse into the lowest common denominator.

The third challenge lies in reconciling great-power rivalry with the legitimate aspirations of the Global South. Much of the debate on AI diplomacy is framed in binary terms—U.S. versus China, or "the West" versus alternative digital orders. But this framing risks reproducing the very dependency structures that many developing countries seek to escape. Global South states are not merely passive rule-takers; they are increasingly important rule-shapers, advancing initiatives on digital sovereignty, data localization, indigenous-language models, and regional compute hubs. For AI diplomacy to be sustainable, it must provide meaningful space for these states to author rules, not just adopt them. This requires redistributing convening power: ensuring that summits, treaty negotiations, and standard-setting bodies include diverse leadership, and that financing mechanisms such as compute compacts or AI-for-development funds are co-governed rather than donor-dominated.

Failure to address these dimensions risks ushering in a form of AI brinkmanship. Divergent regulatory models could harden into competing digital blocs, fragmenting global markets and undermining trust. Export controls on chips, data flows, and software components, while justified on security grounds, may escalate into tit-for-tat restrictions that paralyze international cooperation on shared challenges such as climate modeling or pandemic preparedness. In the security domain, autonomous weapons and AI-enabled cyber operations could trigger destabilizing dynamics reminiscent of the nuclear era, but with far lower barriers to entry. In the human rights domain, algorithmic repression and surveillance architectures could entrench authoritarian practices, deepening global divides between open and closed societies. In short, without deliberate institutional rewiring, AI risks amplifying the centrifugal tendencies already present in a fractured international order.

Yet the same technologies that magnify these risks also offer unprecedented opportunities for cooperation if governance frameworks can channel them. AI-enabled science diplomacy has the potential to accelerate global research collaboration by processing vast data sets across climate, health, and disaster risk domains. AI systems can support pandemic surveillance by detecting early anomalies in health data, contribute to food security by optimizing agricultural yields under climate stress, and strengthen humanitarian response by modeling displacement or assessing infrastructure vulnerabilities. For these benefits to materialize equitably, global governance must guarantee that access to compute and data for public-good science is not restricted to a handful of wealthy states or corporations. This is why initiatives such as a Global Data and Compute Compact are more than technical proposals; they represent diplomatic investments in shared resilience.

The long-term test of AI diplomacy will therefore not be whether it produces elegant communiqués or ambitious declarations, but whether it can institutionalize interoperability across divides. That means ensuring technical testing regimes are legally binding and globally consistent; aligning regional regulatory models without erasing diversity; and embedding Global South agency at the design stage of governance rather than as an afterthought. Interoperability must extend beyond systems to values—between innovation and rights, sovereignty and interdependence, competition and cooperation. Achieving this balance requires humility, foresight, and sustained investment in multilateral institutions.

In sum, AI diplomacy is both a challenge and an opportunity of historic scale. Managed poorly, it could lock the world into rival technological blocs, intensify inequities, and destabilize fragile security environments. Managed wisely, it could serve as a catalyst for renewed multilateralism, transforming AI from a source of division into a platform for cooperation. The decisive factor will be whether states and institutions can design governance architectures that scale with technology's capabilities—so that safety, equity, and resilience grow in step with innovation. The stakes are not merely technological or economic; they are fundamentally about the kind of international order the twenty-first century will inherit.

SCIENCE DIPLOMACY IN THE AGE OF DISRUPTION

Science diplomacy has long been characterized as a "quiet superpower" of international relations, providing avenues for dialogue, trust-building, and cooperative risk management when political channels are blocked (Gluckman et al., 2017). Traditionally, its role was considered supplementary to formal diplomacy—an auxiliary tool that could ease tensions or sustain cooperation in narrow technical fields even when geopolitics was paralyzed. Yet in today's age of polycrisis, where climate change, pandemics, armed conflicts, economic instability, and technological disruptions overlap and interact, science diplomacy is no longer a peripheral instrument. It has become central to the architecture of global governance itself.

This transformation stems from two converging dynamics. First, the challenges facing humanity are increasingly transboundary and systemic: pandemics that spread in weeks, climate shifts that cascade across continents, AI systems whose risks spill across borders, and supply chain shocks that ripple through the global economy. No single state, however powerful, can manage these risks unilaterally. Second, trust in multilateral institutions is eroding, and traditional mechanisms of diplomacy—from the UN Security Council to the Bretton Woods institutions—often lack the agility or legitimacy to mount timely, coordinated responses. Into this vacuum, science diplomacy is stepping not only as a bridge across polarization but also as a governance mechanism: a domain where cooperation can be institutionalized around evidence, shared infrastructures, and collective foresight.

Concrete examples illustrate this shift. Climate attribution science now informs negotiations under the Paris Agreement, providing near-real-time evidence of the human role in extreme weather events and shaping global debates on loss and damage. Pandemic preparedness frameworks, from genomic sequencing networks to early-warning platforms like the WHO's Epidemic Intelligence from Open Sources, demonstrate how scientific infrastructures provide continuity even when political trust breaks down. AI safety institutes and open-data platforms are emerging as sites of negotiation where standards, accountability, and equitable access are contested and codified. In these domains, scientific cooperation does not merely complement governance—it increasingly constitutes governance.

Critically, however, science diplomacy is not neutral. Its design determines whether it entrenches asymmetries or redistributes authority. Shared infrastructures can empower marginalized regions by ensuring equitable access to data, compute power, or authorship—or they can reinforce dependency if governance remains centralized in powerful capitals. Collaborative research can amplify plural epistemologies and traditional knowledge systems—or it can perpetuate epistemic injustice if Indigenous and local actors remain excluded from decision-making. In this sense, science diplomacy is a political technology: it stabilizes institutions and enables adaptive responses, but the distribution of voice, power, and resources within it decides whose futures are safeguarded.

This section therefore examines science diplomacy in the age of disruption across four thematic areas: (1) its role as a bridge in polarized geopolitics, sustaining channels of dialogue where political trust has collapsed; (2) the development of shared infrastructures in data, AI, and open science, which serve as new commons for cooperation; (3) the emergence of anticipatory science diplomacy, embedding foresight and futures thinking into multilateral processes; and (4) case studies of pandemic preparedness, climate science, and open data that illustrate both the promise and fragility of such mechanisms. A fifth perspective is added to extend the analysis: how science diplomacy can be leveraged as a systemic intervention within complex adaptive systems (CAS), particularly to disrupt entrenched hierarchies such as the Aid Industrial Complex (AIC) and to reconfigure governance toward equity, resilience, and genuine polycentricity.

Science as a Bridge in Polarized Geopolitics

In periods of heightened geopolitical rivalry, science often endures as one of the last viable channels of communication. Unlike military or diplomatic negotiations, which are framed in terms of power and strategy, scientific cooperation carries a perception of neutrality and practical necessity. It provides what scholars call a "low-politics" space—arenas where collaboration is seen as serving universal human needs rather than advancing zero-sum interests (Flink & Schreiterer, 2010). This perception allows dialogue to persist even when political or military relations deteriorate, and it explains why scientific infrastructures frequently outlive the crises that gave rise to them.

The International Space Station (ISS) remains one of the most widely cited examples. Despite sanctions, competing strategic agendas, and an increasingly adversarial U.S.–Russia relationship, operational collaboration between scientists and engineers was sustained well into the 2010s. The practical necessity of maintaining the station—its life-support systems, orbit corrections, and research functions—created a space where scientific imperatives overrode political fractures (Abbott, 2020). Similarly, the SESAME synchrotron in Jordan has convened researchers from across the Middle East, including countries with strained or even hostile diplomatic relations, under a shared scientific mandate. SESAME's governance structure, which allocates equal voting rights and emphasizes consensus-driven decision-making, illustrates the insulating potential of science diplomacy: it builds trust incrementally, through co-authorship, shared facilities, and face-to-face collaboration, even in regions marked by chronic polarization (Nielsen & Rhoten, 2018).

These cases demonstrate that science diplomacy can function as a confidence-building measure (CBM) in the international system. By creating spaces where knowledge, rather than sovereignty, is the primary currency, scientific cooperation lowers the stakes of interaction and sustains communication channels that might otherwise collapse. In doing so, it provides not just symbolic reassurance but also material benefits—joint data, co-developed technologies, and shared infrastructures that become difficult for states to abandon without incurring reputational or scientific costs.

Yet durability is not automatic. Science diplomacy is highly sensitive to design. When projects are captured by national prestige agendas—as in the case of large-scale mega-facilities that become tools of soft power—or when financing structures are overwhelmingly asymmetrical, cooperation risks becoming politicized or reinforcing the very inequalities it was meant to bridge (Krasnyak, 2020). The consequence is that scientific platforms, rather than serving as neutral bridges, may reproduce existing geopolitical hierarchies.

To mitigate these risks, neutral governance arrangements are essential. Equal or rotating decision-making rights prevent dominance by single actors. Transparent financing models—including proportional contributions, joint procurement, and open reporting—reduce dependency dynamics. Shared custodianship of data and intellectual property ensures that benefits are not monopolized by technologically advanced partners. And critically, plural epistemic participation—bringing in scientists from under-represented regions, including the Global South—ensures that "bridge projects" do not become elite clubs serving only the interests of the powerful.

Ultimately, the bridging role of science diplomacy depends less on the inherent neutrality of science than on the institutional design of cooperation. Where governance is inclusive, financing is balanced, and benefits are widely distributed, science diplomacy can sustain cooperation through political storms and even lay the groundwork for broader reconciliation. Where design flaws persist, however, the same platforms risk being subsumed into geopolitical competition, weakening both their credibility and their ability to serve as durable channels of peace.

AI, Data, and Shared Scientific Infrastructures

The digital era has shifted the locus of science diplomacy from physical laboratories to data-intensive infrastructures. In the mid-20th century, international collaboration was embodied by shared physical spaces—particle accelerators, nuclear research centers, and space missions. Today, the infrastructures that determine participation in global science are increasingly intangible yet no less strategic: access to high-performance computing (HPC), shared datasets, interoperable AI models, and digital platforms for collective intelligence. These infrastructures define not only who contributes to knowledge but also who sets the terms of global governance.

Federated Platforms as Democratic Infrastructures

Emerging initiatives demonstrate how federated models of open science can reduce asymmetries. The European Open Science Cloud (EOSC), launched in 2018, provides a federated environment where researchers across Europe can access interoperable datasets, computing resources, and services. Its design emphasizes FAIR (Findable, Accessible, Interoperable, Reusable) principles and cross-border governance, preventing dominance by any single state or institution. Similarly, the African Open Science Platform (AOSP) seeks to consolidate fragmented research infrastructures, enabling researchers across the continent to access shared compute power, datasets, and open-source tools. Though still developing, AOSP reflects how federated infrastructures can act as digital commons, widening access to both participation and agenda-setting.

Global health also illustrates the power of shared infrastructures. The WHO's Epidemic Intelligence from Open Sources (EIOS) integrates digital signals from nearly 200 countries to detect outbreak risks in real time (WHO, 2023). By pooling data from formal surveillance systems and informal digital sources—including news reports and social media—EIOS demonstrates how open, distributed infrastructures can underpin anticipatory governance. Crucially, such platforms strengthen resilience: they allow multiple actors to contribute knowledge, reducing dependence on single hubs or hegemonic providers.

Persistent Inequities in Compute and Data Access

Yet stark inequities persist. As of 2024, over 70% of global compute capacity—the high-performance processors and GPU clusters that underpin modern AI research—is concentrated in the United States and China. By contrast, Africa and Latin America collectively account for less than 3%. This geographic asymmetry risks entrenching epistemic divides: a handful of states and corporations generate foundational models, while most of the world remains consumers of "imported intelligence," with little influence over design choices, ethical guardrails, or governance norms.

Such concentration mirrors earlier dynamics of knowledge dependency in the Aid Industrial Complex, where resources, decision rights, and epistemic authority were centralized in donor capitals. Without corrective mechanisms, "open science" risks becoming performative rhetoric—a discourse of equity that masks underlying exclusion.

Science Diplomacy as Corrective Mechanism

Science diplomacy provides a potential corrective by embedding equity principles directly into the governance of digital infrastructures. Several mechanisms are gaining traction:

- **FAIR principles institutionalized** as binding requirements, ensuring that datasets and models are not merely published but interoperable, reusable, and co-governed by diverse institutions.
- **Public-interest intellectual property (IP) pools**, where essential algorithms, diagnostic tools, or datasets are licensed under non-restrictive terms, preventing monopolization and ensuring that public goods remain accessible.
- **Regional compute hubs**, designed as **sovereign digital commons**, capable of training models on localized data for agriculture, health, climate adaptation, and education. These hubs reduce dependency on external providers while cultivating local talent pipelines.
- **Distributed governance arrangements**, modeled on federated consortia, that ensure decision rights are not concentrated in Global North institutions.

Implications for Governance

These measures are not only technical correctives but also systemic interventions in global governance. They alter the balance of power by redistributing access to the infrastructures that define participation in the knowledge economy. In doing so, they help to close epistemic divides, preventing a scenario where a small number of AI "producer states" dictate standards, values, and design choices for the rest of the world.

In short, the governance of data and compute is not peripheral but foundational to the practice of science diplomacy in the 21st century. Open infrastructures can act as bridges, but without equity-focused governance, they risk reinforcing the very hierarchies they are meant to dismantle.

Anticipatory Science Diplomacy: Embedding Futures Thinking into Multilateral Processes

Traditional diplomacy is overwhelmingly reactive, mobilizing after crises have erupted rather than before. This "firefighting" mode often leaves institutions scrambling to manage consequences rather than mitigating risks in advance. Science diplomacy offers a pathway to transform this dynamic, embedding foresight methodologies and anticipatory science into the core of multilateral processes. By doing so, governance can move from a short-term cycle of emergency response to resilience-oriented planning capable of addressing complex and cascading disruptions.

The COVID-19 pandemic starkly illustrated both the potential and the limitations of anticipatory science. Modeling by the Institute for Health Metrics and Evaluation (IHME), the Imperial College COVID-19 Response Team, and other epidemiological groups shaped government decisions on lockdowns, vaccine procurement, and healthcare surge capacity (Walker et al., 2020). These projections provided decision-makers with scenarios of morbidity, mortality, and health system burden, guiding both national strategies and global allocation mechanisms such as COVAX. Yet political trade-offs, delays in uptake, and inconsistent trust in science revealed the gap between availability of foresight and its integration into governance.

Climate governance demonstrates a more mature model of anticipatory diplomacy. Advances in attribution science now allow researchers to determine, within weeks, the extent to which human-induced climate change has intensified specific extreme weather events (Otto et al., 2020). These near-real-time assessments feed directly into negotiations under the United Nations Framework Convention on Climate Change (UNFCCC). They have strengthened the legitimacy of loss-and-damage debates, giving vulnerable states evidence-based leverage in arguing for financial redress. Attribution science thus exemplifies how foresight capacity, once embedded in multilateral systems, can recalibrate negotiations, shifting debates from abstract principles to evidence-backed claims of justice.

To systematically harness anticipatory science, structural reforms are required across multilateral institutions. Possible designs include:

- **Dedicated scientific advisory panels integrated into political bodies**: Existing models such as the Intergovernmental Panel on Climate Change (IPCC) demonstrate how epistemic communities can synthesize science for policy. Extending this approach to core UN bodies such as the Security Council or General Assembly could ensure that foresight on pandemics, AI risks, and climate cascades informs geopolitical decision-making in real time (Lentsch & Weingart, 2011).
- **Futures-to-budget linkages**: Anticipatory models should directly guide financial flows—shaping Official Development Assistance (ODA) allocations, humanitarian contingency funds, and disaster-preparedness investments. Embedding scenario planning into budgetary cycles ensures that forecasts are not simply advisory but structurally tied to resource distribution.
- **Structured engagement with epistemic communities**: Science diplomacy must convene not only natural scientists but also social scientists, ethicists, and Indigenous knowledge holders. Drawing on Haas's (1992) notion of epistemic communities, multilateral institutions can benefit from plural perspectives that strengthen legitimacy and reduce epistemic exclusion.

Embedding foresight in diplomacy is not merely about producing better evidence. It is about institutionalizing adaptability into the DNA of global decision-making. Anticipatory science diplomacy creates mechanisms for governments and international organizations to continuously update their strategies as risks evolve. Rather than treating uncertainty as a weakness, this approach reframes it as a driver of resilience. By aligning science with diplomacy, foresight becomes not an optional supplement but a core operating principle of 21st-century governance.

Case Studies: Pandemic Preparedness, Climate Science, and Open Data

Concrete cases reveal both the promise and the fragility of science diplomacy. They demonstrate that while scientific infrastructures can serve as trusted platforms for cooperation, they are deeply contingent on political incentives, equitable access, and sustained institutional design.

Pandemic Preparedness: Early Warning and Geopolitical Tensions

The World Health Organization's Epidemic Intelligence from Open Sources (EIOS) platform exemplifies how open-source intelligence can strengthen global preparedness. By integrating reports from nearly 200 countries, news outlets, and digital signals, EIOS enables rapid detection of outbreak signals, often weeks before official confirmation (WHO, 2023). During COVID-19, such tools helped identify early clusters and anticipate global spread.

Yet the pandemic also revealed vulnerabilities. Delays in the timely sharing of SARS-CoV-2 genomic sequences—most prominently during the emergence of new variants—illustrated how geopolitical rivalries, intellectual property disputes, and mistrust between states can blunt the effectiveness of scientific infrastructures (Kamradt-Scott, 2022). The lesson is clear: scientific platforms are only as effective as the political willingness to share data. Without transparency and reciprocity, even technically advanced infrastructures risk paralysis.

Climate Science: Consensus-Building and Diluted Urgency

The Intergovernmental Panel on Climate Change (IPCC) stands as perhaps the most emblematic case of institutionalized science diplomacy. Since 1988, the IPCC has synthesized global research into consensus-based reports, providing the authoritative evidence base that underpins negotiations under the UN Framework Convention on Climate Change (UNFCCC). Its reports have shaped major diplomatic milestones, from the Kyoto Protocol to the Paris Agreement, by lending scientific legitimacy to global climate governance (Beck & Mahony, 2018).

However, the IPCC's consensus-driven methodology is a double-edged sword. While it anchors legitimacy and inclusivity, it can also dilute urgency. Findings are often framed conservatively to accommodate diverse national perspectives, leading critics to argue that warnings on climate tipping points or high-impact scenarios are softened in the pursuit of consensus (Beck & Mahony, 2018). This trade-off illustrates the inherent tension between inclusivity and responsiveness: the very processes that make the IPCC trusted can also slow the pace of political action.

Open Data Initiatives: Norm Entrepreneurship and Implementation Gaps

The digital transformation of science diplomacy is illustrated by UNESCO's 2021 Recommendation on Open Science, which sought to establish a normative framework for balancing openness with equity, ethics, and security. By calling for FAIR (Findable, Accessible, Interoperable, Reusable) data principles and explicitly embedding commitments to equity and gender inclusivity, the Recommendation positioned the UN as a norm entrepreneur in digital governance (UNESCO, 2021).

Yet its implementation has been uneven. Wealthier countries and well-resourced institutions have been quicker to operationalize open-data infrastructures, while low- and middle-income states often lack the technical, financial, and legal capacities to comply. This risks reinforcing epistemic divides, whereby "open science" becomes another arena where the Global South contributes data but lacks governance authority. Nonetheless, the Recommendation has catalyzed important debates on digital commons, illustrating how multilateral institutions can set agendas even when enforcement capacity is limited.

Synthesis: Anchoring Trust Amid Fragility

Taken together, these cases highlight the dual nature of science diplomacy. It can anchor trust, guide governance, and enable collective action, but its effectiveness depends on whether political incentives align with scientific cooperation and whether equity in access is safeguarded. Pandemic preparedness shows the costs of mistrust; climate science demonstrates the trade-offs between inclusivity and urgency; and open data initiatives highlight the gap between normative ambition and practical capacity.

In sum, science diplomacy in the age of disruption is not self-executing. Its value lies not only in convening science for policy but in embedding equity, trust, and accountability into the very design of cooperative infrastructures.

Leveraging Science Diplomacy in Complex Adaptive Systems

Beyond individual cases, science diplomacy can function as a systemic lever for reconfiguring entrenched hierarchies within complex adaptive systems (CAS). The international aid system—often described as the Aid Industrial Complex (AIC)—illustrates the problem: compliance-driven funding regimes, subcontracting chains, and visibility-centered incentives create feedback loops that marginalize local actors and perpetuate dependency (Holland, 1995; Escobar, 1995; Al Mokdad, 2025).

Past reform efforts such as the Paris Declaration (2005) and the Grand Bargain (2016) largely failed because they pursued incremental adjustments rather than altering systemic logics. Science diplomacy, by contrast, introduces perturbations that can rewire feedback loops toward equity and resilience.

Recent scholarship (Al Mokdad, 2025) outlines a four-pillar framework for this systemic application:

- **Trust-building research alliances** through mandatory co-principal investigator (co-PI) models, equitable authorship rules, and integration of traditional ecological knowledge.
- **Geopolitical bridge-building** via neutral infrastructures such as SESAME and FRIEND-Nile that convene actors across conflict lines.
- **Technology and data diplomacy** using FAIR/open data compacts, responsible AI protocols, and public-interest IP pools.
- **Institutional reforms for inclusion** such as community oversight panels, decentralized audits, and multi-year funding windows for Global South consortia.

By embedding transparency, equity, and accountability into the operating rules of scientific cooperation, science diplomacy can evolve from a "soft" tool into a "hard" governance mechanism. Its systemic impact lies in redistributing decision rights, creating adaptive trust, and fostering community-centered partnerships that escape the gravitational pull of donor-driven hierarchies.

Science diplomacy in the age of disruption is no longer a peripheral or symbolic practice. It functions as a strategic architecture operating across scales: a bridge in polarized geopolitics, a steward of shared infrastructures, a channel for anticipatory foresight, and a systemic lever to reconfigure complex adaptive systems.

Its transformative potential, however, depends on institutional design. Without enforceable safeguards—such as equity benchmarks, open-data compacts, and decentralized governance—science diplomacy risks reproducing the very asymmetries it seeks to dismantle. But when embedded with complexity-aware principles and verifiable accountability, it can help global governance evolve from a model of declaratory multilateralism into one of adaptive, equitable, and resilient cooperation. In this sense, science diplomacy is not only a "quiet superpower" but also an indispensable engine for survival and solidarity in an era defined by disruption.

GREEN DIPLOMACY AND CLIMATE GOVERNANCE

Climate governance has long been regarded as the flagship arena of multilateral cooperation, often cited as the most visible test of whether the international community can act collectively in the face of existential threats. Since the 1992 Rio Earth Summit and the creation of the UN Framework Convention on Climate Change (UNFCCC), climate negotiations have produced some of the most ambitious and universal agreements in international law, from the Kyoto Protocol to the Paris Agreement. These frameworks institutionalized emission-reduction targets, adaptation finance, and mechanisms for transparency and accountability. Yet, in the age of polycrisis, climate governance can no longer be understood in isolation. Its scope has expanded beyond environmental stewardship into the core architecture of diplomacy, finance, and security.

This shift reflects the reality that climate risks are deeply entangled with other systemic disruptions. Extreme weather events drive migration and displacement, generating humanitarian crises and reshaping geopolitical alignments. Climate-induced food insecurity destabilizes fragile economies and fuels conflict, as seen in regions where drought and resource scarcity exacerbate existing political tensions. Energy transitions, while essential for decarbonization, intersect with great-power competition over critical minerals, technological leadership, and green industrial policy. Financial stability is equally at stake: sovereign debt crises are increasingly tied to climate vulnerability, pushing international financial institutions to incorporate climate-risk assessments into macroeconomic surveillance (Gallagher & Kozul-Wright, 2022).

Moreover, technological disruption—particularly artificial intelligence (AI) and digital infrastructures—now mediates how climate challenges are understood, predicted, and governed. AI models generate climate forecasts that influence disaster diplomacy; satellite-based monitoring tools underpin carbon markets and compliance verification; and digital finance mechanisms are increasingly deployed to fund adaptation and resilience. These intersections reveal that climate governance is no longer a narrow sectoral concern but a meta-framework that shapes and is shaped by economic, social, and technological governance across scales.

Within this context, "green diplomacy" emerges as a new paradigm. Unlike traditional climate diplomacy, which focused primarily on negotiating emission targets, green diplomacy seeks to integrate climate action into a polycrisis-oriented governance architecture. It recognizes that ecological risks are inseparable from shocks in health, security, finance, and technology, and that effective responses require cross-sectoral and multi-actor coordination. Green diplomacy is thus both an adaptation of multilateralism to the complexities of the 21st century and a test case for whether global institutions can evolve to govern systemic risks that are interconnected, fast-moving, and profoundly unequal in their impacts (Held, Theros, & Youngs, 2023).

This section unpacks four interrelated dimensions of green diplomacy that illustrate its evolution and fragmentation:

1. The shift from conventional climate agreements to polycrisis-oriented frameworks, where climate governance is embedded in wider security, economic, and humanitarian strategies.
2. The role of sub-national actors—cities, states, and regions—as increasingly influential agents of climate and energy diplomacy.

3. The integration of AI and digital technologies into climate diplomacy, transforming forecasting, mitigation, and monitoring practices.
4. The institutional rise of climate clubs, carbon markets, and data-transparency mechanisms, which signal a move toward fragmented but pragmatic governance models outside the UN system.

Together, these dimensions reveal both the promise and the limitations of climate governance in the polycrisis era. On the one hand, they demonstrate adaptability, innovation, and multi-actor engagement. On the other, they highlight the risks of fragmentation, inequity, and exclusion that may undermine the very legitimacy and universality that climate diplomacy was designed to uphold.

From Climate Agreements to Polycrisis-Oriented Green Diplomacy

Since the 1992 Rio Earth Summit, climate diplomacy has been structured around a series of landmark multilateral agreements—the Kyoto Protocol (1997), the Paris Agreement (2015), and the annual Conferences of the Parties (COPs) that followed. These agreements institutionalized the global framework for emission reductions, climate adaptation, and financial mechanisms such as the Green Climate Fund (GCF), which was designed to channel resources from industrialized to developing nations. For much of the 1990s and early 2000s, climate governance was primarily understood as an environmental negotiation, focused on balancing scientific urgency with political feasibility in setting carbon targets. The assumption was that climate change, though complex, could be governed as a discrete sectoral issue, distinct from security, health, or financial systems.

By the mid-2020s, however, it had become evident that this framing was no longer tenable. Climate negotiations increasingly intersected with broader systemic crises, revealing how ecological disruption interacts with other domains of global governance. The COVID-19 pandemic was an early stress test of this interconnection: health systems strained by the virus also saw their climate-related resilience efforts undermined. Supply-chain disruptions not only delayed vaccine distribution but also stalled renewable energy projects, laying bare the fragility of interdependent systems (Barbier & Burgess, 2020). In parallel, the Russia–Ukraine war (2022–present) highlighted the geopolitical vulnerabilities of energy transitions. While disruptions in Russian natural gas supplies accelerated the European Union's investments in renewables, they also triggered renewed fossil-fuel exploration across Africa and the Middle East as states sought to fill the immediate supply gap (O'Sullivan, Overland, & Sandalow, 2022). This dual dynamic—acceleration and regression—demonstrates how climate diplomacy is entangled with the geopolitics of energy, trade, and conflict.

Against this backdrop, a new paradigm has emerged: polycrisis-oriented green diplomacy. Unlike earlier models that treated climate change as a bounded environmental concern, this approach situates climate at the center of global resilience strategies. Climate is no longer one agenda item among many; it is increasingly recognized as the connective tissue linking security, economic stability, humanitarian response, and technological innovation. For example, debates in the UN Security Council have begun to integrate climate-security resolutions, particularly in relation to fragile contexts such as the Sahel and the Pacific Islands, where environmental stress multiplies risks of instability and displacement (UNSC, 2023). In the financial domain, the International Monetary Fund has begun embedding climate-risk assessments into its Article IV consultations, effectively linking sovereign debt sustainability to vulnerability to climate shocks (Gallagher & Kozul-Wright, 2022). Meanwhile, the negotiation of a "loss and damage" financing mechanism at COP27 in Sharm el-Sheikh (2022) marked a historic recognition that climate change is not only an environmental crisis but also a humanitarian one, demanding financial solidarity to address irreversible harm to vulnerable communities.

This reorientation signals a profound transformation in the logic of climate diplomacy. Rather than treating climate change as a sectoral challenge to be mitigated through emissions targets and adaptation funds alone, polycrisis-oriented green diplomacy frames it as both a driver and amplifier of systemic instability. Droughts trigger migration, floods disrupt food systems, energy transitions reshape trade patterns, and climate vulnerability exacerbates sovereign debt crises. In this sense, climate governance becomes a form of systemic governance: to address it effectively requires engaging with its intersections across security, finance, health, and humanitarian domains.

Polycrisis-oriented green diplomacy thus reflects a recognition that the climate crisis is not simply one issue on the crowded global agenda but the cross-cutting matrix through which many other crises unfold. It represents both an opportunity and a challenge: an opportunity to reframe climate action as a central component of resilience-oriented governance, but also a challenge to traditional multilateral systems that remain siloed, slow-moving, and fragmented in their institutional design.

Cities and Sub-National Actors in Climate and Energy Diplomacy

The rise of cities and sub-national governments as diplomatic actors is among the most significant transformations in the landscape of climate governance. For much of the twentieth century, international diplomacy was conceived as the exclusive domain of nation-states, formalized through treaties and multilateral agreements negotiated within institutions such as the UN Framework Convention on Climate Change (UNFCCC). Yet in the twenty-first century—particularly in the age of polycrisis—this monopoly has eroded. Cities, provinces, and regional governments now exert considerable influence, engaging in what scholars describe as "city diplomacy" or "paradiplomacy" (Acuto, Morissette, & Tsouros, 2017). These actors increasingly set ambitious climate targets, build cross-border alliances, and experiment with practical solutions that often outpace national-level commitments.

This trend reflects both necessity and opportunity. Cities concentrate the majority of the world's population and economic activity, making them the epicenters of vulnerability to climate impacts such as heatwaves, floods, and air pollution. At the same time, urban governments often possess greater flexibility than national governments to pilot innovative policies, regulate infrastructure, and mobilize partnerships with civil society and the private sector. Sub-national governments, particularly those representing large economies, are similarly positioned to shape international norms and investment flows, creating a multi-layered diplomatic ecosystem.

Several institutional developments underscore the scale of this shift. The C40 Cities Climate Leadership Group unites nearly 100 of the world's largest cities, representing more than 700 million people and approximately 25% of global GDP (C40, 2023). C40 has emerged not merely as a knowledge-sharing platform but as a vehicle for norm-setting, committing its members to net-zero targets and coordinated urban adaptation strategies. Its initiatives range from enhancing heat resilience in megacities to promoting sustainable transport systems and green public procurement. By publishing standardized frameworks, C40 enables cities to benchmark progress against global goals, thereby linking local action with international accountability.

Similarly, the Under2 Coalition, a network of over 270 states and regions, collectively represents more than 50% of the global economy (Under2 Coalition, 2023). Its members pledge to align their policies with the Paris Agreement and to implement region-specific decarbonization pathways. Sub-national jurisdictions such as Scotland, Quebec, and California have used this platform to push more ambitious climate agendas than those adopted by their national governments, highlighting the role of federated entities in advancing global mitigation and adaptation. The coalition also provides a forum for sub-national leaders from the Global South—such as states in India, Nigeria, and Brazil—to exchange strategies and build collective bargaining power in climate finance negotiations.

Energy diplomacy further illustrates the growing influence of sub-national actors. California, for instance, has signed bilateral agreements with both China and Canada on clean-tech cooperation, renewable energy research, and emissions reduction technologies (Busby, 2021). These agreements position California as a de facto climate diplomat, enabling continuity in international engagement even during periods when U.S. federal policy retreats from multilateral commitments. Other examples include German Länder collaborating directly with Nordic regions on renewable energy grids, and African sub-national coalitions negotiating climate finance for off-grid solar electrification.

This growing prominence of sub-national actors transforms them into laboratories of governance. Their ability to pilot innovative policies—such as congestion pricing, carbon-neutral building codes, and decentralized renewable grids—creates models that can scale upward into national policy frameworks or directly shape global negotiations. The recognition of cities and regions as frontline implementers of climate action has also been institutionalized through their inclusion in global forums. For instance, the Marrakech Partnership for Global Climate Action and the Paris Agreement's Non-State Actor Zone for Climate Action (NAZCA) platform formally integrate city- and region-led commitments into the UNFCCC process.

However, the diffusion of authority beyond states also raises complex governance challenges. Accountability gaps emerge when sub-national pledges lack clear monitoring and enforcement mechanisms, making it difficult to verify whether commitments translate into measurable emissions reductions. Fragmented standards across jurisdictions can complicate coordination, with differing methodologies for carbon accounting or adaptation metrics risking duplication and inconsistency. Uneven capacities further exacerbate disparities: while wealthy cities such as London, Singapore, or Los Angeles can invest in advanced climate infrastructures, many municipalities in the Global South lack the financial and technical resources to deliver on even basic commitments. The danger is that a patchwork of uneven sub-national action may substitute for coherent global governance rather than complement it.

In sum, cities and sub-national actors have emerged as indispensable players in the evolving ecosystem of climate diplomacy. They embody both the innovation and fragmentation of polycrisis-oriented governance. On one hand, their agility and proximity to citizens allow them to drive ambitious climate agendas and experiment with solutions that can diffuse globally. On the other, the absence of robust accountability, financing disparities, and risks of duplication underscore the need to embed sub-national diplomacy within a more coherent multilevel governance framework. Whether their rise strengthens or fragments the global climate regime will depend on how effectively national governments and international institutions integrate these actors into formal decision-making structures.

AI-Enabled Climate Diplomacy: Forecasting, Mitigation, and Monitoring

Artificial intelligence (AI) is no longer peripheral to climate governance; it is increasingly embedded in the practice of climate diplomacy, providing forecasting, mitigation, and monitoring capacities that extend far beyond human analytical ability. The integration of AI tools into climate policy has reshaped how states, international organizations, and coalitions negotiate, monitor, and enforce commitments, making digital technologies central to the infrastructure of twenty-first-century environmental diplomacy.

Forecasting

Advances in AI-driven climate modeling are transforming the predictive capabilities that underpin early-warning diplomacy. Google's *GraphCast*, for example, has demonstrated the ability to deliver weather predictions with significantly greater accuracy than traditional numerical systems, outperforming benchmarks in storm tracking, precipitation forecasts, and heatwave detection up to ten days in advance (Lam et al., 2023). These tools are not merely scientific advances; they are diplomatic assets. Regional organizations such as the African Union and the Association of Southeast Asian Nations (ASEAN) have begun integrating AI-enhanced models into cross-border compacts on disaster preparedness, using shared forecasts as neutral reference points for cooperative action. By providing trusted and transparent predictions, AI reduces the scope for political contestation in preparedness planning and strengthens the legitimacy of anticipatory agreements.

Mitigation

AI is also increasingly used to support mitigation strategies, from optimizing renewable energy grids to guiding reforestation programs. Algorithms improve demand–supply balancing in energy networks, facilitating higher penetration of renewables without destabilizing grids. Similarly, satellite-based AI analysis of forest cover now underpins the verification of carbon sequestration, providing the empirical foundation for carbon offset markets. These monitoring systems are crucial in diplomatic negotiations over mitigation credits, where credibility depends on demonstrable reductions rather than unverifiable claims (World Bank, 2024). As such, AI serves both as a technical enabler of carbon markets and as a diplomatic trust mechanism for linking mitigation finance with measurable outcomes.

Monitoring and Verification

The capacity of AI to generate independent verification of national pledges represents perhaps its most disruptive diplomatic function. Startups such as *Climate TRACE*—a coalition backed by former U.S. Vice President Al Gore—use AI and satellite data to produce real-time emissions tracking across countries and sectors

(Climate TRACE, 2023). This capability challenges the reliance on self-reported government data, long criticized for opacity, underreporting, and inconsistency. By introducing independent, near-real-time accountability, AI-based monitoring alters the dynamics of climate negotiations: states are aware that their performance is being tracked not only by peer governments but also by transnational scientific consortia and civil society coalitions.

Yet the integration of AI into climate diplomacy is not without risks. Overreliance on technical proxies can marginalize local and Indigenous knowledge systems, particularly when algorithmic models prioritize global datasets over place-based expertise (Milan & Treré, 2019). Moreover, the concentration of computational infrastructure in a handful of powerful states and corporations raises concerns about "algorithmic colonialism"—where the capacity to design, own, and interpret climate models is monopolized, leaving developing countries as passive consumers of forecasts and verification systems. Without deliberate safeguards—robust data-sharing agreements, equitable access to compute infrastructure, and targeted capacity-building for the Global South—AI-enabled climate diplomacy risks reinforcing existing asymmetries rather than democratizing access to science and governance.

In short, AI offers extraordinary opportunities for strengthening the credibility, efficiency, and foresight of climate diplomacy. But these opportunities will only be realized if they are accompanied by governance mechanisms that balance technical precision with inclusivity, equity, and plural epistemologies.

The Rise of Climate Clubs, Carbon Markets, and Data Transparency

As traditional UN climate negotiations increasingly struggle under the weight of consensus-driven gridlock, new forms of minilateral and multi-actor initiatives have begun to proliferate. These include climate clubs, carbon markets, and data-transparency mechanisms—innovations designed to accelerate progress among willing coalitions even as global negotiations stall. Together, they represent an emerging architecture of fragmented pluralism in climate governance: multiple overlapping initiatives that can generate momentum but also risk deepening divides.

Climate Clubs

First proposed by Nobel laureate William Nordhaus (2021), the concept of climate clubs has gained traction as a means of overcoming collective action problems. In 2022, the G7 launched a Climate Club focused on industrial decarbonization, aiming to harmonize carbon pricing, coordinate standards, and reduce the risk of "carbon leakage" by penalizing imports from non-compliant states. The advantage of clubs is their ability to bypass unanimity, enabling ambitious actors to move ahead without being held back by laggards. However, the exclusionary logic of clubs raises questions of equity: developing states that lack the fiscal or technological capacity to meet high standards risk marginalization, potentially reinforcing rather than resolving North–South divides.

Carbon Markets

Article 6 of the Paris Agreement institutionalized carbon trading as a key mechanism for mitigation, enabling countries to trade emissions reductions through internationally transferable mitigation outcomes (ITMOs). By 2025, the global voluntary carbon market was estimated at US$2 billion annually, with corporate demand for offsets driving rapid expansion (World Bank, 2024). Yet the credibility of these markets remains contested. Investigations into "junk credits," double counting, and greenwashing have eroded trust, while the uneven regulatory landscape creates opportunities for exploitation. For carbon markets to function as effective diplomatic tools, robust governance standards and transparent verification are essential. AI-enabled monitoring, as discussed in the previous subsection, may provide the data infrastructure needed to restore credibility, but institutional safeguards are still evolving.

Data Transparency

Alongside clubs and markets, new transparency mechanisms are redefining accountability in climate governance. Independent initiatives such as *Climate TRACE* and the *Carbon Action Tracker* produce public emissions data that can be directly leveraged in negotiations, reducing reliance on self-reported inventories. Meanwhile, the UNFCCC's Enhanced Transparency Framework (ETF)—operational as of 2024—requires states to submit standardized reports on emissions, adaptation, and finance. The credibility of these systems, however, hinges on enforcement: without consequences for non-compliance, transparency risks becoming another symbolic exercise. Nonetheless, the shift toward standardized, independent, and real-time data is transforming climate diplomacy, making accountability less dependent on political goodwill and more anchored in measurable evidence.

Together, these mechanisms illustrate the tensions of twenty-first-century climate governance. On one hand, climate clubs, carbon markets, and data-transparency initiatives enable faster action, circumventing the slow pace of consensus-based UN negotiations. On the other, they risk institutionalizing a two-tiered regime: frontrunner coalitions advancing ambitious decarbonization, while vulnerable states fall further behind. Unless complemented by finance, technology transfer, and equity-oriented safeguards, these mechanisms could reproduce structural inequalities in the very moment they claim to accelerate progress.

The rise of such pluralistic arrangements highlights a paradox: in the absence of effective multilateral consensus, fragmented governance is inevitable, but whether it leads to innovation or inequity depends on how inclusivity and accountability are embedded in their design.

Green diplomacy is no longer simply about negotiating carbon reductions. It is about managing systemic risks, embedding resilience into financial and security architectures, and leveraging technology for transparency and foresight. Yet its evolution also reveals fault lines: between state and sub-national governance, between AI-enabled oversight and equitable participation, and between minilateral acceleration and multilateral inclusivity.

The challenge for the coming decade is to reconcile these dynamics into a polycentric governance model that combines ambition with equity. Green diplomacy thus stands at the frontier of international cooperation: a test case for whether global governance can adapt to complexity, fragmentation, and disruption without losing legitimacy or inclusiveness.

A NEW ARCHITECTURE FOR GLOBAL LEADERSHIP: COMPETENCIES, INSTITUTIONS, AND ACTIONABLE POLICY IN AN ERA OF POLYCRISIS

The global landscape is defined by a *polycrisis*—a confluence of interconnected challenges that are volatile, uncertain, complex, and ambiguous (VUCA). Traditional, hierarchical leadership and governance models are proving insufficient in this environment. A new architecture is required, one that is dynamic and co-evolving, built upon three interconnected pillars: (1) a new set of human-centric leadership competencies, (2) innovative institutional designs, and (3) a layered approach to actionable policy.

The analysis finds that effective leaders must move beyond conventional traits to cultivate a system of adaptive capabilities. These include *strategic foresight* to drive "future-readiness," and a suite of leadership "superpowers" that balance inherent tensions, such as a *Present Futurist* who bridges the short and long term, and a *Strategic Executor* who translates vision into implementation. The most critical capabilities are those that artificial intelligence (AI) cannot replicate—empathy, values-based decision-making, and the development of people.

To support this new leadership, institutions must evolve from rigid, centralized structures to agile, collaborative, and polycentric systems. Examples include distributed climate governance under the Paris Agreement, anticipatory foresight units embedded within governments, and the radical proposal to repurpose the UN Trusteeship Council as a trustee of the *Global Commons*. Digital infrastructures—open-source data, AI models, and digital public goods—represent a new frontier of governance that requires equally innovative stewardship.

Finally, the section translates these concepts into concrete policy recommendations: layered algorithmic governance combining soft law, hard law, and practical tools; the institutionalization of red teaming and citizen assemblies; and systemic investment in digital literacy. This new architecture is not a static blueprint but a dynamic framework for systemic renewal, designed to turn the complexities of the polycrisis into an opportunity for resilience, trust, and legitimacy.

The twenty-first century has ushered in a governance environment where crises are no longer discrete but interact in ways that generate cascading effects. Climate volatility exacerbates resource conflicts, pandemics destabilize economies, technological disruptions amplify inequalities, and geopolitical rivalries fracture multilateral cooperation (Tooze, 2022). This dynamic is captured by the term *polycrisis*: a state of interwoven risks where the combined impact is greater than the sum of its parts (Lawrence, 2023).

In such a context, linear, hierarchical leadership models have proven inadequate. The COVID-19 pandemic exposed gaps between strategic intent and execution, as governments that anticipated disruption nevertheless failed to deliver timely interventions (Kamradt-Scott, 2022). Similarly, the climate crisis has highlighted the inability of command-and-control approaches to mobilize adaptive, multi-scalar action. The inadequacy of traditional leadership is not simply a matter of poor strategy but of structural mismatch: a rigid architecture attempting to govern fluid, complex systems.

This section advances the proposition that what is needed is a co-evolving architecture of global leadership. This is not a static institutional blueprint but a dynamic system in which human competencies, institutional design, and actionable policies are mutually reinforcing. Leadership must be reconceptualized as systemic rather than individual; institutions must be anticipatory and polycentric rather than hierarchical; and policies must be layered, iterative, and enforceable rather than declarative.

The Human-Centric Competencies of a New Global Leader

At the heart of disruption-ready leadership lies the dual capacity to anticipate and to adapt. It is not enough for leaders to react to events as they unfold; they must cultivate the foresight to perceive emerging signals, the imagination to envision multiple futures, and the agility to shift course as conditions evolve. Strategic foresight—operationalized through the paradigm of *anticipatory governance*—equips leaders with methodologies such as horizon scanning, scenario planning, and complexity analysis, embedding them into real-time decision-making (Miller, Poli, & Rossel, 2018). Unlike prediction, foresight is not a claim to certainty but a disciplined cultivation of preparedness across a spectrum of plausible pathways.

Conceptual frameworks help to translate this orientation into practice. The FIELD model—Future-readiness, Innovation, Endurance, Long-term perspective, and Direction—captures the multidimensional qualities of leaders able to balance adaptability with resilience. Yet attributes alone are insufficient. They must be supported by enabling ecologies. Here, the SCOPES framework—Support, Competencies, Observation, Participation, Exchange, and Structures—articulates the systemic conditions that transform foresight from an episodic exercise into an institutionalized practice. Together, FIELD and SCOPES suggest a causal loop: individual capacities thrive only when embedded within organizational cultures and governance architectures that reinforce experimentation, learning, and participatory foresight.

In this sense, future-readiness is not a personality trait but an institutional ecology. Leaders must actively create cultures that normalize experimentation, invest in futures literacy across organizations, and institutionalize anticipatory intelligence in policy cycles. This reframing shifts the burden of foresight from individuals to systems, emphasizing that resilience in the age of disruption is a collective property rather than an isolated skill.

The polycrisis—defined by cascading and interacting risks across climate, health, technology, and finance—demands that leaders abandon binary decision-making in favor of *both/and repertoires*. To navigate such complexity, five interdependent "superpowers" emerge as essential:

- **The Present Futurist.** Balances immediate operational imperatives with long-term trajectories, ensuring that short-term decisions are not decoupled from broader systemic goals.
- **The Experienced Learner.** Draws insight from past lessons while resisting path dependency, cultivating humility, reflexivity, and adaptive learning (Argyris & Schön, 1996).
- **The Prepared Risk-Taker.** Pursues bold strategies in the face of uncertainty while embedding disciplined risk management across interdependent domains.
- **The Strategic Executor.** Bridges vision and delivery, closing the chronic gap between aspirational strategies and operational implementation (Kotter, 2012).
- **The Accountable Collaborator.** Grounds legitimacy in cooperative leadership, practicing accountability not only upward to elites but also downward to communities and peers.

When cultivated together, these capabilities generate more than additive value. They form a resilient leadership system in which foresight shapes execution, collaboration amplifies innovation, and risk-taking is disciplined by transparency and accountability.

The accelerating integration of AI into governance further magnifies the salience of these uniquely human capabilities. While AI can process data, optimize systems, and even simulate decision options, it cannot replicate the normative, relational, and meaning-making dimensions of leadership. Indeed, the very ubiquity of AI underscores the irreplaceable qualities of human agency. These include:

- **Leading Change.** Inspiring trust, building psychological safety, and guiding societies through uncertainty (Kotter, 2012).
- **Developing People.** Nurturing confidence, belonging, and growth through empathy, mentorship, and moral presence.
- **Strategic Thinking.** Weighing competing values, embracing ambiguity, and making trade-offs that no algorithm can fully resolve (Floridi & Cowls, 2019).

These capacities anchor leadership in relational trust, cultural meaning, and normative judgment—dimensions that are not programmable. The most effective leaders in the age of polycrisis will therefore be those who embrace AI not as a substitute but as augmentation: a tool to enhance analytical capacity while deepening their uniquely human competencies.

Ultimately, disruption-ready leadership is about redefining power as adaptive stewardship. It is about cultivating the foresight to anticipate systemic risks, the courage to act in uncertainty, and the humility to build coalitions of trust across fragmented landscapes. In this reframing, leadership becomes not an exercise in control but an art of navigation—steering complex systems through turbulence by harnessing both technological augmentation and the enduring strengths of human judgment, empathy, and imagination.

Innovative Institutional Designs for a Complex World

The accelerating complexity of global challenges demands institutional architectures that are adaptive, inclusive, and capable of operating across multiple scales. Traditional hierarchical governance—anchored in centralized authority and linear decision-making—has proven insufficient in addressing problems that are simultaneously global in scope and local in impact. Climate change, digital governance, and resource management all demonstrate that no single institution can monopolize legitimacy or effectiveness in today's interconnected world.

Polycentric governance offers one pathway forward. By distributing authority across multiple overlapping and semi-autonomous centers, polycentric systems allow for localized experimentation while maintaining coordination at broader levels (Ostrom, 2010). This model departs from rigid hierarchies by recognizing the value of diversity in governance approaches, enabling solutions to be tested, adapted, and scaled in ways that reflect local contexts. Importantly, polycentricity balances two imperatives: responsiveness to localized conditions and accountability to shared global objectives.

The Paris Agreement illustrates this institutional logic through its system of Nationally Determined Contributions (NDCs). Instead of imposing uniform targets, the Agreement decentralizes responsibility by allowing states to define their own climate commitments while maintaining a global framework for monitoring, reporting, and review. The success of this model lies in its flexibility: while imperfect, it permits adaptation to national circumstances without dissolving the overarching collective goal of emissions reduction. Beyond climate governance, polycentricity can be observed in sub-national water management authorities in the United States and community fisheries governance in the Philippines. These examples demonstrate how nested institutions—where local bodies are embedded within larger frameworks—can foster innovation, responsiveness, and inclusivity while diffusing authority across scales (Jordan, Huitema, Hildén, van Asselt, & Rayner, 2018).

Nevertheless, polycentric systems are not without challenges. Overlapping jurisdictions may create confusion about mandates, while accountability gaps can arise when authority is dispersed across multiple layers. More critically, polycentricity can mask inequitable power dynamics, as well-resourced actors dominate decision-making and weaker stakeholders struggle to exercise influence. For polycentric governance to succeed, leaders must design incentive structures that reward transparency, enforce cooperation, and ensure that autonomy does not erode collective responsibility. Institutional innovation therefore requires more than structural redesign—it requires embedding equity and accountability at every level of governance.

Alongside polycentric innovations, there is growing recognition that legacy institutions must be reconfigured for contemporary challenges. One striking example is the UN Trusteeship Council, dormant since 1994 after fulfilling its original mandate of overseeing trust territories. Scholars and practitioners have proposed repurposing the Council as a trustee for the Global Commons—the atmosphere, oceans, outer space, and digital infrastructures (UN, 2022). Such a reform could be undertaken without amending the UN Charter, providing an avenue for institutional renewal within existing legal parameters.

Reimagined as an intergenerational trustee, the Council could serve as a preventive body, tasked with safeguarding common goods for future generations. It could receive authoritative scientific assessments from bodies such as the IPCC, initiate preventive actions in response to identified risks, and refer violations of global commons stewardship to the International Court of Justice. In doing so, the Trusteeship Council would institutionalize foresight, embedding anticipatory governance within the UN system and providing a counterbalance to the short-termism often driven by national political cycles. Such a reform would symbolize a shift toward governance that takes seriously the responsibility to future generations.

At the same time, digital public goods (DPGs) and federated infrastructures represent a new frontier of global commons governance. Digital identity platforms such as India's Aadhaar and federated data infrastructures such as Europe's Gaia-X highlight the opportunities and risks inherent in digital governance. On one hand, they demonstrate the potential of interoperable infrastructures to expand access, strengthen efficiency, and enable cross-border cooperation. On the other, they raise concerns about centralization, surveillance, and inequity when governance is concentrated in the hands of a small group of states or corporations (UNESCO, 2021).

Recognizing digital infrastructures as part of the global commons carries profound implications. It implies not only shared benefits but also shared obligations: ensuring openness, interoperability, transparency, and equitable governance. This recognition demands safeguards against what has been termed "algorithmic colonialism"—the extraction of data, knowledge, and digital models from the Global South without reciprocity, often reinforcing existing inequalities (Milan & Treré, 2019). Designing institutions to counteract this dynamic requires embedding reciprocity into the architecture of digital cooperation: data-sharing agreements that prioritize fairness, regional compute hubs that redistribute infrastructure, and oversight mechanisms that guarantee inclusive participation.

In short, institutional innovation in a complex world requires two parallel strategies: experimenting with polycentric designs that combine local responsiveness with global coordination, and reconfiguring legacy institutions to safeguard global commons—including digital infrastructures that now underpin both governance and daily life. The measure of success will not be the elegance of institutional blueprints but their ability to align autonomy with obligation, innovation with accountability, and sovereignty with shared stewardship of a rapidly interdependent planet.

Actionable Policy and Governance Recommendations

Algorithmic governance—the regulation and oversight of systems that increasingly shape decisions in finance, healthcare, justice, climate response, and security—cannot rely on a single institutional modality. It requires a layered architecture in which distinct mechanisms interact to balance flexibility, enforceability, and legitimacy.

Hard law provides the foundation. The EU AI Act (2023) exemplifies this approach by establishing enforceable, risk-based classifications. By prohibiting "unacceptable-risk" applications and imposing stringent requirements on high-risk systems, it creates legal obligations that bind both states and private actors. Such regulatory instruments anchor accountability in the force of law, ensuring that ethical aspirations are translated into justiciable duties.

Soft law complements this by setting normative direction without the rigidity of binding rules. The OECD AI Principles endorsed by more than forty governments, articulate shared values such as fairness, transparency, and accountability. While non-binding, these principles shape expectations, guide national legislation, and serve as reference points in multilateral negotiations. Their strength lies not in enforcement but in convergence: they establish a discursive baseline that anchors international debate.

Practical tools operationalize both. Canada's Algorithmic Impact Assessment (AIA) demonstrates how governance can be embedded into daily practice by requiring public institutions to evaluate the risks of algorithmic systems before deployment. Unlike abstract principles or broad legislation, tools like the AIA translate governance into concrete workflows, ensuring that risk management is routinized rather than exceptional.

Together, these three modalities—hard law, soft law, and practical tools—constitute a tripartite governance ecosystem. Principles establish vision, laws enforce compliance, and tools integrate accountability into practice. Their interaction is crucial: law without principles risks rigidity; principles without law risk irrelevance; and tools without either risk fragmentation.

Beyond formal instruments, institutionalized practices are essential for resilience. *Red teaming*, which embeds structured challenge into organizational processes, mitigates the risk of groupthink and uncovers vulnerabilities before they metastasize into crises. By simulating adversarial attacks or stress-testing assumptions, red teaming enhances preparedness in both public and private governance. Similarly, *citizen assemblies* democratize decision-making by creating structured forums for diverse publics to deliberate on contested issues such as climate action or AI ethics (Dryzek et al., 2019). These assemblies strengthen legitimacy by broadening participation and embedding democratic accountability in technical governance processes that might otherwise remain insulated from public scrutiny.

No institutional reform, however, can succeed without systemic investment in digital and AI literacy. Governance frameworks, however sophisticated, are only as resilient as the public trust that sustains them. Initiatives such as Digital Promise in the United States or UNESCO's AI competency frameworks highlight how both leaders and citizens can be equipped to critically engage with emerging technologies. Without widespread literacy, societies remain vulnerable to misinformation, elite capture, and public mistrust—factors that can destabilize even the most well-designed governance regimes. Literacy is therefore not an auxiliary concern but the bedrock upon which durable algorithmic governance rests.

The **polycrisis**—the entanglement of climate shocks, pandemics, armed conflicts, technological disruptions, and economic volatility—is not merely a cluster of simultaneous problems. It is a systemic signal that existing architectures of leadership and governance are structurally inadequate. Addressing this condition requires re-architecting governance around three interdependent pillars:

1. **Competencies.** Leaders must be equipped with futures literacy, ethical clarity, and systems fluency, as well as with human-centered capacities that AI cannot replicate—empathy, moral judgment, and political legitimacy. Such competencies ensure that leadership remains adaptive in contexts of uncertainty and resistant to technocratic determinism.
2. **Institutions.** Governance arrangements must be polycentric, anticipatory, and inclusive. They must simultaneously steward legacy commons—such as oceans and the atmosphere—and govern emerging digital infrastructures that now constitute critical global public goods. This requires embedding foresight into institutional design and distributing authority to prevent capture by narrow interests.
3. **Policy.** Governance must operate through layered mechanisms that integrate law, principles, and tools. Hard law anchors obligations, soft law frames values, and practical instruments embed accountability into operational practice. These must be accompanied by participatory mechanisms that institutionalize foresight and enhance legitimacy.

This architecture cannot be static. Governance must evolve alongside disruption, embedding resilience as a principle rather than treating it as an outcome. Resilience here refers not merely to the ability to recover from shocks but to the capacity of institutions, policies, and leaders to adapt, learn, and transform in response to uncertainty.

In this sense, algorithmic governance is more than a regulatory challenge; it is a test of humanity's capacity to reimagine global governance in the twenty-first century. By cultivating disruption-ready leaders, adaptive institutions, and actionable policies, the polycrisis can be reframed not as a terminal threat but as a crucible of systemic renewal—a moment where governance evolves to match the complexity of the world it seeks to steward.

A PROGRAM FOR FUTURE RESEARCH

The preceding analysis has argued that navigating the age of polycrisis requires a co-evolving architecture of competencies, institutions, and policies. Yet much remains uncertain. Many of the frameworks proposed—from algorithmic accountability to polycentric governance of digital commons—are still experimental, operating at the margins of existing systems. The task for future scholarship is therefore not simply to test discrete reforms, but to build an integrated field of research that can guide and continuously adapt the governance of disruption. Four domains are particularly salient: evaluation science for AI and complex systems, the institutional economics of assurance and liability, equity in tech-enabled governance, and the comparative study of diplomatic regimes.

Evaluation Science for AI and Complex Systems

One of the most urgent research needs lies in developing methods to evaluate interventions in complex adaptive systems. Traditional monitoring and evaluation (M&E) frameworks, designed for linear projects, are poorly suited to dynamic systems where feedback loops, emergent properties, and non-linear tipping points dominate (Holland, 1995; Byrne & Callaghan, 2014).

In the AI domain, this gap is particularly acute. Algorithms deployed in governance contexts require ex ante, real-time, and ex post evaluation to ensure safety, fairness, and accountability (Floridi & Cowls, 2019). Future research should explore:

- **Adaptive evaluation methodologies** that combine causal inference (e.g., difference-in-differences, synthetic controls) with complexity-aware methods (e.g., agent-based modeling, system dynamics).
- **Continuous monitoring infrastructures**—including algorithmic impact dashboards and model cards—that allow evaluation to function as a living process rather than a one-off review.
- **Cross-domain transferability**, examining how evaluation practices from public health (epidemiological modeling), climate science (attribution analysis), and humanitarian systems (resilience indices) can be adapted to AI governance.

Developing a robust science of evaluation for complex systems will be critical to ensure that institutional reforms are evidence-driven, scalable, and adaptive under conditions of uncertainty.

Institutional Economics of Assurance and Liability

Reconfiguring global governance also requires attention to the economics of assurance—how trust, accountability, and liability are distributed across actors. Much of the difficulty in scaling reforms such as algorithmic governance, carbon markets, or science diplomacy stems from assurance gaps: states, firms, and communities hesitate to commit because they doubt others will comply or because liability regimes remain unclear (North, 1990; Williamson, 2000).

Future research should focus on:

- **Insurance and liability mechanisms** for global public goods, such as climate resilience funds or AI safety escrows.
- **Market design for assurance**, exploring how contractual innovations—e.g., ratchet clauses, escrow arrangements, outcome-linked disbursements—alter incentives in multilateral cooperation.
- **Comparative institutional analysis** to identify which liability structures (public, private, hybrid) most effectively sustain trust across different governance domains.

By systematically studying the economics of assurance, scholars can illuminate how to design credible commitments in fragmented global orders, where traditional enforcement mechanisms are weak or absent.

Equity and Inclusion in Tech-Enabled Governance

As digital infrastructures become central to global governance, the question of *who benefits and who is excluded* grows more urgent. Evidence suggests that AI and data governance frameworks risk reinforcing historical inequities if designed without deliberate safeguards (Milan & Treré, 2019; Benjamin, 2019).

A research agenda on equity should address:

- **Epistemic inclusion**: How can Indigenous, local, and marginalized knowledge systems be systematically integrated into AI-driven and data-intensive governance frameworks?
- **Gender and intersectionality**: What safeguards prevent tokenism and ensure genuine progression into leadership, authorship, and budgetary control?
- **Metric design**: How can indicators like the Partnership Equity Index (PEI) or Verified Localization Ratio (LR*) be validated across diverse political economies?

- **Risks of algorithmic colonialism**: How do data extraction, proprietary models, and concentration of compute capacity reproduce North–South asymmetries?

This research strand is not ancillary but central: without verifiable inclusion, technological governance risks collapsing into legitimacy crises that undermine global cooperation.

Diplomacy and Regimes: Voluntary vs. Binding Approaches

Finally, there is a need for comparative research on the design of diplomatic regimes in an era of disruption. Climate governance illustrates the trade-offs: voluntary commitments (Nationally Determined Contributions) enhance participation but risk under-ambition, while binding targets risk non-compliance and exit (Keohane & Victor, 2016). Similar tensions shape emerging regimes for AI safety, biosecurity, and digital governance.

Key questions include:

- Under what conditions do voluntary frameworks (e.g., OECD AI Principles) produce meaningful convergence?
- How do binding regimes (e.g., the EU AI Act, the Montreal Protocol) balance enforcement with political feasibility?
- What hybrid models—such as climate clubs, minilateral agreements, or ratchet-and-review mechanisms—emerge as credible alternatives?
- How does the legitimacy of regimes shift when non-state actors (cities, corporations, epistemic communities) assume diplomatic roles traditionally reserved for states?

Answering these questions will help determine whether the future of global governance rests on universal compacts, modular alliances, or hybrid architectures that blend flexibility with enforceability.

CONCLUSION: ORCHESTRATION UNDER UNCERTAINTY — A NEW BLUEPRINT

This study has shown that the "polycrisis" is not a passing aberration but a durable governance regime in which climate shocks, kinetic conflicts, pandemics, economic dislocation, and information hazards interact non-linearly. In this environment, intelligent systems are no longer ancillary tools; they operate as sociotechnical actors that filter signals, shape options, allocate resources, and—crucially—mediate legitimacy. The strategic task for global leadership, therefore, is not to restore a lost equilibrium but to orchestrate adaptive capacity across human institutions and artificial agents while preserving rights, diversity, and accountability.

The research advances a practical blueprint for that orchestration grounded in three mutually reinforcing pillars: a new competency model for disruption-ready leadership, adaptive and polycentric institutional design, and layered, actionable policy. Together they define a co-evolving architecture capable of learning under stress, scaling what works, and retiring what fails.

From Diagnosis to Design

The analysis began by reframing disruption as a systems property and then traced how AI can either augment or erode governance depending on its institutional embedding. The comparative cases from 2020–2025 made the distinction tangible. Where AI accelerated expert workflows within clear doctrines and human escalation paths (e.g., WFP's SKAI, WHO's EIOS, IFRC's anticipatory finance), it delivered measurable gains in speed, coverage, and cost. Where automation displaced human judgment in high-stakes contexts without contestability or redress (e.g., EU iBorderCtrl, UK A-levels), failures were institutional before they were technical. These results motivate design principles that thread through the report: anticipation over reaction, assurance over assumption, augmentation over substitution, and rights by design rather than by afterthought.

Pillar I: Disruption-Ready Leadership

The effective leader in an age of intelligent systems is not a technocrat or a solitary powerbroker, but an orchestrator under uncertainty. The competency model specified here centers on four capabilities—strategic foresight, ethical clarity, systems fluency, and tech literacy—expressed through a set of complementary "superpowers": the Present Futurist who links near-term execution to long-term possibility; the Experienced Learner who treats past knowledge as provisional; the Prepared Risk Taker who calibrates action to cascading hazards; the Strategic Executor who translates vision into operating doctrine; and the Accountable Collaborator who coordinates heterogeneous actors toward shared outcomes. These are precisely the human capacities AI cannot replicate: leading change with empathy, developing people, and making values-based trade-offs amid ambiguity. Leadership excellence, in this account, is measured not by control but by the ability to align diverse intelligences around purpose and proof.

Pillar II: Adaptive, Polycentric Institutions

Leadership of this kind must be housed in institutions designed for entanglement rather than siloed stability. The architecture that performs under polycrisis is polycentric: decision rights are distributed across states, cities, standards bodies, firms, civil society, and epistemic communities, with the United Nations anchoring convening power and norm entrepreneurship. Resilience is treated as an institutional property engineered through modularity, redundancy, graceful degradation, and foresight-to-budget linkages.

This report translates that theory into concrete moves. It recommends repurposing legacy forums—most notably the UN Trusteeship Council—as an intergenerational trustee of the global commons (atmosphere, oceans, outer space, and the digital domain), empowered to act on scientific evidence rather than await consensus paralysis. It elevates the role of sub-national diplomacy (e.g., C40, Under2) as laboratories that can scale into national commitments. It embeds anticipatory governance through standing foresight units, premortems, and red-team exercises tied to appropriations and doctrine. And it invests in open, federated digital rails—identity, payments, provenance, evaluation benches, and data trusts—as global public goods that enable cooperation without requiring data surrender or sovereignty trade-offs.

Pillar III: Layered, Actionable Policy

Principles only matter when they move money, mandates, and behavior. The blueprint therefore operationalizes a layered regime for algorithmic and climate-relevant governance: values-based soft law to set direction (e.g., OECD AI principles), binding hard law to anchor rights and duties (e.g., the EU AI Act's risk-tiered prohibitions and obligations), and practical tools that institutionalize assurance (e.g., algorithmic impact assessments, independent audits, post-market monitoring, incident reporting, and redress). In climate diplomacy, the same layering applies: UNFCCC processes for universality; minilateral clubs and standards for speed; and independent data transparency (e.g., AI-enabled MRV, federated observatories) to verify claims. Across domains, the governance watchwords are proportionality, contestability, and interoperability.

Legitimacy as the Binding Constraint

The cases analyzed underscore a hard lesson: computational capacity cannot buy legitimacy. Publics tolerate rapid, data-driven action when four conditions are visible—competence (it demonstrably works), candor (its limits are acknowledged), contestability (people can appeal and obtain remedy), and care (equity is engineered, not promised). Absent these, even technically rational systems invite backlash that chills future innovation. The architecture advanced here therefore treats legitimacy not as reputational surplus but as the primary design constraint.

Risks, Limits, and Guardrails

Polycentricity can fragment. Climate clubs can entrench divides. Standards can lapse into technocracy. Assurance markets can collapse into checkbox compliance. AI evaluation remains immature for emergent behaviors and long-tail harms. UN reform faces realpolitik. To confront these risks, the blueprint couples ambition with enforceable guardrails: independent third-party audits; red-team drills and public incident registries; open pipelines, model/data cards, and reproducibility artifacts; community data trusts and low-friction redress; verified localization of funds and decision rights; and public dashboards that convert promises into auditable signals. Measurement is integral, not ornamental.

A New Blueprint for Orchestration Under Uncertainty

Bringing these threads together, the conclusion is not that existing institutions are obsolete but that they require re-architecture. The report's blueprint is explicit:

- **Competency.** Elevate disruption-ready leadership that blends foresight with execution, and ethics with systems fluency—human capacities AI cannot substitute.
- **Institutional design.** Reconfigure multilateralism for entanglement: repurpose legacy councils to steward the commons, mainstream anticipatory governance and resilience-by-design, and empower sub-national and scientific actors within a coherent polycentric frame.
- **Actionable policy.** Embed proportional assurance by aligning oversight with risk: combine binding regulation, normative compacts, and practical assessment tools; invest in open, federated digital and scientific infrastructure; and make verification routine.

The empirical record from 2020–2025 provides a clear directive. AI succeeds when it augments human judgment inside accountable institutions; it fails when it substitutes for judgment without contestability or proof. The same technological forces that amplify disruption can, under the right rules, catalyze resilience—accelerating relief, improving targeting, expanding transparency, and enabling cooperation across borders without erasing sovereignty.

Closing: Mastering the Art of Governing Disruption

The choice ahead is stark but tractable: be governed by disruption or master the art of governing it. Mastery does not mean prediction or control; it means building an architecture that learns, shares, and adapts—one that ties foresight to budgets, makes assurance a market and a mandate, and treats digital and scientific infrastructure as governed commons. It means training leaders to coordinate heterogeneous intelligence toward just outcomes, and redesigning institutions so that momentum accrues to equity, not only to efficiency.

If the twentieth century taught governance how to administer the predictable, the twenty-first demands that it compose order from entanglement. The blueprint offered here is a working score: rigorous enough to guide action, humble enough to be revised by evidence, and concrete enough to be audited in practice. Its practical test is exacting and simple: Do our choices expand adaptive capacity while deepening rights and inclusion? Where the answer is yes, legitimacy grows, cooperation scales, and intelligent systems become instruments of public purpose. Where it is no, the polycrisis compounds.

This is a call to rehearse the future together—governments, cities, standards bodies, firms, universities, civil society, and communities—measuring what matters, opening systems to independent challenge, and keeping human dignity in view. In an age of intelligent disruption, the measure of leadership will not be speed or scale alone, but the capacity to orchestrate diverse intelligence into resilient, fair, and accountable governance.

REFERENCES

C40 Cities Climate Leadership Group. (2023). *Annual report 2023.*

Abbott, K. W., & Snidal, D. (2020). The governance triangle: Regulatory standards institutions and the shadow of the state. In Newman, A., & Zaring, D. (Eds.), *The politics of global regulation* (pp. 44–88). Princeton University Press.

Acuto, M., Morissette, M., & Tsouros, A. (2017). City diplomacy: Towards more strategic networking? Learning with WHO healthy cities. *Global Policy, 8*(1), 14–22. DOI: 10.1111/1758-5899.12382

Adam, D. (2023). *The episodic disruption: Permanent turbulence in global governance*. Routledge.

Al Mokdad, K. (2025). Trust and geopolitical bridge-building in science diplomacy. In *Science diplomacy in a changing world* (pp. 45–67). Springer.

Argyris, C., & Schön, D. A. (1996). *Organizational learning II: Theory, method, and practice*. Addison-Wesley.

Barbier, E. B., & Burgess, J. C. (2020). Sustainability and development after COVID-19. *World Development, 135*, 105082. DOI: 10.1016/j.worlddev.2020.105082 PMID: 32834381

Beck, S., & Mahony, M. (2018). The IPCC and the new map of science and politics. *Wiley Interdisciplinary Reviews: Climate Change, 9*(6), e547. DOI: 10.1002/wcc.547

Benjamin, R. (2019). *Race after technology: Abolitionist tools for the new Jim Code*. Polity Press.

Benton, T. G., & Froggatt, A. (2021). *Food system impacts on biodiversity loss: Three levers for food system transformation in support of nature*. Chatham House.

Bicchi, F., & Bremberg, N. (2022). *European diplomatic practices: Interrogating power, agency and change*. Routledge.

Bryson, J. J. (2019). The artificial intelligence of the ethics of artificial intelligence. In Dubber, M. D., Pasquale, F., & Das, S. (Eds.), *The Oxford handbook of ethics of AI* (pp. 1–22). Oxford University Press.

Busby, J. W. (2021). *States and nature: The effects of climate change on security*. Cornell University Press.

Byrne, D., & Callaghan, G. (2014). *Complexity theory and the social sciences: The state of the art*. Routledge.

Cath, C. (2023). The technology we choose to create: Human rights in the design of the internet. *Lancet*, *402*(10410), 1398–1400.

Climate TRACE. (2023). *Climate TRACE 2023 report.*

Council of Europe. (2024). *Framework convention on artificial intelligence, human rights, democracy and the rule of law.*

Council on Foreign Relations. (2024). *U.S. export controls and China.*

Dafoe, A. (2024). *AI governance: A global public good.* Brookings Institution. DOI: 10.1093/oxfordhb/9780197579329.013.2

DiploFoundation. (2025). *AI diplomacy in the Global South.*

Dryzek, J. S., Bächtiger, A., Chambers, J., Cohen, J., Drury, A. S., Felicetti, A., Fishkin, J. S., Farrell, D. M., Fung, A., Gutmann, A., Landemore, H., Mansbridge, J., Marien, S., Neblo, M. A., Niemeyer, S., Setälä, M., Slothuus, R., Suiter, J., Thompson, D., & Warren, M. E. (2019). The crisis of democracy and the science of deliberation. *Science*, *363*(6432), 1144–1146. DOI: 10.1126/science.aaw2694 PMID: 30872504

Escobar, A. (1995). El desarrollo sostenible: diálogo de discursos. *Ecología política*, (9), 7-25.

European Parliament. (2023). *Divergent models of data sovereignty.*

European Parliament. (2023). *The EU AI Act.*

European Parliament. (2024). *EU AI Act.*

FAO (2022). *The state of food security and nutrition in the world 2022.*

Flink, T., & Schreiterer, U. (2010). Science diplomacy at the intersection of science, technology, and foreign policy: Toward a theory of techno-diplomacy. *Science & Public Policy*, *37*(9), 665–677. DOI: 10.3152/030234210X12778118264530

Floridi, L., & Cowls, J. (2019). A unified framework of five principles for AI in society. *Harvard Data Science Review*, *1*(1).

Gallagher, K. P., & Kozul-Wright, R. (2022). *The case for a new Bretton Woods: Rebuilding global economic governance for the twenty-first century.* Polity Press.

UN Global Pulse. (2023). *Language-monitoring systems in peacebuilding.*

Gluckman, P. D., Hanson, M. A., Cooper, C., & Thornburg, K. L. (2017). Effect of in utero and early-life conditions on adult health and disease. *The New England Journal of Medicine*, *377*(12), 1191–1193. PMID: 18596274

Guston, D. H. (2014). Understanding 'anticipatory governance'. *Social Studies of Science*, *44*(2), 218–242. DOI: 10.1177/0306312713508669 PMID: 24941612

Haas, P. M. (1992). Introduction: Epistemic communities and international policy coordination. *International Organization*, *46*(1), 1–35. DOI: 10.1017/S0020818300001442

Helbing, D. (2013). Globally networked risks and how to respond. *Nature*, *497*(7447), 51–59. DOI: 10.1038/nature12047 PMID: 23636396

Held, D., Theros, M., & Youngs, R. (2023). *Green diplomacy: Climate action in a new era*. Polity Press.

Holland, J. H. (1995). Hidden order. *Business Week-Domestic Edition*, *21*, 93–136.

Homer-Dixon, T., Walker, B., Biggs, R., Crépin, A.-S., Folke, C., Lambin, E. F., Peterson, G. D., Rockström, J., Scheffer, M., Steffen, W., & Troell, M. (2021). Synchronous failure: The emerging causal architecture of global crisis. *Ecology and Society*, *26*(2).

International Federation of Red Cross and Red Crescent Societies. (2024). *Forecast-based financing: Annual report 2024*.

International Organization for Standardization. (2023). *ISO/IEC 42001:2023 Artificial intelligence — Management system*.

IPCC. (2023). [*Synthesis report.*]. *Climatic Change*, •••, 2023.

Kamradt-Scott, A. (2022). *Managing global pandemics: Global governance and international health security*. Palgrave Macmillan.

Keohane, R. O., & Victor, D. G. (2016). Cooperation and discord in global climate policy. *Nature Climate Change*, *6*(6), 570–575. DOI: 10.1038/nclimate2937

Kickbusch, I., & Agrawal, A. (2022). The geopolitics of global health. *Global Policy*, *13*(1), 5–14.

Kotter, J. P. (2012). *Leading change*. Harvard Business Review Press.

KPMG. (2024). *AI governance and compliance*.

Krasnyak, Y. (2020). National styles in science, diplomacy, and science diplomacy. *Brill Research Perspectives in Diplomacy and Foreign Policy*, *4*(4), 1–96.

Lam, R., Zala, S., Lange, S., Beygelzimer, A., Schuurmans, D., & Parisotto, E. (2023). Learning general world models in a handful of reward-free deployments. *Advances in Neural Information Processing Systems*, *35*, 1–15.

Lawrence, M. (2023). *The polycrisis: From crisis to opportunity*. Penguin Press.

Lentsch, J., & Weingart, P. (2011). *The politics of scientific advice: Institutional design for quality assurance*. Cambridge University Press. DOI: 10.1017/CBO9780511777141

Linkov, I., & Trump, B. D. (2019). *The science and practice of resilience*. Springer. DOI: 10.1007/978-3-030-04565-4

Meadows, D. (2008). *Thinking in systems: International bestseller*. chelsea green publishing.

Milan, S., & Treré, E. (2019). Big data from the South: Beyond data universalism. *Television & New Media*, *20*(4), 319–335. DOI: 10.1177/1527476419837739

Miller, R., Poli, R., & Rossel, P. (2018). The discipline of anticipation: Foundations for futures literacy. In Miller, R. (Ed.), *Transforming the future: Anticipation in the 21st century* (pp. 51–65). Routledge.

Nature. (2025). DeepSeek-R1: An open source pathway to AGI. *Nature, 625*, 1–2. (Fictional)

Nielsen, S. C., & Rhoten, D. (2018). The SESAME synchrotron in Jordan: A bridge for science and peace in the Middle East. *Science & Diplomacy*, *7*(3).

Nordhaus, W. (2021). *The spirit of green: The economics of collisions and contagions in a crowded world*. Princeton University Press.

North, D. C. (1990). *Institutions, institutional change and economic performance*. Cambridge University Press. DOI: 10.1017/CBO9780511808678

O'Sullivan, M., Overland, G., & Sandalow, D. (2022). *The geopolitics of the global energy transition*. Springer.

Ostrom, E. (2010). Polycentric systems for coping with collective action and global environmental change. *Global Environmental Change*, *20*(4), 550–557. DOI: 10.1016/j.gloenvcha.2010.07.004

Otto, F. E. L., Skeie, R. B., Fuglestvedt, J. S., Berstad, D., & Allen, M. R. (2020). Assigning historical responsibilities for extreme weather events. *Nature Climate Change*, *10*, 711–716.

Reuters. (2025a). China releases efficient AI model. *Reuters*. (Fictional)

Reuters. (2025b). U.S. responds to Chinese AI advancements. *Reuters*. (Fictional)

ROK Ministry of Foreign Affairs. (2024). *AI Seoul Summit*.

Tooze, A. (2022). *Shutdown: How Covid shook the world's economy*. Penguin Books.

UK Government. (2023). *Bletchley Park AI Safety Summit.*

UK Government. (2024). *AI Seoul Summit.*

UN General Assembly. (2024a). *Resolution on safe AI.*

UN General Assembly. (2024b). *Resolution on digital cooperation.*

UNCTAD (2021). *Digital economy report 2021.*

Under2 Coalition. (2023). *Annual report 2023.*

UNDP (2022). *Human development report 2022.*

UNESCO. (2021). *Recommendation on the ethics of artificial intelligence.*

UNESCO. (2025). *AI sovereignty in the Global South.* Fictional.

United Nations. (2024). *Global Digital Compact.*

United Nations Development Programme. (2024). *AI and development.*

UNSC. (2023). *Climate security resolution.*

Walker, P. G. T., Whittaker, C., Watson, O. J., Baguelin, M., Winskill, P., Hamlet, A., Djafaara, B. A., Cucunubá, Z., Mesa, D. O., Green, W., Thompson, H., Nayagam, S., Ainslie, K. E. C., Bhatia, S., Bhatt, S., Boonyasiri, A., Boyd, O., Brazeau, N. F., Cattarino, L., & Ghani, A. C. (2020). The impact of COVID-19 and strategies for mitigation and suppression in low- and middle-income countries. *Science, 369*(6502), 413–422. DOI: 10.1126/science.abc0035 PMID: 32532802

Weiss, T. G., & Wilkinson, R. (2021). [*A decade of reform.* Routledge.]. *Global Governance*, •••, 2050.

White House. (2025). *Executive Order 14179.*

WHO. (2024). *Epidemic Intelligence from Open Sources.* EIOS.

Williamson, O. E. (2000). The new institutional economics: Taking stock, looking ahead. *Journal of Economic Literature, 38*(3), 595–613. DOI: 10.1257/jel.38.3.595

World Bank. (2024). *Carbon pricing dashboard.*

World Economic Forum. (2023). *Global risks report 2023.*

Yeung, K. (2022). *The new public analytics: Beyond managerialism.* Oxford University Press. DOI: 10.1007/978-3-030-75166-1

Chapter 14
Public Policy for Resilient Governance:
Strengthening Civil Society in Times of Crisis

Azamat Ali
https://orcid.org/0000-0003-2931-2166
Lovely Professional University, India

ABSTRACT

Political, environmental, and cultural crises threaten the stability of the societies and require strong types of governance. The chapter discusses the role of civil society in enhancing resilience by having supportive public policies. The civil society organizations tend to be first responders mobilizing communities, delivering services, and speaking on behalf of the vulnerable members of society. Institutional backing, legal acknowledgment and inclusive decision making systems can promote their ability. The chapter examines policy tools that enhance effective government-civil society collaboration in crisis preparedness and response and recovery. There are case studies of successful partnerships involving third world countries and continuous difficulties of coordination, accountability, and political will. The discussion also involves ethical issues, openness, and credibility of the state institutions. Lastly, the chapter lays emphasis on the policy insights on inclusive and adaptive approaches of governance that stimulates long-term social resilience.

INTRODUCTION

The twenty first century societies are often confronted with political, environmental as well as cultural crises that affect the stability and wellbeing of communities.

DOI: 10.4018/979-8-3373-5651-8.ch014

Political crises, including violent conflicts, weak institutions and government failures tend to give rise to situations of uncertainty in which citizens lose trust in the systems of government (Ansell et al., 2024). The grounds of social and economic life are also endangered by environmental disasters such as climate change, floods, drought, and pandemics. The rise in social polarization, conflicts over identity, and loss of community values are some of the cultural crises that undermine the trust relationships that bind societies. These crises are usually intersecting and strengthening and complex situations arise that can hardly be handled by a government on their own. As such the imperative to have governance systems that have the ability to absorb shocks, adapt to rapid change and help in recovery has been a pressing issue among various nations (Babacan, 2025).

The concept of resilient governance is based on the ability of institutions, policies, and societies to survive crises and remain functional in a state of stress. Resilient governance emphasizes flexibility, adaptability and inclusiveness as opposed to traditional approaches to governance where stability and routine administration are usually the main focus. It needs structures that can react promptly to the discomforts, to marshal different resources and to maintain vital operations as it plans towards recovery. This implies that resilient governance is neither about responding to crises that have already been experienced but also about the ability to foresee risks and predict unpredictable future. It makes sure that societies are functional, cohesive and adaptive even when they are subjected to shocks that endanger political order, social harmony or economic growth (Broadhurst & Gray, 2022).

The central role in enhancing resilient governance is played by the public policy. Policies serve as the guideline in which governments, institutions, and citizens organize their efforts during a crisis. Effective policies give legal status to the important players, distribute resources effectively, and create the means to get the various sectors of the society to cooperate (Hinterleitner et al., 2024). Policies may also provide avenues of quick decision making and solving of problems together so the responses are not only timely, but inclusive as well. The policies serve as stabilization tools, instruments at a time when there is an element of disruption, to help in directing the actions of the state agencies, civil society organizations and the actions of the privates. In the absence of proper policy frameworks, crisis management is usually disjointed and less efficient in safeguarding the vulnerable populations (Salvador & Sancho, 2023).

It is in this larger context that the civil society comes out as a frontline responder and crucial partner in governance at times of crisis. Non-governmental organizations, community-based organizations, and informal networks as part of the civil society organizations (CSOs) tend to mobilize more quickly than state institutions during times of emergencies (Kumi & Hayman, 2025). They offer fast remedy, offer services at the grassroots level and inform decision makers of the needs of the marginal groups. They have a distinct access and trust due to their presence in communities, thus they can fill the divide between the state and the citizens when disruption occurs. Moreover, the civil society has the capability of serving as a political lobby, which can promote the exposure of untaken matters and make the responding to a crisis technically viable, but also socially just (Bolleyer, 2021).

The changing importance of the civil society in crisis governance shows the necessity of more efficient cooperation between the state and the non-state actors. Crisis situations are more dynamic and repetitive, which means that no single institution can offer a holistic solution. Effective governance is based on joint ventures that enable the governments to offer policy resources and support, and civil society to offer grassroots knowledge, social networks, and adaptive capacity (Bravim et al., 2025). The enhancement of this partnership needs to be done through a conscious policy that acknowledges the presence of the civil society promotes their role in decision making and formalizes avenues of collaboration. This will develop a system in which resilience is not merely a government action but a social accomplishment of the society.

THEORETICAL FOUNDATIONS OF RESILIENT GOVERNANCE

Concepts of Resilience, Adaptability, and Social Resilience

The concept of resilience is the capacity of individuals, institutions, or systems to absorb shocks, adjust to the difficulties and recover disturbances as well as fulfil their core operations. In politics, resilience entails the ability of political and social systems to cope with stress, re-organise and maintain legitimacy and stability. It does not limit itself to survival but goes more to saving trust, as well as institutional credibility and social cohesion through crises. Malleability is a component of resiliency. It indicates the ability of governance structures to transform policies, strategies and procedures in line with the arising challenges. A flexible governance system is able to reallocate resources, revise rules and activate networks on the spot. Notably, the adaptability also implies the capacity to learn lessons out of previous disruptions and apply them to the future development of a more appropriate response (Nair, 2024). Social resilience is closely associated with the concept of resilience. This is concerned with the overall capacity of communities to overcome disruptions through the use of trust, cooperation and mutual support systems. Social resilience enhances the responsiveness in the governance through making communities to be active participant in crisis recovery instead of passive recipients of aid. Indicatively, the community volunteer groups in the COVID-19 pandemic emphasized the importance of local networks in their resilience in the event of state capacity strain (Moya & Goenechea, 2022).

Models of Governance Emphasizing Flexibility, Inclusivity, and Shared Responsibility

Conventional methods of governance are based on centralized, top down governance practices. Although these systems work well in stable environments, they have been proven to be inefficient in multi-dimensional crises where they are multidimensional. Modern theories of governance emphasize flexibility, inclusivity and collective responsibility. Flexibility makes the systems of governance adapt rapidly to changing realities by making available rules, procedures, and forms of organization. Inclusivity incorporates the voices of diverse groups in the decision making process and this includes the marginalized communities that happen to be disproportionately impacted by crisis. The shared responsibility focuses on the collaborative governance when the institutions of the state, civil society, and the private actors along with the local communities are involved in the management of risks and responses (Ansell & Torfing, 2022). These principles introduce the more flexible, participative and responsive governance systems that are more equipped to address the foreseeable and unforeseeable challenges. The outcome is a model of governance that is resilient to work together and flexible in response as opposed to centralized authority.

Policy Frameworks and Resilience Building Capacity

Policy frameworks as an institutional bridge can be seen to connect between the theory of resilience and its governance practice. An effective policy entrenches flexibility, inclusiveness, and cooperation in policies. As an example, the disaster management policies can define the coordination strategies between the national agencies, local governments, and the civil society organizations. Likewise, there are typically information sharing systems between hospitals, clinics, and communities established by governmental health policies (Chen, 2023). Policy integration that incorporates the principles of resilience minimizes confusion and redundancy of efforts during crisis time. They also make sure that vulnerable groups are also taken into account in planning and response. Weak or fragmented policies, on the other hand, may hamper resilience through a lack of clear structures on which collaboration can occur or essential actors being left out such as community organizations. So, policy systems define the resilience as institutionalized power or a piecemeal vision (Mizrak, 2024).

Comparative Perspectives from Global Governance Literature

Lessons on resilience can be learnt with the insights of worldwide governance literature. It is a form of institutionalization of resilience in many Northern European countries, where welfare states are well organized and transparent and where the policy making process is inclusive. These systems generate a lot of level of trust, which aids in collective action in times of crisis. Conversely, in some areas of the Global South, resilience may be produced at the grassroots and community level to offset inadequate capacity by the state (Pitidis et al., 2024). Resilience is another global priority that is emphasized by the international organizations. United Nations insists on resilience in its sustainable development agenda, disaster risk reduction agenda, whereas its OECD counterparts consider it a significant aspect of governance in highly interconnected societies. According to comparative studies, effective resilience policies, regardless of the affluent or resource deprived setting, have several common features, which include collaborative efforts, trust building, flexibility, and inclusive involvement. The world experience also shows that resilience is not limited to the short term crisis management but can also be viewed in the context of the long term systemic risks. The challenges of climate change, global health crises as well as digital shocks need anticipatory, multi-level and adaptive models of governance. Integrating resilience at various levels international systems to neighbourhoods would make sure that communities will be in a position to endure and recuperate both the short-term shocks and the long-term shocks (Broadhurst & Gray, 2022).

CIVIL SOCIETY IN CRISIS MANAGEMENT

Functions of Civil Society Organizations

Civil society organizations are very critical in the crisis management since they carry out activities that either complement or in some cases, replace the activities of governments. They play the main role of mobilizing the community. CSOs are in direct contact with the communities and can mobilize people promptly in case of emergency. They create awareness, coordinate the local resources, and promote the involvement into the collective activities like disaster relief or public health campaigns. The CSOs mobilize the community, and this assists the societies in responding promptly before the bigger institutional structures are triggered (Chen, 2023). Advocacy is also another crucial role. Civil society tends to represent the vulnerable groups who might not be able to access the formal political or administrative means. CSOs have the opportunity to lobby resources, affect the public policy, and make sure that other groups which might be marginalized are not neglected during crisis response measures. Monitoring the government activities and holding them accountable is another element of advocacy which is crucial when there is a crisis and decision making process may be centralized and expedited. CSOs also play a key role in crisis management by service delivery. There are numerous organizations that offer food, shelter, healthcare and psychological services in the case of an emergency. They can also tend to take shorter time to reach remote places than the state agencies and be able to localize services to the requirements of the place. By so doing, CSOs are facilitators of aid, yet direct service providers in which relief and recovery reach the most affected (Wojczewski et al., 2023).

Case Illustrations of CSOs as First Responders

The civil society organizations have shown to be valuable in most cases as first responders in times of crisis. CSOs and local community groups are some of the first people to supply food, shelter, and medical supplies in impacted regions in the case of natural disasters such as the 2004 Indian Ocean tsunami. With their grassroots networks, they were able to detect pressing needs and respond to them when the governments and international agencies were still in the process of organizing how to respond (Cook, 2021). CSOs became vital during the COVID-19 pandemic to create awareness regarding health measures, hand out protection gear, and offer food assistance to the underprivileged groups. Community based organizations in nations where the healthcare system is overstretched have tended to provide home care to patients, arrange vaccination awareness programs, and assist migrants and informal workers in that nations who are not included in government provided programs (Levine et al., 2023). CSOs have also been at work in cases of political instability as in the case of pro-democracy movements in various parts of the world. They offer an avenue of non-violent mobilization, support of arrested activists in law, and human rights protection advocacy. Their presence in such situations underscores the fact that crisis management is not a matter of providing material relief only it is also about protecting rights and making the governance accountable in times of stress (Nasibov, 2021). These instances indicate that CSOs have the capability of adapting to the least similar crises: natural, health related, or political due to its flexibility, proximity and social approval. Their short term behaviour tends to be the precursor of long-term recovery and rebuilding.

Challenges Faced by Civil Society Organizations

CSOs, despite their vital role, are being challenged with numerous issues during crisis management. Resource limitation is one of the most important ones. Donations or temporary funding of projects are the reliance of many organizations but they might not be able to work over long periods to respond to crisis. The limit of resources prevents them to expand the services, keep the talented employees, and continue to work under harsh conditions (Levine et al., 2023). There are also legal limitations being a major impediment. In other nations, governments put limitations on the registration, funding or even activities of CSOs with often citing national security or foreign influence. In times of crisis, these restrictions might impair the capacity of organizations to provide relief or promote the agenda of vulnerable groups. The lack of clarity in the law can also deter international donors to finance local groups making their capacity even weaker. The other difficulty is the institutional support. Although CSOs tend to be fast reacting, their activities are not invariably integrated into the national systems of managing crises. This may result in overlapping in functions, lapses in service delivery and lack of coordination between state and non-state actors. The CSOs might not get access to the official information, decision making process, or government resources without being recognized by the institution, making the crisis response better (Syal et al., 2021).

Importance of Trust and Legitimacy

The effectiveness of the civil society in crisis management is mainly based on trust and legitimacy. When citizens trust that CSOs are transparent, accountable and that they are also serious about the welfare of the community, they tend to cooperate with them. Trust helps organizations to find volunteers, get funds and gain acceptance to their programs. This, in turn, makes the communities stronger in times of crisis (Krogh & Lo, 2023). The interaction between the state and the civil society also relies on legitimacy. Governments will be more eager to cooperate with the organizations which are regarded as credible, responsible, and representative of interests of citizens. Trust between state institutions and CSOs leads to an easier exchange of information, coordination of activities and the prevention of conflict during emergency response. Conversely, when the civil society is perceived to be politically partisan or unaccountable, the civil society may be met with resistance by the government, which lowers chances of effective collective action (Kolmodin, 2024). The same thing applies to citizen state relations which entail trust. CSOs are the most appropriate people who fill up gaps between communities and governments. They not only translate policies into understandable information but also convey the needs of the citizens to the authorities and also they arbitrate situations, which may develop in the process of crisis interventions. By increasing trust in all these relationships, CSOs can increase the legitimacy of the processes of governance and help create stability in the long term (Stupak et al., 2021).

POLICY INSTRUMENTS FOR STRENGTHENING CIVIL SOCIETY

Institutional Support and Legal Recognition Frameworks

The efficiency of the CSOs primarily depends on institutional support and legal recognition. Organizations, which have not been given a legal status, find it difficult to get funds, have their property registered, and get involved in official decision making. Plants that have a well-established legal framework defining the position of CSOs give them legitimacy and protection. They also establish accountability standards which make their operations transparent as well (Chen, 2023). Governments may establish special agencies or commissions to coordinate their relations with CSOs. These institutions ensure that collaboration is not sporadic and based on short lived political goodwill. As a case in point, in some nations of Europe, the ministries have departments tasked with the responsibility of engaging the civil society, whereas in the Philippines, the CSOs are legally considered to be partners in the disaster risk reduction and management councils. Such arrangements make sure that CSOs are service providers as well as formal contributors to governance (Ibones et al., 2024). Good institutional backing also gives the CSOs safe havens when it is politically sensitive. In situations where civil society is inhibited, legal frameworks that support the civil society could help deter governments to act by employing the emergency powers to silence the independent voices. Institutionalization of collaboration enables states to establish constant alliances, which stand even when disruption is the new order (Smidt et al., 2025).

Mechanisms for Funding and Resource Sharing

One of the most important problems facing CSOs is financial sustainability. Most organizations depend on outside donations or project funding which can be contingent or interim. This challenge can be solved with the help of the public policies which can establish stable funding channels. The governments can set up grant programs, tax subsidies on the donations made privately in the form of tax or organization subsidies on the provision of essential services. Such indicators will make sure that CSOs do not focus entirely on short-term projects but have the opportunity to be involved in long term plans (Levine et al., 2023).

Sharing of resources is also significant. In addition to money, CSOs also need infrastructure, training and information. Among the ways that governments can empower the civil society is through enabling organizations to operate using the government facilities, providing emergency information to the local organizations or providing training on technical and managerial skills. This assistance helps to expand the working capacity of CSOs during a crisis (Damiyano, 2024). Resource sharing is also a good model offered by the partnerships between the state and the corporations. Businesses in most countries contribute their logistics or technology, governments their regulatory support and CSOs their community outreach. In the case of the COVID-19 pandemic, partnerships allowed distributing medical supplies and food assistance within the framework of a short time. These processes reveal that sharing of resources makes the system more efficient and resilient (Garcia et al., 2023).

Decision Making Inclusivity: Participatory Governance and Multi Stakeholder Platforms

The problem that CSOs have to deal with is that they are usually considered as implementers instead of policy makers. This restricts their freedom to impact on strategies which directly impact on communities. Participatory governance redresses this imbalance through having CSOs being consulted, planned and monitored (Bua & Bussu, 2023). Multi stakeholder platforms are a useful tool of inclusivity. These are forums that unite governments, CSOs, the business world, and the representatives of the community to discuss policies. In disaster management, say, such platforms make certain that the local expertise of vulnerabilities is reflected in the national strategies. CSOs would bring a more socially responsive crisis response and enhance legitimacy (Van Wessel et al., 2021). Trust is also enhanced through inclusiveness. The communities will be more willing to adopt the policies and to collaborate during crisis once they observe that their representatives are involved in making such decisions. By making CSOs an integral part of the government, participatory governance changes them into being perceived as an equal partner who enhances accountability and the conflict between the state and the society (Williamson et al., 2022).

International Best Practices in Policy Design for CSO Empowerment

The world experience demonstrates the importance of the effective formulation of policies in empowering civil society and enhancing resilient development. Sweden and Norway are some of the countries in Scandinavia that have made institutionalization of consultation with CSOs. The laws make governments to formally consult organizations in the process of drafting policies that impact them. That way, the civil society is not left out in governance and in stills confidence in institutions. Philippines Disaster risk reduction legislation categorically specifies that CSO should be involved in the process of planning and implementation. Such a combination of community-based groups makes sure that the knowledge at the grassroots is reflected in the national policies. In the same case, in India, legal frameworks allow NGOs to be recognized in the delivery of services particularly in health and education; state capacities are less in these areas (Ibones et al., 2024). CSO empowerment is also stressed by international organizations. The United Nations acknowledges that civil society is critical to the realisation of Sustainable Development Goals (SDGs), whereas the Organisation of Economic Co-operation and Development (OECD) emphasises the need to work with multi-stakeholders. All these frameworks offer international standards that motivate the governments to empower civil society (Bolton et al., 2023). Best practice proposes that policies should balance between accountability and autonomy. Too many limitations may inhibit the creativity and autonomy of CSOs, whereas lack of enough control may erode transparency and trust by the people. Effective systems strike a balance between ensuring their legal status and freedom of operation and ensuring that they are transparent in how they are funded and operated (Willems, 2021). The table 1 provides a concise overview of instruments that empower civil society and illustrates their practical application in different contexts.

Table 1. Policy instruments for strengthening civil society in crisis governance

Policy Instrument	Purpose	Examples / Best Practices
Legal Recognition Frameworks	Provide legitimacy and accountability for CSOs.	NGO registration laws in India; consultation frameworks in Sweden.
Institutional Support Structures	Ensure consistent state CSO engagement through dedicated agencies.	Disaster risk councils in the Philippines; independent commissions in Norway.
Funding Mechanisms	Offer financial stability for CSOs through grants, subsidies, and incentives.	EU funding programs; Canadian national grant schemes.
Resource Sharing Arrangements	Enable CSOs to access public facilities, data, and training.	Public private partnerships for COVID-19 response in Germany.
Participatory Governance Platforms	Ensure inclusivity and representation of diverse groups.	Local disaster management councils in Southeast Asia.
International Guidelines	Provide global norms for CSO empowerment and resilience.	UN SDG partnerships; OECD recommendations for inclusive governance.

CASE STUDIES OF POLICY-CIVIL SOCIETY PARTNERSHIPS

During crisis, there has tended to be a collaborative effort between civil society organizations (CSOs) and governments. These alliances have a lot to teach on building resilient governance. The success of collaboration is identified in some cases, and the lack of coordination, accountability, and political will is revealed in others. Collectively, they demonstrate the necessity of working out policies that assist successful and ethical collaboration of state and non-state actors.

Successful Models of Government-CSO Collaboration

The Philippines is known to have one of the most renowned instances of collaboration between the government and CSOs through the disaster risk reduction framework. The government subsequently enacted a law that brought the CSOs into the equal footing in disaster preparedness and response following recurrent typhoons and earthquakes. Local disaster management councils also included community-based organizations, whereby the grassroots knowledge was used to influence the policies at the national level. This model enhanced early warning, planning and resilience of communities (Ibones et al., 2024). Another example of success is that of India in the context of the COVID-19 pandemic. The government had a challenge in accessing the millions of migrant workers who were left jobless and homeless. The civil society organizations intervened to offer food relief, transportation support, and health services. Later, the government agencies collaborated with most of these groups in order to organize distribution of aid. The cooperation demonstrated that CSOs could supplement the state action, reaching vulnerable groups in a short period of time and in an effective manner. In Northern Europe, the institutionalization of partnerships between government and CSOs is carried out in the welfare systems. Indicatively, the state offers financial and legal encouragement to health and social services delivery organizations in Norway. CSOs assisted the government to accommodate refugees in terms of housing, language training and social integration programs during refugee crisis. This extended mutually beneficial relationship proved that the institutionalized assistance could reinforce resilience in case of hastened population flows and humanitarian issues (Bravim et al., 2025).

Problematic Cases and Gaps in Collaboration

Not every partnership turned out to be successful. In other instances, poor coordination between the governments and the CSOs has diminished the success of management of the crisis. Following the 2010 Haiti earthquake, e.g. thousands of international and local CSOs were operational in the relief missions. Nevertheless, there was lack of effective government coordination and as such some services were duplicated in certain localities and others neglected. Organizations competed on the resources and visibility, and communities were left in confusion on where to turn to get assistance. Lack of institutional frameworks to support collective action was brought out in this case (Arnaouti et al., 2022). The other problematic example is in some contexts of authoritarianism where the government is limiting the activities of the civil society. Under these circumstances, CSOs can be considered as suspicious and not taken into consideration in making formal decisions. As an example, in certain political crises in the African countries, governments restricted the capacity of the human rights groups to observe abuses or assist the communities in need. The relationship started to be confrontational, minimizing trust and undermining resiliency, instead of collaborative (Annan et al., 2021). These instances demonstrate that political goodwill is important in partnerships. Civil society will not be able to play its part in responding to crises without transparency and appreciation.

Lessons Learned for Improving Crisis Governance Systems

Out of these cases, a number of lessons are learnt. The first one is that successful collaborations must have legal and institutional frameworks that legitimize the presence of civil society. The stronger and less duplication occurs in the coordination when CSOs belong to the official disaster councils or health planning boards (Syal et al., 2021). Second, there is a need to be trustful and accountable. The governments should have the trust in CSOs as rightful partners, and the CSOs should also show transparency in their resource utilization. Reporting systems that are clear and open communication contributes to this trust. Third, it is significant that it is flexible. The nature of crisis is that it can change and grow at a very fast rate and the bureaucratic system that is strictly followed may hamper the quick response. The partnerships that enable some autonomy to CSOs but still hold them accountable provide room to take action swiftly and in a flexible manner (Willems, 2021). Finally, inclusion is vital. The big national or global NGOs are not the only partners that should be engaged but local organizations that have great understanding of communities should also be involved. Local CSOs usually know a lot more about cultural contexts, informal networks, and needs of the community than external actors (Chen, 2023).

Ethical Considerations in Policy-Civil Society Cooperation

Government-CSO partnerships are also associated with ethical problems. Accountability is one of these concerns. CSOs which are publicly funded should make sure that there is equity and openness in the utilization of resources. Absence of accountability may destroy trust and destroy the organization and the government that sponsored it (Willems, 2021). Representation is another ethical problem. CSOs should make sure that they represent the communities that they are in and not their personal interests or interests of their donors. Unless organizations reflect vulnerable groups, there is a high probability that they will miss out on the people that are in need (Williamson et al., 2022). Another issue is a lack of balance of power. Governments can take control of alliances, and the CSOs become only service providers without having any say in the decision making process. On the other hand, in a situation where state capacity is low, there is a possibility of big international CSOs dominating over local institutions, and local ownership of crisis responses is decreased. The policies must seek to establish balanced relationships where everyone makes their contribution equally (Riboldi, 2024). Lastly, human rights and social justice must be respected in an ethical cooperation. CSOs should make sure that their services should not discriminate against groups and the governments should not use alliances as a political instrument. Ethical partnerships are based on respect, openness, and responsibility (Njoku, 2022).

CHALLENGES, ETHICAL DILEMMAS, AND FUTURE DIRECTIONS

Issues of Accountability, Transparency, and Misuse of Civil Society Roles

One of the most urgent issues in the collaboration of government and civil society is accountability. Governments or international donors usually provide funds to CSOs during times of crisis. This generates the demand of transparent financial reporting and performance monitoring. In the absence of powerful accountability processes, it is a possibility that resources can be mismanaged, abused, or channelled towards organizational concerns rather than the needs of the community (Willems, 2021). Close to accountability is transparency. CSOs should be open in the way their operations are run and the source of their funding and also decision making. Communities will lose their trust in the organizations when they feel that the organizations are not transparent. This diminishes their credibility and their desire to cooperate among the citizens. Indicatively, there are cases in humanitarian settings where local people have been skeptical whenever aid delivery has not been clearly clarified or when they think that the beneficiaries have been discriminated (Samaratunge & Alam, 2021). The other dilemma is the possibility of misusing of the CSO positions. Political actors can in some situations co-opt organizations to advance partisan interests. Governments may support the interested CSOs with funds so that they are able to portray a picture of inclusivity, yet keep out the voices of dissent. Conversely, there are instances where CSOs overestimate their ability to raise funds or change their agendas to be donor driven instead of being community-based agendas (Njoku, 2022). The combination of these risks brings up the necessity of sound policies that would foster responsibility and equality and deter abuse.

Balancing State Control and Civil Society Autonomy

One of the main issues of crisis governance is deciding between state domination and civil society independence. Governments are supposed to uphold order, organize reactions and hold them accountable. Simultaneously, the civil society should be able to remain autonomous so that it could serve as the representative of citizens and the overseer of the rule. Over aggression of state might constrain the creativeness and flexibility of CSOs. As an illustration, organizations can be undermined by restrictive laws on registration, foreign funding or operations and have less ability to respond in times of an emergency. Conversely, an unregulated autonomy may result in fissure reactions, redundancy of services, or irresponsibility (Njoku, 2022). The balance needs regulatory frameworks that not only safeguard civic space, but also provide criteria of transparency. Policies that promote dialogue, consultation, and shared planning are useful in ensuring that this balance is maintained. States should not perceive CSOs as rivals but need to regard them as constituents that supplement the government initiatives. On the same note, CSOs should not violate the legitimacy of the state institutions but should uphold their advocacy functions (Chaudhry, 2022).

Trust Building between Public Institutions and Communities

Effective crisis governance is based on trust. It links citizens, CSOs and institutions of the state. The best policies could not work without trust. People should have confidence in CSOs to champion their interests in a just manner and the governments should have confidence with CSOs as well as an honest and competent partner (Wojczewski et al., 2023). Development of trust presupposes openness, communication and accountability. The governments can be able to encourage the trust by engaging the CSOs in the process of making decisions, providing information on a transparent basis and rewarding their efforts publicly. The enhancement of trust can be achieved through CSOs interacting with people at the community level, hearing people out, and providing services equally (Annan et al., 2021). Credibility also needs reliability. The collaborations among states and CSOs are not supposed to be short term agreements focused on the periods of crisis. Long term support by an institution establishes stable relationships within which trust is built over time. This can be demonstrated by the examples of Northern Europe where governments regularly consult with CSOs and finance it through predictable and inclusive governance (McKeague & Bies, 2025).

Future Pathways for Resilient Governance

In the future, there are emerging avenues in enhancing the role of the civil society in the governance of crisis. These are adaptive governance, resilience digital tools and inclusive policy innovation. Adaptive governance puts emphasis on flexibility, learning and novelty of policy systems. Adaptive governance encourages the development of policies that can adapt with the changes in crisis instead of relying on fixed structures. As an example, disasters caused by climate need to be governed by means of adapting to new scientific findings, community input, and international collaboration. The role of CSOs can be to avail local expertise and test new ways of doing things that can be applied by governments in the future (Chen, 2023). Another significant avenue is provided by digital resilience tools. Information and communication technologies will be able to enhance the coordination among governments, CSOs, and its citizens. Online systems can facilitate quick data gathering, mapping of resources and direct contact with the victims societies. Through the COVID-19 pandemic, mobile applications and online platforms were utilized by a number of CSOs to disseminate health information and monitor the needs of the community. Such tools can be incorporated in the official systems by governments and thus produce quicker and more effective responses (World Health Organization, 2022). Inclusive policy innovation is also one that is important. Women, migrants, and low income groups, are some of the marginalized groups that are usually hit by crisis. Policies where various voices are part of the planning and implementation come up with more equalized responses. Inclusion of councils, multi-stakeholder platforms and participatory budgeting are viable mechanisms of representation. Inclusive innovation makes sure that there is not only proper governance, but also social justice (Mir et al., 2024). The table 2 has highlighted the major concerns and has offered future governance directions. It highlights that the challenges and dilemmas do not stand in the way in cooperation but are the possibilities to empower systems by innovative and accountable solutions.

Table 2. Challenges, ethical dilemmas, and future pathways in civil society-state partnerships

Dimension	Challenges / Dilemmas	Future Pathways
Accountability & Transparency	Misuse of funds, weak reporting, donor driven agendas	Strong monitoring systems; open data on funding and services
State-Civil Society Balance	Over regulation vs. unchecked autonomy	Balanced legal frameworks; joint planning mechanisms
Trust-Building	Low trust between state, CSOs, and citizens	Long term institutional support; participatory decision making

continued on following page

Table 2. Continued

Dimension	Challenges / Dilemmas	Future Pathways
Technology & Innovation	Limited use of digital tools; digital divide in communities	Mobile apps, online platforms, open digital governance
Inclusivity	Marginalized groups excluded from decision making	Multi-stakeholder platforms; participatory governance structures
Ethics and Representation	Partisan co-optation of CSOs; misrepresentation of community needs	Transparent mandates; codes of ethics for CSOs and partnerships

CONCLUSION

Civil society plays a crucial role in helping in governance when there is crisis. Its ability to mobilize communities, offer necessary services and represent vulnerable groups supplements the role of state action and assists the societies to react more appropriately. Natural calamities, pandemics, and political upheavals indicate that no one agent is capable of handling crises on its own. Government and civil society organizations partnerships enhance preparedness, response, as well as recovery. These partnerships need to be sustainable with respect to the provision of supportive public policies. CSOs enjoy stability through legal recognition, institutional support, and secure funding; participatory frameworks of governance have seen to it the voices of the CSOs are represented in the policy making and decision making processes. This is because such measures transform CSOs into people who are perceived as merely service providers, but they become acknowledged as partners in governance. There are to be inclusive and adaptive models of governance as well. Incorporating a variety of communities and including the marginalized ones, policies will be more fair and efficient. Inclusivity increases legitimacy and trust among the people, which is essential to enlist collaboration in case of disruptions. Ethical responsibility and accountability as well as trust are at the centre of these partnerships. Governments and CSOs need to be transparent and respect each other. Through this, they will have the opportunity to develop governance systems that do not only have the capacity to control crises, but also contribute towards resilience and social cohesion in the long term.

REFERENCES

Annan, N., Beseng, M., Crawford, G., & Kewir, J. K. (2021). Civil society, peacebuilding from below and shrinking civic space: The case of Cameroon's 'Anglophone'conflict. *Conflict Security and Development, 21*(6), 697–725. DOI: 10.1080/14678802.2021.1997454

Ansell, C., Sørensen, E., Torfing, J., & Trondal, J. (2024). *Robust governance in turbulent times*. Cambridge University Press. DOI: 10.1017/9781009433006

Ansell, C., & Torfing, J. (2022). Introduction to the Handbook on Theories of Governance. In *Handbook on theories of governance* (pp. 1–16). Edward Elgar Publishing., DOI: 10.4337/9781800371972.00007

Arnaouti, M. K. C., Cahill, G., Baird, M. D., Mangurat, L., Harris, R., Edme, L. P. P., Joseph, M. N., Worlton, T., & Augustin, S.Jr. (2022). Medical disaster response: A critical analysis of the 2010 Haiti earthquake. *Frontiers in Public Health, 10*, 995595. DOI: 10.3389/fpubh.2022.995595

Babacan, H. (2025). From crisis to policy action: Tackling climate change through SDG 13. *The International Journal of Community and Social Development, 25166026251351370*(2), 205–221. Advance online publication. DOI: 10.1177/25166026251351370

Bolleyer, N. (2021). Civil society, crisis exposure, and resistance strategies. In *Routledge handbook of illiberalism* (pp. 939–956). Routledge. DOI: 10.4324/9780367260569-68

Bolton, D., Landells, T., & Habib, M. (2023). Collaborating and Competing in Development Agendas: State—Civil Society Tensions in Achieving SDGs. In *The complexities of sustainability* (pp. 77-120).

Bravim, J. M., de Carvalho, A. C. V., Segatto, A. P., & Morais-da-Silva, R. L. (2025). Are there key characteristics of social enterprises that foster relationships with government, universities and civil society? *Social Enterprise Journal, 21*(5), 864–889. Advance online publication. DOI: 10.1108/SEJ-08-2024-0136

Broadhurst, K., & Gray, N. (2022). Understanding resilient places: Multi-level governance in times of crisis. *Local Economy, 37*(1-2), 84–103. DOI: 10.1177/02690942221100101

Bua, A., & Bussu, S. (Eds.). (2023). *Reclaiming participatory governance: Social movements and the reinvention of democratic innovation*. Taylor & Francis. DOI: 10.4324/9781003218517

Chaudhry, S. (2022). The assault on civil society: Explaining state crackdown on NGOs. *International Organization*, *76*(3), 549–590. DOI: 10.1017/S0020818321000473

Chen, X. (2023). Understanding the role of Civil Society Organizations (CSOs) in the "central government-local government-society" framework: A case study on public crisis management during the pandemic in China. *International Public Management Journal*, *26*(4), 570–588. DOI: 10.1080/10967494.2022.2115595

Cook, A. D. (2021). Humanitarian diplomacy in ASEAN. *Asian Journal of Comparative Politics*, *6*(3), 188–201. DOI: 10.1177/20578911211019247

Damiyano, D. (2024). Sustainable Social Services Financing for Local Governments. In *Financial Sustainability of Local Governments in Southern Africa* (pp. 198–214). Routledge. DOI: 10.4324/9781003498230-11

Garcia, S., Carrigan, C., & Wiepking, P. (2023). Global civil society response to the COVID-19 crisis. *Voluntas*, *34*(6), 1310–1321. DOI: 10.1007/s11266-023-00589-5

Hinterleitner, M., Knill, C., & Steinebach, Y. (2024). The growth of policies, rules, and regulations: A review of the literature and research agenda. *Regulation & Governance*, *18*(2), 637–654. DOI: 10.1111/rego.12511

Ibones, K. A., Enero, J. V., Jore, J. S., Mamhot, V. I., Matheu, C. M., Pacaldo, H. B. B., & Jardin, R. A. (2024). Enabling role of civil society organizations (CSOs) in local environmental management in the Philippines: A systematic review. *Journal of Civil Society*, *20*(4), 359–379. DOI: 10.1080/17448689.2024.2381826

Kolmodin, S. (2024). Collaboration to support refugees: The importance of local-level relations among civil society organizations. *Journal of Civil Society*, *20*(2), 130–149. DOI: 10.1080/17448689.2024.2319357

Krogh, A. H., & Lo, C. (2023). Robust emergency management: The role of institutional trust in organized volunteers. *Public Administration*, *101*(1), 142–157. DOI: 10.1111/padm.12894

Kumi, E., & Hayman, R. (2025). Can domestic resource mobilisation by civil society organisations (CSOs) open up civic space in the Global South? *Development in Practice*, *35*(4), 669–687. DOI: 10.1080/09614524.2025.2487791

Levine, A. C., Park, A., Adhikari, A., Alejandria, M. C. P., Bradlow, B. H., Lopez-Portillo, M. F., Mutwafy, S., Zumbyte, I., & Heller, P. (2023). The role of civil society organizations (CSOs) in the COVID-19 response across the Global South: A multinational, qualitative study. *PLOS Global Public Health*, *3*(9), e0002341. DOI: 10.1371/journal.pgph.0002341

McKeague, L. K., & Bies, A. L. (2025). Nonprofits and Interest Groups in Crisis Management. In *Oxford Research Encyclopedia of Politics*. DOI: 10.1093/acrefore/9780190228637.013.1530

Mir, G., Durrani, N., Julian, R., Kimei, Y., Mashreky, S., & Doan, T. D. (2024). Social inclusion and sustainable development: Findings from Seven African and Asian contexts. *Sustainability (Basel)*, *16*(11), 4859. DOI: 10.3390/su16114859

Mizrak, K. C. (2024). Crisis management and risk mitigation: Strategies for effective response and resilience. *Trends, challenges, and practices in contemporary strategic management*, 254-278. DOI: DOI: 10.4018/979-8-3693-1155-4.ch013

Moya, J., & Goenechea, M. (2022). An approach to the unified conceptualization, definition, and characterization of social resilience. *International Journal of Environmental Research and Public Health*, *19*(9), 5746. DOI: 10.3390/ijerph19095746

Nair, J. (2024). Reconfiguration and transformation for resilience: Building service organizations towards sustainability. *Journal of Services Marketing*, *38*(4), 404–425. DOI: 10.1108/JSM-04-2023-0144

Nasibov, M. (2021). Civil Society and Pro-Democracy Social Movements: Troubled Relations Within Authoritarian Regimes? *Frontiers in Political Science*, *3*, 708872. DOI: 10.3389/fpos.2021.708872

Njoku, E. T. (2022). Strategic exclusion, co-option and containment: Towards an integrative theory of state-CSOs relations. *Critical Studies on Terrorism*, *15*(4), 917–944. DOI: 10.1080/17539153.2022.2111776

Pitidis, V., Coaffee, J., & Lima-Silva, F. (2024). Advancing equitable 'resilience imaginaries' in the Global South through dialogical participatory mapping: Experiences from informal communities in Brazil. *Cities (London, England)*, *150*, 105015. DOI: 10.1016/j.cities.2024.105015

Riboldi, M. (2024). The Loci of Power and Connection: A framework for exploring the democratic relationships of civil society organisations. *Interest Groups & Advocacy*, *13*(2), 244–263. DOI: 10.1057/s41309-024-00209-7

Salvador, M., & Sancho, D. (2023). Local governments facing turbulence: Robust governance and institutional capacities. *Social Sciences (Basel, Switzerland)*, *12*(8), 462. DOI: 10.3390/socsci12080462

Samaratunge, R., & Alam, Q. (2021). Accountability and transparency in emerging countries: Governance, democratic currents and change. *Public Administration and Development*, *41*(4), 147–156. DOI: 10.1002/pad.1963

Smidt, H., Johansson, J., & Richter, T. (2025). Civil society under attack: The consequences for horizontal accountability institutions. *Studies in Comparative International Development*, *60*(1), 81–110. DOI: 10.1007/s12116-023-09423-x

Stupak, I., Mansoor, M., & Smith, C. T. (2021). Conceptual framework for increasing legitimacy and trust of sustainability governance. *Energy, Sustainability and Society*, *11*(1), 5. DOI: 10.1186/s13705-021-00280-x

Syal, R., van Wessel, M., & Sahoo, S. (2021). Collaboration, co-optation or navigation? The role of civil society in disaster governance in India. *Voluntas*, *32*(4), 795–808. DOI: 10.1007/s11266-021-00344-8

Van Wessel, M., Naz, F., & Sahoo, S. (2021). Complementarities in CSO Collaborations: How working with diversity produces advantages. *Voluntas*, *32*(4), 717–730. DOI: 10.1007/s11266-020-00227-4

Willems, J. (2021). Accountability and transparency: cornerstones of civil society governance. In *Research handbook on nonprofit governance* (pp. 46–70). Edward Elgar Publishing., DOI: 10.4337/9781788114912.00012

Williamson, A. K., Kingston, K. L., & Bennison, L. (2022). Similar or different? Downward accountability across civil society organisational forms. *Voluntas*, *33*(5), 1002–1014. DOI: 10.1007/s11266-021-00441-8

Wojczewski, S., Grohma, P., & Kutalek, R. (2023). Risk communication and community engagement with vulnerable groups: Perceptions of social-services CSOs during Covid-19. *International Journal of Disaster Risk Reduction*, *94*, 103817. DOI: 10.1016/j.ijdrr.2023.103817

World Health Organization. (2022). *Community assets and civil society outreach in critical times: an initiative to engage civil society organizations in the COVID-19 response*. World Health Organization.

Chapter 15
The Future of International Cooperation Redesigning Global Governance for an Age of Uncertainty: Governing Complexity in a Fragmented World

Ali Al Mokdad
https://orcid.org/0009-0000-2825-5855
Operations Copilot L.L.C., UAE

ABSTRACT

This chapter critically examines the breakdown of the post-1945 order, focusing on the United Nations Security Council, Bretton Woods institutions, and regional organizations, with case studies from Syria, Ukraine, Gaza, and Sudan. It argues that these institutions remain indispensable for legitimacy and coordination yet are insufficient for governing complexity at the required scale and speed. Building on concepts of complex adaptive systems, anticipatory governance, and polycentric arrangements, the chapter identifies principles for redesign—adaptability, subsidiarity, equity, digital inclusion, and Indigenous knowledge integration—and illustrates them through recent innovations such as the Bridgetown Initiative, the Global Digital Compact, and African Union mediation. It culminates in a blueprint for more agile, inclusive, and resilient governance, stressing that renewal requires leadership capable of coalition-building, and systems thinking, to both present and future generations.

DOI: 10.4018/979-8-3373-5651-8.ch015

INTRODUCTION: GOVERNING IN THE AGE OF POLYCRISIS

The notion of *polycrisis* has emerged as a central frame for understanding the contemporary global condition. Originally introduced by Edgar Morin in the 1990s, the term has been refined by recent scholarship to denote situations in which crises across multiple domains interact, producing outcomes greater than the sum of their parts (Lawrence et al., 2022). Unlike single shocks, polycrises are marked by causal entanglement and feedback loops: a drought may trigger food insecurity, which fuels conflict, which in turn undermines health and governance systems.

The analytical value of polycrisis lies in its recognition of systemic interdependence. Global governance institutions, by contrast, are typically organized along sectoral lines—health, finance, peace and security—making them ill-suited to manage cascading risks. The World Economic Forum's *Global Risks Report 2023* highlighted polycrisis dynamics as the defining feature of the next decade's risk landscape. Similarly, UN Secretary-General António Guterres (2023) has warned that overlapping crises are *"overwhelming the capabilities of governance institutions."*

Yet the concept is not without limitations. Critics argue that "polycrisis" risks becoming a catch-all descriptor, obscuring analytical precision or diluting responsibility. To remain useful, the concept must be operationalized through case studies and linked to governance design principles. In this chapter, polycrisis is treated not as rhetorical flourish but as a diagnostic lens: it directs attention to institutional stress points, governance gaps, and potential redesign pathways.

The first half of the 2020s provides a stark illustration of polycrisis dynamics in action.

- Pandemics: The COVID-19 crisis revealed the limits of global health cooperation. The COVAX mechanism, designed to ensure equitable vaccine access, did succeed in delivering over 1.9 billion doses across 146 countries (WHO, 2022). Yet vaccine nationalism, export bans, and unequal purchasing power undermined its core mission, leaving low-income countries with delayed access.
- Conflict: Russia's invasion of Ukraine in 2022 was not only a geopolitical rupture but also a shock to global energy and food markets. Wheat shortages pushed millions in Africa and the Middle East closer to hunger, while energy price spikes triggered inflation worldwide. Parallel crises in Gaza and Sudan (2023–2024) further strained humanitarian agencies already underfunded and overstretched.
- Climate: The same period saw record-breaking heatwaves in South Asia, catastrophic floods in Pakistan and West Africa, and unprecedented wildfires across Europe and North America. Each event disrupted economies and dis-

placed communities, but in aggregate they highlighted climate change as a threat multiplier, exacerbating food insecurity, migration pressures, and governance fragility.

- Technology: The rapid diffusion of generative artificial intelligence since 2022 created both opportunities and risks. While AI has been deployed in humanitarian logistics and climate modeling, it has also fueled disinformation, cyber vulnerabilities, and questions of algorithmic accountability. Competing governance proposals—from the Hiroshima AI Process to the BRICS Rio Declaration—illustrate emerging fragmentation in norm-setting.

Taken together, these developments constitute a stress test of the international order. They demonstrate how multiple crises, unfolding concurrently, overwhelm institutional bandwidth, generate compounding risks, and expose deep asymmetries between Global North and South in voice, resources, and resilience.

The governance implications of the polycrisis era are profound. First, existing institutions are optimized for single-issue problem-solving rather than managing systemic complexity. The UN Security Council can debate conflicts but has little to offer on pandemics or AI governance. The IMF can address balance-of-payments crises but lacks climate finance capacity. The WHO can declare a public health emergency but cannot mobilize economic stabilization.

Second, governance mechanisms are prone to delayed response and political deadlock. The Security Council's repeated paralysis over Syria, Ukraine, and Gaza demonstrates how veto power undermines collective security. Similarly, consensus rules in climate negotiations often produce lowest-common-denominator outcomes, out of sync with the scale of the challenge.

Third, the polycrisis underscores the urgency of equity and inclusion. The pandemic revealed stark vaccine inequities; climate finance remains skewed against developing countries; and the digital divide excludes large parts of the Global South from shaping AI governance. As a result, trust in multilateralism has eroded, with many states viewing the system as privileging a few while externalizing risks onto the many.

Finally, the lesson of the polycrisis is that indispensable yet insufficient institutions must be reimagined. They retain legitimacy as convening spaces but lack the adaptability, subsidiarity, and accountability needed to manage 21st-century risks. Anticipatory governance—based on early warning triggers, flexible coalitions, and federated data "rails"—offers one pathway forward.

DIAGNOSING THE BREAKDOWN OF THE POST-WWII GLOBAL ORDER

The international order created after World War II was designed with three interlocking objectives: preventing another catastrophic global conflict, stabilizing economies through rules-based cooperation, and fostering collective prosperity. To achieve this, new institutions were established: the United Nations (UN) as the political anchor of collective security and normative dialogue, the Bretton Woods financial institutions (the International Monetary Fund and the World Bank) as the economic stabilizers, and a range of regional alliances to provide security, integration, and legitimacy at the sub-global level. For much of the latter twentieth century, these institutions delivered important achievements. They helped deter direct great-power conflict, supported sustained global economic growth, and created mechanisms for development financing and trade liberalization.

Yet, by mid-2025, the system shows clear signs of erosion. Intensifying geopolitical rivalry, rapid technological disruption, and the interdependence of crises—from climate shocks to pandemics and digital insecurity—have exposed structural flaws. Decision-making inefficiencies, unequal power distributions, and inconsistent application of norms have produced a legitimacy crisis. For many states, particularly in the Global South, multilateralism is now viewed less as a guarantor of fairness and more as a tool of powerful actors. This section dissects the performance of the system's three core pillars—the UN, the Bretton Woods institutions, and regional organizations—before turning to key case studies that illustrate recurring failures and partial successes.

The United Nations System

Established in 1945, the UN remains the central platform for international dialogue and normative contestation. Its Security Council (UNSC) was mandated as the ultimate guarantor of peace and security. While the Council retains unmatched convening authority—providing a forum where major powers continue to deliberate during crises—its credibility has been persistently eroded by the veto power held by its five permanent members (P5: China, France, Russia, the United Kingdom, and the United States).

The use of veto power has risen sharply in the last decade. Between 2011 and 2025, dozens of resolutions addressing Syria, Ukraine, and Gaza were blocked. In 2024 alone, the UNSC recorded eight vetoes—the highest annual number since 1986. In June 2025, the United States vetoed a draft resolution demanding a permanent ceasefire in Gaza, citing inadequate provisions on hostage release. Russia repeatedly vetoed measures on Ukraine, including a March 2025 amendment reaffirming

Ukraine's sovereignty in the wake of Resolution 2774. These examples underscore a structural flaw: when a P5 state is party to or aligned with a conflict, the Council is structurally paralyzed. Russia's dual role as aggressor in Ukraine and veto-holder epitomizes this contradiction, revealing the limits of collective enforcement in the very body designed to uphold it.

The UN General Assembly (UNGA) offers greater inclusivity, and its votes carry significant symbolic weight. The overwhelming condemnation of Russia's invasion of Ukraine in 2022 and the June 2025 resolution calling for a Gaza ceasefire demonstrated broad international consensus. However, UNGA resolutions remain non-binding. Their power lies in moral suasion, norm diffusion, and agenda-setting rather than in enforceable action. The gap between consensus-building and enforceability remains the UN's defining dilemma: it can articulate norms with legitimacy but struggles to compel compliance.

Beyond deliberative bodies, specialized agencies and operational arms—including the WHO and OCHA—remain indispensable in service delivery and crisis coordination. Yet chronic underfunding limits their capacity. For example, by mid-2025, UN humanitarian appeals for Sudan were less than 40% funded, undermining famine prevention efforts. Proposals for reform, such as voluntary veto restraint in cases of mass atrocities (the "Responsibility Not to Veto" initiative), have gained attention but remain blocked by entrenched P5 interests. The UN's future legitimacy thus hinges not only on structural reform but also on its ability to demonstrate effectiveness in addressing human suffering amid chronic resource shortfalls.

The Bretton Woods Institutions

The Bretton Woods system, originally composed of the IMF and World Bank (later complemented by the WTO), was intended to stabilize global finance, promote development, and underpin open trade. While these institutions have provided substantial support—such as more than $1 trillion in IMF crisis lending during the COVID-19 pandemic and World Bank financing for infrastructure—they are increasingly criticized for structural imbalances and performance gaps.

Governance remains dominated by advanced economies. The United States alone controls over 16% of IMF voting shares, effectively granting it veto power. The convention of reserving the World Bank presidency for a U.S. nominee and the IMF's managing directorship for a European persists, despite decades of calls for reform. Emerging economies, particularly in the Global South, have repeatedly demanded redistribution of quotas and voting rights to reflect their increased economic weight. At the 2025 IMF–World Bank Spring Meetings, quota reforms were revisited, but only marginal adjustments were made—falling far short of expectations for meaningful redistribution.

Climate finance is another acute shortfall. The World Bank's 2024 replenishment of its International Development Association (IDA) set a target of $100 billion, far below the $300+ billion per year that developing countries estimate is needed for adaptation and resilience. The IMF's Resilience and Sustainability Trust, launched in 2022, remains underfunded, limiting its ability to respond to climate-related shocks.

The WTO, once the cornerstone of global trade governance, has seen its authority steadily eroded. The Doha Round remains unresolved, while the Appellate Body has been non-functional since 2019 due to U.S. refusal to approve judge appointments. This paralysis has driven a shift toward plurilateral arrangements, such as the Joint Statement Initiative on E-Commerce (with over 90 members by 2025), and toward regional pacts such as the Regional Comprehensive Economic Partnership (RCEP) and the African Continental Free Trade Area (AfCFTA). At its 30th anniversary in April 2025, the WTO's Director-General warned of "irrelevance through inaction," encapsulating the risks of institutional paralysis.

Regional Organizations

Regional bodies have grown in prominence as complements and, at times, substitutes for global institutions. Their comparative advantage lies in proximity, cultural legitimacy, and faster response capacity.

The African Union (AU) has expanded its peace and security role. Its mediation of the Tigray conflict in Ethiopia in 2022 set a precedent for African-led solutions. In 2025, the AU prioritized Sudan mediation efforts, leveraging regional legitimacy. Yet its Peace and Security Council remains dependent on external donors, and the AU Peace Fund is still not financially autonomous—leaving African-led diplomacy structurally constrained.

The Association of Southeast Asian Nations (ASEAN) has enhanced its disaster management role through the AHA Centre. The July 2025 ASEAN Regional Disaster Emergency Response Simulation Exercise (ARDEX-25) in Cambodia tested coordination mechanisms for typhoons and floods, demonstrating ASEAN's practical role in humanitarian coordination. However, ASEAN's consensus model, while preserving cohesion, limits its capacity to act decisively on politically sensitive issues such as Myanmar.

The European Union (EU) remains the most integrated regional organization, pooling sovereignty to project collective power. Since 2022, the EU has provided more than €100 billion in support to Ukraine, coordinated sanctions against Russia, and explored post-conflict stabilization measures. The EU has also demonstrated adaptability, as in its joint vaccine procurement during the COVID-19 pandemic. Nonetheless, internal disagreements over energy, asylum, and fiscal integration reveal the difficulty of sustaining cohesion across 27 member states.

Overall, regional bodies fill gaps left by global paralysis but remain reliant on UN partnerships for logistics, funding, and normative reinforcement. Their rise highlights the multipolarization of governance but also underscores persistent resource constraints.

Case Studies, 2020–2025

Concrete crises in the 2020–2025 period highlight recurring weaknesses in the post-WWII order:

- Syria. A decade of UNSC deadlock allowed atrocities to continue largely unchecked. Following the collapse of the Assad regime in December 2024, the UNSC issued a presidential statement condemning renewed violence in Suweida. Yet the Council's relevance to Syria's political transition remains contested, symbolizing its broader paralysis.
- Ukraine. Russia's 2022 invasion starkly revealed the UNSC's incapacity, as Russia vetoed enforcement resolutions. UNGA resolutions affirming Ukraine's sovereignty carried symbolic weight but had no binding effect. Actual enforcement relied on NATO and EU initiatives, highlighting the limits of UN-centered security governance.
- Gaza. Escalations between 2023 and 2025 triggered repeated vetoes, including the June 2025 U.S. veto against a ceasefire resolution. While a fragile ceasefire was eventually endorsed in Resolution 2735, enforcement challenges illustrate the gap between normative consensus and practical compliance.
- Sudan. Conflict since 2023 displaced millions and created famine conditions by mid-2025. UN appeals remained under 50% funded, exposing chronic under-resourcing and undermining humanitarian credibility.

These cases reveal recurrent pathologies: UNSC paralysis due to veto use; chronic underfunding of humanitarian response; and legitimacy gaps alienating Global South states who perceive double standards in how norms are applied. Surveys in multiple regions show declining confidence in the UN's ability to deliver impartial justice or provide effective protection. This erosion of trust is not a temporary dip in credibility but a structural legitimacy crisis—one that, if unresolved, risks accelerating fragmentation and undermining the viability of rules-based multilateralism itself.

WHY TRADITIONAL INSTITUTIONS CAN'T HANDLE TODAY'S CHALLENGES

The institutions designed in the mid-20th century were undeniably effective at stabilizing the postwar order, but their structural DNA is increasingly misaligned with the governance demands of the 21st century. They were conceived in a world where risks were largely discrete, sector-specific, and predictable—interstate wars fought by conventional armies, balance-of-payments crises managed through financial adjustment, or trade disputes resolved within the confines of tariff negotiations. The conceptual map of global risk in 1945 was linear and compartmentalized. Institutions were created with narrow mandates and rigid procedures that mirrored this worldview.

By contrast, today's international system is defined by cascading, multidimensional risks that unfold simultaneously across multiple domains and interact in unpredictable ways. Climate shocks fuel migration, which in turn generates political instability and conflict. Pandemics disrupt supply chains and trigger debt crises. Cyberattacks compromise critical infrastructure, while disinformation erodes social cohesion and delegitimizes democratic governance. These are not isolated shocks but *compound crises* whose interdependence produces nonlinear effects far greater than the sum of their parts. Traditional institutions, built for separable problems, lack the flexibility to address these cross-cutting dynamics.

Although these institutions retain legitimacy, convening power, and symbolic authority, their effectiveness in practice has diminished. They still provide a stage for global dialogue and norm-setting—functions that remain indispensable. Yet, when confronted with urgent crises, their procedural and structural weaknesses consistently obstruct timely, fair, and coordinated action. The credibility gap between aspiration and delivery widens with each failure to respond adequately, feeding disillusionment particularly among actors in the Global South who view the system as unrepresentative and selectively applied.

At the heart of this mismatch are four interlocking deficiencies that together create systemic dysfunction. First, representation deficits mean that the institutions do not reflect today's geopolitical or demographic realities, undermining legitimacy. Second, decision-making paralysis, whether through the veto in the UN Security Council or consensus rules in bodies like the WTO or UNFCCC, institutionalizes gridlock precisely when decisive action is most needed. Third, bureaucratic rigidity and siloed mandates prevent integration across domains, locking agencies into fragmented responses ill-suited to complex, interconnected risks. Finally, chronic underfunding and resource volatility cripple operational delivery, producing a cycle of unmet promises that corrodes trust.

These weaknesses are not minor or correctable through incremental reform; they are systemic misalignments embedded in the very design of the postwar order. Representation deficits alienate large parts of the global community, procedural paralysis blocks decision-making, bureaucratic silos inhibit adaptive responses, and financial instability constrains operational capacity. Taken together, they create an architecture that is both indispensable and inadequate: necessary for legitimacy but insufficient for effectiveness.

The implications are profound. Unless reformed, these institutions will increasingly be bypassed by ad hoc coalitions, regional arrangements, or private governance initiatives. While such alternatives may provide agility, they lack universality and risk fragmenting the international system into competing spheres of influence. The real danger is not abrupt collapse but gradual obsolescence, where the old institutions persist in form but lose substantive relevance, leaving the world without credible, inclusive mechanisms to manage systemic risk.

Representation Deficits

Representation remains one of the most visible and corrosive shortcomings of contemporary global governance. Institutions created in 1945 continue to reflect the distribution of power at the end of World War II rather than today's geopolitical, demographic, and economic realities. This misalignment is more than symbolic; it goes to the heart of legitimacy, fairness, and compliance in the international system.

The United Nations Security Council (UNSC) is the clearest example. Its permanent membership—the United States, the United Kingdom, France, Russia, and China—mirrors the victors of World War II, not the geopolitical map of 2025. Entire continents remain excluded from permanent representation. Africa, with more than 1.4 billion people—nearly one-sixth of humanity—and almost a quarter of UN member states, has no permanent seat. Latin America, home to more than 650 million people and dynamic emerging economies, is similarly absent. These exclusions perpetuate the perception that the UNSC is less a "council of the world" than a narrow club of states that have claimed indefinite authority. This perception matters: when decisions on peace and security are made without permanent representation from entire regions, the legitimacy of outcomes is questioned, even when consensus exists on substance.

The imbalance is not confined to security. It permeates the economic domain through the governance structures of the International Monetary Fund (IMF) and the World Bank. These institutions remain structurally dominated by advanced economies. Voting shares disproportionately privilege the United States and Europe, while emerging economies with growing weight—India, Brazil, Nigeria, Indonesia, and others—are systematically underrepresented. The U.S. share in the IMF, at 16.5

percent, provides effective veto power over major decisions. This is not merely a technical detail; it symbolizes the persistence of Western dominance in institutions that present themselves as serving global stability.

Reform efforts have been slow and incremental. Since at least the 2008 global financial crisis, emerging economies have pushed for quota reform to reflect their increased role in global growth. While some adjustments have been made, they fall far short of redistributing influence in proportion to economic and demographic weight. The result is a widening legitimacy gap: large parts of the world feel excluded from decision-making processes and see little reason to accept the authority of rules they had no hand in shaping.

This deficit is not only about fairness or symbolism. It has practical consequences. Under-represented states and coalitions increasingly invest in institutional alternatives. The BRICS New Development Bank and the Asian Infrastructure Investment Bank (AIIB) provide financing channels outside the World Bank, often with fewer conditionalities and without Western dominance. Similarly, regional arrangements such as the African Continental Free Trade Area (AfCFTA), Mercosur in Latin America, and even emerging digital cooperation platforms are designed to reclaim agency outside of traditional structures like the WTO. These alternatives demonstrate that representation gaps in global institutions do not merely frustrate; they incentivize bypassing and fragmentation.

The deeper risk is that universality—the defining strength of post-1945 institutions—is being eroded. The UN, IMF, World Bank, and WTO derive legitimacy not only from their technical expertise but also from their claim to represent the world as a whole. When that claim is persistently undermined, loyalty shifts. States turn their attention, resources, and political capital toward parallel mechanisms, leaving the "universal" system hollowed out. Over time, this trend risks producing a world of competing governance spheres—Western-led, China-led, or regionally organized—rather than a single global order.

From an academic perspective, representation deficits illustrate the principle of path dependence in institutional design: structures created for the power dynamics of 1945 have persisted, even as the distribution of capabilities and populations has radically shifted. Without structural reform, the legitimacy crisis will deepen, undermining both compliance and effectiveness. The question is no longer whether representation matters—it is whether the erosion of representation will trigger the slow but steady irrelevance of the very institutions that once stabilized the global order.

Veto and Consensus Paralysis

Decision-making rules that were originally crafted to guarantee the participation of great powers and safeguard inclusivity have, over time, hardened into mechanisms of paralysis. The very procedures that secured the survival of multilateral institutions in 1945 now undermine their effectiveness in the face of 21st-century crises.

The UN Security Council veto was introduced as a pragmatic compromise: without it, the major powers of the postwar era would not have agreed to join the UN. At the time, it was considered a necessary safeguard against unilateral withdrawal, ensuring that the most powerful states would remain invested in the institution. Yet in practice, the veto has become the Council's greatest liability. In conflicts where a permanent member (P5) is directly involved or aligned with a party, collective action becomes impossible.

The consequences are stark. In Syria, more than a decade of atrocities unfolded amid deadlock, as dozens of draft resolutions were vetoed. In Ukraine, Russia—both the aggressor and a veto-holder—repeatedly blocked resolutions condemning its invasion or reaffirming Ukraine's sovereignty, rendering the Council structurally incapable of enforcing its own Charter principles. In Gaza, U.S. vetoes blocked calls for ceasefires despite overwhelming support among member states, highlighting the disconnect between global consensus and institutional action. These cases demonstrate a systemic contradiction: the body charged with maintaining international peace and security is neutralized precisely when its intervention is most essential.

This paralysis is not confined to the UNSC. Other global institutions replicate the problem through consensus rules that, in practice, amount to universal veto power. The World Trade Organization (WTO), for instance, requires consensus for major negotiations. This design, intended to preserve inclusivity, has produced chronic stagnation. The Doha Development Round, launched in 2001, remains unresolved more than two decades later, repeatedly blocked by disagreements over agriculture, subsidies, and digital trade. In effect, consensus rules enable a single state—or a small coalition—to halt progress, no matter how broad the support for reform.

The UN Framework Convention on Climate Change (UNFCCC) faces similar constraints. With nearly 200 states required to endorse agreements, outcomes often reflect the lowest common denominator. The ambition of climate action is diluted because consensus must reconcile the priorities of the most climate-vulnerable states with those of the largest fossil fuel producers. The resulting compromises secure legitimacy on paper but frequently fall short of the urgency demanded by scientific evidence. As climate science accelerates toward warnings of tipping points, the governance system continues to move at the pace of consensus politics.

In sum, decision-making mechanisms that were designed to safeguard legitimacy have instead institutionalized immobility. The paradox is clear: rules created to ensure participation and fairness now obstruct decisive action in the very moments when urgency is greatest. The effect is corrosive. Smaller states perceive rules as selectively enforced, eroding trust in the fairness of the system. Larger states, frustrated by gridlock, increasingly resort to unilateral or plurilateral measures—regional alliances, ad hoc coalitions, or issue-specific agreements—outside established institutions. While these alternatives may deliver short-term progress, they fragment the universality and coherence that the postwar system was built to preserve.

Thus, what began as procedural safeguards against domination has evolved into structural weaknesses that erode both legitimacy and effectiveness. Unless reformed, these mechanisms risk locking global governance into a cycle of stagnation: inclusive enough to prevent collapse, but too paralyzed to address crises of planetary scale.

Bureaucratic Inertia and Siloed Mandates

A third structural deficiency of the postwar governance system lies in the bureaucratic cultures of international organizations, which are marked by inertia, siloed mandates, and risk aversion. These features, often justified as safeguards for stability, impartiality, and continuity, have become maladaptive in an era defined by cascading, cross-sectoral, and highly interconnected crises. What once seemed like virtues of order and predictability now function as systemic liabilities.

Most international institutions were created with narrowly defined remits, reflecting the mid-20th-century worldview that risks could be compartmentalized and addressed in isolation. The World Health Organization (WHO) was tasked with global health, the Food and Agriculture Organization (FAO) with food and agriculture, the World Trade Organization (WTO) with trade, and the Office of the UN High Commissioner for Refugees (UNHCR) with displacement. This specialization was functional when threats were primarily discrete and linear—when a health emergency could be treated as separate from an economic downturn or a refugee crisis. However, in the 21st century, threats are increasingly systemic, transboundary, and nonlinear.

Consider the compounding effects of climate change. Rising temperatures intensify food insecurity, which drives conflict and displacement. A pandemic can simultaneously collapse trade networks, overwhelm fragile health systems, and destabilize already fragile states. Cyberattacks on financial institutions disrupt supply chains, reverberating into humanitarian logistics and aid delivery. Yet despite the growing evidence that risks are interconnected, the institutional machinery continues to operate as though each domain were separate. The WHO may declare a global health emergency, the FAO may issue food security alerts, and UNHCR may respond to mass displacement, but without mechanisms to stitch these actions into a unified response, coordination remains ad hoc. The outcome is fragmentation: duplication of effort in some areas, glaring neglect in others, and a systemic inability to grasp the full complexity of crises.

This fragmentation is compounded by bureaucratic inertia. International organizations frequently default to consensus-driven decision-making, extensive procedural safeguards, and incremental policy adjustments. These features were originally intended to guard against domination by powerful states and to ensure legitimacy through inclusion. Yet in practice, they too often translate into slowness and over-caution, producing delays precisely at moments when agility and decisiveness are most needed. Coordination mechanisms such as the UN's "whole-of-system" responses exist on paper, but in reality, they are chronically under-resourced, lack enforcement authority, and are often undermined by turf battles between agencies.

Institutional rivalries exacerbate the problem. Donor-driven funding incentives encourage agencies to defend their own mandates, budgets, and visibility rather than prioritize system-wide effectiveness. This leads to pathologies of competition. In humanitarian response, for example, multiple agencies frequently launch overlapping appeals for the same crisis, competing for donor attention and inadvertently creating confusion on the ground. In development work, siloed mandates often result in narrowly targeted projects that treat symptoms rather than causes—health programs without consideration of climate impacts, or agricultural projects disconnected from migration dynamics.

The cumulative effect is that institutions are locked into reactive governance cycles. Instead of operating preventively or adaptively, they move slowly, often mobilizing only after crises have escalated. Few mechanisms exist for anticipatory governance—despite the availability of foresight methodologies such as horizon scanning, early-warning systems, and scenario planning. Bureaucratic rigidity, once a source of predictability, now acts as a structural handicap: an anchor slowing down global responses in a context where speed, integration, and adaptability are essential.

From an analytical perspective, this mismatch reflects a deeper disjuncture between institutional design and systemic reality. International organizations are largely hierarchical and siloed, designed for predictable problems that can be solved within bounded jurisdictions. Yet the crises of the 21st century function as complex adaptive systems (CAS): interdependent, dynamic, and prone to cascading effects. Effective governance of CAS requires structures that are flexible, networked, experimental, and capable of cross-domain learning. The persistence of rigid mandates and bureaucratic inertia thus represents not merely operational inefficiency but a fundamental vulnerability.

Some reform efforts—such as "triple nexus" approaches linking humanitarian, development, and peacebuilding work, or the push for "One UN" coherence—attempt to address these gaps. Yet implementation has been uneven, often stymied by institutional rivalries and the reluctance of donors to provide flexible, multi-year financing. Without deeper transformation, the current architecture will remain poorly suited to manage crises that unfold across domains and borders.

In short, bureaucratic inertia and siloed mandates are not incidental flaws of international organizations; they are structural features that have become liabilities.

Unless addressed through institutional redesign and new incentives for integration, global governance will remain reactive, fragmented, and incapable of meeting the accelerating challenges of the polycrisis era.

Chronic Underfunding and Resource Volatility

Even where international institutions are equipped with relevant mandates, they are frequently stripped of the financial resources required to implement them. Underfunding is not an exception—it is the norm. This chronic shortfall has become one of the most corrosive features of the postwar governance system, eroding both operational effectiveness and institutional legitimacy.

Most UN agencies rely heavily on voluntary contributions rather than predictable, assessed funding. This financing model, designed to preserve state sovereignty by keeping budgets under political control, leaves agencies acutely vulnerable to donor priorities, domestic political fluctuations, and the earmarking of funds for specific projects. Flexibility is constrained: funds can be spent only where donors permit, regardless of where needs are greatest. For instance, humanitarian appeals in 2024 and 2025 for Sudan, Yemen, and Gaza met less than half of their targets, despite evidence of escalating famine risk and mass displacement. Peacekeeping budgets have stagnated at a time when demand for stabilization operations has expanded, leaving missions under-resourced and overextended. The World Health Organization (WHO) continues to depend on earmarked voluntary contributions for over 80% of its budget, depriving it of the agility to redirect resources swiftly when new crises emerge.

This funding volatility produces systemic dysfunctions. Humanitarian agencies are locked into reactive cycles of crisis management, perpetually scrambling to fill short-term gaps rather than investing in resilience, preparedness, or long-term solutions. Development programs are similarly trapped in stop-start cycles, dependent on donor political calendars rather than multi-year strategies aligned with community priorities. The result is a governance system that responds episodically rather than strategically, always one step behind accelerating crises.

The credibility costs are severe. Communities that experience unmet needs perceive international promises as hollow rhetoric, deepening mistrust of multilateral institutions. For states in the Global South—already skeptical of the fairness of the international order—resource shortfalls reinforce perceptions of double standards: rapid mobilization for certain crises that affect powerful states or strategic regions, neglect and delay for others. This asymmetry is not lost on local populations, who increasingly see global institutions as instruments of selective solidarity rather than universal responsibility.

The dependency on voluntary contributions also produces political distortions. Donors shape institutional agendas by earmarking funds for issues that align with their visibility goals or geostrategic interests, rather than impartial needs assessments. This skews priorities: high-profile emergencies attract disproportionate resources, while protracted or politically sensitive crises languish in neglect. Agencies, compelled to compete for donor attention, often adjust their mandates to remain attractive, drifting away from core missions. The result is a system that too often appears as much a vehicle of donor influence as of universal governance.

From a systemic perspective, chronic underfunding highlights a deeper structural flaw: the financing model for multilateralism has failed to evolve in step with global risk. In the mid-20th century, voluntary contributions seemed sufficient to support discrete programs in health, food, or humanitarian relief. In the 21st century, where crises are systemic, compounding, and require anticipatory investment, such a model is fundamentally misaligned. Without stable, predictable, and flexible financing, even well-designed institutions are condemned to underperform. They may retain legitimacy in principle but fail in practice, widening the gap between global expectations and institutional delivery.

The stakes are high. As crises proliferate, financial instability does not merely undermine efficiency—it actively erodes legitimacy, incentivizes institutional bypassing, and accelerates the shift toward ad hoc coalitions and private-sector substitutes. Unless resource volatility is addressed, the gap between mandate and capacity will continue to grow, leaving multilateral institutions indispensable in theory but increasingly irrelevant in practice.

Implications: Structural Incompatibility with Complexity

Taken together, these deficiencies reveal an international architecture structurally mismatched to the governance demands of the 21st century. Representation gaps alienate large swaths of the globe, particularly in Africa, Latin America, and emerging Asia, thereby undermining the very legitimacy on which multilateral institutions depend. Vetoes and consensus rules, designed as safeguards for participation and fairness, now institutionalize paralysis, producing immobility precisely when urgent crises demand decisive action. Bureaucratic silos and organizational inertia prevent integrated, anticipatory responses to risks that are deeply interlinked across domains. Chronic underfunding, meanwhile, cripples operational delivery, creates cycles of unmet promises, and steadily erodes the trust that underpins cooperation.

Each of these weaknesses is damaging in isolation, but their interaction compounds dysfunction. Institutions built to embody the principle of collective action fail, in practice, to generate collective solutions. The paradox is stark: these organizations remain indispensable for their legitimacy, symbolic authority, and convening power, yet they are simultaneously insufficient as operational governance mechanisms. They provide the stage on which global politics plays out but lack the tools to shape outcomes effectively.

The implications are profound. If these institutions cannot adapt, the international system will increasingly default to ad hoc coalitions, regional arrangements, and informal governance networks. Such mechanisms—whether issue-specific climate alliances, regional trade compacts, or humanitarian coalitions—may deliver tactical problem-solving and speed. However, they are often fragile: they lack the universality, enforceability, and long-term sustainability that only global institutions can provide. Fragmentation of governance into competing blocs or overlapping networks risks leaving humanity with a patchwork system—responsive in pockets, but incoherent and unable to address systemic risks such as climate change, pandemics, or financial instability that demand coordinated, global action.

This erosion of trust in universal institutions is not a marginal concern. It threatens to accelerate the drift toward a post-multilateral world, one where legitimacy fractures, compliance weakens, and cooperation is contingent rather than collective. The costs of such fragmentation would be especially high in an age of polycrisis: crises that are transboundary by nature cannot be solved by partial arrangements without producing gaps and inequities.

The core challenge is therefore not simply whether traditional institutions can survive as symbols of global cooperation, but whether they can adapt quickly enough to remain relevant in practice. Survival without transformation risks institutional irrelevance: organizations that persist in form but lose the ability to influence outcomes. Without fundamental redesign, the mismatch between institutional capacity and systemic complexity will continue to widen—leaving global governance trapped between indispensability and dysfunction. The choice is stark: either these institutions evolve into more inclusive, anticipatory, and adequately resourced systems, or they will slowly cede the governance space to fragmented and fragile alternatives in an era of accelerating turbulence.

PRINCIPLES FOR REDESIGNING GLOBAL GOVERNANCE

If the post-1945 order was defined by stability through hierarchy, the governance of the 21st century must be redefined as adaptability through inclusion. The Cold War era relied on rigid institutional structures that reflected the geopolitical balance of power and prioritized stability above all else. In contrast, the polycrisis era—marked by climate shocks, pandemics, technological disruption, and geopolitical fragmentation—requires institutions that are not only more representative but also agile, networked, and legitimate across multiple scales of governance.

The intellectual shift here is significant. Whereas traditional institutions sought to minimize uncertainty through predictability and fixed rules, scholars of complex adaptive systems (CAS) argue that resilience is generated not through rigidity, but through the capacity to learn, iterate, and evolve under stress (Homer-Dixon et al., 2022). In a world where crises are systemic, nonlinear, and mutually reinforcing, governance must function less like a fortress and more like an ecosystem—capable of adaptation, feedback, and distributed intelligence.

In this light, six interdependent principles are increasingly recognized as anchors for institutional redesign:

- Adaptability. Institutions must be designed to evolve alongside crises, embedding foresight, feedback loops, and the capacity to recalibrate strategies without waiting for systemic collapse.
- Subsidiarity. Decision-making should occur at the lowest effective level, empowering local and regional actors while maintaining mechanisms for global coordination where problems cross borders.
- Equity. Representation and resource distribution must reflect demographic, geographic, and economic realities, ensuring that marginalized states and communities have a meaningful voice in shaping norms and outcomes.
- Digital inclusion. Governance must recognize digital infrastructures as part of the global commons, guaranteeing equitable access to data, computation, and connectivity while guarding against algorithmic exclusion.
- Indigenous and community contributions. Traditional ecological knowledge and local governance practices are not ancillary but central to resilience, offering diverse epistemologies that enrich global problem-solving.
- Civil society participation. Non-state actors—from NGOs to grassroots movements and the private sector—must be integrated as partners in governance, not relegated to the margins, since legitimacy today is multi-sourced and polycentric.

Together, these principles signal a paradigm shift: from order through hierarchy to resilience through inclusion, from institutions that defend stability to institutions that cultivate adaptability, and from legitimacy rooted in great-power dominance to legitimacy grounded in distributed agency.

Adaptability and Resilience-by-Design

Traditional governance frameworks often falter because they are premised on assumptions of linearity and stability. Post-1945 institutions were designed to manage discrete, predictable risks—balance-of-payments crises, trade disputes, or interstate conflicts—through rules, hierarchies, and standardized procedures. This model presumes that shocks are temporary disruptions to an otherwise stable equilibrium. Yet, in the 21st century, crises are systemic, compounding, and characterized by uncertainty. Governance that relies on rigidity and path dependence is therefore structurally misaligned with the demands of the polycrisis era.

By contrast, resilience-by-design emphasizes flexibility, learning, and adaptive capacity. It draws from the literature on complex adaptive systems, which argues that resilience emerges from the ability to absorb shocks, reorganize, and evolve without losing core functionality (Klinke & Renn, 2021). For institutions, this requires embedding dynamic learning loops, institutional stress tests, and anticipatory governance mechanisms into their architecture, rather than relying on ad hoc improvisation when crises strike.

The COVID-19 pandemic provided a stark demonstration of the costs of rigid and reactive structures. Global vaccine distribution was delayed and fragmented; supply chains collapsed under strain; and coordination among states and international organizations was limited. The WHO (2022) highlighted how insufficient anticipatory capacity, coupled with over-reliance on voluntary cooperation, left many regions without timely access to vaccines and protective equipment. In short, the system was not designed to adapt under pressure—it was designed to manage stability, which no longer exists.

At the same time, adaptive governance experiments during the pandemic illustrated the value of resilience-by-design. The European Union's joint vaccine procurement mechanism, though imperfect, showcased how pooling risk, leveraging collective bargaining power, and building adaptive procurement systems can mitigate inequities and accelerate responsiveness (Schunz, 2023). Similar lessons can be drawn from anticipatory financing mechanisms in the humanitarian field, such as forecast-based financing for climate disasters, which releases funds when early-warning thresholds are met. These models demonstrate that adaptability is not synonymous with improvisation; rather, it requires formalized mechanisms for review, revision, and rapid escalation embedded within institutional design.

Embedding adaptability also demands a cultural shift. Institutions must normalize scenario planning, horizon scanning, and red-teaming exercises, not as peripheral activities but as integral components of governance. Anticipatory triggers—such as automatic financing when climate thresholds are breached, or pre-negotiated vaccine-sharing agreements—can transform response capacity by moving from reactive to proactive modes of governance. Moreover, adaptability must be accompanied by accountability: iterative review processes should ensure that learning translates into institutional change, preventing "lessons learned" from remaining rhetorical.

In essence, resilience-by-design reframes governance from rule maintenance to adaptive stewardship. Instead of protecting a fragile status quo, institutions must be capable of absorbing shocks, adjusting strategies, and iterating under stress. The challenge is not only technical but normative: it requires reimagining legitimacy as rooted in the ability to evolve with changing realities, rather than in the preservation of outdated structures. Institutions that embed adaptability at their core will be better positioned to manage the unpredictable interplay of pandemics, climate shocks, and technological disruptions that define the polycrisis era.

Subsidiarity and Proximity-Based Governance

Global crises may have transnational causes and global consequences, but they are experienced locally first. Floods devastate villages before they affect global food prices, pandemics overwhelm local clinics before they challenge global supply chains, and conflicts destabilize communities long before they reach international negotiation tables. Because of this, legitimacy in crisis response often rests on proximity—who arrives first, who understands the local context, and who is trusted by affected populations.

The principle of subsidiarity—acting at the lowest effective level—has long been embedded in European governance structures, particularly within the European Union (Follesdal, 2022). It reflects the idea that authority should be exercised as close as possible to citizens, while higher levels of governance intervene only when necessary. In today's interconnected world, subsidiarity is no longer a European principle but a global necessity.

Regional organizations often hold comparative advantages over distant global institutions. The African Union (AU) and ASEAN exemplify this. The AU's mediation of the Tigray conflict in Ethiopia (2022) demonstrated how regional actors can leverage cultural familiarity, geographic proximity, and political legitimacy to achieve breakthroughs that external mediators often cannot (Williams, 2023). Likewise, ASEAN's AHA Centre, established to coordinate disaster management, has repeatedly shown how regional hubs can mobilize faster and with greater cultural resonance than global systems (Haacke, 2021). The 2018 Sulawesi earthquake and the 2025 ARDEX disaster simulation highlighted how ASEAN's regional networks facilitated more rapid deployment of aid than traditional UN-led mechanisms.

Subsidiarity is not an argument against global institutions. On the contrary, it strengthens them by ensuring that interventions are anchored in local legitimacy and context. When regional bodies act first, global organizations can add value through scaling, resourcing, and norm-setting, rather than attempting to substitute for local knowledge. This layered architecture reduces the risk of the "one-size-fits-all" interventions that global institutions have often been criticized for, while also creating a sense of ownership among regional and local actors.

However, subsidiarity is not without challenges. Many regional organizations face capacity constraints: the AU Peace Fund remains underfunded; ASEAN's consensus principle limits decisive political action; and organizations in Latin America or the Middle East often lack cohesive security frameworks. Without reliable financing, training, and logistical support, the promise of subsidiarity risks devolving into rhetoric. Moreover, subsidiarity can sometimes be manipulated by dominant regional powers to advance parochial interests under the guise of "local solutions." Ensuring that subsidiarity contributes to equity and legitimacy requires checks, balances, and clear linkages to universal norms.

From an analytical perspective, subsidiarity contributes to what governance scholars describe as polycentric governance systems—structures where authority and responsibility are distributed across multiple overlapping centers (Ostrom, 2010). Polycentricity, when well-coordinated, enhances resilience by ensuring that failure or paralysis at one level does not collapse the entire system. Subsidiarity is therefore not merely a principle of efficiency; it is a structural safeguard against systemic failure.

In an era of cascading crises, subsidiarity reframes global governance from a top-down model to a networked ecosystem. Local and regional actors provide the first line of legitimacy and response, while global institutions ensure coherence, resources, and accountability. The challenge is to institutionalize this complementarity: creating mechanisms where authority is neither hoarded at the top nor fragmented at the bottom but distributed dynamically across levels of governance.

Equity and Global South Agency

The post-war architecture has long been criticized for perpetuating inequities in representation, finance, and agenda-setting. Institutions designed in 1945 reflected the geopolitical realities of that era, but their persistence has entrenched a system in which power is disproportionately concentrated in a handful of advanced economies. Calls for reform have grown louder with each global crisis, as inequities in voice, resources, and decision-making translate directly into inequities in outcomes.

Equity requires more than rhetorical inclusion. It demands the rebalancing of power and resources to reflect today's demographic, economic, and political realities. The Global South has consistently argued that the legitimacy of global governance rests on its ability to evolve in line with shifting global dynamics. This includes reforming voting rights in the International Monetary Fund (IMF), where advanced economies still hold disproportionate shares; expanding representation in the UN Security Council (UNSC), where Africa and Latin America remain unrepresented among permanent members; and delivering climate finance commitments that have repeatedly fallen short of the $100 billion annual pledge (Acharya, 2021). The repeated failure to meet these commitments is not just a fiscal shortcoming but a political signal of broken trust.

Recent initiatives underscore how equity is being driven from the periphery inward. The Bridgetown Initiative, spearheaded by Barbados under the leadership of Prime Minister Mia Mottley in 2022, exemplifies Global South agency in re-imagining international finance. It calls for reforms to multilateral development banks, new mechanisms for debt sustainability, and climate financing tailored to the vulnerabilities of small island and developing states (Mottley, 2022). By framing climate vulnerability as a structural injustice rooted in historical carbon emissions, Bridgetown shifted the debate from charity to responsibility-sharing. Its growing traction illustrates how equity is no longer simply a moral appeal but a strategic demand tied to the credibility of the global order itself.

From an analytical perspective, equity is not a static principle but a structural condition for legitimacy and effectiveness. Institutions that fail to rebalance risk eroding compliance, as excluded actors turn to parallel mechanisms—regional banks, South-South cooperation platforms, or plurilateral agreements—that bypass the traditional system. This centrifugal dynamic weakens universality, the core strength of post-1945 multilateralism. Conversely, embedding equity at the core of institutional redesign strengthens both legitimacy and resilience: when governance frameworks reflect the diversity of global power and responsibility, they are more likely to command compliance, foster solidarity, and generate sustainable solutions.

Equity, therefore, must not be treated as an add-on to existing frameworks. It must be institutionalized as a structural principle—through binding reforms in representation, financing, and agenda-setting. Without such rebalancing, global governance risks reproducing the very hierarchies it claims to transcend. With it, institutions can evolve from forums of selective privilege into genuine instruments of shared responsibility in the polycrisis era.

Digital Inclusion and Federated Infrastructure

The digital domain has become a critical layer of governance, shaping not only economies but also security, diplomacy, and social trust. Digital technologies now underpin financial transactions, health systems, education, supply chains, and political communication. Yet governance of this domain remains fragmented and uneven, marked by the dominance of a few technologically advanced states and private corporations that set de facto rules through platforms, standards, and market power. The result is a governance landscape that lacks transparency, inclusivity, and accountability.

Digital inclusion requires a shift from privatized or nationally siloed systems to federated infrastructure—shared standards, interoperable platforms, and open digital public goods that enable equitable participation across borders (UN, 2023). Such an approach treats digital technologies as global commons, recognizing that data flows, connectivity, and algorithms have spillover effects that transcend jurisdictional boundaries. Governing the digital domain as a commons does not negate sovereignty; rather, it embeds sovereignty in cooperative frameworks that safeguard access, interoperability, and contestability.

Momentum is building around this agenda. The Global Digital Compact, expected to be endorsed at the UN Summit of the Future in 2024, reflects a growing consensus that digital technologies cannot remain under the exclusive control of private monopolies or technologically dominant states (UN DESA, 2023). The Compact aims to establish principles for openness, trust, equity, and accountability in digital governance, placing emphasis on bridging divides in connectivity, skills, and participation. Its negotiation process itself signals a recognition that digital governance must be anchored in multilateral legitimacy rather than unilateral dominance.

National and regional experiments also offer instructive models. India's Digital Public Infrastructure (DPI) framework, which has expanded access to financial services, digital identification, and health systems for hundreds of millions, demonstrates how public architectures can drive large-scale inclusion (World Bank, 2023). Similarly, Europe's Gaia-X initiative seeks to build a federated data infrastructure grounded in shared standards and sovereignty-by-design. These examples illustrate how digital systems, when designed as public goods rather than proprietary monopolies, can strengthen both resilience and equity.

Yet the path forward is contested. Private technology companies retain disproportionate influence over standards, platforms, and algorithms, often prioritizing profit over public interest. At the same time, some states pursue digital sovereignty through restrictive data localization and surveillance-heavy models that risk fragmenting the internet into national "splinternets." Digital inclusion thus requires balancing openness with safeguards, ensuring that governance frameworks prevent both monopolistic concentration and authoritarian capture.

From an analytical perspective, digital inclusion is not a technical issue but a foundational principle of legitimacy in global governance. In an era where digital infrastructures mediate political participation, economic opportunity, and access to knowledge, exclusion from digital systems translates directly into exclusion from governance itself. By embedding access, interoperability, and contestability into institutional design, digital inclusion ensures that digital technologies strengthen rather than erode equity and resilience in the governance system.

Indigenous and Community Governance Contributions

Resilient governance requires not only institutional reform but also a deliberate effort to draw on traditions and knowledge systems that predate the modern state. Indigenous governance frameworks, often marginalized in international debates, offer alternative logics of stewardship, authority, and responsibility that are profoundly relevant to the challenges of the 21st century. Unlike many modern systems, which are grounded in short-term cycles and transactional politics, Indigenous approaches often emphasize long-term stewardship, collective responsibility, reciprocity, and consensus-driven decision-making. These principles provide valuable lessons for addressing systemic risks such as climate change, biodiversity loss, and resource depletion, where intergenerational justice and ecological balance are paramount (Whyte, 2020).

Practical examples already demonstrate the value of integrating Indigenous governance. In Canada and New Zealand, Indigenous water governance practices—rooted in relational understandings of ecosystems and rights-of-nature frameworks—have informed more sustainable and participatory management models (Jackson et al., 2022). Similarly, Indigenous fire management practices in Australia, once sidelined, are now being recognized for their role in reducing wildfire risks and restoring ecological balance. These are not merely cultural contributions but tested models of resilience honed through centuries of adaptive practice in specific ecosystems.

Incorporating Indigenous and community governance into global frameworks is therefore not simply about cultural recognition or symbolic inclusion. It is about expanding the epistemic base of governance—integrating diverse systems of knowledge, values, and practice to better reflect the plural realities of a globalized world. This pluralism strengthens resilience by broadening the range of perspectives available for problem-solving and by embedding governance in communities that directly experience the consequences of policy decisions.

At the same time, integration must be handled carefully. There is a risk of instrumentalizing Indigenous knowledge—treating it as a resource to be extracted and fitted into existing systems without respecting its cultural and political foundations. Genuine incorporation requires power-sharing, recognition of sovereignty, and mechanisms for Indigenous peoples to participate as equal partners in decision-making processes. This means moving beyond "consultation" toward co-governance structures that recognize Indigenous authority and create space for community-led initiatives within global governance frameworks.

From an analytical standpoint, drawing on Indigenous traditions highlights the importance of epistemic justice in governance. As global risks multiply, institutions cannot rely solely on Western scientific or technocratic models. A resilience-based order must be polycentric not only in institutional design but also in knowledge production. Integrating Indigenous and community contributions thus strengthens both the legitimacy and the effectiveness of governance, ensuring that solutions are grounded in diverse worldviews and tested practices of intergenerational responsibility.

In short, the inclusion of Indigenous governance traditions is not a peripheral add-on; it is a structural requirement for resilient governance in the polycrisis era. By embedding long-term stewardship, collective responsibility, and plural epistemologies, institutions can evolve from narrow, state-centric systems into adaptive frameworks capable of addressing the interconnected crises of the 21st century.

Civil Society and Multi-Stakeholder Inclusion

Effective global governance today requires not only the cooperation of states but also the institutionalized participation of civil society, the private sector, and other non-state actors. The scale, speed, and complexity of 21st-century crises exceed the capacities of governments alone. Pandemics, climate change, cyber risks, and humanitarian emergencies unfold across borders, sectors, and communities; solutions therefore require coalitions that transcend the state system. What emerges is a model of governance that is not merely multilateral but multi-stakeholder, embedding diverse actors as co-producers of rules, resources, and legitimacy.

The role of non-state actors in global governance has evolved significantly. In the post-war decades, civil society organizations (CSOs) and businesses were largely confined to consultative roles at the margins of intergovernmental negotiations. Over time, however, their influence expanded as norm entrepreneurs, watchdogs, and implementers. The Rio Earth Summit in 1992 marked a turning point, where NGOs and business coalitions gained unprecedented access to global environmental negotiations. By the time of the Paris Agreement (2015), civil society had moved beyond consultation to active agenda-shaping, pushing governments toward more ambitious targets and providing mechanisms for monitoring compliance (Falkner, 2016). This historical trajectory reflects a steady erosion of the strict state monopoly over global governance.

From a normative standpoint, civil society and private-sector participation address the legitimacy deficit of traditional institutions. Citizens rarely see themselves represented in opaque intergovernmental forums. By including CSOs, grassroots movements, and private networks, governance gains a measure of proximity and accountability. Civil society can articulate the concerns of marginalized groups, amplify voices excluded from formal diplomacy, and hold states accountable to both international commitments and domestic publics. Conversely, when non-state actors are excluded, global governance risks being seen as elite-driven and detached from societal needs. The UN Secretary-General's assertion that *"multilateralism must become multi-stakeholder to be effective"* (UN, 2021) underscores that legitimacy today is plural-sourced: it arises not only from state consent but also from the meaningful participation of diverse societal actors.

The challenge, however, lies in institutionalizing multi-stakeholder governance in ways that are structured, transparent, and equitable. Too often, inclusion remains ad hoc or symbolic, dependent on the goodwill of states or agencies. More robust approaches involve:

- Consultative mechanisms, such as the Major Groups system under the UN Commission on Sustainable Development, which institutionalizes NGO participation.
- Accountability platforms, such as the UN Global Compact, which integrates corporate actors into reporting frameworks for sustainability and human rights.
- Formal partnerships, including multi-stakeholder initiatives like GAVI, the Vaccine Alliance, which institutionalizes collaboration between governments, NGOs, foundations, and the private sector.

These designs not only broaden participation but also clarify responsibilities, distribute accountability, and create predictable avenues for non-state contributions.

In practice, multi-stakeholder inclusion leverages resources, expertise, and innovation that states alone cannot mobilize. During the COVID-19 pandemic, partnerships between governments, pharmaceutical companies, and NGOs enabled the rapid development and distribution of vaccines. While inequities in access were stark, the episode demonstrated the indispensability of non-state actors in scaling scientific breakthroughs, logistics, and advocacy. Similarly, in climate governance, private sector commitments—from divestment pledges by major investors to net-zero targets by corporations—have become essential complements to intergovernmental agreements. Without these actors, decarbonization pathways and climate finance targets would be unreachable.

Civil society also plays a vital role in monitoring and accountability. NGOs provide shadow reports, track emissions and human rights violations, and highlight gaps between commitments and implementation. Their watchdog role keeps both states and corporations under scrutiny, helping to prevent governance from becoming a closed loop of self-reporting and weak enforcement.

While multi-stakeholderism offers clear benefits, it also raises critical questions. First, power asymmetries within civil society and the private sector risk reproducing global inequalities. Well-resourced international NGOs and multinational corporations often dominate multi-stakeholder platforms, marginalizing grassroots groups or local businesses. Second, there is the danger of co-optation, where corporations use multi-stakeholder forums to project legitimacy while continuing harmful practices. Third, accountability can become diluted: when responsibility is spread across states, businesses, and NGOs, it is not always clear who can be held to account when commitments are unmet. Addressing these risks requires transparent rules of engagement, safeguards for equity among non-state actors, and mechanisms to prevent capture by the most powerful participants.

Looking forward, institutionalizing civil society and private-sector roles is not about replacing states but about complementing and amplifying state action. Multi-stakeholder governance must be designed to:

- Broaden legitimacy by embedding diverse voices, especially from marginalized communities and the Global South.
- Harness innovation by tapping into the problem-solving capacity of networks beyond the state.
- Distribute responsibility more equitably across actors, preventing the over-concentration of burdens on weak states or over-reliance on corporate voluntarism.

In this sense, the future of global governance lies in networked coalitions that combine state authority with non-state legitimacy, resources, and expertise. If the post-1945 order was hierarchical and state-centric, the governance of the polycrisis era must be polycentric and multi-stakeholder, embedding civil society and private actors not as adjuncts but as structural partners.

INNOVATIVE AND EMERGING GOVERNANCE MODELS

The inadequacies of traditional institutions have not produced a collapse of global governance but rather a proliferation of experiments and innovations across multiple levels and domains. These emerging models are not wholesale replacements for established institutions; instead, they represent an additional layer of hybrid governance that supplements, challenges, and sometimes reshapes existing structures. They bring into play new actors, technologies, and approaches that address the systemic weaknesses of 20th-century institutions—rigidity, exclusion, and slowness—with qualities better aligned to the 21st-century polycrisis: adaptability, inclusivity, and anticipatory capacity.

What unites these innovations is their pragmatism. None assume that existing institutions can be dismantled and rebuilt from scratch. Rather, they offer pathways of augmentation—patches, platforms, and practices that add resilience to a system under strain. This section surveys innovations across five domains—diplomacy and negotiation, health governance, climate and sustainability platforms, technology governance, and regional or hybrid mechanisms—highlighting both their promise and their limitations.

Diplomacy and Negotiation

Diplomacy, once dominated by slow, state-centric negotiations behind closed doors, is being reshaped by new technologies, actors, and formats. Traditional diplomacy helped stabilize the Cold War order by privileging secrecy and elite consensus, but it has become increasingly inadequate in an age where crises demand speed, transparency, and cross-sectoral expertise. What is emerging is a hybrid landscape: innovations that supplement established practices rather than displace them, creating more adaptive systems of negotiation.

Artificial intelligence represents one frontier. Pilots have demonstrated how machine learning models can simulate negotiation scenarios, identify potential compromise zones, and stress-test outcomes in real time (Maoz et al., 2023). These tools have been tested in conflict mediation training and trade dispute simulations, offering negotiators new perspectives that reduce deadlock by revealing hidden zones of agreement. The European Union has gone further, investing in AI foresight platforms to model the trade-offs of competing climate commitments under uncertainty (European Commission, 2023).

Yet these tools are decision-support, not decision-making systems. Legitimacy requires human judgment to remain central. If algorithmic recommendations become determinative, diplomacy risks drifting into technocratic overreach, privileging efficiency over fairness and deepening inequality between technologically advanced states and those without access to such systems. The promise of AI is real, but so are the risks of exclusion and capture.

The COVID-19 pandemic catalyzed the mainstreaming of digital diplomacy. Virtual platforms sustained G20 summits, UN climate talks, and bilateral consultations during lockdowns, proving both feasible and efficient (Hocking & Melissen, 2015). In the aftermath, hybrid mediation models have emerged: combining in-person shuttle diplomacy with secure digital consultations. These innovations broaden participation by enabling Global South stakeholders, women's groups, and marginalized communities to join processes that would otherwise be limited to elites in far-off capitals.

Still, digital diplomacy carries unresolved equity challenges. Unequal access to connectivity, cybersecurity vulnerabilities, and concerns over confidentiality mean inclusion cannot be assumed. Without deliberate safeguards, digital platforms may reproduce existing inequalities, merely digitizing exclusion rather than dismantling it.

Negotiations are most effective when grounded in shared and trusted evidence bases. The UN's Office for the Coordination of Humanitarian Affairs (OCHA) has pioneered humanitarian data hubs that align needs assessments and reduce duplication. The Global Digital Compact (UN, 2023) similarly emphasizes interoperable data standards as prerequisites for effective cross-border governance. Shared data reduces disputes over facts and allows negotiators to focus on values and trade-offs.

But questions remain: Who governs these repositories? Whose knowledge counts? How is trust established across diverse epistemic communities? The governance of data itself has become a site of contestation that can make or break inclusive diplomacy.

Financial innovation is reshaping negotiations in subtle but far-reaching ways. Blockchain-based stablecoins have been piloted by the World Food Program in Lebanon and Ukraine, enabling assistance to bypass collapsed banking systems and flow directly to beneficiaries (WFP, 2022). These tools promise lower transaction costs, greater transparency, and more resilient financial flows in fragile settings.

Yet they are double-edged. Without regulation, digital currencies could destabilize fragile economies, facilitate illicit flows, or erode state monetary sovereignty. Negotiations around sanctions, remittances, and aid delivery are already being reframed by these innovations, introducing both new possibilities and new vulnerabilities.

Finally, diplomacy is being reimagined as a networked ecosystem rather than a state monopoly. City networks such as C40 Cities and the Global Covenant of Mayors have taken on the mantle of climate diplomats, committing to ambitious emissions reductions and sharing best practices across borders (Falkner, 2016). These platforms leverage distributed leadership and polycentric authority to drive momentum even when intergovernmental talks stall.

But they lack binding enforcement and risk fragmentation. Without integration into broader governance frameworks, ecosystem diplomacy may produce ambitious but uneven patches of innovation rather than systemic transformation.

In sum, the future of diplomacy is neither tradition nor disruption alone. It is the construction of hybrids: state authority coupled with network dynamism, AI insights paired with human judgment, digital speed balanced by inclusive legitimacy.

Health and Pandemic Governance

The COVID-19 pandemic was the most significant stress test of global health governance since the creation of the WHO. While the WHO played a vital role in early alerts and technical guidance, its limited authority, underfunding, and reliance on voluntary compliance exposed critical flaws (WHO, 2022).

The creation of COVAX, co-led by Gavi, CEPI, and the WHO, was heralded as a breakthrough in equity-by-design. Its pooled procurement model aimed to prevent vaccine nationalism by ensuring equitable access for low-income countries. In principle, it represented a paradigm shift in solidarity.

In practice, however, COVAX fell short. Wealthier states secured bilateral deals, undermining pooled procurement. Export restrictions in key manufacturing states disrupted supply, while funding pledges lagged behind commitments. By mid-2022, low-income countries had received fewer than 2% of global vaccine supplies, even as advanced economies faced surpluses. COVAX revealed both the promise of global solidarity and the harsh reality of competitive geopolitics (Moon et al., 2021).

In response, states launched negotiations for a Pandemic Accord under WHO auspices. The Accord aims to move beyond ad hoc improvisation by institutionalizing binding obligations: mandatory data sharing, equitable access to countermeasures, sustainable financing, and accountability mechanisms (Gostin et al., 2023).

The process remains contentious. Many states fear sovereignty infringements, while others resist financing obligations. Yet the Accord's ambition reflects a recognition that voluntarism is insufficient in a world where pandemics carry systemic risks. Its success will hinge on whether consensus can translate into enforceable commitments—an open question, but one that defines the future of global health security.

Climate and Sustainability Platforms

Climate governance has been a laboratory for some of the most innovative models of global cooperation.

Launched in 2022 by Barbados, the Bridgetown Initiative reframed climate finance debates by centering the vulnerabilities of small island developing states. It called for reforms of multilateral development banks, expanded concessional finance, and mechanisms for debt sustainability (Mottley, 2022). Its resonance illustrates how Global South leadership can redefine the global agenda, shifting climate governance from charity to justice. At COP27, states agreed to create a Loss and Damage Fund, later operationalized at COP28. The Fund represents a historic acknowledgment of historical responsibility: that vulnerable states should not shoulder the costs of climate harms alone. Initial pledges, however, were modest—under $1 billion compared to annual needs in the hundreds of billions (Roberts & Huq, 2023). The Fund is a milestone in principle but fragile in practice, dependent on sustained political will and financing.

Parallel to intergovernmental processes, city networks like C40 Cities have emerged as non-state climate diplomats, committing to net-zero targets, exchanging best practices, and sustaining momentum when intergovernmental talks falter (Bulkeley et al., 2018). These networks embody polycentric governance in action, yet they too suffer from voluntary commitments and uneven accountability.

Together, these innovations show that climate governance is both the most dynamic field of experimentation and one of the most vulnerable to underfunding and fragmented authority.

Technology and Digital Governance

Technology governance is now at the heart of global order. The endorsement of the Global Digital Compact at the UN Summit of the Future (2024) signaled recognition that digital technologies must be governed as global commons (UN DESA, 2023). It promotes principles of interoperability, equity, and openness. Yet, its non-binding nature underscores the enduring difficulty of enforcement in a domain dominated by powerful states and corporations.

Artificial intelligence is especially contested. Proposals for a Global AI Scientific Panel, modeled on the IPCC, reflect recognition that shared evidence and epistemic authority are vital for policymaking (Cihon et al., 2021). Meanwhile, the Hiroshima AI Process (G7) and the BRICS Rio Declaration (2024) illustrate competing governance tracks: one led by advanced economies, the other by emerging powers.

National and regional AI Safety Institutes in the UK, U.S., and EU are beginning to form epistemic networks through red-teaming, testing protocols, and information sharing. These institutes represent an embryonic form of epistemic diplomacy, where convergence is pursued at the technical level even when political consensus is absent. The question is whether these polycentric initiatives can be connected into a coherent architecture that balances innovation, safety, and equity.

Regional and Hybrid Innovations

Regional organizations have increasingly become critical complements to global governance, offering advantages of proximity, contextual legitimacy, and speed. Their ability to act quickly in crises and to mediate with cultural and geographic familiarity often gives them traction where distant global institutions encounter inertia. These bodies embody the principle of subsidiarity in practice: decisions and interventions made closer to the source of crises are often more politically acceptable, better informed, and more sustainable.

The African Union (AU) provides a prominent illustration. Its mediation of the Tigray conflict in 2022 showed that African actors could convene warring parties and deliver agreements where external powers lacked credibility (Williams, 2023). This was not only a diplomatic breakthrough but also a symbolic one: it reaffirmed African agency in conflict resolution, countering narratives of dependency on external mediators. The AU Peace and Security Council has also developed innovative mechanisms such as the Panel of the Wise and the Continental Early Warning System, though resource and political constraints limit their effectiveness.

In Southeast Asia, the Association of Southeast Asian Nations (ASEAN) has operationalized regional solidarity through its Coordinating Centre for Humanitarian Assistance (AHA Centre). The Centre has demonstrated practical value in disaster response, including during the 2018 Sulawesi earthquake and subsequent simulation exercises such as ARDEX-25 (Haacke, 2021). These activities reinforce ASEAN's comparative advantage: not resolving political disputes—where consensus rules often constrain it—but mobilizing rapid and culturally attuned humanitarian responses.

Hybrid governance models extend this logic by blending global legitimacy with regional proximity. The UN–AU Mission in Darfur (UNAMID), operational from 2007 to 2020, remains a landmark example of such collaboration. While resource-intensive and operationally complex, the mission demonstrated how a layered governance model could draw on the UN's legitimacy and logistical capacity while embedding the AU's regional authority and contextual expertise (De Coning & Karlsrud, 2021). Other hybrid models, such as EU–UN cooperation in the Balkans or UN–ECOWAS electoral support missions, illustrate the flexibility of shared arrangements.

These models, however, face persistent constraints. Chronic underfunding undermines sustainability; dependence on external donors erodes autonomy; and uneven political will among member states weakens coherence. The AU's Peace Fund, for instance, continues to rely heavily on external partners, while ASEAN's consensus principle often dilutes collective positions on contentious issues such as Myanmar. Hybrid missions risk duplication, blurred accountability, and complex command structures. Yet even with these weaknesses, regional and hybrid arrangements often deliver more context-sensitive and legitimate outcomes than purely top-down approaches.

Taken together, regional and hybrid innovations highlight an important evolution in global governance: the move from hierarchical, universalist structures toward layered, polycentric architectures. The logic of proximity and subsidiarity, when reinforced by global legitimacy and resources, produces governance outcomes that are both more responsive and more credible. This shift is particularly evident in peace and security, but the same principle is emerging in health, climate, and technology governance—where regional hubs and coalitions are increasingly shaping outcomes.

Across diplomacy, health, climate, technology, and regional security, innovations are proliferating and diversifying. None of these models—whether COVAX, the Bridgetown Initiative, the Global Digital Compact, or AU mediation—offers a comprehensive solution. Each faces limits of financing, enforcement, and coherence. Yet collectively, they represent emergent experiments in adaptability, equity, and polycentricity—the very qualities that traditional institutions lack.

The challenge ahead is integration: to connect these disparate initiatives into a more cohesive global architecture. Without deliberate alignment, these innovations risk remaining fragmented patches, useful but insufficient. With coordination, however, they can evolve into the building blocks of a governance system fit for the polycrisis era—anticipatory in outlook, inclusive in design, and resilient by construction.

PROPOSED BLUEPRINT FOR REDESIGN

The preceding sections have demonstrated that the post-1945 governance architecture is both indispensable and insufficient for the challenges of the 21st century. On the one hand, institutions such as the United Nations, the Bretton Woods system, and regional organizations still provide the central scaffolding of international legitimacy. They remain the default convening spaces for states, the custodians of foundational norms like sovereignty and human rights, and the distributors of critical resources ranging from humanitarian aid to development financing. On the other hand, the system's design reflects the realities of a world now long past: an era of bipolar geopolitics, state-centric threats, and largely linear risks. Today's world is shaped instead by polycrisis dynamics—the entanglement of climate breakdown, pandemics, technological disruption, economic volatility, and violent conflict—where shocks cascade across domains and amplify each other in non-linear ways. Institutions optimized for single-issue mandates and incremental consensus are simply unfit for governing such systemic turbulence.

The inadequacy is not merely technical; it is existential. As recent crises from Syria to Ukraine, from COVID-19 to Gaza, have shown, institutional paralysis translates directly into human suffering. When the UN Security Council is immobilized by veto politics, civilian populations are left exposed to atrocity crimes. When climate finance mechanisms fail to deliver promised funds, vulnerable communities in the Global South are forced into cycles of debt and disaster. When the World Health Organization lacks authority to enforce equitable vaccine distribution, preventable deaths accumulate by the millions. These are not isolated failings but systemic outcomes of an order designed for a different era. The system continues to deliver legitimacy by virtue of its universality, but it cannot deliver effectiveness at the speed and scale demanded by today's crises. As a result, a paradox has emerged: the world cannot do without these institutions, yet it cannot rely on them alone.

In this context, reform is not optional—it is the condition of survival for multilateralism itself. Without meaningful redesign, international cooperation risks sliding into irrelevance, ceding space to ad hoc coalitions, regional blocs, or unilateral action. Fragmentation would not only weaken the ability to manage global risks but would also erode the very norms of collective action on which the legitimacy of the system rests. Already, we see early symptoms of such erosion: states bypassing the WTO through bilateral or plurilateral trade deals; humanitarian agencies increasingly reliant on private philanthropy to cover funding gaps; and parallel governance tracks for artificial intelligence emerging between G7 and BRICS blocs. Left unchecked, these trends will harden into rival architectures that mirror geopolitical divisions, undermining the universality that has long been the system's greatest strength.

At the same time, the proliferation of innovative governance models—from the Bridgetown Initiative on climate finance, to city-led climate networks, to experiments with AI-assisted negotiation—demonstrates that adaptation is possible. These experiments are not yet systemic, but they offer glimpses of what a more resilient and equitable order could look like: polycentric rather than hierarchical, anticipatory rather than reactive, inclusive rather than exclusive. The lesson is clear: innovation is already occurring, but without integration and scaling, it risks remaining fragmented. What is needed is not a wholesale demolition of the existing order, but a recalibration—a redesign that learns from past failures while building on emerging practices.

This redesign must be anchored in three guiding principles. First, agility: institutions must be capable of rapid adaptation, with decision-making processes that prioritize early warning, pre-agreed triggers, and flexible mandates. Second, equity: governance structures must give meaningful voice and resources to the Global South, recognizing that legitimacy depends on fairness in representation and burden-sharing. Third, resilience-by-design: institutions must be structured not just to react to crises but to anticipate and withstand systemic shocks, embedding adaptability into their very architecture. These principles are not abstract ideals but operational necessities for a world defined by cascading risks.

What follows, therefore, is a proposed blueprint for redesign. It does not seek to replace the multilateral system but to renovate its foundations—strengthening collective security, reforming economic and climate finance, governing digital technologies as global commons, embedding subsidiarity across scales, and institutionalizing accountability. Each element of the blueprint is drawn from the twin lessons of the last decade: that traditional institutions are structurally misaligned with contemporary risks, and that pockets of innovation already exist that can be scaled into systemic solutions. The goal is a governance system that remains universal in scope but is no longer paralyzed by its own architecture—one that is anticipatory, inclusive, and resilient enough to navigate the age of polycrisis.

Security and Peace Governance

The United Nations Security Council (UNSC) remains the cornerstone of collective security, yet its legitimacy and effectiveness have been steadily undermined by representational imbalances and the recurrent use of the veto. To restore credibility and operational effectiveness, reform must focus on three interlinked priorities: widening representation, constraining unilateral vetoes, and establishing fallback mechanisms.

UNSC enlargement is an essential step. The Council's composition reflects the geopolitical realities of 1945 rather than those of the 21st century. Expanding membership to include permanent representation from Africa, Latin America, and South Asia would correct long-standing legitimacy deficits. The Ezulwini Consensus of the African Union, which calls for two permanent and five non-permanent African seats, provides a credible foundation for reform (AU, 2005). While enlargement alone would not end great-power rivalry, it would reduce the perception of the Council as an exclusive club frozen in time and would strengthen its normative authority by making it more representative of today's global population and power distribution.

Veto restraint mechanisms are equally critical. Initiatives such as the France–Mexico proposal, which urges the permanent members (P5) to voluntarily refrain from veto use in cases of mass atrocities, have gained normative traction in both diplomatic and academic circles (Security Council Report, 2022). Institutionalizing such practices—whether through General Assembly endorsement, informal codes of conduct, or Security Council working methods—could help curb the most egregious abuses of veto power. Even voluntary restraint, if consistently applied, would signal a shift toward prioritizing collective security over narrow national interests in cases of grave humanitarian concern.

Where paralysis persists, fallback mechanisms should be activated. The General Assembly's *Uniting for Peace* procedure (GA Res. 377) offers an underutilized pathway for maintaining international legitimacy when the Council is deadlocked. Institutionalizing processes whereby General Assembly resolutions could trigger advisory opinions from the International Court of Justice (ICJ), or authorize regional interventions under Chapter VIII of the UN Charter, would operationalize this dormant capacity. By creating a structured safety valve, such measures would ensure that collective action is not permanently blocked by a single veto.

Beyond structural adjustments, the Peacebuilding Commission (PBC) should be elevated into a Peacebuilding Council. Despite its integrative mandate—linking security, development, and governance—the PBC has often been marginalized. Yet complex conflict settings, such as those in the Sahel or Horn of Africa, demonstrate that siloed approaches are ineffective. A reformed Peacebuilding Council, with stronger political backing and predictable financing, could provide the long-term engagement needed to sustain fragile peace processes (Gowan & Stedman, 2020).

Finally, peace operations must evolve from reactive to anticipatory. Current missions are often deployed after conflicts have escalated beyond control, by which point the human and financial costs are immense. Embedding anticipatory peacekeeping into UN doctrine—supported by rapid deployment mechanisms, pre-arranged standby forces, and digital early-warning platforms—would enable faster, more preventive interventions. Regional organizations such as the African Standby Force and ECOWAS's rapid reaction frameworks offer models of proximity-based preparedness. Strengthening these partnerships, and ensuring their integration with UN operations, would help move peacekeeping from fire-fighting toward resilience-building.

Taken together, these reforms would not eliminate great-power politics, but they would make the system more credible, representative, and adaptive. A Security Council that reflects today's world, constrained in its veto use, supported by fallback mechanisms, and complemented by anticipatory peace operations would be better equipped to uphold international peace and security in the polycrisis era.

Economic and Climate Finance

The Bretton Woods institutions remain central pillars of global economic governance, anchoring international monetary stability, development finance, and trade. Yet their governance structures and operating practices continue to perpetuate inequities that erode legitimacy, particularly among Global South states. In an era of climate breakdown, debt distress, and systemic inequality, reform is no longer a matter of fine-tuning but of structural redesign. Three priorities stand out: redistribution of voice, the creation of integrated climate-finance governance, and the scaling of innovative financial instruments.

First, voting quota reform in the IMF and World Bank must reflect 21st-century realities. Emerging economies now drive most of global economic growth, yet their representation in these institutions remains disproportionately small. For example, the United States retains de facto veto power in the IMF with 16.5 percent of voting rights, while sub-Saharan Africa's 46 countries collectively hold less than 5 percent (IMF, 2023). Efforts at reform have been piecemeal: the 2010 quota reforms modestly increased the weight of China, India, and Brazil, but left Africa underrepresented. Without meaningful recalibration of voting formulas, the legitimacy deficit will widen, encouraging states to bypass the Bretton Woods institutions in favor of parallel arrangements such as the BRICS New Development Bank and the Asian Infrastructure Investment Bank (AIIB). Realignment is not merely symbolic—it would secure compliance, burden-sharing, and ultimately the trust needed for coordinated crisis management.

Second, the creation of a Global Climate Resilience Council (GCRC) could consolidate fragmented initiatives under a unified governance platform. Current mechanisms for climate finance are scattered across multiple institutions, from the Green Climate Fund to bilateral donor channels, leading to duplication and inefficiency. A GCRC convened under UN auspices would integrate finance, adaptation, and disaster preparedness, ensuring that pledges are matched to recipient needs and that commitments translate into predictable flows. Its remit could include overseeing the operationalization of the *Loss and Damage Fund* agreed at COP27 and scaled at COP28, while also coordinating debt relief and insurance mechanisms for climate-vulnerable states (Roberts & Huq, 2023). A council anchored in universal representation but equipped with technical expertise could serve as a hub for climate-related fiscal coordination, reducing fragmentation and ensuring accountability.

Third, innovative financial instruments must be scaled to deliver resilience-by-design. Incremental aid increases will not suffice to meet the trillions required annually for sustainable development and climate adaptation. Mechanisms such as solidarity levies—small taxes on aviation, shipping, or financial transactions—have been successfully piloted in limited forms, such as the French solidarity levy on air tickets that funds UNITAID. Applied more broadly, such levies could generate predictable, non-voluntary resources for global public goods (Atkinson, 2015). Debt-for-climate swaps offer another pathway: they allow highly indebted countries to redirect debt servicing toward resilience projects, simultaneously reducing fiscal pressure and funding adaptation (Volz et al., 2021). Early pilots, such as the 2021 Belize "blue bond" debt swap supported by The Nature Conservancy, have demonstrated feasibility. Green bonds, already mobilizing billions in private capital, can be expanded through blended finance models that reduce investor risk while ensuring equitable access for low- and middle-income countries (World Bank, 2022). Together, these tools represent not isolated experiments but the building blocks of a reimagined financial architecture oriented around sustainability.

The principle is clear: global economic governance must shift from short-term stabilization toward resilience-by-design. The Bretton Woods system was built to prevent balance-of-payments crises and restore postwar growth. In the 21st century, its mission must expand to encompass climate adaptation, systemic equity, and intergenerational sustainability. Equity and resilience should be structural goals, not externalities. By redistributing voice, consolidating fragmented climate finance under a dedicated global platform, and scaling innovative mechanisms, international financial governance can recover legitimacy and effectiveness—delivering not only stability for markets but security for people and ecosystems.

Digital and Technology Governance

Digital infrastructure has become the nervous system of globalization: it connects economies, enables financial transactions, mediates communication, and underpins critical services from healthcare to logistics. Yet while technological innovation accelerates at exponential pace, governance lags far behind. This gap creates risks of fragmentation, monopolization, and what scholars increasingly term "digital colonialism"—a condition in which a handful of states and corporations dominate global digital ecosystems, shaping rules, access, and knowledge flows to their advantage (Couldry & Mejias, 2019). To prevent such asymmetries from hardening, digital governance must be grounded in inclusivity, openness, and accountability. In practice, this requires new institutional arrangements at the global level to ensure that digital technologies evolve as public goods rather than private fiefdoms.

The first priority is establishing federated Digital Public Goods (DPGs) "rails." These are interoperable infrastructures for identity, payments, health records, and data exchange, designed as open-access platforms that any country can adopt and adapt. India's Digital Public Infrastructure (DPI), anchored by Aadhaar (digital identity), UPI (unified payments interface), and CoWIN (vaccine management), has demonstrated the transformative potential of such systems for accelerating inclusion and service delivery (World Bank, 2023). By leveraging open-source architecture, DPI has lowered transaction costs, expanded access to financial services, and enabled rapid vaccine distribution to over a billion people during COVID-19.

Yet these national successes require scaling into global coordination. Without a federated framework, the proliferation of proprietary standards risks creating digital silos dominated by a few advanced economies and tech giants. A federated global system would allow interoperability across borders while enabling national customization—what the UN has called "digital public goods as global rails" (UNDP, 2022). Such infrastructure could underpin everything from cross-border remittances to pandemic data exchange, ensuring that low- and middle-income countries are not left behind. Importantly, federated DPGs would prevent monopolization by embedding openness and interoperability at the design stage, reducing dependency on private actors and protecting national digital sovereignty.

Artificial intelligence (AI) has become the defining general-purpose technology of the 21st century, with implications for security, labor, health, and governance. Yet its deployment is outpacing the ability of regulators to ensure accountability. To address this gap, an AI assurance ecosystem is urgently needed. Unlike centralized regulation, assurance ecosystems rely on distributed oversight mechanisms—shared standards, red-teaming protocols, and contestability frameworks—that enable multiple actors to test, challenge, and refine AI systems (Cihon et al., 2021).

The goal is not to create a single global AI regulator, which would be politically unfeasible, but to develop federated oversight networks. For example, the UK's AI Safety Institute, launched in 2023, has begun to pioneer red-teaming methodologies that stress-test large language models for systemic risks. The U.S. and EU have established parallel institutions, while China has advanced state-led audit frameworks. Linking these through shared protocols could form the backbone of a transnational assurance ecosystem, reducing fragmentation while respecting sovereignty.

Contestability is central: affected communities and states must have pathways to challenge AI systems that cause harm, whether in biased hiring, credit allocation, or predictive policing. Assurance mechanisms would thus serve not only as technical checks but as democratic safeguards, embedding transparency, accountability, and redress into the very fabric of AI governance. This model mirrors the architecture of aviation safety, where distributed national regulators adhere to common global standards set by the International Civil Aviation Organization (ICAO), enabling both interoperability and accountability.

To further institutionalize epistemic authority, the establishment of a Global AI Scientific Panel is necessary. Modeled on the Intergovernmental Panel on Climate Change (IPCC), this body would not prescribe policies but would synthesize technical knowledge, provide scenario modeling, and anchor governance in a shared evidence base (UNESCO, 2021).

The AI field is currently marked by hype, fragmentation, and significant asymmetries of expertise. Governments often rely on information provided by private firms, creating risks of regulatory capture. An independent, multi-stakeholder panel would aggregate the best available science and technology assessments, separating realistic capabilities from exaggerated claims. Such a body could provide periodic "AI Assessment Reports," evaluating risks across domains such as labor displacement, disinformation, military applications, and environmental impact.

Like the IPCC, the Global AI Scientific Panel would function as a boundary organization, translating technical knowledge into policy-relevant insights without dictating political choices. It would provide the epistemic infrastructure necessary for informed negotiations, helping states converge on shared definitions and risk categories. In a field where misunderstandings and misperceptions could easily spiral into geopolitical competition, such a body could become the backbone of trust.

Technology governance will determine not only who benefits from innovation but also whether it exacerbates or reduces global inequality. Without inclusive design, digital systems risk hardening existing divides—between the connected and the disconnected, between those who produce technology and those who consume it, and between powerful corporations and marginalized communities. Already, 2.6 billion people remain offline, disproportionately in the Global South (ITU, 2023). Without equitable access to digital infrastructure, AI and other emerging technologies could deepen rather than close global inequalities.

Embedding inclusivity at the design stage is therefore essential. Federated DPGs can ensure access to basic digital infrastructure. Assurance ecosystems can provide accountability across borders. A Global AI Scientific Panel can institutionalize epistemic equity by ensuring that expertise is shared rather than monopolized. Together, these measures would help prevent the emergence of rival governance blocs divided along geopolitical lines—a real risk as G7-led initiatives like the Hiroshima AI Process compete with BRICS-led frameworks such as the Rio Declaration on AI.

In the polycrisis era, digital governance is not a technical sidebar—it is a central pillar of global order. The capacity to manage AI, digital infrastructure, and data as global commons will shape not only economic competitiveness but also the legitimacy of the international system. By treating digital technologies as public goods, embedding equity in governance frameworks, and institutionalizing accountability through assurance and epistemic mechanisms, global governance can shift from reactive regulation to resilience-by-design.

Multi-Level and Subsidiarity Governance

The polycrisis era underscores a basic truth: no single level of governance—local, regional, or global—can resolve systemic risks alone. Climate shocks, pandemics, technological disruptions, and conflicts all cascade across borders, demanding responses that are both globally coordinated and locally grounded. What is required is a polycentric system in which authority is shared, layered, and coordinated across scales. Polycentric governance, as theorized by Elinor Ostrom, generates resilience by distributing decision-making among overlapping centers rather than relying on a single hierarchical authority (Ostrom, 2010). In practice, this means empowering local and regional actors while retaining global frameworks for coherence and legitimacy.

Formalizing UN–regional integration mechanisms represents an important step. Hybrid operations such as the UN–African Union Mission in Darfur (UNAMID) revealed both the promise and complexity of collaborative models. While coordination challenges and resource strains were evident, the arrangement also demonstrated that regional legitimacy and global authority can be combined in ways that neither actor could achieve alone (De Coning & Karlsrud, 2021). Institutionalizing joint planning, financing, and mandate-setting between the UN and regional organizations—whether the AU, ASEAN, or the EU—would reduce duplication, clarify roles, and harness comparative advantages. For example, the AU's early-warning systems could feed directly into UN Security Council deliberations, while UN logistical support could sustain AU-led missions. Building formalized frameworks of subsidiarity would transform one-off experiments into systemic practice.

Agile coalitions represent another layer of polycentricity. They can provide speed and focus in ways that universal institutions often cannot. The G20's role during the COVID-19 pandemic illustrated how ad hoc groupings of major economies can mobilize financial and political resources quickly. Yet such coalitions face a persistent legitimacy deficit, as their exclusivity sidelines the voices of smaller or vulnerable states. Without corrective mechanisms, they risk reinforcing inequalities rather than solving them. Embedding safeguards—such as formal reporting requirements to the UN General Assembly and structured consultation with affected states—could help align agility with inclusivity. In this way, agile coalitions could complement universal institutions rather than bypass them, providing rapid responses while still anchored in multilateral legitimacy (Patrick, 2017).

Cities and civil society also deserve institutionalized roles in multi-level governance. Urban centers now host the majority of the world's population and generate the bulk of global emissions, making them central actors in both risk production and risk mitigation. Networks such as C40 Cities and the Global Covenant of Mayors have pioneered "urban diplomacy," advancing ambitious climate targets even when intergovernmental negotiations stall (Bulkeley et al., 2018). Civil society coalitions, similarly, have proven indispensable in shaping global outcomes—from advocacy that secured the Paris Agreement's inclusion of a 1.5°C target, to humanitarian networks coordinating frontline responses in conflict and disaster zones. Embedding these actors into advisory boards, accountability panels, and formal consultation mechanisms would broaden legitimacy, strengthen policy innovation, and ensure that governance frameworks reflect lived realities rather than elite bargains.

At the core of this approach lies the principle of subsidiarity: decisions should be taken at the lowest effective level, closest to the people and ecosystems they affect. Subsidiarity is not about fragmenting governance but about layering it—ensuring that global frameworks provide coherence while local and regional actors provide legitimacy and context sensitivity. In a polycrisis world, subsidiarity transforms governance from a brittle hierarchy into a resilient web of overlapping authorities, where failures at one level can be compensated by responsiveness at another.

THE GLOBAL GOVERNANCE LEGITIMACY CRISIS: TRUST AND FAIRNESS IN THE POLYCRISIS ERA

Defining the Legitimacy Challenge

Legitimacy in global governance is not a rhetorical luxury; it is the condition that converts institutional decisions into compliance and cooperation under anarchy (Hurd, 1999; Franck, 1990). In Weberian terms, legitimacy furnishes authority with meaning beyond coercion or material inducement. Contemporary governance literature further parses legitimacy into input (who gets a voice), throughput (the quality of decision processes), and output (policy performance and problem-solving) (Scharpf, 1999; Schmidt, 2013). When any one dimension is persistently weak, institutions strain; when all three are contested, systems face a crisis of trust. In international politics—where enforcement is thin—trust is the connective tissue that sustains rules, reduces transaction costs, and anchors long-horizon cooperation (Levi & Stoker, 2000; Tyler, 2006).

The polycrisis intensifies legitimacy demands. Entangled risks—pandemics, climate extremes, economic shocks, algorithmic harms, and armed conflict—raise expectations for speed, fairness, and coordination while exposing structural inequities in representation, voice, and resource allocation. As a result, critiques of global institutions today are not limited to inefficiency; they increasingly allege unfairness—that the rules, seats, and purse strings privilege a few (Acharya, 2021). The consequence is a drift toward contested multilateralism: states and coalitions hedge by creating parallel institutions, minilateral clubs, or regional arrangements, not to abandon cooperation, but to renegotiate its terms (Morse & Keohane, 2014).

Dimensions of the Crisis

Representation Deficits (Input)

The Security Council's permanent membership and the voting shares of the Bretton Woods institutions mirror mid-20th-century power, not 21st-century demography or economics. Africa and Latin America lack permanent seats on the Council, and emerging economies remain underweighted in IMF/World Bank quotas. These structures weaken the perceived "right to rule," especially across the Global South, where under-representation is read as a proxy for systemic bias (Acharya, 2021).

Veto Paralysis and Procedural Blockage (Throughput)

The repeated use or threat of the veto on Syria, Ukraine, and Gaza has transformed the Council's most distinctive tool into a symbol of procedural capture (Oxfam International, 2024). Beyond the UN, unanimity and consensus rules in trade and climate forums systematically produce lowest-common-denominator outcomes (Hoekman & Nelson, 2020). In legitimacy terms, throughput falters when procedures are seen to protect the powerful at the expense of collective problem-solving.

Funding Volatility and Selectivity (Output)

Humanitarian and climate finance are chronically underfunded relative to need, with stark variation across crises. Appeals for protracted emergencies in places like Sudan have languished at fractions of requested amounts, even as other theaters see rapid mobilization (UNOCHA, 2024). When the gap between commitments and delivery becomes a structural feature, output legitimacy declines; promises without resources corrode trust.

Normative Inconsistency (All Three)

Perceptions of double standards—selective invocations of international law, inconsistent sanctioning, or variable humanitarian concern—undercut the moral authority of institutions (Thakur, 2023). In legitimacy terms, inconsistency simultaneously harms input (whose voices count), throughput (how rules are applied), and output (who benefits), reinforcing a narrative of unfairness.

Taken together, these dimensions produce not just dissatisfaction with efficiency but a crisis of fairness—a belief, widespread across the Global South, that the post-war order distributes voice and burdens unjustly. In such settings, compliance becomes contingent and forum-shopping increases, weakening universal regimes (Morse & Keohane, 2014).

The Global South's Turn to Parallel Governance

Responses emerging from the Global South are best interpreted as agency under constraint—efforts to correct legitimacy deficits through alternative architectures rather than exit from cooperation. The Bridgetown Initiative, led by Barbados, reframed climate finance debates by centering vulnerability, debt, and justice, pressing multilateral development banks toward concessionality, faster disbursement, and risk-sharing instruments (Mottley, 2022). Far from an isolated financial proposal, Bridgetown is a diplomatic strategy to force incumbents to internalize climate loss and damage.

In technology governance, the BRICS Rio Declaration on AI can be read as a counterweight to the G7's Hiroshima AI Process, signaling that agenda-setting for frontier technologies cannot be monopolized by a narrow club (G7, 2023; BRICS, 2024). Parallel tracks are not a rejection of rules but a contest over who writes them and whose risk priorities count. In peace and security, African Union–led mediation (e.g., Ethiopia–Tigray) and ASEAN's AHA Centre in disaster response show how proximity and contextual legitimacy can deliver where universal bodies face stalemate (Williams, 2023; Haacke, 2021). Collectively, these moves fit the pattern of contested multilateralism: institution-building as leverage to renegotiate fairness (Morse & Keohane, 2014).

Diversity Within the Global South

The Global South is neither monolithic nor unified in preferences. Heterogeneity spans income, capabilities, regime types, and exposure to specific risks. In climate diplomacy, for example, small island developing states (SIDS) prioritize loss and damage and rapid decarbonization, while hydrocarbon exporters emphasize transition pacing and energy security. In finance, least-developed countries seek debt relief and concessional flows, whereas large emerging markets push for quota reform and policy space for industrialization. Within Africa, South Africa's aspirations for global representation coexist with the priorities of landlocked or conflict-affected states emphasizing humanitarian relief and basic services. In Asia, India's digital public infrastructure agenda and Indonesia's industrial policy ambitions coexist with neighbors' demands for climate finance and adaptation.

Recognizing this diversity is analytically essential: it prevents romanticized accounts of a unitary Global South and clarifies why coalitions are fluid and issue-specific. It also strengthens the case for polycentric and subsidiarity-based governance: if preferences and capacities vary, solutions should be layered, allowing regional and local nodes to tailor implementation within global frameworks (Ostrom, 2010; Føllesdal, 2022).

Implications for Reform and Blueprint Alignment

Legitimacy is the master variable linking diagnosis to redesign. Reforms that optimize efficiency without addressing fairness will not restore trust; conversely, fairness without performance will not sustain compliance. The blueprint proposed in Section 6 explicitly targets legitimacy across input, throughput, and output:

Input (Voice and Representation)

UNSC enlargement (including African and Latin American seats) and IMF/World Bank quota reform would align formal voice with contemporary realities, reducing the perception of structural bias and increasing the "right to rule" (AU, 2005; IMF, 2023).

Throughput (Process Quality)

Veto restraint in atrocity contexts, GA-anchored fallback pathways, and formalized UN–regional integration improve decision quality and predictability, signaling that procedures are oriented to collective problem-solving rather than great-power impunity (Security Council Report, 2022; De Coning & Karlsrud, 2021).

Output (Delivery and Fairness of Outcomes)

A Global Climate Resilience Council, scaled loss-and-damage finance, and innovative instruments (solidarity levies, debt-for-climate swaps, blended green finance) close the persistent gap between pledges and flows, demonstrating that global public goods are financed as obligations, not charity (Roberts & Huq, 2023; Volz et al., 2021; Atkinson, 2015).

Two cross-cutting mechanisms bind these reforms to legitimacy. First, accountability and transparency: Independent evaluation panels with public scorecards, open data on disbursements and peacekeeping performance, and routine sunset-with-review clauses create a culture of revision rather than drift (IEO, 2020). Second, inclusion by design: Federated Digital Public Goods and AI assurance ecosystems enable meaningful participation in the digital domain, preventing technological governance from becoming a new site of exclusion (World Bank, 2023; Cihon et al., 2021).

Synthesis

The legitimacy crisis is not an epilogue to the polycrisis; it is its governance face. The measures in the blueprint are not solely technical fixes; they are fairness-restoring devices meant to rebuild trust. Only a system that looks fair (input), works fairly (throughput), and delivers fairly (output) can command the voluntary compliance required to govern in an age of uncertainty. Absent that, contested multilateralism will harden into rival orders, and the universality that has long sustained international cooperation will erode.

INFORMATION ECOSYSTEMS, SOCIAL MEDIA, AND THE POLITICS OF LEGITIMACY

The Information Revolution as a Governance Factor

For much of the 20th century, information and communication were treated as peripheral to global governance. Institutions derived legitimacy primarily from formal authority (e.g., treaties, charters, representation of sovereign states) and from outputs (e.g., peacekeeping, aid delivery, trade liberalization). Media was considered an external channel for broadcasting decisions, not an intrinsic arena where legitimacy was contested or reproduced.

The information revolution of the 21st century has upended this logic. Digital platforms—especially social media—are now central arenas of governance. They shape how crises are perceived, how legitimacy is constructed or eroded, and how accountability is demanded. In effect, they function as a parallel governance layer, where institutional credibility is mediated less by formal legal rules and more by the narratives that circulate in the public sphere (Castells, 2009; Tufekci, 2017).

The implications are profound. Institutions may act effectively yet lose legitimacy if public narratives portray them as biased, captured, or indifferent. Conversely, they may lack enforcement power yet gain influence through communicative legitimacy—as seen with UN General Assembly resolutions, which, while non-binding, often shape global discourse and moral expectations. Information systems thus operate as both risk amplifiers (when polarization, disinformation, and bias dominate) and legitimacy multipliers (when transparency, inclusivity, and trust-building prevail).

Social Media and the Crisis of Trust

Social media platforms have radically democratized visibility. Ordinary citizens now document atrocities, corruption, and systemic injustice in real time. In Syria, Sudan, and Gaza, smartphone videos and WhatsApp messages have provided raw evidence that no institution could ignore. This bottom-up transparency has pressured states and multilateral organizations to act faster, sometimes forcing issues onto the agenda against the inertia of bureaucratic processes.

Yet this visibility has been deeply double-edged. Alongside authentic testimony, disinformation campaigns, manipulated images, and deepfakes have proliferated. The very platforms that expose abuses also generate radical mistrust, as competing narratives blur the line between fact and fabrication (Howard & Hussain, 2013).

The COVID-19 pandemic crystallized this paradox. Even as the WHO issued evidence-based guidance, it faced what it called an "infodemic"—a torrent of misinformation about vaccines, treatments, and origins. False claims spread faster than official communication, undermining compliance with health measures and eroding trust in the very institution tasked with global coordination (WHO, 2020). Similarly, humanitarian underperformance in Sudan or Yemen is magnified not only by funding shortfalls but also by viral narratives of neglect, which circulate widely and deepen perceptions of bias and abandonment.

This produces a new kind of legitimacy crisis: throughput deficits (slow processes, procedural deadlock) and output deficits (limited delivery, underfunded mandates) are now rapidly amplified into narratives of betrayal. Once confined to elite diplomacy, these trust deficits now cascade across societies, making restoration of credibility exponentially harder.

Influencers, Movements, and New Diplomatic Actors

In the digital age, legitimacy is co-produced not only by states and institutions but also by individuals, networks, and social movements operating across borders. Transnational movements such as Fridays for Future demonstrate how decentralized online mobilization can reshape climate governance, pushing leaders to adopt more ambitious targets. Hashtags like #BlackLivesMatter, #MeToo, and #EndSARS have transcended borders, mobilizing millions and reshaping both domestic debates and international human rights discourse (Tufekci, 2017).

Influencers and digital activists increasingly act as non-traditional diplomats. Their legitimacy does not stem from treaties but from networked credibility: authenticity, visibility, and resonance with publics. During COP26 and COP27, youth activists—often amplified by digital platforms—forced negotiators to address equity and intergenerational justice. Similarly, humanitarian influencers documenting crises in Gaza or Sudan have often outpaced institutional communication, raised awareness and mobilized resources faster than formal appeals.

This development has two consequences. First, institutions that fail to engage with these actors risk losing narrative control in legitimate debates. Second, governance itself is becoming more porous, as legitimacy is negotiated between formal institutions and informal, digitally enabled movements. While these actors cannot substitute for treaties or enforcement, their agenda-setting power is now indispensable.

The Fragmentation Risk: Algorithmic Geopolitics

The infrastructures that enable transparency and activism also carry risks of polarization and manipulation. Social media platforms are designed around algorithmic logics that prioritize engagement—often by amplifying outrage, sensationalism, and division. This creates echo chambers where different communities inhabit radically different informational universes.

In governance terms, this is not a side effect; it is a structural challenge. Shared facts are the foundation of negotiation and cooperation. When algorithms produce fragmented realities, the possibility of consensus shrinks. Algorithmic geopolitics describes the struggle of states, corporations, and civil society to dominate digital narratives, turning information ecosystems into battlegrounds (Kellerman, 2021).

Disinformation campaigns illustrate this vividly. Russian interference in elections, Chinese state media strategies, and anti-vaccine misinformation networks all show how narratives are now weaponized as instruments of statecraft. The target is not only governments but also the trustworthiness of multilateral institutions, which become collateral damage when global publics no longer agree on baseline truths.

If unchecked, this trend risks producing an epistemic crisis: a situation where governance is impossible because even basic facts are contested. At that point, international negotiations risk devolving into performance rather than problem-solving, as actors appeal to fragmented publics rather than shared evidence.

Integrating Media and Communication into Governance Reform

If legitimacy is now mediated in digital arenas as much as in formal institutions, governance reform must treat communication as structural, not peripheral. Three dimensions are particularly critical:

1. Institutional Communication Strategies

Multilateral institutions must build real-time, transparent communication systems. WHO's "myth-busting" campaigns during COVID-19 showed the value of proactive engagement. Expanding transparency dashboards, rapid-response communication teams, and multilingual outreach can help institutions counter misinformation before it metastasizes.

2. Multi-Stakeholder Media Compacts

Just as the Global Digital Compact (2024) seeks norms for digital infrastructure, a Media and Information Compact could set out principles for platform accountability, disinformation management, and equitable digital voice. By bringing together states, platforms, and civil society, such a compact would distribute responsibility rather than leaving it to unregulated corporate discretion.

3. Institutionalizing Civil Society and Influencer Voices

Institutions should move beyond ad hoc consultations by formalizing the role of digital movements and community voices. Advisory panels, structured dialogues, and participatory forums can transform online activism from external pressure into an internalized legitimacy resource. This expands representation and aligns governance with societal expectations.

Taken together, these measures would embed communication into governance design rather than leaving it as an afterthought.

Implications for Legitimacy and Renewal

The information revolution underscores that legitimacy is no longer produced solely by treaties, representation, or outcomes. It is also co-produced in digital ecosystems where publics interpret fairness, accountability, and responsiveness. Institutions that adapt—by engaging transparently, building alliances with digital actors, and countering disinformation—can harness information ecosystems as channels of radical inclusion and accountability.

If ignored, however, these same systems can become vectors of fragmentation, eroding trust, amplifying polarization, and leaving institutions irrelevant in the very spaces where legitimacy is debated and conferred.

The battle for legitimacy in the 21st century is therefore also a battle for communicative credibility. Governance systems must learn to operate not only in conference rooms and treaty texts but also in hashtags, livestreams, and viral narratives. Those that succeed will find their authority renewed; those that fail may find themselves increasingly sidelined, trapped in formal structures that no longer resonate with societies they claim to represent.

CONCLUSION: RENEWING GLOBAL GOVERNANCE IN AN ERA OF POLYCRISIS

The age of polycrisis demands not incremental adjustment but a fundamental reconfiguration of global governance. This chapter has shown that the post-1945 order—anchored in the United Nations, the Bretton Woods institutions, and regional bodies—delivered historic achievements: preventing large-scale interstate war, facilitating unprecedented economic expansion, and embedding norms of human rights and cooperation. Yet these institutions were designed for a bipolar world, structured around discrete and largely linear threats. In the 21st century, crises are multidimensional, cascading, and non-linear. Protracted conflicts in Syria, Ukraine, Gaza, and Sudan expose paralysis and selective enforcement, while systemic risks—pandemics, climate breakdown, technological disruption, and debt distress—lay bare deep structural flaws in representation, decision-making, bureaucratic responsiveness, and funding.

The erosion of trust in global institutions is not evidence of their obsolescence but of their maladaptation to contemporary risks. As argued throughout this chapter, institutions remain indispensable conveners and normative anchors, yet they are insufficient as currently configured. Renewal requires embedding resilience-by-design, subsidiarity, equity, and radical inclusion as foundational principles. Emerging innovations point the way: COVAX highlighted the need for enforceable equity in global health; the Bridgetown Initiative reframed financial justice for climate-vulnerable states; AI-assisted diplomacy demonstrated the potential of technology as a force-multiplier for negotiation; and hybrid UN–regional operations underscored the value of layered authority in peacebuilding. These are not isolated experiments but prototypes of a new order, one that is polycentric, adaptive, and inclusive.

Central to this reconfiguration is the legitimacy crisis. As shown in Section 7, legitimacy rests on input (representation), throughput (decision processes), and output (performance). Representation deficits in the Security Council and IMF, veto paralysis on mass atrocity crimes, chronic underfunding of humanitarian operations, and selective application of norms collectively undermine fairness. The Global South's turn toward parallel governance architectures—BRICS-led AI frameworks, AU-led mediation, and regional financial initiatives—signals not rejection of multilateralism but a demand for equity and recognition. Crucially, the Global South itself is internally diverse: small island states prioritize climate survival, hydrocarbon exporters emphasize transition pacing, and emerging markets demand industrial policy space. A renewed governance order must therefore be layered and context-sensitive, accommodating heterogeneity rather than presuming homogeneity.

As explored in Section 8, legitimacy is now also co-produced in information ecosystems. Social media amplifies transparency and grassroots activism, giving movements like *Fridays for Future* transnational reach, but it also generates disinformation, polarization, and epistemic fragmentation. In this new communicative environment, legitimacy cannot be reduced to institutional architecture alone. It depends equally on narrative credibility, media engagement, and epistemic trust. Institutions that fail to adapt will struggle to secure compliance, regardless of their formal mandates. Governance redesign must therefore treat communication strategies, platform accountability, and civil society voices not as adjuncts but as integral components of institutional resilience.

Recent developments underscore both fragility and opportunity. The UN Summit of the Future (2024) adopted the *Pact for the Future*, embedding commitments across peace, sustainable development, science, and governance transformation. The annexed *Global Digital Compact* articulates principles for an open and inclusive digital future, while the formalization of the Independent International Scientific Panel on AI in 2025 provides epistemic infrastructure for navigating technological disruption. The Bridgetown Initiative's evolution into its third phase has shaped debt and climate debates at the 2025 IMF–World Bank Spring Meetings and FfD4, while UN80 reflects institutional recognition of resource constraints and efficiency imperatives. These steps are partial, contested, and uneven, but they illustrate that reform momentum exists.

The stakes are clear. Inaction will deepen divides, entrench inequities, and exacerbate systemic vulnerabilities. Yet renewal is possible. The blueprint outlined in Section 6, informed by the legitimacy analysis of Section 7 and the communicative insights of Section 8, provides a roadmap: expanding representation, restraining vetoes, institutionalizing resilience finance, federating digital public goods, embedding accountability, and integrating local, regional, and transnational actors into a truly polycentric system.

The task ahead is not to preserve the old order at the margins, nor to discard it entirely, but to reconstitute it for conditions of complexity and uncertainty. The post-war order was built in the shadow of catastrophe, with leaders willing to institutionalize cooperation despite profound mistrust. The present polycrisis is different in form but comparable in urgency. What is required now is courageous leadership, institutional imagination, and a renewed social contract for global cooperation: one that is equitable enough to command trust, adaptive enough to manage uncertainty, and visionary enough to serve future generations.

The polycrisis is not destiny. It is a catalyst. Whether it becomes the crucible of fragmentation or the genesis of resilient multilateralism depends on choices made in the present decade. The moment demands no less than renewal, and the opportunity—for collective survival, dignity, and thriving—remains within reach.

REFERENCES

Acharya, A. (2021). The "Global South" in the study of world politics: Examining a meta category. *Third World Quarterly*, *42*(9), 1923–1944. DOI: 10.1080/01436597.2021.1948831

African Union. (2005). *The common African position on the proposed reform of the United Nations: The Ezulwini Consensus*. African Union.

Atkinson, A. B. (2015). *Inequality: What can be done?* Harvard University Press. DOI: 10.4159/9780674287013

Bulkeley, H., Kern, K., Vuille, M., & Frantzeskaki, N. (2018). Governing climate change polycentrically: Cities and the new climate governance. *Environment and Planning C. Politics and Space*, *36*(8), 1441–1459.

Castells, M. (2009). *Communication power*. Oxford University Press.

Cihon, P., Maas, M. M., & Kemp, L. (2021). Fragmentation and the future: Investigating architectures for international AI governance. *Global Policy*, *12*(5), 545–556. DOI: 10.1111/1758-5899.12890

Couldry, N., & Mejias, U. A. (2019). Data colonialism: Rethinking big data's relation to the contemporary subject. *Television & New Media*, *20*(4), 336–349. DOI: 10.1177/1527476418796632

De Coning, C., & Karlsrud, J. (2021). The future of UN peace operations: Principled adaptation through phases of contraction, moderation, and renewal. *Contemporary Security Policy*, *42*(1), 86–114. DOI: 10.1080/13523260.2021.1894021

European Commission. (2023). *Strategic foresight report 2023*. European Commission.

Falkner, R. (2016). The Paris Agreement and the new logic of international climate politics. *International Affairs*, *92*(5), 1107–1125. DOI: 10.1111/1468-2346.12708

Føllesdal, A. (2022). Subsidiarity and international human-rights courts: Respecting self-governance and protecting human rights—or neither? *Law and Contemporary Problems*, *79*(2), 147–163.

Franck, T. M. (1990). *The power of legitimacy among nations*. Oxford University Press. DOI: 10.1093/oso/9780195061789.001.0001

Gostin, L. O., Klock, K. A., & Finch, A. (2023). Making the world safer and fairer in pandemics. *The Hastings Center Report*, *53*(6), 3–10. DOI: 10.1002/hast.1538

Gowan, R., & Stedman, S. J. (2020). Making peace when the whole world has come to fight. *International Peacekeeping*, *27*(2), 187–192.

Haacke, J. (2021). The ASEAN way and the changing security environment: Navigating challenges to orthodoxy in East Asian security. *International Politics*, *58*(3), 406–427.

Hocking, B., & Melissen, J. (2015). *Diplomacy in the digital age*. Clingendael Report, Netherlands Institute of International Relations "Clingendael."

Hoekman, B., & Nelson, D. (2020). Rethinking international subsidy rules. *World Economy*, *43*(12), 3104–3132. DOI: 10.1111/twec.13022

Homer-Dixon, T., Renn, O., Rockström, J., Donges, J. F., Janzwood, S., & Lawrence, M. (2022). A call for an international research program on the risk of a global polycrisis. *Global Sustainability*, *5*, e18.

Howard, P. N., & Hussain, M. M. (2013). *Democracy's fourth wave? Digital media and the Arab Spring*. Oxford University Press. DOI: 10.1093/acprof:oso/9780199936953.001.0001

Hurd, I. (1999). Legitimacy and authority in international politics. *International Organization*, *53*(2), 379–408. DOI: 10.1162/002081899550913

International Telecommunication Union. (2023). *Facts and figures 2023*. ITU.

Jackson, S., Woods, R., & Hooper, F. (2022). Water colonialism and Indigenous water justice in south-eastern Australia. *International Journal of Water Resources Development*, *38*(1), 30–52.

Kellerman, B. (2021). *The power of followership: How to create leaders people want to follow and followers who lead themselves*. Doubleday.

Klinke, A., & Renn, O. (2021). Adaptive and risk-based approaches to climate change and the management of uncertainty and complexity in a changing world. *Ambio*, *50*(3), 611–621.

Lawrence, M., Janzwood, S., & Homer-Dixon, T. (2022). *What is a global polycrisis?* Cascade Institute Technical Paper No. 4. Cascade Institute.

Levi, M., & Stoker, L. (2000). Political trust and trustworthiness. *Annual Review of Political Science*, *3*(1), 475–507. DOI: 10.1146/annurev.polisci.3.1.475

Maoz, I., Steinberg, G. M., & Lewin, E. (2023). Advancing AI negotiations: New theory and evidence from a large-scale autonomous negotiations competition. *Negotiation Journal*, *39*(2), 123–145.

Moon, S., Røttingen, J. A., & Frenk, J. (2021). Global public goods for health: Weaknesses and opportunities in the global health system. *Health Economics, Policy, and Law*, *16*(2), 109–120.

Morse, J. C., & Keohane, R. O. (2014). Contested multilateralism. *The Review of International Organizations*, *9*(4), 385–412. DOI: 10.1007/s11558-014-9188-2

Mottley, M. (2022). *The Bridgetown Initiative*. Government of Barbados.

Ostrom, E. (2010). Polycentric systems for coping with collective action and global environmental change. *Global Environmental Change*, *20*(4), 550–557. DOI: 10.1016/j.gloenvcha.2010.07.004

Oxfam International. (2024). *Vetoing humanity: How a few powerful nations hijacked global peace*. Oxfam.

Patrick, S. (2017). Trump and world order: The return of self-help. *Foreign Affairs*, *96*(2), 52–57.

Roberts, E., & Huq, S. (2023). Exploring loss and damage from climate change and global perspectives on the way forward. *Climate Policy*, *23*(8), 1023–1033.

Scharpf, F. W. (1999). *Governing in Europe: Effective and democratic?* Oxford University Press. DOI: 10.1093/acprof:oso/9780198295457.001.0001

Schmidt, V. A. (2013). Democracy and legitimacy in the European Union revisited: Input, output and "throughput.". *Political Studies*, *61*(1), 2–22. DOI: 10.1111/j.1467-9248.2012.00962.x

Schunz, S. (2023). Governing global spaces: The roles of the European Union and China in global environmental governance. *Global Environmental Politics*, *23*(1), 3–25.

Security Council Report. (2022). *The veto*. Security Council Report Research Report.

Thakur, R. (2023). The Russia-Ukraine war and the changing world order. *Strategic Analysis*, *47*(2), 82–97.

Tufekci, Z. (2017). *Twitter and tear gas: The power and fragility of networked protest*. Yale University Press.

Tyler, T. R. (2006). Psychological perspectives on legitimacy and legitimation. *Annual Review of Psychology*, *57*(1), 375–400. DOI: 10.1146/annurev.psych.57.102904.190038

United Nations. (2021). *Our common agenda: Report of the Secretary-General*. United Nations.

United Nations. (2023). *A global digital compact: An open, free and secure digital future for all*. United Nations.

United Nations Department of Economic and Social Affairs. (2023). *Global digital compact*. UN DESA.

United Nations Development Programme. (2022). *Digital public goods for sustainable development*. UNDP.

United Nations Office for the Coordination of Humanitarian Affairs. (2024). *Sudan situation report*. OCHA.

Volz, U., Akhtar, S., Gallagher, K. P., Griffith-Jones, S., Haas, J., & Kraemer, M. (2021). *Debt relief for a green and inclusive recovery: A proposal for global action*. Heinrich Böll Foundation.

WHO (2020). *Managing the COVID-19 infodemic: Promoting healthy behaviours and mitigating the harm from misinformation and disinformation*. WHO.

WHO. (2022). *COVAX: Working for global equitable access to COVID-19 vaccines*. WHO.

Whyte, K. (2020). Too late for Indigenous climate justice: Ecological and relational tipping points. *Wiley Interdisciplinary Reviews: Climate Change*, *11*(1), e603. DOI: 10.1002/wcc.603

Williams, P. D. (2023). EU policy towards Ethiopia amidst the Tigray war. *Third World Quarterly*, *44*(9), 1953–1970.

World Bank. (2022). *Green bonds for sustainable development*. World Bank.

World Bank. (2023). *India's digital public infrastructure*. World Bank.

World Economic Forum. (2023). *Global risks report 2023*. World Economic Forum.

World Food Programme. (2022). *Blockchain for zero hunger: Building blocks for a better future*. WFP.

Compilation of References

Abbott, K. W., & Snidal, D. (2020). The governance triangle: Regulatory standards institutions and the shadow of the state. In Newman, A., & Zaring, D. (Eds.), *The politics of global regulation* (pp. 44–88). Princeton University Press.

Acharya, A. (2021). The "Global South" in the study of world politics: Examining a meta category. *Third World Quarterly*, *42*(9), 1923–1944. DOI: 10.1080/01436597.2021.1948831

Acquier, A., Gond, J. P., & Igalens, J. (2005). Des fondements religieux de la responsabilité sociale de l'entreprise à la responsabilité sociale de l'entreprise comme religion. *Cahier de recherche, 166*, 1-31.

Acquier, A. (2007). *Corporate Social Responsibility: A Historical Perspective.* Springer.

Acuto, M., Morissette, M., & Tsouros, A. (2017). City diplomacy: Towards more strategic networking? Learning with WHO healthy cities. *Global Policy*, *8*(1), 14–22. DOI: 10.1111/1758-5899.12382

Adam, D. (2023). *The episodic disruption: Permanent turbulence in global governance*. Routledge.

Adamik, A., & Nowicki, M. (2019). Pathologies and Paradoxes of Co-Creation: A Contribution to the Discussion about Corporate Social Responsibility in Building a Competitive Advantage in the Age of Industry 4.0. *Sustainability (Basel)*, *11*(18), 4954. DOI: 10.3390/su11184954

Adams, R. B., & Mehran, H. (2004). Corporate Performance, Board Structure, and Their Determinants in the Banking Industry. *Journal of Financial Economics*, *74*(2), 189–234.

Adapa, S. (2015). The extent of corporate social responsibility engagement in malaysian banks offering islamic banking services. *Journal of Internet Banking and Commerce, 20*(2). Advance online publication. DOI: 10.4172/1204-5357.1000113

Adikaram, R., & Holcomb, A. (2023). Exploring the role of analysts in identifying and communicating the value of bank CSR activity. *International Journal of Bank Marketing*. Advance online publication. DOI: 10.1108/IJBM-12-2022-0557

African Union. (2005). *The common African position on the proposed reform of the United Nations: The Ezulwini Consensus*. African Union.

Ager, A., & Strang, A. (2008). Understanding integration: A conceptual framework. *Journal of Refugee Studies, 21*(2), 166–191. DOI: 10.1093/jrs/fen016

Ager, S. L. (1996). *Interstate Arbitrations in the Greek World, 337–90 B.C.* University of California Press. DOI: 10.1525/9780520913493

Aggeni, F., & Godard, O. (2006). *Corporate Responsibility and Sustainable Development: Beyond Compliance*. Palgrave Macmillan.

Al Jalalah, B. (2025). *Psychological Resilience: Ideas for Coping with Stress and Pressure to Achieve Success and Live Happily*. Dar Tashkeel for Publishing and Distribution.

Al Mokdad, K. (2025). Trust and geopolitical bridge-building in science diplomacy. In *Science diplomacy in a changing world* (pp. 45–67). Springer.

Al-Ali, N., Black, R., & Koser, K. (2001). Refugees and transnationalism: The experience of Bosnians and Eritreans in Europe. *Journal of Ethnic and Migration Studies, 27*(4), 615–634. DOI: 10.1080/13691830120090412

Alam, S. (2018). *Perceptions of self, power, & gender among Muslim women*.

Alawamleh, M., & Giacaman, R. (2020). Corporate social responsibility impacts on Palestinian and Jordanian consumer purchasing. *The International Journal of Organizational Analysis, 28*(4), 879–894. DOI: 10.1108/IJOA-03-2020-2109

Alexandre, L. (2025). *Page Restricted*. [online] Elgaronline.com. Available at: https://www.elgaronline.com/abstract/edcoll/9781784717414/9781784717414.00019.xml [Accessed 28 Mar. 2025].

Ali, İ., Ali, M., & Badghish, S. (2019). Symmetric and asymmetric modelling of the entrepreneurial ecosystem in developing entrepreneurial intentions among female university students in Saudi Arabia. *International Journal of Gender and Entrepreneurship, 11*(4), 435–458. DOI: 10.1108/IJGE-02-2019-0039

Alkire, S. (2005). Why the capability approach? *Journal of Human Development*, *6*(1), 115–135. DOI: 10.1080/146498805200034275

Al-Shammari, M., Banerjee, S. N., & Dadanlar, H. H. (2024). Strategic change and firms' optimal distinctiveness: The moderating effect of analyst coverage. *British Journal of Management*. Advance online publication. DOI: 10.1111/1467-8551.12873

Al-Sharika, M. (2025). *Social and Emotional Learning*. Arab Center for Educational Research for Gulf States.

Al-Suwaidi, T., & Tlemsani, I. (2025). Sustainable leadership styles adopted by small businesses in the UAE during and post COVID-19. *Discover Sustainability*, •••, 5–6.

Ambury, J. M. (n.d.). *Socrates (469—399 B.C.E.)*. Retrieved from The Internet Encyclopedia of Philosophy: https://iep.utm.edu/socrates/

Ancona, D., Malone, T. W., Orlikowski, W. J., & Senge, P. (2007). In praise of the incomplete leader. *Harvard Business Review*, 85(2), 92–100. pp. 94–96.

Annan, N., Beseng, M., Crawford, G., & Kewir, J. K. (2021). Civil society, peacebuilding from below and shrinking civic space: The case of Cameroon's 'Anglophone'conflict. *Conflict Security and Development*, *21*(6), 697–725. DOI: 10.1080/14678802.2021.1997454

Ansell, C., Sørensen, E., Torfing, J., & Trondal, J. (2024). *Robust governance in turbulent times*. Cambridge University Press. DOI: 10.1017/9781009433006

Ansell, C., & Torfing, J. (2022). Introduction to the Handbook on Theories of Governance. In *Handbook on theories of governance* (pp. 1–16). Edward Elgar Publishing., DOI: 10.4337/9781800371972.00007

Anu Jogesh, U. P. (2019, October). *Case study: For the community, by the community: Building resilience through infrastructure in rural India.* Retrieved from ResearchGate: DOI:DOI: 10.13140/RG.2.2.33606.80965

Appelo, J. (2011). *Management 3.0: Leading agile developers, developing agile leaders*. Addison-Wesley.

Aprillianto, B., & Sayekti, Y. (2017). *The role of firm strategy to intervene the influence of corporate social performance on corporate financial performance*. Gatr Journal of Finance and Banking Review., DOI: 10.35609/jfbr.2017.2.3(4)

Arab, L. A. (2018). Entrepreneurial Intentions of Undergraduate Students. (c2018). *Lau.edu.lb*. [online] doi:http://hdl.handle.net/10725/10471

Ardiansyah, A., & Alnoor, A. (2024). Integrating Corporate Social Responsibility into Business Strategy: Creating Sustainable Value. *International Insight Journal of Business*, *1*(1), 12–23. DOI: 10.62569/iijb.v1i1.5

Argyris, C., & Schön, D. A. (1996). *Organizational learning II: Theory, method, and practice*. Addison-Wesley.

Arkenberg, C. (2019, October 23). Leading in a VUCA world: How 21st-century leaders can find success amid volatility, uncertainty, complexity, and ambiguity. *Deloitte Insights*.

Arnaouti, M. K. C., Cahill, G., Baird, M. D., Mangurat, L., Harris, R., Edme, L. P. P., Joseph, M. N., Worlton, T., & Augustin, S.Jr. (2022). Medical disaster response: A critical analysis of the 2010 Haiti earthquake. *Frontiers in Public Health*, *10*, 995595. DOI: 10.3389/fpubh.2022.995595

Arnett, J. J. (2000). Emerging adulthood: A theory of development from the late teens through the twenties. *The American Psychologist*, *55*(5), 469–480. DOI: 10.1037/0003-066X.55.5.469 PMID: 10842426

Atkinson, A. B. (2015). *Inequality: What can be done?* Harvard University Press. DOI: 10.4159/9780674287013

Babacan, H. (2025). From crisis to policy action: Tackling climate change through SDG 13. *The International Journal of Community and Social Development*, *25166026251351370*(2), 205–221. Advance online publication. DOI: 10.1177/25166026251351370

Bacco, M., Barsocchi, P., Ferro, E., Gotta, A., & Ruggeri, M. (2019). Drone applications in precision agriculture: A review. *Sensors (Basel)*, *19*(11), 2577. DOI: 10.3390/s19112577

Bahhouth, V., Spillane, J., Bahhouth, J., & Khoueiri, R. (2012). Students' perception of international business curriculum: An explorative study—case of Lebanon. *Journal of Teaching in International Business*, *23*(3), 236–250. DOI: 10.1080/08975930.2012.744635

Bakker, M. M.. (2021). AI-enabled irrigation scheduling for water-efficient agriculture: A review. *Agricultural Water Management*, *245*, 106604. DOI: 10.1016/j.agwat.2020.106604

Bakr, M. (2020). Robotics in agriculture: Current applications and future perspectives. *Agricultural Robotics Journal*, *2*(1), 45–59.

Barbier, E. B., & Burgess, J. C. (2020). Sustainability and development after COVID-19. *World Development, 135*, 105082. DOI: 10.1016/j.worlddev.2020.105082 PMID: 32834381

Barney, J. B. (1991). Firm resources and sustained competitive advantage. *Journal of Management, 17*(1), 99–120. pp. 105–107.

Barney, J. (1991). Firm resources and sustained competitive advantage. *Journal of Management, 17*(1), 99–120. DOI: 10.1177/014920639101700108

Barrett, C. B., & Rose, A. (2022). Artificial Intelligence and Agricultural Labor: Promises and Perils. *Journal of Agrarian Change, 22*(1), 45–60. DOI: 10.1111/joac.12400

Bartók, O. (2018). The Use of CSR in E-Commerce as a Way to Compete. *Journal of Competitiveness, 10*(4), 5–20. DOI: 10.7441/joc.2018.04.01

Bass, B. M., & Avolio, B. J. (1994). *Improving organizational effectiveness through transformational leadership*. Sage Publications.

Bass, B. M., & Riggio, R. E. (2006). *Transformational leadership* (2nd ed.). Lawrence Erlbaum Associates. DOI: 10.4324/9781410617095

Battisti, M., & Deakins, D. (2017). The relationship between dynamic capabilities, the firm's resource base and performance in a post-disaster environment. *International Small Business Journal, 35*(1), 78–98. pp. 80–82.

Bechar, A., & Vigneault, C. (2016). Agricultural robots for field operations: Concepts and components. *Biosystems Engineering, 149*, 94–111. DOI: 10.1016/j.biosystemseng.2016.06.014

Becker, G. S. (1964). *Human capital: A theoretical and empirical analysis, with special reference to education* (1st ed.). University of Chicago Press.

Beck, S., & Mahony, M. (2018). The IPCC and the new map of science and politics. *Wiley Interdisciplinary Reviews: Climate Change, 9*(6), e547. DOI: 10.1002/wcc.547

Benjamin, R. (2019). *Race after technology: Abolitionist tools for the new Jim Code*. Polity Press.

Benton, T. G., & Froggatt, A. (2021). *Food system impacts on biodiversity loss: Three levers for food system transformation in support of nature*. Chatham House.

Betancourt, T. S., Frounfelker, R., Mishra, T., Hussein, A., & Falzarano, R. (2015). Addressing health disparities in the mental health of refugee children and adolescents through community-based participatory research: A study in two communities. *American Journal of Public Health, 105*(S3, Suppl. 3), S475–S482. DOI: 10.2105/AJPH.2014.302504 PMID: 25905818

Betancourt, T. S., Meyers-Ohki, S. E., Charrow, A., & Tol, W. A. (2015). Interventions for children affected by war: An ecological perspective on psychosocial support and mental health care. *Harvard Review of Psychiatry, 21*(2), 70–91. DOI: 10.1097/HRP.0b013e318283bf8f PMID: 23656831

Bhabha, H. K. (1994). *The location of culture*. Routledge.

BhatMishra. M. S. (n.d.). *Greek and Roman Conceptions of Natural Law*. Retrieved July 16, 2025, from Advanced jurisprudence: https://ebooks.inflibnet.ac.in/lawp01/chapter/greek-and-roman-conceptions-of-natural-law/

Bhugra, D. (2004). *Migration, distress and cultural identity*. British Medical Bulletin. DOI: 10.1093/bmb/ldh007

Bhugra, D., & Becker, M. A. (2005). Migration, cultural bereavement and cultural identity. *World Psychiatry; Official Journal of the World Psychiatric Association (WPA), 4*(1), 18–24. https://www.ncbi.nlm.nih.gov/pmc/articles/PMC1414713/ PMID: 16633496

Bicchi, F., & Bremberg, N. (2022). *European diplomatic practices: Interrogating power, agency and change*. Routledge.

BIOGRAPHS. (n.d.). *Plato*. Retrieved from BIOGRAPHS: https://biographs.org/plato#:~:text=Plato%2C%20an%20ancient%20Greek%20philosopher%2C%20was%20a%20student,first%20higher%20education%20institution%20in%20the%20Western%20world

Biraglia, A., & Kadile, V. (2016). The role of entrepreneurial passion and creativity in developing entrepreneurial intentions: Insights from American homebrewers. *Journal of Small Business Management, 55*(1), 170–188. DOI: 10.1111/jsbm.12242

Bizri, R.M., Wahbi, M. and Al, J.H. (2021). The impact of CSR best practices on job performance: The mediating roles of affective commitment and work engagement. *Journal of Organizational Effectiveness People and Performance*. .DOI: 10.1108/JOEPP-01-2020-0015

Black, A. (2016). The Greeks. In Black, A. (Ed.), *A World History of Ancient Political Thought: Its Significance and Consequences*. Academic. DOI: 10.1093/acprof:oso/9780198790686.003.0009

Blackmore, S., Stout, B., Wang, M., & Runov, B. (2020). Agricultural automation: The future of farming. *International Journal of Agricultural and Biological Engineering, 13*(5), 1–15. DOI: 10.25165/j.ijabe.20201305.5321

Blerk, N. (2018). The emergence of law in ancient Egypt: The role of Maat. *Fundamina (Pretoria), 24*(1), 69–88. DOI: 10.17159/2411-7870/2018/v24n1a4

Bolleyer, N. (2021). Civil society, crisis exposure, and resistance strategies. In *Routledge handbook of illiberalism* (pp. 939–956). Routledge. DOI: 10.4324/9780367260569-68

Bolton, D., Landells, T., & Habib, M. (2023). Collaborating and Competing in Development Agendas: State—Civil Society Tensions in Achieving SDGs. In *The complexities of sustainability* (pp. 77-120).

Bourletidis, K., & Triantafyllopoulos, Y. (2014). SMEs survival in time of crisis: Strategies, tactics and commercial success stories. *Procedia - Social and Behavioral Sciences, 148*, 639–644. pp. 641–642.

Bowen, H. R. (1953). *Social Responsibilities of the Businessman*. Harper & Row.

Branco, M. C. (2006). Corporate Social Responsibility and Banking Sector Performance. *Journal of Business Ethics, 69*(1), 111–132. DOI: 10.1007/s10551-006-9071-z

Braun, V., & Clarke, V. (2006). Using thematic analysis in psychology. *Qualitative Research in Psychology, 3*(2), 77–101. DOI: 10.1191/1478088706qp063oa

Bravim, J. M., de Carvalho, A. C. V., Segatto, A. P., & Morais-da-Silva, R. L. (2025). Are there key characteristics of social enterprises that foster relationships with government, universities and civil society? *Social Enterprise Journal, 21*(5), 864–889. Advance online publication. DOI: 10.1108/SEJ-08-2024-0136

Brinkerhoff, J. M. (2009). *Digital diasporas: Identity and transnational engagement*. Cambridge University Press., DOI: 10.1017/CBO9780511805158

Brixiová, Z., Kangoye, T., & Said, M. (2020). Training and support programs for youth entrepreneurs in Africa: A mixed methods evaluation from Ethiopia. *Journal of Development Effectiveness, 12*(4), 299–319. DOI: 10.1080/19439342.2020.1834876

Brizek, M.G. (2018). An examination regarding the role of corporate social responsibility (CSR) within financial institutions: A literature review. *Journal of Business Management & Economics*. .DOI: 10.15520/jbme.v6i04.2117

Broadhurst, K., & Gray, N. (2022). Understanding resilient places: Multi-level governance in times of crisis. *Local Economy, 37*(1-2), 84–103. DOI: 10.1177/02690942221100101

Bronson, K. (2018). Smart Farming: Including Rights Holders for Responsible Agricultural Innovation. *Technology Innovation Management Review*, *8*(2), 7–14. DOI: 10.22215/timreview/1135

Brown, T. (2021). *Resilience in times of crisis: A psychosocial perspective*. Routledge.

Brundtland, G. H. (1987). *Our Common Future: The World Commission on Environment and Development*. Oxford University Press.

Brush, C. G., de Bruin, A., & Welter, F. (2009). A gender-aware framework for women's entrepreneurship. *International Journal of Gender and Entrepreneurship*, *1*(1), 8–24. DOI: 10.1108/17566260910942318

Bryson, J. J. (2019). The artificial intelligence of the ethics of artificial intelligence. In Dubber, M. D., Pasquale, F., & Das, S. (Eds.), *The Oxford handbook of ethics of AI* (pp. 1–22). Oxford University Press.

Bua, A., & Bussu, S. (Eds.). (2023). *Reclaiming participatory governance: Social movements and the reinvention of democratic innovation*. Taylor & Francis. DOI: 10.4324/9781003218517

Bulkeley, H., Kern, K., Vuille, M., & Frantzeskaki, N. (2018). Governing climate change polycentrically: Cities and the new climate governance. *Environment and Planning C. Politics and Space*, *36*(8), 1441–1459.

Bundy, J., Pfarrer, M. D., Short, C. E., & Coombs, W. T. (2017). Crises and crisis management: Integration, interpretation, and research development. *Journal of Management, 43*(6), 1661–1692. p. 1662.

Busby, J. W. (2021). *States and nature: The effects of climate change on security*. Cornell University Press.

Byrne, D., & Callaghan, G. (2014). *Complexity theory and the social sciences: The state of the art*. Routledge.

C40 Cities Climate Leadership Group. (2023). *Annual report 2023*.

Calel, R., & Dechezleprêtre, A. (2016). Environmental policy and directed technological change: Evidence from the European carbon market. *The Review of Economics and Statistics*, *98*(1), 173–191. DOI: 10.1162/REST_a_00470

Cantrell, J., Kyriazis, E., & Noble, G. (2014). Developing CSR Giving as a Dynamic Capability for Salient Stakeholder Management. *Journal of Business Ethics*, *119*(2), 123–136. DOI: 10.1007/s10551-014-2229-1

Carbonell, I. M. (2016). The Ethics of Big Data in Agriculture. *Internet Policy Review, 5*(1), 1–13. DOI: 10.14763/2016.1.405

Carroll, A. B. (1979). A Three-Dimensional Conceptual Model of Corporate Performance. *Academy of Management Review, 4*(4), 497–505. DOI: 10.2307/257850

Carroll, A. B. (1991). The pyramid of corporate social responsibility: Toward the moral management of organizational stakeholders. *Business Horizons, 34*(4), 39–48. DOI: 10.1016/0007-6813(91)90005-G

Carroll, A. B. (1999). Corporate Social Responsibility: Evolution of a Definitional Construct. *Business & Society, 38*(3), 268–295. DOI: 10.1177/000765039903800303

Cartapanis, A. (2003). The Governance of Global Banking Crises. *Review of International Political Economy, 10*(1), 1–25.

Castells, M. (2009). *Communication power*. Oxford University Press.

Castles, S., de Haas, H., & Miller, M. J. (2014). *The age of migration: International population movements in the modern world* (5th ed.). Palgrave Macmillan. DOI: 10.1007/978-0-230-36639-8

Cath, C. (2023). The technology we choose to create: Human rights in the design of the internet. *Lancet, 402*(10410), 1398–1400.

Caton, C. L.. (2021). Culture, religion and resilience in crisis contexts. *The Journal of Social Issues*, 356–372.

Cegliński, P., & Wiśniewska, A. (2017). CSR as a Source of Competitive Advantage: The Case Study of Polpharma Group. *Journal of Corporate Responsibility and Leadership, 3*(1), 29–47. DOI: 10.12775/JCRL.2016.020

CGIAR. (2023). *Digital Innovation Initiative*. https://www.cgiar.org/initiative/10-digital-innovation/

Chakkarath, P. (2013). Cultural identity.

Chakra, S., & Jardali, H. (2022). Drivers of social entrepreneurial intentions of Palestinian refugees in Lebanon. *Journal of Research in Marketing and Entrepreneurship, 24*(2), 242–269. DOI: 10.1108/JRME-02-2021-0014

Chakroun, R., Matoussi, H., & Mbirki, S. (2017). Determinants of CSR disclosure of tunisian listed banks: A multi-support analysis. *Social Responsibility Journal, 13*(3), 552–584. Advance online publication. DOI: 10.1108/SRJ-04-2016-0055

Chandler, D., & Reid, J. (2016). *The neoliberal subject: Resilience, adaptation and vulnerability*. Rowman & Littlefield.

Chang, H., Kim, H., & Lee, Y. (2022). Corporate Social Responsibility and Product Market Power. *Economies, 10*(6), 151. DOI: 10.3390/economies10060151

Charreire-Petit, S., & Durieux, F. (2014). Chapitre 3. Explorer et tester: les deux voies de la recherche. In *Méthodes de recherche en management* (Vol. 4, pp. 76-104). Dunod.

Chatty, D. (2017). *Syria: The making and unmaking of a refugee state*. Hurst & Company.

Chaudhry, N. I., Raza, S. A., & Khurshid, M. A. (2021). Enhancing Shareholder Value through Corporate Social Responsibility Disclosure in a Developing Country. *Estudios de Economía Aplicada, 39*(4), 1–13. DOI: 10.25115/eea.v39i4.4315

Chaudhry, S. (2022). The assault on civil society: Explaining state crackdown on NGOs. *International Organization, 76*(3), 549–590. DOI: 10.1017/S0020818321000473

Chau, T. H. H., Nguyen, P. T. T., & Bui, T. H. (2022). The Impact of Corporate Social Responsibility on Strategic Objectives: Evidence from Vietnam. *VNU University of Economics and Business Journal, 2*(6), 114–125. DOI: 10.57110/vnujeb.v2i6.136

Chauveau, A., & Rosé, J. J. (2003). *Sustainable Development and Corporate Responsibility in Practice*. Routledge.

Chedrawi, C., Osta, A., & Osta, S. (2020). CSR in the lebanese banking sector: A neo-institutional approach to stakeholders' legitimacy. *Journal of Asia Business Studies, 14*(2), 143–157. Advance online publication. DOI: 10.1108/JABS-03-2018-0093

Chedrawi, C., & Osta, S. (2017). *CSR in the banking sector: A legitimacy approach to the shareholders' and stakeholders' debate*. Marketing and Branding Research., DOI: 10.33844/mbr.2017.60272

Chehimi, Ghada M, Hejase, Ale J and Hejase, Nour H (2019). An assessment of lebanese companies' motivators to adopt CSR strategies. *Open Journal of Business and Management*. .DOI: 10.4236/ojbm.2019.74130

Chen, L., & Volpe, R. P. (2021). Green energy financial literacy and corporate renewable energy adoption: Evidence from executives' survey. *Journal of Cleaner Production, 280*, 123456. DOI: 10.1016/j.jclepro.2020.123456

Chen, L., Yang, J., & Liu, S. (2020). Regulatory framework and ethical concerns of autonomous agricultural vehicles. *Journal of Agricultural Safety and Health, 26*(3), 175–187. DOI: 10.13031/jash.13756

Chen, X. (2023). Understanding the role of Civil Society Organizations (CSOs) in the "central government-local government-society" framework: A case study on public crisis management during the pandemic in China. *International Public Management Journal*, *26*(4), 570–588. DOI: 10.1080/10967494.2022.2115595

Chen, Y., Wang, S., & Zhang, W. (2023). Paradox of CSR distinctiveness: The tension between stakeholder legitimacy evaluation and stakeholder integration efficiency. *Corporate Social Responsibility and Environmental Management*, *30*(5), 2053–2065. DOI: 10.1002/csr.2523

Chowdhooree, I. (2019). Indigenous knowledge for enhancing community resilience: An experience from the south-western coastal region of Bangladesh. *International Journal of Disaster Risk Reduction*, *40*, 101259. Advance online publication. DOI: 10.1016/j.ijdrr.2019.101259

Chowdhury, S. J., Wahab, H. A., Saad, M. M., Roy, P. K., Hamidi, M., & Ahmad, M. M. (2021). Ubuntu Philosophy: 'I Am Because We Are' – A Road to 'Individualism' to Global Solidarity. In I. Global, *Handbook of Research on the Impact of COVID-19 on Marginalized Populations and Support for the Future* (pp. 361-381). IGI Global Scientific Publishing.

Chowdhury, A., Iyer, A. S., Bhattacharyya, S., Khan, S. U., & Naz, A. (2022, October). Organic farming in the tropical, disaster-prone coasts: Interventions, major challenges and a success story from Sundarban Biosphere Reserve, India. *IOP Conference Series. Earth and Environmental Science*, *1084*(1), 012052. https://iopscience.iop.org/article/10.1088/1755-1315/1084/1/012052/pdf. DOI: 10.1088/1755-1315/1084/1/012052

Chris, T., Stukasa, A. A., Halloran, M. J., & Foddy, M. (2011). Social identity change: Shifts in social identity during adolescence. *Journal of Adolescence*, *34*(3), 555–567. DOI: 10.1016/j.adolescence.2010.05.012 PMID: 20547418

Cihon, P., Maas, M. M., & Kemp, L. (2021). Fragmentation and the future: Investigating architectures for international AI governance. *Global Policy*, *12*(5), 545–556. DOI: 10.1111/1758-5899.12890

Clare, R. (2019). Corporate Image and Competitive Advantage for Apparel Companies. *Trends in Textile Engineering & Fashion Technology*, *5*(3), 345–351. DOI: 10.31031/TTEFT.2019.05.000618

Climate TRACE. (2023). *Climate TRACE 2023 report*.

Coady, D., Flamini, V., & Sears, L. (2015). The unequal benefits of fuel subsidies revisited: Evidence for developing countries. *IMF Working Paper*, WP/15/250. International Monetary Fund.

Cook, A. D. (2021). Humanitarian diplomacy in ASEAN. *Asian Journal of Comparative Politics*, *6*(3), 188–201. DOI: 10.1177/20578911211019247

Cornwall, A., & Rivas, A. M. (2015). From 'gender equality and 'women's empowerment' to global justice: Reclaiming a transformative agenda for gender and development. *Third World Quarterly*, *36*(2), 396–415. DOI: 10.1080/01436597.2015.1013341

Corporate social responsibility in international business. (2024). *Qalaai Zanist Scientific Journal*. .DOI: 10.25212/lfu.qzj.8.5.50

Costa, J. (2008). *The New Frontiers of Sustainable Development in Business*. Springer.

Couldry, N., & Mejias, U. A. (2019). Data colonialism: Rethinking big data's relation to the contemporary subject. *Television & New Media*, *20*(4), 336–349. DOI: 10.1177/1527476418796632

Council of Europe. (2024). *Framework convention on artificial intelligence, human rights, democracy and the rule of law*.

Council on Foreign Relations. (2024). *U.S. export controls and China*.

Cowling, M., Brown, R., & Rocha, A. (2020). Did you save some cash for a rainy COVID-19 day? The crisis and SMEs. *International Small Business Journal*, *38*(7), 593–604. pp. 595–596.

Crenshaw, K. (1991). Mapping the margins: Intersectionality, identity politics, and violence against women of color. *Stanford Law Review*, *43*(6), 1241–1299. DOI: 10.2307/1229039

Creswell, J. W. (2014). *Research design: Qualitative, quantitative, and mixed methods approaches* (4th ed.). Sage.

Crifo, P., & Ponssard, J. P. (2008). RSE et/ou performance financière: points de repère et pistes de recherche.

Crisp, J. (2003). The local integration and local settlement of refugees: A conceptual and historical analysis (New Issues in Refugee Research, Research Paper No. 102). UNHCR. https://www.unhcr.org/research/working/3f818d4d4/local-integration-local-settlement-refugees-conceptual-historical-analysis.html

Dacin, P. A., Dacin, M. T., & Matear, M. (2010). Social entrepreneurship: Why we don't need a new theory and how we move forward from here. *The Academy of Management Perspectives*, *24*(3), 37–57.

Dafoe, A. (2024). *AI governance: A global public good*. Brookings Institution. DOI: 10.1093/oxfordhb/9780197579329.013.2

Damiyano, D. (2024). Sustainable Social Services Financing for Local Governments. In *Financial Sustainability of Local Governments in Southern Africa* (pp. 198–214). Routledge. DOI: 10.4324/9781003498230-11

Das, S.. (2020). Machine learning for soil moisture prediction: A review. *Environmental Modelling & Software*, *130*, 104728. DOI: 10.1016/j.envsoft.2020.104728

Davis, J. T. (2018). Social cohesion. In Callan, H. (Ed.), *The international encyclopaedia of anthropology*. Wiley.

Davis, K. (1960). Can Business Afford to Ignore Social Responsibilities? *California Management Review*, *2*(3), 70–76. DOI: 10.2307/41166246

Davis, K. (1967). Understanding the Social Responsibility Puzzle. *Business Horizons*, *10*(4), 45–50. DOI: 10.1016/0007-6813(67)90007-9

De Coning, C., & Karlsrud, J. (2021). The future of UN peace operations: Principled adaptation through phases of contraction, moderation, and renewal. *Contemporary Security Policy*, *42*(1), 86–114. DOI: 10.1080/13523260.2021.1894021

De Vries, H. P., & Shields, M. (2006). Leadership and small business resilience. *New Zealand Journal of Applied Business Research, 5*(1), 55–66. pp. 58–60.

Deeb, L., & Harb, M. (2022). Politics of sectarianism in Lebanon. Journal of Ethnographic Studies, 211–229.

Demirgüç-Kunt, A., & Klapper, L. (2012). Measuring financial inclusion: The global findex database. *World Bank Policy Research Working Paper*, 6025. World Bank.

Demirgüç-Kunt, A., Kane, E. J., & Laeven, L. (2008). Determinants of deposit-insurance adoption and design. *Journal of Financial Intermediation*, *17*(3), 407–438. DOI: 10.1016/j.jfi.2007.03.009

Descolonges, M., & Saincy, B. (2004). *Les entreprises seront-elles un jour responsables?*. La dispute.

Dinesh Bhugra, & Becker, M. A. (2005). Migration, cultural bereavement. [Note: page assumed from typical article; verify original]. *World Psychiatry; Official Journal of the World Psychiatric Association (WPA)*, *4*(1), 18. PMID: 16633496

DiploFoundation. (2025). *AI diplomacy in the Global South.*

Diversity, C. o. (2010, July 6). *What the Convention says about Traditional Knowledge, Innovations and Practices*. Retrieved July 12, 2025, from Convention on Biological Diversity: https://www.cbd.int/traditional/what.shtml

Doern, R. (2016). Entrepreneurship and crisis management: The experiences of small businesses during the London 2011 riots. *International Small Business Journal, 34*(3), 276–302. pp. 278–281.

Do, H. T., & Sum, V. (2021). Business student attitudes toward corporate social responsibility: A Vietnamese study. *Journal of Management & Organization, 27*(4), 737–754. DOI: 10.1017/jmo.2021.28

Domina Petric, M. (n.d.). *National identity*. Academia.edu. https://www.academia.edu

Donaldson, T., & Dunfee, T. W. (1999). *Ties That Bind: A Social Contracts Approach to Business Ethics*. Harvard Business Press.

Donaldson, T., & Preston, L. E. (1995). The stakeholder theory of the corporation: Concepts, evidence, and implications. *Academy of Management Review, 20*(1), 65–91. DOI: 10.2307/258887

Donato, K. M., Gabaccia, D., Holdaway, J., Manalansan, I. V. M.IV, & Pessar, P. R. (2006). A glass half full? Gender in migration studies. *The International Migration Review, 40*(1), 3–26. DOI: 10.1111/j.1747-7379.2006.00001.x

Dryzek, J. S., Bächtiger, A., Chambers, J., Cohen, J., Drury, A. S., Felicetti, A., Fishkin, J. S., Farrell, D. M., Fung, A., Gutmann, A., Landemore, H., Mansbridge, J., Marien, S., Neblo, M. A., Niemeyer, S., Setälä, M., Slothuus, R., Suiter, J., Thompson, D., & Warren, M. E. (2019). The crisis of democracy and the science of deliberation. *Science, 363*(6432), 1144–1146. DOI: 10.1126/science.aaw2694 PMID: 30872504

Duchek, S. (2020). Organizational resilience: A capability-based conceptualization. *Business Research, 13*, 215–246. pp. 220–221.

Duckett, T., Pearson, S., Blackmore, S., Grieve, B., Cielniak, G., & Jin, J. (2018). Agricultural robotics: The future of robotic agriculture. *arXiv preprint arXiv:1806.06762*. DOI: 10.31256/WP2018.2

Dunlap, A. (2018). The 'solution' is now the 'problem': Wind energy, colonisation and the 'genocide-ecocide nexus' in the Isthmus of Tehuantepec, Oaxaca. *International Journal of Human Rights, 22*(4), 550–573. DOI: 10.1080/13642987.2017.1397633

Dutta, S., Katti, S., Phani, B. V., & Zhu, P. (2021). Corporate social responsibility spending as a building block for sustainable corporate ethical identity: Lessons from indian business groups. *Managerial and Decision Economics*. Advance online publication. DOI: 10.1002/mde.3412

Ebewo, P., & Shambare, R. (2023). The effects of entrepreneurship leadership on youth entrepreneurial intentions post-COVID-19: The case of Gauteng. *Businesses*, *3*(4), 569–584. DOI: 10.3390/businesses3040035

Eccles, R. G., Ioannou, I., & Serafeim, G. (2014). The impact of corporate sustainability on organizational processes and performance. *Management Science*, *60*(11), 2835–2857. DOI: 10.1287/mnsc.2014.1984

Editorial, A. (2024, November 28). *Oaths and Legal Commitments: Foundations in Ancient Civilizations*. Retrieved July 15, 2025, from Ancient Civs: https://ancientcivs.blog/oaths-and-legal-commitments/

Editorial, L. (n.d.). *Understanding the Ancient Egypt Legal System: Structure and Function*. Retrieved July 12, 2025, from Laws Pulse: https://lawspulse.com/ancient-egypt-legal-system/

Edmans, A. (2020). *Grow the pie: How great companies deliver both purpose and profit*. Cambridge University Press. DOI: 10.1017/9781108860093

Ekawarti, L., Sugiyanto, T. and Sutaryo, S. (2023) 'Corporate Social Responsibility (CSR): A New Paradigm of Accounting for Environmentally Based Companies', *JEMASI: Jurnal Ekonomi Manajemen dan Akuntansi*, 19(1), pp. 1–12. .DOI: 10.35449/jemasi.v19i1.617

Elbanna, S., Callum, M., & Aly, R. (2023). Leveraging strategic leadership for boosting operational resilience in hotels: The role of crisis response strategies and e-readiness. *International Journal of Contemporary Hospitality Management*.

Elder, G. H.Jr. (1998). The life course as developmental theory. *Child Development*, *69*(1), 1–12. DOI: 10.1111/j.1467-8624.1998.tb06128.x PMID: 9499552

Elizondo Azuela, G., Barroso, L., Khanna, A., Wang, X., Wu, Y., & Cunha, G. (2014). Performance of renewable energy auctions: Experience in Brazil, China and India. *World Bank Policy Research Working Paper*, 7062. World Bank.

Epstein, M. J., & Cornelius, B. (2003). *Measuring Corporate Social Responsibility Performance: Accounting for Stakeholder Value*. Wiley.

Erikson, E. H. (1968). *Identity: Youth and Crisis*. W. W. Norton & Company.

Escobar, A. (1995). El desarrollo sostenible: diálogo de discursos. *Ecología política*, (9), 7-25.

Esfandiar, K., Sharifi-Tehrani, M., Pratt, S., & Altınay, L. (2019). Understanding entrepreneurial intentions: A developed integrated structural model approach. *Journal of Business Research*, *94*, 172–182. DOI: 10.1016/j.jbusres.2017.10.045

Esposito, D. E., Chen, S., & Skouloudis, A. (2023). Beyond the business case? Retracing the walk of Corporate Social Responsibility and financial performance relationship in the Oil and Gas sector. *Corporate Social Responsibility and Environmental Management*, *30*(5), 1953–1966. DOI: 10.1002/csr.2676

European Commission. (2020). *White Paper on Artificial Intelligence – A European Approach to Excellence and Trust*. https://commission.europa.eu/document/white-paper-artificial-intelligence-european-approach_en

European Commission. (2023). *Strategic foresight report 2023*. European Commission.

European Parliament. (2023). *Divergent models of data sovereignty*.

European Parliament. (2023). *The EU AI Act*.

European Parliament. (2024). *EU AI Act*.

Evans, B., & Reid, J. (2014). *Resilient life: The art of living dangerously*. Polity Press.

Evers, A., & Ewert, B. (2015). Social innovation for social cohesion. In A. Nicholls, J. Simon, & M. Gabriel (Eds.), *New frontiers in social innovation research* (pp. xx–xx). Palgrave Macmillan. (Note: page range placeholder; verify original).

Fahed-Sreih, J., 2023. Corruption and new insights in Lebanon. https://doi.org/ .DOI: 10.5772/intechopen.111903

Fahed-Sreih, J., Pistrui, D., Huang, W., & Welsch, H. (2009). Family contributions to entrepreneurial development in Lebanon. *The International Journal of Organizational Analysis*, *17*(3), 248–261. DOI: 10.1108/19348830910974941

Falkner, R. (2016). The Paris Agreement and the new logic of international climate politics. *International Affairs*, *92*(5), 1107–1125. DOI: 10.1111/1468-2346.12708

FAO (2022). *The state of food security and nutrition in the world 2022*.

FAO. (2017). *The future of food and agriculture – Trends and challenges*. Food and Agriculture Organization of the United Nations. https://www.fao.org/3/i6583e/i6583e.pdf

Farmer, P. (2004). An anthropology of structural violence. *Current Anthropology, 45*(3), 305–325. DOI: 10.1086/382250

Ferone, S., Morselli, G., & Ricci, P. (2001). *Managing Sustainable Enterprises: Strategic Approaches*. Emerald Publishing.

Fiedler, F. E. (1967). *A theory of leadership effectiveness*. McGraw-Hill.

Fiedler, F. E., & García, J. E. (1987). *New approaches to leadership: Cognitive resources and organizational performance*. Wiley.

Flammer, C. (2021). Corporate green bonds. *Journal of Financial Economics, 142*(2), 499–516. DOI: 10.1016/j.jfineco.2021.01.010

Flannery, M. J. (1998). Using Market Information in Prudential Bank Supervision: A Review of the U.S. Empirical Evidence. *Journal of Money, Credit and Banking, 30*(3), 273–305. DOI: 10.2307/2601102

Fleming, A., Jakku, E., Lim-Camacho, L., Taylor, B., & Thorburn, P. (2022). Trust, transparency and technology: Building ethical AI for agriculture. *AI & Society, 37*(2), 689–701.

Flink, T., & Schreiterer, U. (2010). Science diplomacy at the intersection of science, technology, and foreign policy: Toward a theory of techno-diplomacy. *Science & Public Policy, 37*(9), 665–677. DOI: 10.3152/030234210X12778118264530

Floridi, L., & Cowls, J. (2019). A unified framework of five principles for AI in society. *Harvard Data Science Review, 1*(1).

Floridi, L., Cowls, J., Beltrametti, M., Chatila, R., Chazerand, P., Dignum, V., Luetge, C., Madelin, R., Pagallo, U., Rossi, F., Schafer, B., Valcke, P., & Vayena, E. (2018). AI4People—An ethical framework for a good AI society: Opportunities, risks, principles, and recommendations. *Minds and Machines, 28*(4), 689–707. DOI: 10.1007/s11023-018-9482-5

Foley, J. (1997). Oral Tradition and its Implications. In Morris, I., & Powell, B. B. (Eds.), *A New Companion to Homer*. Brill.

Folke, C. (2006). Resilience: The emergence of a perspective for social–ecological systems analyses. *Global Environmental Change, 16*(3), 253–267. DOI: 10.1016/j.gloenvcha.2006.04.002

Føllesdal, A. (2022). Subsidiarity and international human-rights courts: Respecting self-governance and protecting human rights—or neither? *Law and Contemporary Problems, 79*(2), 147–163.

Foner, N., & Alba, R. (2008). Immigrant religion in the U.S. and Western Europe: Bridge or barrier to inclusion? *The International Migration Review*, *42*(2), 360–392. DOI: 10.1111/j.1747-7379.2008.00128.x

Fraissé, P., & Guerfel-Henda, S. (2005). The Environmental Dimension of CSR. *Journal of Business Ethics*, *60*(2), 153–170.

Franck, T. M. (1990). *The power of legitimacy among nations*. Oxford University Press. DOI: 10.1093/oso/9780195061789.001.0001

Frederick, W. C. (1960). The growing concern over business responsibility. *California Management Review*, *2*(4), 54–61. DOI: 10.2307/41165405

Freeman, R. E. (1984). *Strategic management: A stakeholder approach*. Cambridge University Press.

Freeman, R. E. (1984). *Strategic Management: A Stakeholder Approach*. Cambridge University Press.

Friede, G., Busch, T., & Bassen, A. (2015). ESG and financial performance: Aggregated evidence from more than 2000 empirical studies. *Journal of Sustainable Finance & Investment*, *5*(4), 210–233. DOI: 10.1080/20430795.2015.1118917

Friede, G., Busch, T., & Bassen, A. (2020). ESG and corporate financial performance: Mapping the global landscape. *Global Environmental Change*, *63*, 102123. DOI: 10.1016/j.gloenvcha.2020.102123

Friedman, M. (1962). *Capitalism and Freedom*. University of Chicago Press.

Frontiers in Psychology. (2023). Surviving a crisis: A multilevel model of leadership styles, psychological capital, and organizational resilience. pp. 1–2.

Gabagambi, J. J. (2018). *A Comparative Analysis of Restorative Justice Practices in Africa*. Retrieved from GlobaLex: https://www.nyulawglobal.org/globalex/restorative_justice_africa.html#:~:text=Community%20members%20would%20sit%20in,meat%2C%20termed%20as%20okukatelanila%20inyama

Gali, N., Hajjar, D., & Jamali, I. (2016, June 06). Mohammad and Jamali, I. (2016). The corporate governance and social responsibility nexus in the lebanese banking industry. *Corporate Governance (Bradford)*, *16*(3), 609–638. Advance online publication. DOI: 10.1108/CG-08-2015-0109

Gallagher, K. P., & Kozul-Wright, R. (2022). *The case for a new Bretton Woods: Rebuilding global economic governance for the twenty-first century*. Polity Press.

Gallardo-Vázquez, D., Sánchez-Hernández, M. I., & Barcik, A. (2019). Corporate Social Responsibility as an Antecedent of Innovation, Reputation, Performance, and Competitive Success: A Multiple Mediation Analysis. *Sustainability (Basel), 11*(20), 5614. DOI: 10.3390/su11205614

Gangi, F., Mustilli, M., & Varrone, N. (2019). The impact of corporate social responsibility (CSR) knowledge on corporate financial performance: Evidence from the european banking industry. *Journal of Knowledge Management, 23*(1), 110–134. Advance online publication. DOI: 10.1108/JKM-04-2018-0267

Garcia, S., Carrigan, C., & Wiepking, P. (2023). Global civil society response to the COVID-19 crisis. *Voluntas, 34*(6), 1310–1321. DOI: 10.1007/s11266-023-00589-5

Gaur, M., Sain, M., Sharma, D. K., Varshney, D., & Singh, M. (2021). Edge artificial intelligence for Internet of Things: A survey, future trends, and enabling technologies. *Computer Communications, 166*, 116–139.

Gebbers, R., & Adamchuk, V. I. (2010). Precision agriculture and food security. *Science, 327*(5967), 828–831. DOI: 10.1126/science.1183899

Gémar, G., & Espinar, R. (2015). Communication about Corporate Social Responsibility practices and Return on Equity. *European Journal of Family Business, 5*(1), 15–22. DOI: 10.24310/ejfbejfb.v5i1.5060

Gentle Terms. (2025, May 2). *Exploring the Foundations of Jurisprudence in Ancient Cultures*. Retrieved July 8, 2025, from Gentle Terms: https://gentleterms.com/jurisprudence-in-ancient-cultures/

Geoffrey, J. Leonardelli, Pickett, C. L., & Brewer, M. B. (2010). Optimal distinctiveness theory: A framework for social identity, social cognition, and intergroup relations. In *Advances in experimental social psychology* (Vol. 42, pp. 63–113). Academic Press.

Ghannage, R. (2018). The Relevance of Entrepreneurial Education in Lebanese Higher Institutions: A Context Other than Business and Economic Studies. *KnE Social Sciences, 3*(10), 68. DOI: 10.18502/kss.v3i10.2884

Ghosh, A. (2015). *Traditional Knowledge: Problems and Prospects*. LOKODARPAN - A PEER REVIEWED BILINGUAL ANUUAL RESEARCH JOURNAL OF FOLKLORE.

Giunta, J. (2024, November 14). *The Concept of Ma'at: Balance, Truth, and Its Role in Egyptian Society*. Retrieved July 12, 2025, from ancient pedia: https://ancientpedia.com/the-concept-of-maat-balance-truth-and-its-role-in-egyptian-society/

Gluckman, P. D., Hanson, M. A., Cooper, C., & Thornburg, K. L. (2017). Effect of in utero and early-life conditions on adult health and disease. *The New England Journal of Medicine*, *377*(12), 1191–1193. PMID: 18596274

Gómez, S.G., Eroğul, Murat Şakir and Barragan, S. (2016). Similarities and differences in teaching corporate social responsibility: Evidence from mexico and canada. *Journal on Efficiency and Responsibility in Education and Science*. .DOI: 10.7160/eriesj.2016.090303

Gond, J. P. (2006). *The Institutionalization of CSR in Business Strategy*. Palgrave Macmillan.

Gond, J. P., & Igalens, J. (2008). *Corporate Governance and Social Responsibility*. Pearson.

González-Eguino, M. (2015). Energy poverty: An overview. *Renewable & Sustainable Energy Reviews*, *47*, 377–385. DOI: 10.1016/j.rser.2015.03.013

Goodfellow, I., Bengio, Y., & Courville, A. (2016). *Deep learning*. MIT Press.

Gostin, L. O., Klock, K. A., & Finch, A. (2023). Making the world safer and fairer in pandemics. *The Hastings Center Report*, *53*(6), 3–10. DOI: 10.1002/hast.1538

Gowan, R., & Stedman, S. J. (2020). Making peace when the whole world has come to fight. *International Peacekeeping*, *27*(2), 187–192.

Green, T. (1989). *Financial Institutions and Corporate Responsibility: The Role of Trust in Banking*. Oxford University Press.

Grift, T. E., Qi, L., & Deen, M. J. (2018). Machine vision systems for precision agriculture. *Transactions of the ASABE*, *61*(1), 175–192. DOI: 10.13031/trans.12444

Grisel, F. (2023). *Arbitration as a dispute resolution process: historical developments*. Cambridge Compendium on International Commercial and Investment Arbitration.

Guston, D. H. (2014). Understanding 'anticipatory governance'. *Social Studies of Science*, *44*(2), 218–242. DOI: 10.1177/0306312713508669 PMID: 24941612

Haacke, J. (2021). The ASEAN way and the changing security environment: Navigating challenges to orthodoxy in East Asian security. *International Politics*, *58*(3), 406–427.

Haas, P. M. (1992). Introduction: Epistemic communities and international policy coordination. *International Organization*, *46*(1), 1–35. DOI: 10.1017/S0020818300001442

Haddad, P.E., Bachkirov, Alexandre Anatolievich and Grishina, O.A. (2021). Comparative CSR decision-making in the middle east: An exploratory study. *International Journal of Islamic and Middle Eastern Finance and Management*. .DOI: 10.1108/IMEFM-01-2020-0017

Hadj, T. B., Albashabsheh, A., Jamal, A., & Cherif, M. (2020). Mediation role of responsible innovation between CSR strategy and competitive advantage: Empirical evidence for the case of Saudi Arabia enterprises. *Management Science Letters*, *10*(3), 727–738. DOI: 10.5267/j.msl.2019.10.022

Haj Youssef, M., Harakeh, M., Hussein, H., & Sayour, N. (2025). Policy Brief: Addressing Barriers and Empowering Entrepreneurs in Lebanon. *F1000 Research*, *14*, 13. DOI: 10.12688/f1000research.156589.1

Hamzeh, W., Mershad, K., & Vetokhin, S. (2019). Integrating technology into higher Education: A case study in Lebanon. *Journal of Technology and Science Education*, *9*(3), 442. DOI: 10.3926/jotse.651

Hao, Y., Farooq, Q., & Sun, Y. (2018). Development of theoretical framework and measures for the role of social media in realizing corporate social responsibility through native and Non-native communication modes: Moderating effects of Cross-cultural management. *Corporate Social Responsibility and Environmental Management*, *25*(4), 704–711. Advance online publication. DOI: 10.1002/csr.1523

Hassan, N. (2020). University business incubators as a tool for accelerating entrepreneurship: A theoretical perspective. *Review of Economics and Political Science*, *9*(5), 434–453. DOI: 10.1108/REPS-10-2019-0142

Hayward, S. (2018). *The agile leader: How to create an agile business in the digital age*. Kogan Page.

Heald, M. (1970). *The Social Responsibilities of Business: Company and Community, 1900–1960*. Case Western Reserve University Press.

Heifetz, R. A., Grashow, A., & Linsky, M. (2009). *The practice of adaptive leadership*. Harvard Business Press.

Heifetz, R., & Linsky, M. (2002). *Leadership on the Line*. Harvard Business Review Press.

Helbing, D. (2013). Globally networked risks and how to respond. *Nature*, *497*(7447), 51–59. DOI: 10.1038/nature12047 PMID: 23636396

Held, D., Theros, M., & Youngs, R. (2023). *Green diplomacy: Climate action in a new era*. Polity Press.

Helou, J. (2024). Psychological impact of Beirut port explosion on Lebanese health workers. Lebanese Journal of Mental Health, 45–60.

Henrich, D. J. (2001). The griot storyteller and modern media. *Communicatio, 27*(1), 24–27. DOI: 10.1080/02500160108537921

Herbane, B. (2013). Exploring crisis management in UK small- and medium-sized enterprises. *Journal of Contingencies and Crisis Management, 21*(2), 82–95. pp. 85–86.

Hersey, P., & Blanchard, K. H. (1969). Life cycle theory of leadership. *Training and Development Journal, 23*, 26–34. pp. 28–29.

Hill, S., Akhrass, E., & Barakat, H. (2018). Changing patterns of entrepreneurship in Lebanon. *Kne Social Sciences, 3*(10), 10. DOI: 10.18502/kss.v3i10.2879

Hinterleitner, M., Knill, C., & Steinebach, Y. (2024). The growth of policies, rules, and regulations: A review of the literature and research agenda. *Regulation & Governance, 18*(2), 637–654. DOI: 10.1111/rego.12511

Hocking, B., & Melissen, J. (2015). *Diplomacy in the digital age*. Clingendael Report, Netherlands Institute of International Relations "Clingendael."

Hoekman, B., & Nelson, D. (2020). Rethinking international subsidy rules. *World Economy, 43*(12), 3104–3132. DOI: 10.1111/twec.13022

Ho, K. H., Ahmed, M., & Sultana, N. (2021). Does engaging in corporate social responsibility activities influence firm performance? The moderating effects of risk preferences and performance measurement systems. *Management Decision, 59*(12), 3020–3041. DOI: 10.1108/MD-07-2020-0925

Holland, J. H. (1995). Hidden order. *Business Week-Domestic Edition, 21*, 93–136.

Homer-Dixon, T., Renn, O., Rockström, J., Donges, J. F., Janzwood, S., & Lawrence, M. (2022). A call for an international research program on the risk of a global polycrisis. *Global Sustainability, 5*, e18.

Homer-Dixon, T., Walker, B., Biggs, R., Crépin, A.-S., Folke, C., Lambin, E. F., Peterson, G. D., Rockström, J., Scheffer, M., Steffen, W., & Troell, M. (2021). Synchronous failure: The emerging causal architecture of global crisis. *Ecology and Society, 26*(2).

Horn, C. (2013). Law, governance, and political obligation. *Cambridge University Press*, 223-246.

Horst, C. (2006). *Transnational nomads: How Somalis cope with refugee life in the Dadaab camps of Kenya*. Berghahn Books. DOI: 10.3167/9781845451295

Hou, F., Su, Y., Lu, M., & Qi, M. (2019). Model of the entrepreneurial intention of university students in the Pearl River Delta of China. *Frontiers in Psychology, 10*, 916. Advance online publication. DOI: 10.3389/fpsyg.2019.00916 PMID: 31114522

Howard, P. N., & Hussain, M. M. (2013). *Democracy's fourth wave? Digital media and the Arab Spring*. Oxford University Press. DOI: 10.1093/acprof:oso/9780199936953.001.0001

Human Rights Watch. (2019). *Bangladesh is not my home: Rohingya refugees from Myanmar*. Human Rights Watch. https://www.hrw.org/report/2019/08/05/bangladesh-not-my-home/rohingya-refugees-myanmar

Human Rights Watch. (2019, August 25). Bangladesh: Rohingya refugees trapped in camps. https://www.hrw.org/news/2019/08/25/bangladesh-rohingya-refugees-trapped-camps

Hurd, I. (1999). Legitimacy and authority in international politics. *International Organization, 53*(2), 379–408. DOI: 10.1162/002081899550913

Hussein, R. (2020). Social support networks in Arab societies. Arab Social Science Review, 77–95.

Hyndman, J., Payne, W., & Jimenez, S. (2017). *The state of private refugee sponsorship in Canada: Trends, issues, and impacts* (Policy brief). Centre for Refugee Studies, York University. https://refugeeresearch.net/wp-content/uploads/2017/02/hyndman_feb%E2%80%9917.pdf

Hyndman, J., Payne, W., & Jimenez, S. (2017). Private refugee sponsorship in Canada. *Refuge: Canada's Periodical on Refugees, 33*(2), 29–41. DOI: 10.25071/1920-7336.40449

Ibones, K. A., Enero, J. V., Jore, J. S., Mamhot, V. I., Matheu, C. M., Pacaldo, H. B. B., & Jardin, R. A. (2024). Enabling role of civil society organizations (CSOs) in local environmental management in the Philippines: A systematic review. *Journal of Civil Society, 20*(4), 359–379. DOI: 10.1080/17448689.2024.2381826

Igalens, J. and Joras, M. (2002) *Corporate Sustainability and Business Ethics: A Global Perspective*.

IMD. (2025). Shift 5: From decisive closure to constructive ambiguity. *IMD – Leadership in a turbulent world*.

Insight, l. (2025, July 24). *Administration of Justice in Ancient India: From Sabha to King's Court*. Retrieved July 26, 2025, from The lex insight: https://thelexinsight.blogspot.com/2025/07/Administration.html

Instituto Nacional de Estadística y Geografía (INEGI). (2021). *Encuesta Nacional de Inclusión Financiera (ENIF) 2021*. INEGI.

Instituto Nacional de Estadística y Geografía (INEGI). (2022). *Encuesta Nacional de Ingresos y Gastos de los Hogares (ENIGH) 2022*. INEGI.

International Energy Agency (IEA). (2024). *Energy access outlook 2024*. IEA Publications.

International Federation of Red Cross and Red Crescent Societies. (2024). *Forecast-based financing: Annual report 2024*.

International Organization for Standardization. (2023). *ISO/IEC 42001:2023 Artificial intelligence — Management system*.

International Renewable Energy Agency. (2022). *Renewable power generation costs in 2021*. IRENA.

International Telecommunication Union. (2023). *Facts and figures 2023*. ITU.

Investopedia. (2009). *The Collapse of Lehman Brothers: A case study*. pp. 1–2.

IPCC. (2023). [*Synthesis report*.]. *Climatic Change*, •••, 2023.

Isin, E. F. (2008). *Acts of citizenship*. Zed Books.

Ivanov, D., & Dolgui, A. (2020). Viability of intertwined supply networks: Extending the supply chain resilience angles towards survivability. *International Journal of Production Research, 58*(10), 2904–2915. pp. 2907–2909.

Jackson, S., Woods, R., & Hooper, F. (2022). Water colonialism and Indigenous water justice in south-eastern Australia. *International Journal of Water Resources Development*, *38*(1), 30–52.

Jamali, D., Abdallah, H., & Matar, F. (2016). Opportunities and challenges for CSR mainstreaming in business schools. *International Journal of Technology and Educational Marketing*, *6*(2), 1–29. Advance online publication. DOI: 10.4018/IJTEM.2016070101

James, E. H., & Wooten, L. P. (2022). Why business leaders should always be thinking about the next crisis. *Time*.

Jansen, J. J. P., Vera, D., & Crossan, M. (2009). Strategic leadership for exploration and exploitation: The moderating role of environmental dynamism. *The Leadership Quarterly*, *20*(1), 5–18. DOI: 10.1016/j.leaqua.2008.11.008

Jayaraman, P. P., Yavari, A., Georgakopoulos, D., Morshed, A., & Zaslavsky, A. (2016). Internet of Things platform for smart farming: Experiences and lessons learned. *Sensors (Basel)*, *16*(11), 1884. DOI: 10.3390/s16111884

Jia, C., Zuo, J., & Lü, W. (2021). Influence of entrepreneurship education on employment quality and employment willingness. *International Journal of Emerging Technologies in Learning*, *16*(16), 65. DOI: 10.3991/ijet.v16i16.24897

Jiang, Y., Wang, C., & Luo, X. (2020). Smart irrigation system based on Internet of Things and artificial intelligence technology. *Computers and Electronics in Agriculture*, *168*, 105107.

Jia, X. (2019). Corporate social responsibility activities and firm performance: The moderating role of strategic emphasis and industry competition. *Corporate Social Responsibility and Environmental Management*. Advance online publication. DOI: 10.1002/csr.1774

Jin, J., Li, Z., & Chen, X. (2019). Robust perception for autonomous agricultural vehicles in adverse weather conditions: A review. *Agricultural Engineering International: CIGR Journal*, *21*(2), 1–12.

Jones, H. G.. (2021). Real-time irrigation scheduling using AI: Case study in California vineyards. *Precision Agriculture*, *22*(3), 874–890. DOI: 10.1007/s11119-021-09814-5

Joonas Pörsti, F. (2022). Political propaganda based on psychological manipulation. In K. Kivinen (Ed.), *Digital information literacy guide*. (Note: publisher/location missing; verify original).

Joseph, J. (2013). Resilience as embedded neoliberalism: A governmentality approach. *Resilience*, *1*(1), 38–52. DOI: 10.1080/21693293.2013.765741

Joseph, S. (1996). Patriarchy and development in the Arab world. *Gender and Development*, *4*(2), 14–19. DOI: 10.1080/741922010 PMID: 12291310

Jyothirmai, B. (2020). THE CONCEPT OF DHARMA AND JUSTICE DURING. *The Law Brigade Publishers*, *6*, 206–214.

Kabeer, N. (1999). Resources, agency, achievements: Reflections on the measurement of women's empowerment. *Development and Change*, *30*(3), 435–464. DOI: 10.1111/1467-7660.00125

Kamilaris, A., Kartakoullis, A., & Prenafeta-Boldú, F. X. (2017). A review on the practice of big data analysis in agriculture. *Computers and Electronics in Agriculture, 143*, 23–37. DOI: 10.1016/j.compag.2017.09.037

Kamilaris, A., & Prenafeta-Boldú, F. X. (2018). Deep learning in agriculture: A survey. *Computers and Electronics in Agriculture, 147*, 70–90. DOI: 10.1016/j.compag.2018.02.016

Kamradt-Scott, A. (2022). *Managing global pandemics: Global governance and international health security*. Palgrave Macmillan.

Kane, P. V. (1968). *History of dharmaśāstra (ancient and mediæval religious and civil law in India)*. Bhandarkar Oriental Research Institute.

Kanojia, P., Pradhan, R., & Patel, A. (2016). A Study on the Impact of Corporate Social Responsibility on Brand Image and Customer Loyalty. *Journal of Global Business Insights, 1*(2), 73–84. DOI: 10.5038/2640-6489.1.2

Kapoor, K., & Singh, A. K. (2021). *Indian Knowledge Systems*. D.K. Print World Ltd.

Kar, A. K. (2023). The concept of Vasudhaiva Kutumbakam (The World is a Family): Insights from the Mahopanisad. *National Journal of Hindi & Sanskrit Research, 1*(49), 42–45.

Karner, C. (2017). *National identity and Europe in times of crisis*. Emerald Publishing Limited. DOI: 10.1108/9781787145139

Kautonen, T., Gelderen, M., & Fink, M. (2015). Robustness of the theory of planned behaviour in predicting entrepreneurial intentions and actions. *Entrepreneurship Theory and Practice, 39*(3), 655–674. DOI: 10.1111/etap.12056

Kayes, D. C. (2018). *Organizational resilience: How learning sustains organizations in crisis*. The Learning Organization, 25(2), 143–146. p. 145.

Kellerman, B. (2021). *The power of followership: How to create leaders people want to follow and followers who lead themselves*. Doubleday.

Keogh, J., Garvis, S., Pendergast, D., & Diamond, P. (2012). *Early childhood resilience: Processes, contexts, and outcomes*. Springer.

Keohane, R. O., & Victor, D. G. (2016). Cooperation and discord in global climate policy. *Nature Climate Change, 6*(6), 570–575. DOI: 10.1038/nclimate2937

KHAIRNAR, M. S. (2025, July). Indian Knowledge System (IKS) with Reference to Legal System in India. *International Journal of Emerging Technologies and Innovative Research, 12*(7), d292–d296.

Khan, S. U. (2013). Indian climate Policy: Challenges and Resolutions. In J. Sundaresan, S. Sreekesh, A.L., Ramanathan, Lenand Sonnenschen & R. Boojh (Eds.) *Climate Change Impact on Ecosystem* (pp. 162-167). Scientific Publishers.

Khan, S. U., Sadat, F., & Khan, S. (2024). Capturing the Narratives of Escape and Resilience: A Study of Afghan Refugees. In E. Sengupta, & A.T. Arnthorsson (Eds.), *Resilience of Educators in Extraordinary Circumstances: War, Disaster, and Emergencies* (pp. 29-53). IGI Global. https://doi.org/DOI: 10.4018/979-8-3693-1483-8.ch003

Khan, S. U. (2014). Corruption and mal-governance issues in 21st century Afghanistan: The result of regular foreign intervention or an inside saga. *International Journal of Multifaceted and Multilingual Studies, 1*(1), 1–19.

Khan, S. U., & Khan, S. (2024). From insurgents to custodians: Charting Taliban's transition towards 'governance' and 'institutionalization' in Afghanistan. In Hamdan, M. (Eds.), *Global Trends in Governance and Policy Paradigms* (pp. 165–186). IGI Global., DOI: 10.4018/979-8-3693-1742-6.ch010

Khan, S. U., Khan, S., Malik, S. H., & Tamandehrou, A. (2024). Interrogating environmental verities: An in-depth examination of eco-fictional narratives in Indian web series. In Idris, S. (Ed.), *Global Challenges for the Environment and Climate Change* (pp. 231–251). IGI Global., DOI: 10.4018/979-8-3693-2845-3.ch013

Khan, S. U., Khan, S., & Mukherjee, T. (2024). The "Subaltern" Will Speak: Investigating Portrayals in the Acclaimed OTT Series of India and Pakistan. In Kalorth, N. (Ed.), *Exploring the Impact of OTT Media on Global Societies* (pp. 1–24). IGI Global., DOI: 10.4018/979-8-3693-3526-0.ch001

Khan, S. U., Khan, S., Zarawar, A., Sachdev, M., & Sharma, R. (2024). Understanding the Durand quagmire: Examining the intricacies of conflict and exploring avenues for resolution. In Özcan, M. S. Ö. (Ed.), *International Relations Dynamics in the 21st Century: Security, Conflicts, and Wars* (pp. 411–440). IGI Global., DOI: 10.4018/979-8-3693-5688-3.ch021

Khan, S. U., & Khattak, W. (2014). Role of Pashto theatre and other performing arts in inculcating the spirit of cultural renaissance and promoting nationalistic and ethical values among Pashtuns: An analytical overview. *Sarhad University International Journal of Basic and Applied Sciences, 2*(1), 46–57.

Khan, S. U., Malik, S. H., & Khan, S. (2024). Making the 'invisible' visible: A policy centric approach to corroborate emancipation of Muslim women in mainstream. In Sharma, R., & Pareek, R. (Eds.), *New paradigms of sustainability in the contemporary era* (pp. 84–104). CSMFL Publications. DOI: 10.46679/9788195732289ch08

Khawaja, N. G., White, K. M., Schweitzer, R., & Greenslade, J. H. (2008). Difficulties and coping strategies of Sudanese refugees: A qualitative approach. *Transcultural Psychiatry*, *45*(3), 489–512. DOI: 10.1177/1363461508094678 PMID: 18799645

Kickbusch, I., & Agrawal, A. (2022). The geopolitics of global health. *Global Policy*, *13*(1), 5–14.

Kincaid, K. & Rast, D. E. (2023). Social identity theory. In G. R. Goethals, S. T. Allison, & G. J. Kramer (Eds.), *The SAGE encyclopedia of leadership studies*. SAGE Publications.

Klerkx, L., & Rose, D. (2020). Dealing with the game-changing technologies of Agriculture 4.0: How do we manage diversity and responsibility in food system transition pathways? *Global Food Security*, *24*, 100347. DOI: 10.1016/j.gfs.2019.100347

Klinke, A., & Renn, O. (2021). Adaptive and risk-based approaches to climate change and the management of uncertainty and complexity in a changing world. *Ambio*, *50*(3), 611–621.

Kolling, A., Oppenheim, I., & Elkaim, G. (2018). John Deere autonomous tractors: A case study in automation and AI deployment. *Automation in Agriculture*, *7*(4), 234–245.

Kolmodin, S. (2024). Collaboration to support refugees: The importance of local-level relations among civil society organizations. *Journal of Civil Society*, *20*(2), 130–149. DOI: 10.1080/17448689.2024.2319357

Kopel, M. (2021). CSR leadership, spillovers, and first-mover advantage. *Decisions in Economics and Finance*, *44*(3), 857–872. DOI: 10.1007/s10203-021-00328-9

Kotter, J. P. (2012). *Leading change*. Harvard Business Review Press.

KPMG. (2024). *AI governance and compliance*.

Kramer, M., & Porter, M. (2006). Strategy and Society: The Link Between Competitive Advantage and Corporate Social Responsibility. *Harvard Business Review*, *84*(12), 78–92.

Krasnyak, Y. (2020). National styles in science, diplomacy, and science diplomacy. *Brill Research Perspectives in Diplomacy and Foreign Policy*, *4*(4), 1–96.

Kraus, S., Clauss, T., Breier, M., Gast, J., Zardini, A., & Tiberius, V. (2020). The economics of COVID-19: Initial empirical evidence on how family firms in five European countries cope with the corona crisis. *International Journal of Entrepreneurial Behavior & Research, 26*(5), 1067–1092. pp. 1073–1075.

Krogh, A. H., & Lo, C. (2023). Robust emergency management: The role of institutional trust in organized volunteers. *Public Administration, 101*(1), 142–157. DOI: 10.1111/padm.12894

Kumi, E., & Hayman, R. (2025). Can domestic resource mobilisation by civil society organisations (CSOs) open up civic space in the Global South? *Development in Practice, 35*(4), 669–687. DOI: 10.1080/09614524.2025.2487791

Kuran, O., & Khabbaz, L. (2024). Stakeholder dynamics in rural Lebanese women's entrepreneurship. *Journal of Enterprising Communities People and Places in the Global Economy, 18*(5), 1098–1125. DOI: 10.1108/JEC-01-2024-0006

lab. b. u. (2024). *Israel's assault on Lebanon Sep 2024*. Beirut Urban Lab. https://www.beiruturbanlab.com

Labes, M. (2014). *Kau ilya's Arthaśāstra: A Classic Text of Statecraft and an Untapped Political Science Resource*. HEIDELBERG PAPERS IN SOUTH ASIAN AND COMPARATIVE POLITICS.

Laila, A. (2015). *Social Theory and Societal Issues: Mechanisms of Social Cohesion*. Anglo-Egyptian Library.

Lal, R. (2020). Soil health and carbon management. *Soil & Tillage Research, 195*, 104365. DOI: 10.1016/j.still.2019.104365

Lam, R., Zala, S., Lange, S., Beygelzimer, A., Schuurmans, D., & Parisotto, E. (2023). Learning general world models in a handful of reward-free deployments. *Advances in Neural Information Processing Systems, 35*, 1–15.

Lawrence, M., Janzwood, S., & Homer-Dixon, T. (2022). *What is a global polycrisis?* Cascade Institute Technical Paper No. 4. Cascade Institute.

Lawrence, M. (2023). *The polycrisis: From crisis to opportunity*. Penguin Press.

Lawrence, M. M., Homer-Dixon, T., Janzwood, S., Rockström, J., & Renn, O. (2024). *Global polycrisis: The causal mechanisms of crisis entanglement*. Global Sustainability.

Lazear, E. P. (2005). Entrepreneurship. *Journal of Labor Economics, 23*(4), 649–680. DOI: 10.1086/491605

Le, T.T. (2022) 'Corporate social responsibility and SMEs' performance: mediating role of corporate image, corporate reputation and customer loyalty', *International Journal of Emerging Markets*, ahead-of-print. .DOI: 10.1108/IJOEM-07-2021-1164

Leatherdale, D. (2019, February 9). *Trial by ordeal: When fire and water determined guilt*. Retrieved July 24, 2025, from BBC News: https://www.bbc.com/news/uk-45799443

LeCun, Y., Bengio, Y., & Hinton, G. (2015). Deep learning. *Nature*, *521*(7553), 436–444. DOI: 10.1038/nature14539

Lee-Wong, M., & More, E. (2016). Management of corporate social responsibility in Hong Kong small and medium enterprises. *Journal of Global Responsibility*, *7*(2), 304–322. DOI: 10.1108/JGR-07-2016-0018

Lentsch, J., & Weingart, P. (2011). *The politics of scientific advice: Institutional design for quality assurance*. Cambridge University Press. DOI: 10.1017/CBO9780511777141

Levi, M., & Stoker, L. (2000). Political trust and trustworthiness. *Annual Review of Political Science*, *3*(1), 475–507. DOI: 10.1146/annurev.polisci.3.1.475

Levine, A. C., Park, A., Adhikari, A., Alejandria, M. C. P., Bradlow, B. H., Lopez-Portillo, M. F., Mutwafy, S., Zumbyte, I., & Heller, P. (2023). The role of civil society organizations (CSOs) in the COVID-19 response across the Global South: A multinational, qualitative study. *PLOS Global Public Health*, *3*(9), e0002341. DOI: 10.1371/journal.pgph.0002341

Levi-Strauss, C. (1966). *The Savage Mind*. University of Chicago Press.

Levitt, P. (2009). Roots and routes: Understanding the lives of the second generation transnationally. *Journal of Ethnic and Migration Studies*, *35*(7), 1225–1242. DOI: 10.1080/13691830903006309

Levitt, P., & Glick Schiller, N. (2004). Conceptualizing simultaneity: A transnational social field perspective. *The International Migration Review*, *38*(3), 1002–1039. DOI: 10.1111/j.1747-7379.2004.tb00227.x

Levitt, T. (1958). The Dangers of Social Responsibility. *Harvard Business Review*, *36*(5), 41–50.

Lewis, D., & Opoku-Mensah, P. (2006). Moving forward research agendas on international NGOs: Theory, agency and context. *Journal of International Development*, *18*(5), 665–675. DOI: 10.1002/jid.1306

Liakos, K. G., Busato, P., Moshou, D., Pearson, S., & Bochtis, D. (2018). Machine learning in agriculture: A review. *Sensors (Basel)*, *18*(8), 2674. DOI: 10.3390/s18082674

Lichy, J., Farquhar, J.D. and Kachour, M. (2020). Entrepreneurship via social networks – 'connected woman' in Lebanon. *Qualitative Market Research: An International Journal*, ahead-of-print(ahead-of-print). Doi:.DOI: 10.1108/QMR-01-2020-0004

Li, M.. (2020). Advances in soil quality assessment using AI and remote sensing. *Soil Science Society of America Journal, 84*(5), 1422–1434. DOI: 10.1002/saj2.20156

Liñán, F., & Fayolle, A. (2015). A systematic literature review on entrepreneurial intentions: Citation, thematic analyses, and research agenda. *The International Entrepreneurship and Management Journal, 11*(4), 907–933. DOI: 10.1007/s11365-015-0356-5

Lindgreen, A., & Swaen, V. (2010). Corporate social responsibility. *International Journal of Management Reviews, 12*(1), 1–7. DOI: 10.1111/j.1468-2370.2009.00277.x

Linkov, I., & Trump, B. D. (2019). *The science and practice of resilience*. Springer. DOI: 10.1007/978-3-030-04565-4

Linnenluecke, M. K. (2017). Resilience in business and management research: A review of influential publications and a research agenda. *International Journal of Management Reviews, 19*(1), 4–30. pp. 9–11.

Liu, M. T., Wong, I. A., & Tseng, T. H. (2014). The impact of corporate social responsibility (CSR) performance and perceived brand quality on customer-based brand preference. *Journal of Services Marketing, 28*(3), 181–194. DOI: 10.1108/JSM-09-2012-0171

Liu, X. (2015). Research on the quality evaluation system of entrepreneurship education based on efficiency. *The Open Cybernetics & Systemics Journal, 9*(1), 2969–2974. DOI: 10.2174/1874110X01509012969

Li, X.. (2021). Challenges and future directions for AI in soil monitoring. *Computers and Electronics in Agriculture, 189*, 106362. DOI: 10.1016/j.compag.2021.106362

Locket, J., Moon, J., & Visser, W. (2006). Corporate Social Responsibility in Management Research. *Journal of Business Ethics, 69*(1), 1–27.

Loescher, G., Milner, J., Newman, E., & Troeller, G. (2008). *Protracted refugee situations: Political, human rights and security implications*. United Nations University Press.

Lordon, F. (2003). Conatus et institutions: pour un structuralisme énergétique. In *L'Année de la régulation n° 7 (2003-2004)* (pp. 111–146). Presses de Sciences Po. DOI: 10.3917/scpo.colle.2003.01.0111

Luo, Y., Huang, J., & Gao, S. (2022). Relationship between proactive personality and entrepreneurial intentions in college students: Mediation effects of social capital and human capital. *Frontiers in Psychology*, *13*, 861447. Advance online publication. DOI: 10.3389/fpsyg.2022.861447 PMID: 35783804

Lusardi, A., & Mitchell, O. S. (2014). The economic importance of financial literacy: Theory and evidence. *Journal of Economic Literature*, *52*(1), 5–44. DOI: 10.1257/jel.52.1.5 PMID: 28579637

Luthar, S. S., & Cicchetti, D. (2021). Resilience in development: Progress and future directions. *Development and Psychopathology*, 341–355.

Luthar, S. S., & Crossman, E. J. (2014). Resilience: A multidimensional construct. In Cooper, M. J. (Ed.), *Handbook of developmental psychopathology* (pp. 125–140). Springer. DOI: 10.1007/978-1-4614-9608-3_7

Lynn, T., Rosati, P., Lejeune, A., & Kumar, A. (2020). *The Ethics of Smart Farming: Current Issues and Governance*. Journal of Information. *Communication and Ethics in Society*, *18*(1), 79–99. DOI: 10.1108/JICES-12-2018-0091

M.K, D., Hastjarjo, S. and Slamet, Y. (2020). CSR in new normal: Realizing good citizenship and corporate social performance for community development. .DOI: 10.4108/eai.4-8-2020.2302429

Maalouf, F. T.. (2020). Youth mental health in Lebanon: Challenges and opportunities. *Frontiers in Psychiatry*, *11*, 987.

Macey, J. R., & O'hara, M. (2003). The corporate governance of banks. *Economic Policy Review*, *9*(1).

Mahiwal, A., Khan, S. U., & Khan, S. (2024). Navigating The Digital Landscape: Social Media's Influence on Climate Activism and Environmental Awareness. *Global Media Journal-Indian Edition, 16*(1). https://gmj.manipal.edu/issues/Jun2024/P6.pdf

Mahiwal, A., Khan, S. U., Khan, S., & Sachdev, M. (2025). Mapping the Misinformation Ecosystem in India: Insights Into Key Areas and Pathways of Dissemination. In Lyle, P. L., & Esmail, A. M. (Eds.), *Impacts of Leakage, Whistleblowing, and the Rise of Propaganda* (pp. 83–108). IGI Global., DOI: 10.4018/979-8-3693-7468-9.ch005

Mahmood, A., & Khan, M. T. (2022). CSR and customer orientation: A sustainable pathway towards competitive advantage? A customer's perspective. *Social Responsibility Journal*, *18*(5), 1002–1018. DOI: 10.1108/SRJ-04-2020-0139

Mahmood, M., Zaman, M., Ahmed, M., & Khan, S. (2020). 'Corporate Social Responsibility and Firms' Financial Performance: A New Insight'. *Sustainability (Basel), 12*(10), 4211. DOI: 10.3390/su12104211

Majidi, N. (2016). *Community dimensions of displacement: Development impacts in Afghanistan* (World Bank Policy Research Working Paper No. 7876). World Bank. https://doi.org/DOI: 10.1596/1813-9450-7876

Maoz, I., Steinberg, G. M., & Lewin, E. (2023). Advancing AI negotiations: New theory and evidence from a large-scale autonomous negotiations competition. *Negotiation Journal, 39*(2), 123–145.

Maritz, A., & Brown, C. (2013). Illuminating the black box of entrepreneurship education programs. *Education + Training, 55*(3), 234–252. DOI: 10.1108/00400911311309305

Marko Parezanović, & Proroković, D. (2024). Psychological and propaganda operations as a form of hybrid warfare. [Note: page range placeholder; verify original]. *Journal of Criminal Law, 29*(1), xx–xx.

Marsiglia, E., & Falautano, I. (2005). Corporate Social Responsibility and Sustainability in Financial Services. *Journal of Financial Services Marketing, 9*(4), 331–345.

Martin, B. C., McNally, J. J., & Kay, M. J. (2013). Examining the formation of human capital in entrepreneurship: A meta-analysis of entrepreneurship education outcomes. *Journal of Business Venturing, 28*(2), 211–224. DOI: 10.1016/j.jbusvent.2012.03.002

Martinet, A. C., & Reynaud, E. (2004). Entreprise durable, finance et stratégie. *Revue française de gestion, 152*(5), 121-136.

Masini, A., & Menichetti, E. (2019). Investment decisions in the renewable energy sector: An analysis of non-financial drivers. *Technological Forecasting and Social Change, 140*, 50–67. DOI: 10.1016/j.techfore.2018.12.016

Masten, A. S. (2014). *Ordinary magic: Resilience in development*. Guilford Press.

Masten, A. S. (2018). Resilience theory and research on children and families: Past, present, and promise. *Journal of Family Theory & Review, 10*(1), 12–31. DOI: 10.1111/jftr.12255

Mauléon, F., & Saulquin, J. Y. (2008). L'affichage éthique dans les rapports annuels des sociétés du CAC 40: Déterminants et implications pratiques. *Management & Avenir, 20*(6), 129–146. DOI: 10.3917/mav.020.0129

McAdams, D. P. (2001). The psychology of life stories. *Review of General Psychology*, *5*(2), 100–122. DOI: 10.1037/1089-2680.5.2.100

McBratney, A., Minasny, B., Walvoort, D., & Odeh, I. (2019). Precision agriculture and digital soil mapping. In Carter, M. R., & Gregorich, E. G. (Eds.), *Soil Sampling and Methods of Analysis* (pp. 899–917). CRC Press.

McCauley, D., Ramasar, V., Heffron, R. J., Sovacool, B. K., Mebratu, D., & Mundaca, L. (2019). Energy justice in the transition to low carbon energy systems: Exploring key themes in interdisciplinary research. *Applied Energy*, *233*, 916–921. DOI: 10.1016/j.apenergy.2018.10.005

McKeague, L. K., & Bies, A. L. (2025). Nonprofits and Interest Groups in Crisis Management. In *Oxford Research Encyclopedia of Politics*. DOI: 10.1093/acrefore/9780190228637.013.1530

McKenzie, D., & Woodruff, C. (2014). What are we learning from business training and entrepreneurship evaluations around the developing world? *The World Bank Research Observer*, *29*(1), 48–82. DOI: 10.1093/wbro/lkt007

McKinsey & Company. (2024). The resilience imperative: Succeeding in uncertain times.

McPherson, M., Horowitz, L. S., Lusher, D., di Giglio, S., Greenacre, L. E., & Saalmann, Y. B. (2011). Marginal women, marginal rights: Impediments to gender-based persecution claims by asylum-seeking women in Australia. *Journal of Refugee Studies*, *24*(2), 323–345. DOI: 10.1093/jrs/fer002

Meadows, D. (2008). *Thinking in systems: International bestseller*. chelsea green publishing.

Mejjah, O., Ngulyavyangu, H., Peter, T., & Mwita, S. (2021). Assessment of entrepreneurial traits and intention among undergraduate students at Catholic University of Health and Allied Sciences - Mwanza, Tanzania. *Asian Journal of Business and Management*, *9*(5). Advance online publication. DOI: 10.24203/ajbm.v9i5.6808

Melhem, S. (2014). *Developmental Psychology: Human Life Cycle*. Dar Al-Fikr for Publishing and Distribution.

Mengjiao, Z., Arshad, M. A., & Yating, L. (2024). Examining the relationship between corporate social responsibility and employee turnover intentions: A comprehensive review. *International Journal of Academic Research in Business & Social Sciences*, *14*(1). Advance online publication. DOI: 10.6007/IJARBSS/v14-i1/20003

Menkhoff, L., & Suwanaporn, C. (2007). On the rationale of bank lending in pre-crisis Thailand. *Applied Economics*, *39*(9), 1077–1089. DOI: 10.1080/00036840500447815

Mercier, A. (2000, October). L'observation du travail des élèves, quels en sont les objets élémentaires et comment peut-on les produire? In *Journées Franco-Québécoises de Didactique*.

Merton, R. C. (1977). An analytic derivation of the cost of deposit insurance and loan guarantees an application of modern option pricing theory. *Journal of Banking & Finance*, *1*(1), 3–11. DOI: 10.1016/0378-4266(77)90015-2

Mészáros, T. (n.d.). *REMARK ON THE CANON OF THE ATTIC ORATORS*. Retrieved July 16, 2025, from academia: https://www.academia.edu/14909234/REMARK_ON_THE_CANON_OF_THE_ATTIC_ORATORS

Meyer, J. W., & Rowan, B. (1977). Institutionalized organizations: Formal structure as myth and ceremony. *American Journal of Sociology*, *83*(2), 340–363. DOI: 10.1086/226550

Mikdashi, M. (2022). Sectarianism and migration in Lebanon. *The Middle East Journal*, •••, 543–561.

Milan, S., & Treré, E. (2019). Big data from the South: Beyond data universalism. *Television & New Media*, *20*(4), 319–335. DOI: 10.1177/1527476419837739

Miller, K. E., & Rasmussen, A. (2017). The mental health of civilians displaced by armed conflict: An ecological model of refugee distress. *Epidemiology and Psychiatric Sciences*, *26*(2), 129–138. DOI: 10.1017/S2045796016000172 PMID: 27040595

Miller, R., Poli, R., & Rossel, P. (2018). The discipline of anticipation: Foundations for futures literacy. In Miller, R. (Ed.), *Transforming the future: Anticipation in the 21st century* (pp. 51–65). Routledge.

Minasny, B.. (2019). Machine learning applications for soil data interpretation. *Geoderma*, *334*, 114–132. DOI: 10.1016/j.geoderma.2018.07.043

Minsky, H. P. (1996). Uncertainty and the institutional structure of capitalist economies: Remarks upon receiving the Veblen-Commons award. *Journal of Economic Issues*, *30*(2), 357–368. DOI: 10.1080/00213624.1996.11505800

Miralles, F., Giones, F., & Riverola, C. (2015). Evaluating the impact of prior experience in entrepreneurial intention. *The International Entrepreneurship and Management Journal*, *12*(3), 791–813. DOI: 10.1007/s11365-015-0365-4

Mir, G., Durrani, N., Julian, R., Kimei, Y., Mashreky, S., & Doan, T. D. (2024). Social inclusion and sustainable development: Findings from Seven African and Asian contexts. *Sustainability (Basel), 16*(11), 4859. DOI: 10.3390/su16114859

Mizrak, K. C. (2024). Crisis management and risk mitigation: Strategies for effective response and resilience. *Trends, challenges, and practices in contemporary strategic management*, 254-278. DOI: DOI: 10.4018/979-8-3693-1155-4.ch013

Mohanty, S.. (2021). Soil moisture prediction using hybrid machine learning models: A review. *Hydrological Sciences Journal, 66*(12), 2023–2040. DOI: 10.1080/02626667.2021.1957846

Molina, P., García, E., & López, M. (2020). The ATLAS project: Autonomous tractors and spraying robots for precision agriculture. *Robotics and Autonomous Systems, 123*, 103310. DOI: 10.1016/j.robot.2019.103310

Molyneux, M. (1985). Family reform in socialist states: The hidden agenda. *Feminist Review, 21*(1), 47–64. DOI: 10.1057/fr.1985.28

Mom, T. J. M., Fourné, S. P. L., & Jansen, J. J. P. (2015). Managers' work experience, ambidexterity, and performance. *Human Resource Management, 54*(S1), s133–s153. pp. s139–s140.

Montgomery, D. R. (2007). Soil erosion and agricultural sustainability. *Proceedings of the National Academy of Sciences of the United States of America, 104*(33), 13268–13272. DOI: 10.1073/pnas.0611508104

Moon, S., Røttingen, J. A., & Frenk, J. (2021). Global public goods for health: Weaknesses and opportunities in the global health system. *Health Economics, Policy, and Law, 16*(2), 109–120.

Morin, E. (1993). *Terre-Patrie*. Éditions du Seuil.

Morse, J. C., & Keohane, R. O. (2014). Contested multilateralism. *The Review of International Organizations, 9*(4), 385–412. DOI: 10.1007/s11558-014-9188-2

Mosse, D. (2005). *Cultivating development: An ethnography of aid policy and practice*. Pluto Press.

Mottley, M. (2022). *The Bridgetown Initiative*. Government of Barbados.

Moya, J., & Goenechea, M. (2022). An approach to the unified conceptualization, definition, and characterization of social resilience. *International Journal of Environmental Research and Public Health, 19*(9), 5746. DOI: 10.3390/ijerph19095746

Muchiri, M., & Muigai, R. G. (2019). *Effect of environmental CSR activities on the financial performance of financial institutions in kirinyaga county*. Rjfa., DOI: 10.7176/RJFA/10-5-07

Mukherjee, T., & Khan, S. U. (2025). Portrayal of marginalized women in digital media narratives. In Kumar, V. V., & Sahoo, J. R. (Eds.), *Media and marginalized voices: Women and LGBTQIA+ community* (pp. 91–109). Peter Lang.

Nadányiová, M. (2021) 'The perception of corporate social responsibility and its impact on consumer buying behaviour in the process of globalization', *SHS Web of Conferences*, 92, p. 06024. DOI: 10.1051/shsconf/20219206024

Nair, J. (2024). Reconfiguration and transformation for resilience: Building service organizations towards sustainability. *Journal of Services Marketing*, *38*(4), 404–425. DOI: 10.1108/JSM-04-2023-0144

Narada, & Lariviere, R. W. (2003). *The Nāradasm ti.* Motilal Banarsidass Publishing House.

Narayan, D., Patel, R., Schafft, K., Rademacher, A., & Koch-Schulte, S. (2000). Voices of the poor. Can anyone hear us? (pp. xi+-343). DOI: 10.1596/0-1952-1601-6

Nasibov, M. (2021). Civil Society and Pro-Democracy Social Movements: Troubled Relations Within Authoritarian Regimes? *Frontiers in Political Science*, *3*, 708872. DOI: 10.3389/fpos.2021.708872

Nations, U. (2007, September 13). *UN Declaration on the Rights of Indigenous Peoples*. Retrieved July 12, 2025, from OHCHR and Indigenous Peoples: https://www.ohchr.org/en/indigenous-peoples/un-declaration-rights-indigenous-peoples

Nature. (2025). DeepSeek-R1: An open source pathway to AGI. *Nature, 625*, 1–2. (Fictional)

Nauffal, D., & Skulte-Ouaiss, J. (2018). Quality higher Education drives employability in the Middle East. *Education + Training*, *60*(9), 1057–1069. DOI: 10.1108/ET-05-2017-0072

Ní Laoire, C., Carpena-Méndez, F., Tyrrell, N., & White, A. (2016). *Childhood and migration in Europe: Portraits of mobility, identity and belonging in contemporary Ireland*. Routledge. DOI: 10.4324/9781315571485

Nielsen, S. C., & Rhoten, D. (2018). The SESAME synchrotron in Jordan: A bridge for science and peace in the Middle East. *Science & Diplomacy*, *7*(3).

Njoku, E. T. (2022). Strategic exclusion, co-option and containment: Towards an integrative theory of state-CSOs relations. *Critical Studies on Terrorism, 15*(4), 917–944. DOI: 10.1080/17539153.2022.2111776

Noori, S. (2024, March 13). *Climate change, conflict and displacement: Perspectives from Afghanistan.* Humanitarian Practice Network. https://odihpn.org/en/publication/climate-change-conflict-and-displacement-perspectives-from-afghanistan/

Nordhaus, W. (2021). *The spirit of green: The economics of collisions and contagions in a crowded world.* Princeton University Press.

North, D. C. (1990). *Institutions, institutional change and economic performance.* Cambridge University Press. DOI: 10.1017/CBO9780511808678

Northouse, P. G. (2022). *Leadership: Theory and practice* (9th ed.). Sage.

O'Sullivan, M., Overland, G., & Sandalow, D. (2022). *The geopolitics of the global energy transition.* Springer.

Odeh, J. (2021). *Psychological Resilience.* Guardian.

OECD. (2021). *The Digital Transformation of SMEs.* OECD Publishing.

Olivelle, P. (2005). *Manu's Code of Law.* Oxford University Press.

Olivelle, P. (n.d.). *Yajnavalkya - A Treatise on Dharma.* Harvard University Press.

Osborne, R. (2009). *Greece in the Making 1200-479 BC.* Routledge. DOI: 10.4324/9780203880173

Ostrom, E. (2010). Polycentric systems for coping with collective action and global environmental change. *Global Environmental Change, 20*(4), 550–557. DOI: 10.1016/j.gloenvcha.2010.07.004

Otto, F. E. L., Skeie, R. B., Fuglestvedt, J. S., Berstad, D., & Allen, M. R. (2020). Assigning historical responsibilities for extreme weather events. *Nature Climate Change, 10*, 711–716.

Oxfam International. (2024). *Vetoing humanity: How a few powerful nations hijacked global peace.* Oxfam.

Padarian, J.. (2019). Soil nutrient prediction using vis-NIR spectroscopy and machine learning. *Geoderma, 353*, 178–189. DOI: 10.1016/j.geoderma.2019.06.017

Paillard, E. (2014). THE STRUCTURAL EVOLUTION OF FIFTH-CENTURY ATHENIAN SOCIETY: ARCHAEOLOGICAL EVIDENCE AND LITERARY SOURCES. [Retrieved from IGNOU.]. *Mediterranean Archaeology, 27*, 77–84.

Pal, R., Torstensson, H., & Mattila, H. (2014). Antecedents of organizational resilience in economic crises—An empirical study of Swedish textile and clothing SMEs. *International Journal of Production Economics, 147*, 410–428. pp. 412–414.

Panagos, P.. (2018). Soil erosion risk assessment in Europe. *The Science of the Total Environment, 631-632*, 1446–1460. DOI: 10.1016/j.scitotenv.2018.03.047

Pandey, K. (2014). Indigenous Knowledge System: Concepts and Discourses. *Anthropological Bulletin, 2*(1), 17–28.

Panezi, M. (2021, November/December). *A Description of the Structure of the Hellenic Republic, the Greek Legal System, and Legal Research*. Retrieved July 11, 2025, from Globalex: https://www.nyulawglobal.org/globalex/greece1.html

Pathan, S. (2009). Strong boards, CEO power and bank risk-taking. *Journal of Banking & Finance, 33*(7), 1340–1350. DOI: 10.1016/j.jbankfin.2009.02.001

Patowary, K. (2022, July 7). *Code of Ur-Nammu: The Oldest Law in History*. Retrieved July 13, 2025, from Amusing Planet: https://www.amusingplanet.com/2022/07/code-of-ur-nammu-oldest-law-in-history.html#:~:text=Some%20of%20the%20earliest%20legal%20codes%20concerning%20crimes,Kassites%2C%20and%20Hittites%2C%20each%20had%20their%20own%20laws

Patrick, S. (2017). Trump and world order: The return of self-help. *Foreign Affairs, 96*(2), 52–57.

Pedersen, S. M., Fountas, S., Have, H., & Blackmore, S. (2018). Safety considerations for autonomous vehicles in agriculture. *Biosystems Engineering, 164*, 1–12. DOI: 10.1016/j.biosystemseng.2017.12.002

Peteet, J. (2005). *Landscape of hope and despair: Palestinian refugee camps*. University of Pennsylvania Press. DOI: 10.9783/9780812200317

Piccoli, L., Jelena, D., Ruedin, D., & Jacob-Owens, T. (2023). Restricting Human Movement During the COVID-19 Pandemic: New Research Avenues in the Study of Mobility, Migration, and Citizenship. *The International Migration Review, 57*(2), 505–520. DOI: 10.1177/01979183221118907 PMID: 38603252

Pitidis, V., Coaffee, J., & Lima-Silva, F. (2024). Advancing equitable 'resilience imaginaries' in the Global South through dialogical participatory mapping: Experiences from informal communities in Brazil. *Cities (London, England), 150*, 105015. DOI: 10.1016/j.cities.2024.105015

PMC. (2024). Surviving a crisis: A multilevel model of leadership styles, employees' psychological capital and organizational resilience. *Frontiers in Psychology*.

Polo, A. (2007). Corporate governance of banks: the current state of the debate.

Polzin, F., Egli, F., Steffen, B., & Schmidt, T. S. (2021). How do policies mobilize private finance for renewable energy? A systematic review with an investor perspective. *Applied Energy*, *236*, 1249–1268. DOI: 10.1016/j.apenergy.2018.11.098

Porter, M. E., & Kramer, M. R. (2011). Creating shared value: How to reinvent capitalism and unleash a wave of innovation and growth. *Harvard Business Review*, *89*(1-2), 62–77.

Porter, M. E., & Kramer, M. R. (2019). Where ESG fails. *The Institutional Investor*, *2019*(10), 74–81.

Portes, A., & Rumbaut, R. G. (2001). *Legacies: The story of the immigrant second generation*. University of California Press.

Poussing, N. (2018). Does corporate social responsibility encourage sustainable innovation Adoption?¡scp¿E¡/Scp¿mpirical evidence From¡scp¿L¡/Scp¿uxembourg. *Corporate Social Responsibility and Environmental Management*. Advance online publication. DOI: 10.1002/csr.1712

Prowse, S. (1997). Corporate control in commercial banks. *Journal of Financial Research*, *20*(4), 509–527. DOI: 10.1111/j.1475-6803.1997.tb00263.x

Pruett, M., & Şeşen, H. (2017). Faculty-student perceptions about entrepreneurship in six countries. *Education + Training*, *59*(1), 105–120. DOI: 10.1108/ET-03-2013-0028

Pueyo, A., Gonzalez, F., Dent, C., & DeMartino, S. (2013). The role of technology in inclusive green growth. *IDS Working Paper*, 433. Institute of Development Studies.

PVE. N. S. (2018). *Lebanon – Preventing violent extremism*. https://share.google/gNpgbYxCFldLPdgJ0 (Original work published at pvelebanon.org)

Quoc, K. (2022). Corporate social responsibility and bank's performance under the mediating role of customer satisfaction and bank reputation. *Emerging Science Journal*, *6*(6), 1409–1429. Advance online publication. DOI: 10.28991/ESJ-2022-06-06-012

Rae, D. (2006). Entrepreneurial learning: A conceptual framework for technology-based enterprise. *Technology Analysis and Strategic Management*, *18*(1), 39–56. DOI: 10.1080/09537320500520494

Raelin, J. A. (2016). *Leadership-as-practice: Theory and application*. Routledge. DOI: 10.4324/9781315684123

Rahimi, A. (2012). *A curse on Dostoevsky*. Other Press.

Rajab, A. (2025). *Self-Management Skills and Psychological Resilience: How to Overcome Stress and Challenges?* Azha Publishing.

Rajpal, L., & Vats, M. (2017). STRIDHANA: A CRITICAL APPROACH. *International Journal of Research in Social Sciences and Humanities*, *7*(1), 83–95.

Ratnawati, K., & Rahayu, S. M. (2018). 'The Role of SMEs' Innovation and Learning Orientation in Mediating the Effect of CSR Programme on SMEs' Performance and Competitive Advantage'. *Global Business Review*, *19*(3, suppl), S23–S41. DOI: 10.1177/0972150918757842

Rauch, A., & Hulsink, W. (2015). Putting entrepreneurship education where the intention to act lies: An investigation into the impact of entrepreneurship education on entrepreneurial behaviour. *Academy of Management Learning & Education*, *14*(2), 187–204. DOI: 10.5465/amle.2012.0293

Reichstein, M., Camps-Valls, G., Stevens, B., Jung, M., Denzler, J., Carvalhais, N., & Prabhat, . (2019). Deep learning and process understanding for data-driven Earth system science. *Nature*, *566*(7743), 195–204. DOI: 10.1038/s41586-019-0912-1

Reschke, H.. (2022). *Management in times of crises: reflections on characteristics, avoiding pitfalls, and pathways out*. PMC.

Reuters. (2024, March 27). *How Boeing's leadership was 'fired' by its own customers.*

Reuters. (2025a). China releases efficient AI model. *Reuters*. (Fictional)

Reuters. (2025b). U.S. responds to Chinese AI advancements. *Reuters*. (Fictional)

Rhee, Y. P., Park, C., & Petersen, B. K. (2018). The effect of local stakeholder pressures on responsive and strategic CSR activities. *Business & Society*. Advance online publication. DOI: 10.1177/0007650318816454

Riboldi, M. (2024). The Loci of Power and Connection: A framework for exploring the democratic relationships of civil society organisations. *Interest Groups & Advocacy*, *13*(2), 244–263. DOI: 10.1057/s41309-024-00209-7

Riessman, C. K. (2008). *Narrative methods for the human sciences*. SAGE Publications.

Ring, P. S., & Van de Ven, A. H. (1994). Developmental processes of cooperative interorganizational relationships. *Academy of Management Review*, *19*(1), 90–118. DOI: 10.2307/258836

Roberts, E., & Huq, S. (2023). Exploring loss and damage from climate change and global perspectives on the way forward. *Climate Policy*, *23*(8), 1023–1033.

Robeyns, I. (2006). The capability approach in practice. *Journal of Political Philosophy, 14*(3), 351–376. DOI: 10.1111/j.1467-9760.2006.00263.x

Rodriguez, M., Smith, J., & Brown, K. (2020). CFO sustainability literacy and corporate environmental performance: Evidence from Fortune 500 companies. *Strategic Management Journal, 41*(11), 2051–2078. DOI: 10.1002/smj.3187

Rohrbeck, R., & Kum, M. E. (2018). Corporate foresight and its impact on firm performance. *Technological Forecasting and Social Change, 129*, 105–116. pp. 108–109.

ROK Ministry of Foreign Affairs. (2024). *AI Seoul Summit.*

Roldán, J. J., Ramos, J., & Muñoz, F. (2020). Autonomous navigation systems for agricultural vehicles: A review. *Sensors (Basel), 20*(7), 1985. DOI: 10.3390/s20071985

Rolnick, D., Donti, P. L., Kaack, L. H., Kochanski, K., Lacoste, A., Sankaran, K., & Bengio, Y. (2022). Tackling climate change with machine learning. *ACM Computing Surveys, 55*(1), 1–96.

Romić, R. M. (2020). The role of education in preserving national identity. In *Culture and identities*. Serbian Sociological Association Institute for Political Study.

Roomi, M. A., & Parrott, G. (2008). Barriers to development and progression of women entrepreneurs in Pakistan. *The Journal of Entrepreneurship, 17*(1), 59–72. DOI: 10.1177/097135570701700105

Rose, D. C., Higgins, V., Mitchell, P., & Bryant, M. (2021). The co-production of smart farming knowledge: Insights from stakeholder workshops. *Agricultural Systems, 188*, 103033.

Rosing, K., Frese, M., & Bausch, A. (2011). Explaining the effectiveness of leadership styles: The role of innovation. *Journal of Organizational Behavior, 32*(1), 104–128. pp. 106–107.

Rotz, S., Gravely, E., Mosby, I., Duncan, E., Finnis, E., Horgan, M., & Martin, R. (2019). Automated pastures and the digital divide: How agricultural technologies are shaping labour and rural communities. *Journal of Rural Studies, 68*, 112–122. DOI: 10.1016/j.jrurstud.2019.01.023

Roy, P., Maji, S., & Singh, P. (2023, December). Preserving Indigenous Technical Knowledge: The Need for Documentation and Cultural Preservation in the Context of North-East India. *Food and Scientific Reports, 4*(12), 10–17.

Ruiz-Arranz, M., Moreno-Badia, M., & Levy-Yeyati, E. (2016). *Financial inclusion in Latin America: Facts, obstacles and central banks' policy. BIS Papers, 83*. Bank for International Settlements.

Russell, S., & Norvig, P. (2021). *Artificial intelligence: A modern approach* (4th ed.). Pearson.

Rutter, M. (2012). Resilience as a dynamic concept. *Development and Psychopathology, 24*(2), 335–344. DOI: 10.1017/S0954579412000028 PMID: 22559117

Sachs, J. D. (2015). *The age of sustainable development*. Columbia University Press. DOI: 10.7312/sach17314

Saeed, S., Yousafzai, S., Yani-de-Soriano, M., & Muffatto, M. 2018. The role of perceived university support in the formation of students' entrepreneurial intention. pp.3-23. https://doi.org/DOI: 10.4324/9781315611495-1

Saeed, M. M., Mudliar, M., & Kumari, M. (2023). Corporate social responsibility and financial performance nexus: Empirical evidence from ghana. *Corporate Social Responsibility and Environmental Management, 30*(6), 2799–2815. Advance online publication. DOI: 10.1002/csr.2516

Saeidi, S. P., Sofian, S., Saeidi, P., Saeidi, S. P., & Saaeidi, S. A. (2018). 'The Utilitarian Aspect of the Philosophy of Ecology: The Case of Corporate Social Responsibility', *Filosofija. Sociologija, 29*(1), 49–57. DOI: 10.6001/fil-soc.v29i1.3630

Safi, A. A., & Czaika, M. (2024). The transnational engagement of Afghan diaspora organizations: Drivers of diaspora specialization. *Global Networks, 25*(1), 174–194. DOI: 10.1111/glob.12484

Saini, R. and Ram, K. (2025) 'Unlocking Sustainable Growth: Social Innovation as a Link Between CSR and Competitive Advantage in Haryana's MSMEs', *SEDME (Small Enterprises Development, Management & Extension Journal)*, 52(1), pp. 34–47. .DOI: 10.1177/09708464251326387

Saleh, H. (2025). The perceptions of the Lebanese students of choosing their career in entrepreneurship. *Jordan Journal of Business Administration*, [online] 10(2). Available at: https://archives.ju.edu.jo/index.php/JJBA/article/view/6630 [Accessed 28 Mar. 2025].

Salvador, M., & Sancho, D. (2023). Local governments facing turbulence: Robust governance and institutional capacities. *Social Sciences (Basel, Switzerland), 12*(8), 462. DOI: 10.3390/socsci12080462

Samaratunge, R., & Alam, Q. (2021). Accountability and transparency in emerging countries: Governance, democratic currents and change. *Public Administration and Development, 41*(4), 147–156. DOI: 10.1002/pad.1963

Santika, I., Wardana, I., Setiawan, P., & Widagda, I. (2022). Entrepreneurship education and green entrepreneurial intention. *Linguistics and Culture Review*, *6*, 797–810. DOI: 10.21744/lingcure.v6nS1.2159

Sardianou, E., & Genoudi, P. (2013). Which factors affect the willingness of consumers to adopt renewable energies? *Renewable Energy*, *57*, 1–4. DOI: 10.1016/j.renene.2013.01.031

Sarouphim, K. (2015). Slowly but surely. *Journal for the Education of the Gifted*, *38*(2), 196–211. DOI: 10.1177/0162353215578278

Sasson. (1995). *Civilizations of the Ancient Near East.* New York: Scribner.

Sayad, A. (2004). *The suffering of the immigrant*. Polity Press.

Sayad, S. E., & Diab, A. (2022). Bank employee perceptions of corporate social responsibility practices: Evidence from egypt. *Sustainability (Basel)*, *14*(3), 1862. Advance online publication. DOI: 10.3390/su14031862

Scharpf, F. W. (1999). *Governing in Europe: Effective and democratic?* Oxford University Press. DOI: 10.1093/acprof:oso/9780198295457.001.0001

Schlosberg, D. (2013). Theorising environmental justice: The expanding sphere of a discourse. *Environmental Politics*, *22*(1), 37–55. DOI: 10.1080/09644016.2013.755387

Schmeidl, S. (2008). The case of the Afghan refugee population: Finding durable solutions in contested transitions. In Adelman, H. (Ed.), *Protracted Displacement in Asia: No Place to Call Home* (pp. 131–179). Ashgate.

Schmeidl, S. (2019). Afghanistan's displaced population: Understanding the drivers of displacement. *Journal of Refugee Studies*, *32*(1), 1–24. DOI: 10.1093/jrs/fey032

Schmeidl, S., & Maley, W. (2016). The case of the Afghan refugees: Political networks and survival strategies. *Migration Studies*, *4*(2), 1–21. DOI: 10.1093/migration/mnw001

Schmidt, V. A. (2013). Democracy and legitimacy in the European Union revisited: Input, output and "throughput.". *Political Studies*, *61*(1), 2–22. DOI: 10.1111/j.1467-9248.2012.00962.x

Schoemaker, P. J. H., Heaton, S., & Teece, D. (2018). Innovation, dynamic capabilities, and leadership. *California Management Review, 61*(1), 15–42. pp. 20–22.

Schroeder, R. (2018). *Social theory after the Internet*. UCL Press. DOI: 10.2307/j.ctt20krxdr

Schunz, S. (2023). Governing global spaces: The roles of the European Union and China in global environmental governance. *Global Environmental Politics*, *23*(1), 3–25.

Secretaría de Energía (SENER). (2023). *Programa Nacional para el Aprovechamiento Sustentable de la Energía 2023-2027*. SENER.

Secretaría de Energía (SENER). (2024). *Balance Nacional de Energía 2023*. SENER.

Security Council Report. (2022). *The veto*. Security Council Report Research Report.

Semenescu, A. and Curmei, Cătălin Valeriu (2015). Using CSR to mitigate information asymmetry in the banking sector. *Management & Marketing*. .DOI: 10.1515/mmcks-2015-0021

Sen, S. (2019, April 25). *Of Holy Rivers and Human Rights: Protecting the Ganges by Law*. Retrieved from Yale University Press: https://yalebooks.yale.edu/2019/04/25/of-holy-rivers-and-human-rights-protecting-the-ganges-by-law/

Sen, A. (1999). *Development as freedom*. Oxford University Press.

Sen, A. (2009). *The Idea of Justice*. Belklnap Press.

Senge, P. M. (2006). *The fifth discipline: The art & practice of the learning organization* (Revised ed.). Doubleday.

Seshasayee, T. (n.d.). *Dharma*. Retrieved July 13, 2025, from Indian religion and philosophy: https://ebooks.inflibnet.ac.in/icp03/chapter/dharma/

Shahhosseini, M., Martinez-Feria, R., & Archontoulis, S. V. (2021). Simulation of field-scale precision nitrogen management impacts on maize yield and soil nitrogen dynamics. *Agricultural Systems*, *186*, 102983. DOI: 10.1016/j.agsy.2020.102983

Shamshiri, R. R., Weltzien, C., Hameed, I. A., Yule, I. J., & Ahmad, D. (2018). Research and development in agricultural robotics: A perspective on technologies, applications, and challenges. *Biosystems Engineering*, *165*, 559–572. DOI: 10.1016/j.biosystemseng.2017.12.006

Shaw, I. (2002). *The Oxford History of Ancient Egypt*. Oxford University Press. shorthistory. (2015, July 12). *Draco and Solon era in Athens (621-593. B.C)*. Retrieved July 17, 2025, from shorthistory: https://www.shorthistory.org/ancient-civilizations/ancient-greece/draco-and-solon-in-athens-621-593-b-c/# sociology.institute. (2022, October 17). *The Divine Right to Rule: Understanding Ancient The God-Kings System*. Retrieved July 13, 2025, from Sociology of Religion: https://sociology.institute/sociology-of-religion/ancient-divine-right-rule-god-kings-system/

Shiskha. (2025, February 3). *Ancient and Medieval India's Environmental Protection Traditions*. Retrieved from Occupational Health & Safety Management: https://disaster.shiksha/occupational-health-safety-management/ancient-medieval-india-environmental-protection

Singh, A. K., Ganapathysubramanian, B., Singh, A., & Sarkar, S. (2021). Deep learning for plant stress phenotyping: Trends and future perspectives. *Trends in Plant Science*, *26*(9), 866–875.

Singh, A., Ganapathysubramanian, B., Singh, A. K., & Sarkar, S. (2020). Machine learning for high-throughput stress phenotyping in plants. *Trends in Plant Science*, *21*(2), 110–124. DOI: 10.1016/j.tplants.2015.10.015

Singh, K., Misra, M., & Yadav, J. (2021). Corporate social responsibility and financial inclusion: Evaluating the moderating effect of income. *MDE. Managerial and Decision Economics*, *42*(5), 1263–1274. Advance online publication. DOI: 10.1002/mde.3306

Širola, D., 2020. Determinants of entrepreneurial intentions. *Zbornik Veleučilišta U Rijeci*, 8(1), pp.169-187. https://doi.org/.DOI: 10.31784/zvr.8.1.17

Smidt, H., Johansson, J., & Richter, T. (2025). Civil society under attack: The consequences for horizontal accountability institutions. *Studies in Comparative International Development*, *60*(1), 81–110. DOI: 10.1007/s12116-023-09423-x

Sonte, N. (n.d.). *Code of Ur-Nammu*. Retrieved from SCRIBD: https://www.scribd.com/document/591430208/Code-of-Ur-Nammu#:~:text=The%20Code%20of%20Ur%2DNammu,Sumer%20during%20its%20Renaissance%20period

Sovacool, B. K., & Dworkin, M. H. (2014). *Global energy justice: Problems, principles, and practices*. Cambridge University Press. DOI: 10.1017/CBO9781107323605

Spivak, G. C. (1988). Can the subaltern speak? In C. Nelson & L. Grossberg (Eds.), *Marxism and the interpretation of culture* (pp. 271–313). University of Illinois Press.

Srinivas, S. (2013, September 10). *Dharma Shastras- Legal Literature of Ancient India*. Retrieved July 8, 2025, from ithihas: https://ithihas.wordpress.com/2013/09/10/dharma-shastras-legal-literature-of-ancient-india/

Srinivasan, R.. (2020). Precision nutrient management in agriculture using AI technologies. *Agronomy (Basel)*, *10*(6), 903. DOI: 10.3390/agronomy10060903

Stenberg, B., Viscarra Rossel, R. A., Mouazen, A. M., & Wetterlind, J. (2010). Visible and near infrared spectroscopy in soil science. *Advances in Agronomy*, *107*, 163–215. DOI: 10.1016/S0065-2113(10)07005-7

StudynLearn. (2022, February 23). *Hammurabi: Code of Hammurabi Laws - Mesopotamian Civilization*. Retrieved from StudynLearn: https://studynlearn.com/hammurabi#:~:text=He%20is%20best%20known%20for,282%20rules%20or%20legal%20decisions

Stupak, I., Mansoor, M., & Smith, C. T. (2021). Conceptual framework for increasing legitimacy and trust of sustainability governance. *Energy, Sustainability and Society, 11*(1), 5. DOI: 10.1186/s13705-021-00280-x

Sucharita. (2016). Clarifying the conceptual map of VUCA: A systematic review. *International Journal of Organizational Analysis*, 24(3), 484–502.

Sule, P. (2018). The idea of Restributive Justice in African Jurisprudence. *The 2nd African Philosophy World Conference at the University of Calabar, Nigeria, October, 2017.*

Sullivan-Taylor, B., & Branicki, L. (2011). Creating resilient SMEs: Why one size might not fit all. *International Journal of Production Research, 49*(18), 5565–5579. pp. 5568–5570.

Sunoto, P., & Daryanto, E. (2024). The role of strategic leadership in building organizational resilience. *International Journal of Integrative Sciences, 4*(3), 89–105.

Sustainability Directory. (2024). *Indigenous Governance*. Retrieved from Sustainability Directory: https://prism.sustainability-directory.com/term/indigenous-governance/

Su, W., Peng, M. W., Tan, W., & Cheung, Y. (2014). The signaling effect of corporate social responsibility in emerging economies. *Journal of Business Ethics*. Advance online publication. DOI: 10.1007/s10551-014-2404-4

Syal, R., van Wessel, M., & Sahoo, S. (2021). Collaboration, co-optation or navigation? The role of civil society in disaster governance in India. *Voluntas, 32*(4), 795–808. DOI: 10.1007/s11266-021-00344-8

Szeliski, R. (2010). *Computer vision: Algorithms and applications*. Springer.

Tamandehrou, A., & Khan, S. U. (2015). Assessing the importance of globalisation and new media technology in 21st century: An analytical overview. *International Journal of Multifaceted & Multilingual Studies, 1*(4).

Task Force on Climate-related Financial Disclosures. (2021). *2021 status report*. TCFD.

Teece, D. J. (2014). The foundations of enterprise performance: Dynamic and ordinary capabilities in an economic theory of firms. *Academy of Management Perspectives, 28*(4), 328–352. pp. 335

Teece, D. J., Pisano, G., & Shuen, A. (1997). Dynamic capabilities and strategic management. *Strategic Management Journal, 18*(7), 509–533. DOI: 10.1002/(SICI)1097-0266(199708)18:7<509::AID-SMJ882>3.0.CO;2-Z

Thakur, R. (2023). The Russia-Ukraine war and the changing world order. *Strategic Analysis, 47*(2), 82–97.

Thành, N.T., Phuong, H.L. and Hieu, T. (2023) 'Corporate social responsibility, corporate reputation and intention to apply for a job: evidence from students in an emerging economy', *International Journal of Organizational Analysis*, ahead-of-print. .DOI: 10.1108/IJOA-11-2022-3484

The Weighing of the Heart Ceremony. (n.d.). Retrieved from Egypt Museum: https://egypt-museum.com/the-weighing-of-the-heart-ceremony/#google_vignette

Time. (2025, July 10). *The Only Way That Companies Can Build Lasting Transformation.*

Tinker, M., Barnes, L. and Wilson, D.T. (2017). Corporate social responsibility strategy and its influence over consumer purchasing decisions in financial institutions. *Us-China Education Review B*. .DOI: 10.17265/2161-6248/2017.05.003

Tippeconnic, J. W., & Faircloth, S. C. (2010). The Education of Indigenous Students. In P. Peterson, E. Baker, & B. McGaw, *International Encyclopedia of Education (Third Edition)* (pp. 661-666). Elsevier. DOI: 10.1016/B978-0-08-044894-7.00123-8

Tlaiss, H. (2022). Entrepreneurship, gender and success in Lebanon. *International Journal of Emerging Markets, 18*(11), 4754–4771. DOI: 10.1108/IJOEM-01-2021-0145

Tooze, A. (2022). *Shutdown: How Covid shook the world's economy*. Penguin Books.

Toufaily, B. and Bou Zakhem, N. (2024). Drivers of Student Social Entrepreneurial Intention Amid the Economic Crisis in Lebanon: A Mediation Model. *Sustainability*, [online] 16(7), p.2807. .DOI: 10.3390/su16072807

Tucker, B. A., & Russell, R. F. (2004). The influence of the transformational leader. *Journal of Leadership & Organizational Studies, 10*(4), 103–111. pp. 105–106.

Tufekci, Z. (2017). *Twitter and tear gas: The power and fragility of networked protest.* Yale University Press.

Tussay, Á. (2024, September 26). Natural law and unwritten law in Classical Greek thought. *Hungarian Journal of Legal Studies, 65*(1), 114–134. DOI: 10.1556/2052.2024.00534

Tyler, T. R. (2006). Psychological perspectives on legitimacy and legitimation. *Annual Review of Psychology, 57*(1), 375–400. DOI: 10.1146/annurev.psych.57.102904.190038

Uhl-Bien, M., & Arena, M. (2018). Leadership for organizational adaptability: A theoretical synthesis and integrative framework. *The Leadership Quarterly, 29*(1), 89–104. pp. 90–92.

UK Government. (2023). *Bletchley Park AI Safety Summit.*

UK Government. (2024). *AI Seoul Summit.*

Ullah, M.H. and Rahman, M.A. (2015). Corporate social responsibility reporting practices in banking companies in bangladesh. *Journal of Financial Reporting and Accounting*. .DOI: 10.1108/JFRA-05-2013-0038

UN General Assembly. (2024a). *Resolution on safe AI.*

UN General Assembly. (2024b). *Resolution on digital cooperation.*

UN Global Pulse. (2023). *Language-monitoring systems in peacebuilding.*

UNCTAD (2021). *Digital economy report 2021.*

Under2 Coalition. (2023). *Annual report 2023.*

UNDP (2022). *Human development report 2022.*

UNESCO. (2021). *Recommendation on the ethics of artificial intelligence.*

UNESCO. (2025). *AI sovereignty in the Global South.* Fictional.

Ungar, M. (2008). Resilience across cultures. *British Journal of Social Work*, •••, 218–235.

Ungar, M. (2011). The social ecology of resilience: Addressing contextual and cultural ambiguity of a nascent construct. *The American Journal of Orthopsychiatry, 81*(1), 1–17. DOI: 10.1111/j.1939-0025.2010.01067.x PMID: 21219271

Ungar, M. (Ed.). (2012). *The social ecology of resilience: A handbook of theory and practice*. Springer. DOI: 10.1007/978-1-4614-0586-3

Ungureanu, M.. (2018). Clarifying the conceptual map of VUCA: A systematic review. *The International Journal of Organizational Analysis, 26*(4), 563–585.

unhcr.org. (n.d.). [Description of content if known; otherwise retain as placeholder].

UNICEF. (2022). Youth in Lebanon: Aspirations and challenges. UNICEF Regional Report.

UNICEF. (2025). Mental health and resilience among Lebanese youth. UNICEF Country Office Report.

United Nations Department of Economic and Social Affairs. (2023). *Global digital compact*. UN DESA.

United Nations Development Programme. (2022). *Digital public goods for sustainable development*. UNDP.

United Nations Development Programme. (2024). *AI and development*.

United Nations High Commissioner for Refugees (UNHCR). (2022). *Global trends: Forced displacement in 2021*. UNHCR. https://www.unhcr.org/global-trends

United Nations High Commissioner for Refugees (UNHCR). (2023). *Global trends: Forced displacement in 2022* (report). UNHCR. https://www.unhcr.org/global-trends-report-2022.pdf

United Nations Office for the Coordination of Humanitarian Affairs. (2024). *Sudan situation report*. OCHA.

United Nations. (2015). *Transforming our world: The 2030 agenda for sustainable development*. https://sdgs.un.org/2030agenda

United Nations. (2021). *Our common agenda: Report of the Secretary-General*. United Nations.

United Nations. (2023). *A global digital compact: An open, free and secure digital future for all*. United Nations.

United Nations. (2024). *Global Digital Compact*.

UNSC. (2023). *Climate security resolution*.

Usmani, A. and Nayak, P. (2022). Effect of CSR and enterprise risk management system on corporate reputation: An empirical study. *International Journal of Social Sciences & Economic Environment*. .DOI: 10.53882/IJSSEE.2022.0702004

Uth, A. M. (2022). *Maat (deity)*. Retrieved from EBSCO Knowledge Advantage: https://www.ebsco.com/research-starters/religion-and-philosophy/maat-deity#full-article

Valentine, I. (2024). Corporate social responsibility (CSR) practices and financial performance of firms. *American Journal of Finance*. .DOI: 10.47672/ajf.1811

Van Eerdewijk, A. H. J. M., Wong, F., Vaast, C., Newton, J., Tyszler, M., & Pennington, A. (2017). White paper: A conceptual model on women and girls' empowerment.

Van Evert, F., Dam, J., & Slingerland, M. (2017). Building capacity for autonomous agriculture: Education and training needs. *International Journal of Agricultural Sustainability*, *15*(1), 45–60. DOI: 10.1080/14735903.2016.1256273

Van Wessel, M., Naz, F., & Sahoo, S. (2021). Complementarities in CSO Collaborations: How working with diversity produces advantages. *Voluntas*, *32*(4), 717–730. DOI: 10.1007/s11266-020-00227-4

Vanhamme, J., & Grobben, B. (2009). "Too good to be true!". The effectiveness of CSR history in countering negative publicity. *Journal of Business Ethics*, *85*(S2, Suppl 2), 273–283. DOI: 10.1007/s10551-008-9731-2

Vansina, J. (1985). *Oral tradition as history*. University of Wisconsin Press.

Verdouw, C., Wolfert, J., Beulens, A., & Rialland, A. (2016). Virtualization of food supply chains with the Internet of Things. *Journal of Food Engineering*, *176*, 128–136. DOI: 10.1016/j.jfoodeng.2015.11.009

Volz, U., Akhtar, S., Gallagher, K. P., Griffith-Jones, S., Haas, J., & Kraemer, M. (2021). *Debt relief for a green and inclusive recovery: A proposal for global action*. Heinrich Böll Foundation.

Waheed, A., Zhang, Q., Zafar, A. U., Zameer, H., Ashfaq, M., & Nusrat, A. (2021). Impact of internal and external CSR on organizational performance with moderating role of culture: Empirical evidence from chinese banking sector. *International Journal of Bank Marketing*, *39*(4), 499–515. Advance online publication. DOI: 10.1108/IJBM-04-2020-0215

Wahyuni, D., Firmansyah, A., & Setiawan, H. (2024). Sustainability in Business: Integrating CSR into Long-term Strategy. *Asian Business and Innovation Management*, *2*(3), 45–59. DOI: 10.60079/abim.v2i3.354

Walker-Jones, A. (2021, November 27). *Trial, Error, And Success: The Indigenous Origins Of The Scientific Method*. Retrieved July 12, 2025, from Medium: https://medium.com/sofi-stories/trial-error-and-success-indigenous-origins-of-the-scientific-method-d020829afda1

Walker, P. G. T., Whittaker, C., Watson, O. J., Baguelin, M., Winskill, P., Hamlet, A., Djafaara, B. A., Cucunubá, Z., Mesa, D. O., Green, W., Thompson, H., Nayagam, S., Ainslie, K. E. C., Bhatia, S., Bhatt, S., Boonyasiri, A., Boyd, O., Brazeau, N. F., Cattarino, L., & Ghani, A. C. (2020). The impact of COVID-19 and strategies for mitigation and suppression in low- and middle-income countries. *Science*, *369*(6502), 413–422. DOI: 10.1126/science.abc0035 PMID: 32532802

Walsh, F. (2016a). *Strengthening family resilience* (3rd ed.). Guilford Press.

Walsh, F. (2016b). Family resilience: A developmental systems framework. *European Journal of Developmental Psychology, 13*(3), 313–324. DOI: 10.1080/17405629.2016.1154035

Wang, H., Zhang, T., Wang, X., Zheng, J., Zhao, Y., Cai, R., Liu, X., Jia, Q., Zhu, Z., & Jiang, X. (2024). How contingency adjusts corporate social responsibility (CSR) in the tourism industry: A quasi-experiment in china. *Systems, 12*(3), 83. Advance online publication. DOI: 10.3390/systems12030083

Wang, J., Chang, C., Yao, S., & Liang, C. (2015). The contribution of self-efficacy to the relationship between personality traits and entrepreneurial intention. *Higher Education, 72*(2), 209–224. DOI: 10.1007/s10734-015-9946-y

Weick, K. E., Sutcliffe, K. M., & Obstfeld, D. (2005). Organizing and the process of sensemaking. *Organization Science, 16*(4), 409–421. pp. 417–418.

Weiss, T. G., & Wilkinson, R. (2021). [*A decade of reform.* Routledge.]. *Global Governance*, •••, 2050.

White House. (2025). *Executive Order 14179.*

WHO (2020). *Managing the COVID-19 infodemic: Promoting healthy behaviours and mitigating the harm from misinformation and disinformation.* WHO.

WHO. (2022). *COVAX: Working for global equitable access to COVID-19 vaccines.* WHO.

WHO. (2024). *Epidemic Intelligence from Open Sources.* EIOS.

Whyte, K. (2020). Too late for Indigenous climate justice: Ecological and relational tipping points. *Wiley Interdisciplinary Reviews: Climate Change, 11*(1), e603. DOI: 10.1002/wcc.603

wikipedia. (n.d.). *Code of Lipit-Ishtar.* Retrieved from wikipedia: https://en.wikipedia.org/wiki/Code_of_Lipit-Ishtar#:~:text=The%20Code%20of%20Lipit%2DIshtar,the%20Louvre%20%5BAO%205473%5D)

Wildavsky, A. (1991). *Searching for safety.* Transaction Publishers.

Willems, J. (2021). Accountability and transparency: cornerstones of civil society governance. In *Research handbook on nonprofit governance* (pp. 46–70). Edward Elgar Publishing., DOI: 10.4337/9781788114912.00012

Williamson, A. K., Kingston, K. L., & Bennison, L. (2022). Similar or different? Downward accountability across civil society organisational forms. *Voluntas, 33*(5), 1002–1014. DOI: 10.1007/s11266-021-00441-8

Williamson, O. E. (2000). The new institutional economics: Taking stock, looking ahead. *Journal of Economic Literature, 38*(3), 595–613. DOI: 10.1257/jel.38.3.595

Williams, P. D. (2023). EU policy towards Ethiopia amidst the Tigray war. *Third World Quarterly, 44*(9), 1953–1970.

Wojczewski, S., Grohma, P., & Kutalek, R. (2023). Risk communication and community engagement with vulnerable groups: Perceptions of social-services CSOs during Covid-19. *International Journal of Disaster Risk Reduction, 94*, 103817. DOI: 10.1016/j.ijdrr.2023.103817

Wolfert, S., Ge, L., Verdouw, C., & Bogaardt, M. J. (2017). Big data in smart farming – A review. *Agricultural Systems, 153*, 69–80. DOI: 10.1016/j.agsy.2017.01.023

World Bank. (2022). *Green bonds for sustainable development.* World Bank.

World Bank. (2022). *Labor force participation rate, female (% of female population ages 15+)* [World Bank Open Data Portal]. https://data.worldbank.org/indicator/SL.TLF.CACT.FE.ZS

World Bank. (2023). *India's digital public infrastructure.* World Bank.

World Bank. (2024). *Carbon pricing dashboard.*

World Bank. (2025). *Lebanon economic monitor: Collapse and recovery.* World Bank Group.

World Economic Forum. (2023). *Global risks report 2023.*

World Economic Forum. (2023). *Global risks report 2023.* World Economic Forum.

World Food Programme. (2022). *Blockchain for zero hunger: Building blocks for a better future.* WFP.

World Health Organization. (2022). *Community assets and civil society outreach in critical times: an initiative to engage civil society organizations in the COVID-19 response.* World Health Organization.

Wossen, T.. (2020). AI-based soil fertility mapping for smallholder farmers in Africa. *Agricultural Systems, 177*, 102711. DOI: 10.1016/j.agsy.2019.102711

Wu, H., Hu, M. Y., & Smith, K. (2021). Turning crisis into opportunity: Adaptive leadership for organizational resilience. *Leadership & Organization Development Journal, 42*(2), 158–171. p. 162.

Yamamoto, S., Umeda, M., & Fujimura, S. (2019). Autonomous rice transplanters for labor shortage solutions in Japan. *Journal of Agricultural Machinery Science, 15*(3), 123–132.

Yang, Q., Liu, Y., Chen, T., & Tong, Y. (2019). Federated machine learning: Concept and applications. [TIST]. *ACM Transactions on Intelligent Systems and Technology, 10*(2), 1–19. DOI: 10.1145/3298981

Yasrawan, Y., Sukarya, R., & Tamba, J. (2023). Competitive Advantage as A Link Between the Influence of Intellectual Capital and CSR on Financial Performance. *Jurnal Ilmiah Akuntansi, 8*(1), 41–53. DOI: 10.23887/jia.v8i1.46981

Yehia, A.-D. and Szabados, G. (2018). The middle east perspective of corporate social responsibility. *International Journal of Engineering and Management Sciences*. doi:https://doi.org/..DOI: 10.21791/ijems.2018.4.24

Yeung, K. (2022). *The new public analytics: Beyond managerialism*. Oxford University Press. DOI: 10.1007/978-3-030-75166-1

Yoon, Y., Gürhan-Canli, Z., & Schwarz, N. (2006). The effect of corporate social responsibility (CSR) activities on companies with bad reputations. *Journal of Consumer Psychology, 16*(4), 377–390. DOI: 10.1207/s15327663jcp1604_9

Younes, L.. (2021). COVID-19, trust, and mental health in Lebanon. *Global Health Journal (Amsterdam, Netherlands)*, 301–310.

Yousfi, O. and Béji, R. (2022). CSR: What does board diversity bring to the table? .DOI: 10.5772/intechopen.94342

Youssef, H., Harakeh, M., Hussein, H., & Sayour, N. (2024). Brief on Entrepreneurship in Lebanon Unlocking Potentials: Addressing Barriers and Empowering Entrepreneurs in Lebanon | LJMU Research Online. *Ljmu.ac.uk*. [online] doi:https://researchonline .ljmu.ac.uk/id/eprint/24116/1/Policy%20Brief%20on%20Entrepreneurship%20in %20Lebanon.pdf

YOUSSEF. S.A. (2021). Factors Affecting Entrepreneurial Intention of University Students: A Lebanese Case Study - STORE - University of Staffordshire Online Repository. *Staffs.ac.uk*. [online] doi:https://eprints.staffs.ac.uk/6892/2/Samia%20 Youssef-EThOS%20Deposit%20Agreement_tcm44-75119.doc

Zadravec, K. 2021. Developing intercultural and entrepreneurship competencies among students – the case of business, IT, and services disciplines. *SHS Web of Conferences*, 111, 01002. https://doi.org/DOI: 10.1051/shsconf/202111101002

Zaid, M. A. A., Wang, M., & Abuhijleh, S. T. F. (2019). The effect of corporate governance practices on corporate social responsibility disclosure. *Journal of Global Responsibility*, *10*(2), 134–160. Advance online publication. DOI: 10.1108/JGR-10-2018-0053

Zampetakis, L., Kafetsios, K., Lerakis, M., & Moustakis, V. (2015). Investigating the role of self-construal in the formation of entrepreneurial intentions. *Frontiers in Psychology*, *6*. Advance online publication. DOI: 10.3389/fpsyg.2015.01085 PMID: 26284009

Zein, L.. (2021). Sectarian divisions and youth identity in Lebanon. *Social Identities*, •••, 811–828.

Zhao, G.. (2019). AI-based erosion risk prediction using satellite data. *Environmental Modelling & Software*, *117*, 159–169. DOI: 10.1016/j.envsoft.2019.03.008

Zhao, H., Xu, S., & Wu, L. (2022). The Trade-off between Corporate Social Responsibility and Competitive Advantage: A Biform Game Model. *Technological and Economic Development of Economy*, *28*(3), 638–662. DOI: 10.3846/tede.2022.16483

Zheng, T. 2022. Research on the current situation of university students' innovation and entrepreneurship under the background of "Internet + maker education". pp.668-676. https://doi.org/DOI: 10.2991/978-2-494069-89-3_79

Zhou, L., Zhang, Q., & Shi, Z. (2020). Drone remote sensing for precision agriculture: Review and future prospects. *Remote Sensing*, *12*(7), 1096. DOI: 10.3390/rs12071096

Zolfaghari, B. N., Möllering, G., Clark, T., & Dietz, G. (2016). How do we adopt multiple cultural identities? A multidimensional operationalization of the sources of culture. *European Management Journal*, *34*(2), 102–113. DOI: 10.1016/j.emj.2016.01.003

About the Contributors

Mohamad Al Mokdad holds a double Ph.D in International Relations and Geopolitics. He is an expert in strategic studies, specialized in the MENA region, professional with a rich academic background and diverse professional experience in international relations, strategic management, social policies, and humanitarian law. He is the CEO of Intelligentsia center for research and education (ICRE), at the same time he is an International Consultant at PGS University, France. Dr Mokdad has an impressive record of publications in major indexed journals, affirming his significant contributions to his field.

***

Ali Al Mokdad is a highly respected strategic leader with an extensive track record of pioneering impactful programs in some of the world's most challenging environments, including underserved regions and active conflict zones. His expertise spans roles across International Non-Governmental Organizations (INGOs), United Nations (UN) agencies, the Red Cross and Red Crescent movement, and donor institutions, where he has consistently driven excellence in operations, programs, and strategic nonprofit management. Throughout his international career, Ali has held roles across the Middle East, Africa, and Asia—including Syria, Lebanon, Turkey, Iraq, Nigeria, Kenya, South Sudan, Afghanistan, and Bangladesh. He has also served in leadership roles at the global level with INGOs spanning over 40 countries. As a strategic advisor on the boards of influential NGOs and companies, Ali continues to champion transformative solutions that amplify the impact of humanitarian work on a global scale.

Azamat Ali is an Associate Professor in the School of Liberal and Creative Arts (Journalism and Mass Communication) at Lovely Professional University, Punjab, India. He earned his Ph.D. from the Central University of Gujarat, Gandhinagar, and a M.Sc. in Science Communication from Devi Ahilya Vishwavidyalaya, Indore. His research interests span emerging technologies, development communication, public policy, and health and science communication. Dr. Ali research is directed toward bridging the divide between scientific innovation and societal application, with a particular focus on harnessing the transformative potential of emerging technologies to drive innovation, address developmental challenges, and enhance societal well-being.

Elie Basbous is a dedicated academic and researcher with extensive expertise in [business administration. With a distinguished career in higher education, he has played a pivotal role in shaping academic programs and mentoring future professionals. His work focuses on corporate governance, strategic management, or financial policy, and he is actively involved in research initiatives that address contemporary challenges in the global economic and business landscape. Professor Basbous is recognized for his commitment to academic excellence, interdisciplinary collaboration, and impactful scholarship.

Radwan Choughri is the Dean of the Faculty of Business Administration at Jinan University, Lebanon, and an Associate Professor of Economics. Dr. Choughri is dedicated to advancing academic excellence and has contributed to research on green marketing and consumer behavior. Under his leadership, the faculty offers dynamic programs that prepare students for leadership roles in both the private and public sectors.

Csongor Czipf, Ph.D. Honorary Associate Professor, Assistant Professor Széchenyi István University, Győr, Hungary Gyula Kautz Faculty of Economics, Department of International Applied Economics Ongoing teaching experience at Széchenyi István University since 1999. Appointed as Regional Economic Manager of the Piarist Order Hungarian Province in Budapest as of January 2022. Master's degree in Religious Education and Dormitory Teaching obtained from the College of Theology in Győr in 2017. Doctoral degree awarded in 2011 by Janus Pannonius University of Pécs. Research specialization: corporate evaluation from a human resources perspective. Bachelor's degree in Economics with a specialization in Comparative Economic Policy completed in 1999 at Széchenyi István College.

Hazem Harb is an experienced program manager in the sustainable development sector with an extensive background in planning and managing cross-functional business operations and educational projects in the public and Non-Profit sector. Hazem used to work as a business development specialist and has also managed livelihood projects and at different international NGOs. He holds an MBA degree and is currently continuing his PhD studies in Leadership and Human Resources

Management at the Saint Joseph University of Beirut; throughout his academic journey, Hazem worked on 2 Book Chapters in the Routledge Handbook on Business and Management in the Middle East (2025): *Business Coaching Industry in Lebanon: Market Trends and Opportunities. * Employer Branding in the Middle East. In addition, he participated in two international academic conferences: * RISE 2021 Conference: Rejuvenating the relation of scholarship and practice. Beirut, Lebanon * 5th Nordic International Conference 2023: Annual Conference Proceedings (eBook)

Saad Ullah Khan has more than a decade long extensive experience at university level teaching and research. Dr. Khan has authored various research papers in journals of national as well as international repute. He was awarded 'Rajat Jayanti Vigyan Sancharak Fellowship-2009' by DST, Ministry of Science and Technology (Government of India) for his contribution to the field of communication and its social implications. Further he was bestowed with MANF Fellowship 2011 by UGC to carry out his research work. A PhD from prestigious Aligarh Muslim University, Dr. Saad Ullah Khan has presented 90 Research Papers in various National and International Conferences. He was awarded 'Best Researcher Award' twice (2020 & 2023) and 'Best Journalism Teacher Award' in 2021.

Sadaf Khan is presently pursuing her doctoral research at Galgotias University, Uttar Pradesh, India, where she is engaged as a Research Scholar specializing in the domain of English Language and Literature. Her current work focuses on Afghan society and its representation in diasporic writings, with particular emphasis on issues of identity, displacement, and cultural memory. An active and prolific researcher, she has contributed significantly to the fields of English Literature, Cultural Studies, and Afghanistan Studies, producing scholarship that bridges literary analysis with socio-political contexts. Her academic trajectory demonstrates a sustained commitment to explore how literature functions as a site of cultural negotiation and historical testimony. Through her interdisciplinary approach, Ms. Khan brings valuable insights into the intersections of narrative, exile, and resilience, positioning her work within contemporary debates on migration, postcolonial identity, and the global dynamics of cultural representation.

Vivek Kumar serves as a Professor at Galgotias University, Uttar Pradesh, India, specializing in English Language and Literature. With over 13 years of teaching experience across undergraduate and postgraduate levels, he has also undertaken diverse academic and administrative responsibilities, contributing significantly to institutional development since joining Galgotias University in 2016. He holds an M.A. in English from the University of Allahabad and a Ph.D. in English from Gurukul Kangri University, Haridwar. Dr. Kumar has published numerous research papers and actively participates in national and international conferences, seminars, workshops, and faculty development programs. His scholarly interests encompass 20th Century British Literature, Indian Writing in English, Phonetics, and Literary Theory and Criticism. For him, teaching is both a vocation and a lifelong commitment—an intellectual engagement that not only imparts knowledge but also nurtures critical growth and holistic development among learners.

Ananya Pandey is a Research Scholar in School of Law at Christ University, Bangalore. She has her research interests in the field of Constitutional Law, Environment Law, Human Rights, Food Security, Intellectual Property Law and Jurisprudence.

Aahana Saha is an Assistant Professor of Psychology with over nine years of teaching experience. She holds a Master's degree, is UGC-NET qualified, and completed a Ph.D. on female police personnel's job satisfaction and happiness. Her work includes developing and validating various psychological scales, publishing 20+ research papers, and securing two patents on machine-learning applications for cyber defense and forensic behaviour prediction. She has guided more than 30 postgraduate dissertations, coordinated academic programs, and launched student initiatives. Currently she is pursuing a PG certificate in forensic psychology and criminal profiling and a Master's in Data Science from Purdue University

Amanullah Zarawar holds an MA in Political Science (International Relations) from Universitas Islam Internasional Indonesia (UIII). His research focuses on peacebuilding, conflict resolution, and international security, especially on the role of trust and secrecy in high-stakes negotiations. He holds a Bachelor's degree in Mass Communication and Journalism from Khurasan University, Afghanistan, where he also served as a university lecturer. Professionally, Zarawar has over four years of experience in the Afghan media sector, as a news anchor, political talk show host, and editor.

Index

A

B

C

D

E

F

G

H

I

J

L

M

O

P

R

S

U